Separations Division 2020

Held at the 2020 AIChE Annual Meeting

Online
16 – 20 November 2020

ISBN: 978-1-7138-2297-4

Printed from e-media with permission by:

Curran Associates, Inc.
57 Morehouse Lane
Red Hook, NY 12571

CURRAN ASSOCIATES INC. proceedings.com

Some format issues inherent in the e-media version may also appear in this print version.

Printed with permission by Curran Associates, Inc. (2021)

For permission requests, please contact AIChE
at the address below.

AIChE
120 Wall Street, FL 23
New York, NY 10005-4020

Phone: (800) 242-4363
Fax: (203) 775-5177

www.aiche.org

Additional copies of this publication are available from:

Curran Associates, Inc.
57 Morehouse Lane
Red Hook, NY 12571 USA
Phone: 845-758-0400
Fax: 845-758-2633
Email: curran@proceedings.com
Web: www.proceedings.com

TABLE OF CONTENTS

606173 An Improved Method for Predicting Autoignition Temperatures Based on First Principles

Abstract:

Methods for predicting autoignition temperatures (AIT) have been historically inaccurate and are rarely based on the underlying physical phenomena leading to observed AIT. Previous attempts at predicting AIT have been developed based on Quantitative Structure-Property Relationships (QSPR) and additive group contribution methods. This work presents an improved method for predicting AIT based on the method originally developed by the late Dr. William H. Seaton. The method of Seaton is described in detail including its use of first principles to predict AIT. Improvements on the method of Seaton and underlying principles are presented and discussed.

611821 Water Washing of Startup Crude Oil

Abstract:

During startup of cold sub-sea wells, methanol is injected at the wellhead to prevent the formation of hydrates in the pipeline/riser system. Methanol injection continues until the fluid arrival temperature exceeds hydrate-formation temperature at the operating pressure, and the typical duration is a few days. Methanol contamination is problematic, and hence the methanol specification of the crude oil is usually less than 20 ppmw. Water or brine wash is often used to reduce the methanol content of the startup crude oil. This paper studies the design of these systems using process simulation. The enabling need is for a reliable and accurate thermodynamic model. The model must describe the three-phase system containing the aqueous and oil liquid phases and the vapor phase, and the key needed feature is description of the distribution coefficient of methanol between the oil and aqueous phases. A model has been developed based upon the ElectrolyteNRTL model in Aspen Plus, and the presentation describes the key data sources and the parameter development. The model enables design of an optimum process flowsheet, and a two-stage countercurrent has been determined to be the optimum option.

607019 Universal Correlation for the Binary Interaction Parameter of Hydrogen with Petroleum Fractions in Refinery Hydroprocessing Applications

Abstract:

Hydrotreating and hydrocracking applications in oil refineries use large quantities of hydrogen. Hydrotreating removes hetero-atoms of sulfur, nitrogen and oxygen and can saturate aromatic ring structures, producing more desirable transportation fuels, such as ultra-low sulfur diesel. Hydrocracking is a more severe reaction, which breaks carbon-carbon bonds and improves the ratio of hydrogen to carbon in the fluid, again, making more suitable transportation fuels. Hydrogen is expensive and is either made on-site, perhaps through steam reforming of methane, or is bought from outside as a utility.

To design new hydroprocessing facilities and to improve the performance of existing plant, it is essential to be able to predict the solubility of hydrogen in typical hydrocarbon fluids. The hydrogen has to be separated from the product streams in high and low pressure separators and is then recycled to the reactor feed, where it is combined with hydrogen make-up.

Although Detailed Hydrocarbon Analysis (DHA) enables us to define the composition of reactor feedstocks in terms of a larger slate of defined chemical components, it is still common practice to represent the heavier portion of the feed in terms of undefined components, or petro fractions, characterized by narrow cut boiling points and densities from a petroleum assay.

Cubic equations of state (EoS), such as Soave-Redlich-Kwong and Peng-Robinson are the workhorse thermodynamic methods for calculating vapor liquid equilibrium (VLE) in refinery processes. From the study of the VLE of binary systems of hydrogen as a solute in a defined hydrocarbon solvent, it is evident that a significant correction to the basic Van der Waals mixing rule in the EoS is required if we are to predict accurately the composition of hydrogen in the liquid phase. Moreover, we can show that this correction, through a binary interaction parameter (k_{ij}) needs to be temperature-dependent. While we can correlate k_{ij} for hydrogen binaries with defined components through regression of experimentally measured VLE data, the problem is how we can predict the interaction between hydrogen and any petro fraction? We observe that increasingly large values of k_{ij} are required as the molecular weight of the hydrocarbon increases and any petro fractions are likely to be in the higher molecular weight range. Ignoring the interaction parameter between hydrogen and a petro fraction will result in significant inaccuracy in any prediction of solubility.

In this study, we will show how the wealth of binary VLE data for hydrogen with defined hydrocarbon solvents can be reduced through regression and conversion to Henry's Law constants that correlate with the reduced temperature and critical pressure of the solvent [1,2]. We propose a new universal correlation for these Henry's Law constants that will allow the subsequent prediction for k_{ij} between hydrogen and a petro fraction, based only on knowledge of the critical temperature, critical pressure and acentric factor of the petro fraction that can be calculated from well-established estimation methods. The validity of this approach will be shown by comparing the predicted solubility of hydrogen in a well-analyzed diesel stream with a base case using methods currently described in the Technical Data Book of the American Petroleum Institute [3].

[1] Trinh TKH, de Hemptinne JC, Lugo R, Ferrando N, Passarello JP, "Hydrogen Solubility in Hydrocarbon and Oxygenated Organic Compounds." J. Chem. Eng. Data 2016,61,119-34

[2] Fernández-Prini R, Alvarez JL, Harvey AH, "Henry's Constants and Vapor–Liquid Distribution Constants for Gaseous Solutes in and at High Temperatures "J. Phys. Chem. Ref. Data, 3012, 32, 903-912

[3] API Tech Data Book 10, 2019

600821 Distillation Curve and Vapor Pressure Predictions of Alcohol-Gasoline Mixtures Using an Advanced Equation of State

Abstract:

Due to federal environmental regulations, such as the Energy Policy Act of 2005 in the United States, ethanol is frequently blended into gasoline in an effort to reduce tailpipe emissions. Moreover, there is a shift towards increasing the use of more sustainable energy resources such as biofuels, which includes ethanol, as transportation fuels. A fundamental understanding of how blending alcohol into a gasoline stream affects the properties of said stream is of critical importance. Specifically, the change in the distillation curve and vapor pressure is essential from a manufacturing and engine performance perspective. In the plant setting, there exists empirical correlations to predict these changes. However, these are typically very limited in scope as a correlation is only applicable to a specific plant gasoline mixed with a specific alcohol in a certain blend ratio.

Process simulators of petroleum process rely on equation of state (EoS) predictions of the phase behavior of petroleum pseudo-components defined by normal boiling point T_b, specific gravity *SG,* and *MW*. For gasoline blending these pseudo-components are typically generated from D86 curves. Hence, any method to be widely used in the prediction of alcohol / gasoline blending in refining processes should be accurate for petroleum pseudo-components.

In this work, we present a fundamental methodology for alcohol / gasoline blend property prediction using the polar PC-SAFT EoS[1] combined with the recently developed petroleum parameterization methodology, TAPPS (formerly referred to as EMPETRO)[2-6]. TAPPS is a highly accurate methodology used to determine hydrocarbon PC-SAFT parameters for petroleum species defined solely by T_b, *SG,* and *MW*. In addition to highly accurate predictions for pure component vapor pressures and densities, TAPPS accounts for polar features of petroleum such as the fact that aromatic species receive hydrogen bonds. This makes it well suited to describe alcohol / gasoline blending.

The approach presented here employs the base gasoline D86, average gasoline SG, and dry vapor pressure equivalent (DVPE) to generate gasoline pseudo-components. Using these pseudo-components, it is demonstrated the PC-SAFT / TAPPS approach allows for the quantitative prediction of the D86 and DVPE of alcohol / gasoline mixtures. We validate the approach by reproducing both literature[7,8] and ExxonMobil refinery data[9] for various blends. This work demonstrates the strength of PC-SAFT / TAPPS in predicting the physical properties of non-traditional blends of petroleum streams and lays the foundation for the development of a predictive blending tool that will be valuable when considering emerging biofuel markets.

References

[1] J. Gross; G. Sadowski, *Ind. Eng. Chem. Res.*, **40** (2001) *1244.*

[2] B. D. Marshall; C. P. Bokis, *Fluid Ph. Equilibria*, **489** (2019) *83.*

[3] B. D. Marshall; C. P. Bokis, *Fluid Ph. Equilibria*, **478** (2018) *34.*

[4] B. D. Marshall, *Fluid Ph. Equilibria*, **493** (2019) *153.*

[5] B. D. Marshall, *Fluid Ph. Equilibria*, **497** (2019) *79.*

[6] B. D. Marshall, *Fluid Ph. Equilibria*, **507** (2020) *112420.*

[7] V. F. Andersen; J. E. Andersen; T. J. Wallington; S. A. Mueller; O. J. Nielsen, *Energy Fuels,* **24** (2010) *2683.*

[8] V. F. Andersen; J. E. Andersen; T. J. Wallington; S. A. Mueller; O. J. Nielsen, *Energy Fuels,* **24** (2010) *3647.*

[9] J. R. Vella; B. D. Marshall, *Ind. Eng. Chem. Res.* (2020) *accepted.*

606540 Development of Mean Ionic Activity Coefficients for Modification of the Krichevsky – Ilinskeya Equation for Methylimidazolim – Type Ionic Liquids

Abstract:

Removal of H_2S from oil and gas streams has been a priority, oftentimes a necessity, for upstream crude oil/gas facilities for decades. However, the advances in production, namely horizontal drilling and hydraulic fracking, have produced oils and natural gases that contain higher concentrations of H_2S than previously seen. Crude oils containing > 0.5 % sulfur are considered sour and must undergo separation and/or conversion processes to lower the H_2S content prior to entering the midstream pipelines and railcars. H_2S scavengers can be categorized into two types: non-regenerative and regenerative. Non-regenerative scavengers, such as triazines, use an irreversible reaction to convert H_2S forming sulfur-containing hydrocarbons. With regenerative scavengers, such as alkanolamines, absorb H_2S using reversible absorption with stripping recovery processes.

Ionic Liquids (ILs) are a new family of absorbent materials that are being sought to replace amine-type reversible absorbents due to lower production of volatile organic carbons (VOCs) and reduced costs of the regenerative stripping phase. The ILs that have been chosen for this work were 1-butyl-3-methylimidazolium hexafluorophosphate, 1-butyl-3-methylimidazolium chloride, and 1-butyl-3-methylimida-zolium tetrafluoroborate. In this work, the mean ionic activity coefficient and the adjustable parameters for the Pitzer Ion Interaction Model have been determined. An infinite dilution procedure has been employed where a concentrated mixture of H_2S in IL was stripped using helium gas. The decrease in H_2S concentration (measured through gas chromatography) over time was used to determine the pertinent thermodynamic parameters.

In addition, the Henry's constant and partial molar volume of H_2S in IL were determined. The Krichevsky – Kasarnovsky equation has been shown to be applicable over a wide range of temperatures, pressures, and solute/solvent systems. However, in some cases (i.e. where T is near T_c) the molar volume ( ) is not constant, and there are strong deviations between the measured solubility and the calculated Henry's constants. From preliminary estimates, the Krichevsky – Ilinskeya equation (a modification of the Krichevsky – Kasarnovsky equation) does not model the phenomenon well. Ultimately, this work seeks to establish a new thermodynamic model that replaces the 2-suffix Margules component of the Krichevsky – Ilinskeya equation with the mean ionic activity coefficient found in Pitzer Ion Interaction model. The experimental results and model development will be presented.

609303 An Integrated Process Automation System in API Manufacturing: Key Characteristics and Case Study

Abstract:

Safety is always first priority and a critical concern in the chemical industry. Jiangsu Provincial Emergency Management Administration (JSEMA) in China released <Basic Requirements for Diagnosis and Improvement of Instinct Safety> (Su Emer. NO.53) in June 2019, which specifies the requirements for the main aspects of chemical production. The concept of "integrated process automation" in Su Emer. No. 53 poses a great challenge to API manufacturing. It emphasizes mandatory safety and intensified automatic implementation. This paper illustrates the key characteristics in the integrated process automation of API manufacturing and gives some case studies.

Such automation system includes sophisticated solution and cutting-edge tools. According to the recipe and process of the product, the consultant optimizes the control strategic with expertise to enhance the safety and meet the regulation requirements. The process control modules or batch control program are deployed in DCS system for the purpose of multifunctional and flexible manufacturing. In case study it gives example of the mixing control strategic in the reactors which is critical and should consider all concerned factors. Usually the geared motor electricity is ignored to monitor, the absence may interfere the cooling system activation in time, its real time monitoring and interlock with cooling media are necessary; another example is the utility facilities configuration and interlock with automatic systems which should be tuned duly to eliminate the hidden trouble. The integrity of safety instrumented functions (SIFs) should be determined through risk assessment tools and rating tools, such as HAZOP (Hazard and Operability Analysis) and LOPA (Layer of Protection Analysis). The related case study reports sharing control components between SIS and DCS will increase the possibility of failure, which should be validated. The allocation of center control room and controller assignment should be concerned with HSE guideline. It also discusses the roadmap of information technology and MES (Manufacture Execution System) for API manufacturing. Our implementation shows that the integrated process automation system can help to reach instinct safety, which increase the productivity and improve the quality of products with regulatory compliance.

605455 A Quantitative Assay of Sodium Triacetoxyborohydride

Abstract:

Sodium triacetoxyborohydride (STAB) is a common reducing agent for reductive amination reactions. Due to sensitivities to water and air, the potency of this reagent tends to degrade over time. Traditional analytical assays for STAB fail to accurately assess the content of the active borohydride species, and often report assay purities of 97% or greater, when the true potency is lower. An incorrect or unknown potency of the STAB being used may lead to significant challenges when scaling up reductive aminations. An excess of STAB may be required and reactions may not achieve full conversion with the amount of STAB charged, requiring additional charges. The use of an excess of STAB comes with greater raw material costs and the thermal and H_2 evolution hazards associated with quenching the excess reagent.

This presentation outlines a simple assay, based on an aldehyde reduction, that was developed to determine the active borohydride content of STAB. The HPLC assay yield of a salicylaldehyde reduction has been shown to accurately determine this potency and has been validated against an H_2 evolution method, as well as yields obtained from a reductive amination.

607708 Delivery Strategies for Live Therapeutic Bacteria

Abstract:

The microbiome is a collection of bacteria, viruses and fungi that perform essential functions for human health. Dysbiosis, or the dysregulation of the microbiome, can contribute to chronic inflammation, metabolic disorders, neurological diseases, and cancer. Clinically, one of the only methods for altering the composition of the microbiome is with broad-spectrum antibiotics, which contribute to dysbiosis and create selective pressure for antibiotic resistant pathogens. As such, live therapeutic bacteria (LTB) have been identified as an attractive alternative to antibiotics, as they can modulate the composition of the microbiome without eliminating commensal species. However, LTBs are commonly delivered orally and experience significant challenges during gastrointestinal (GI) transit, including survival in the harsh conditions of the stomach, adherence to the intestinal lumen, and navigation of interpatient variability in the microbiome due to factors such as diet. Few engineered drug delivery platforms have been rationally developed to address these challenges. In this work, we describe two complementary platforms that separately adapt LTBs to improve viability during gastric transit and modify LTB surfaces to aid in adhesion along the GI tract. In the first system, we sought to understand the influence of factors such as diet on LTB adaptation and colonization. We delivered an LTB to female BALB/c mice on two distinct diets and found that diet significantly influenced both the appearance of a small colony variant of the LTB, as well as its distribution along the GI tract. As adaptation appeared to significantly affect LTB colonization, we developed an approach to control the adaptation of LTBs to specific environmental insults during GI transit. We found that growth in an *in vitro* adaptation media can prime LTBs for the acidic conditions of the stomach, significantly improving their viability during a challenge in simulated gastric fluid. Following transit through the stomach, LTBs must adhere to the intestinal lumen to persist and colonize in the GI tract. Therefore, for our second system, we developed a platform to chemically conjugate targeting ligands to the surface of LTBs to improve their adhesion to the GI mucosa. This platform significantly improves both the rate of colonization, as well as the concentration of LTBs in the GI tract of female BALB/c mice. Collectively, this work represents a simple approach to adapt LTBs to the intestinal environment during *in vitro* growth, while enabling subsequent modification of the LTB surface for improved intestinal adherence. Together, these delivery techniques for LTBs may help overcome challenges associated with GI transit, colonization and adaptation.

609030 In Vitro Biomimetic Tumor Model with Cancer-Specific Gene Regulators and Green Fluorescence Reporter for High Throughput Drug Screening

Abstract:

Although much scientific effort has been expended for cancer treatment, this disease is still one of the leading causes of death around the world. The limitations and shortcomings of current therapies highlight the urgent need for more effective therapeutic strategies. So far, most of the chemotherapy drugs act negatively on the cell division without selection, leading to serious adverse effect on normal cells. The current mainstream of anti-cancer drug development is to find the drugs that have specific targets (DNA, mRNA or proteins). Those targets are overexpressed in cancerous cells and are critical for carcinogenesis. In this study, we engineered breast and lung cancer cells to express the enhanced green fluorescent protein (EGFP) under cancer-specific gene promoters for three known carcinogenesis genes - survivin, Bcl-2 and hTERT, which enable online high-throughput screening of drugs that may up- or down-regulate these cancer genes as indicative of efficacy in inhibiting breast and lung cancers. Additionally, the mechanism of drugs and the crosstalk between cancer genes can be elucidated at the early stage of drug discovery. Another issue in the pharmaceutical industry is the high drug failure rate owing to poor predictability of early drug assays in a two-dimensional cell culture system. In such traditional assay, multiple cell-cell and cell-matrix interactions are deficient, leading to incomprehensive cellar responses to drug agents. To improve the drug assay predictability, our assay has focused on the construction of three-dimensional biomimetic cancer cell models with the recapitulation of multiple elements of tumor microenvironment. While developing the new drug screening assay, we also combined the rapid cell detection approach to achieve the high-throughput capability. Our optimized drug screening process can help to discover more anti-cancer therapies.

609939 Preparation and Use of Fine Grade Engineered Excipients for Direct Compression of Binary Blends of Cohesive Drug Powders

Abstract:

Novel, surface modified excipients are prepared using a fluid energy mill (FEM) to simultaneously mill and dry coat microcrystalline cellulose (MCC) with nano-silica to produce different sizes of fine, well flowing MCC excipient [1]. The purpose of these new class of excipients is to enable direct compression (DC). DC route is one of the most efficient and economic methods for tablet manufacturing since only blending and compression operations are required, greatly simplifies the process-train. However, the popularity of direct compression is limited by its stringent requirements, such as good flowability, high packing density, and excellent compaction for excipient-active pharmaceutical ingredient (API) blends. When the blend has high drug loading with a poorly flowing API, the flowability and tabletability of blends is generally decreased sharply. In a recent work, it has been shown that surprisingly, fine excipients of 20 mm size that have improved flow and bulk density after silica dry coating, can greatly improve direct compression performance [2]. Unfortunately, it could not work for very fine cohesive APIs such as 10 mm acetaminophen (APAP) at higher drug loading. In this work, binary blends of several fine sized dry coated MCC-based excipients are considered with three different fine, cohesive APIs at high drug loadings. Bulk density, FFC, and tablet compactability for all these blends are measured and compared to determine which excipient sizes provide the best overall performance. The results are analyzed for the relative sizes of the excipients and APIs to better understand the interplay between the components of the blends and assessment of the extent of silica transfer from coated excipient powders to API powders and its effect on blend bulk density and flowability. It is expected that using such novel surface engineered fine excipients in blends with fine APIs may also help in reduced segregation tendency during overall processing. Hence, these results are expected to provide industry relevant guidelines for the use of finer surface modified excipients in DC tableting.

596229 Applying Machine Learning to Predict Therapeutic Antibody Viscosity

Abstract:

A general machine learning feature selection protocol to predict antibody stabilities is proposed. These features only require information from sequences and molecular structures of antibodies. In addition, the physically relevant features are easily interpretable and help to design experiments to understand the underlying mechanisms of antibody stabilities. Two case studies are demonstrating using this protocol. First, we applied this framework to predict aggregation behaviors of 21 marketed monoclonal antibodies at high concentration (150 mg/mL), yielding an accuracy of 97% on validation tests with only three features. Models trained using structural features, rather than sequence features, correlated better with aggregation. Second, the concentration dependent viscosity behavior of 27 FDA approved antibody drugs was measured in concentrated solutions at high concentration (150 mg/mL). Combining molecular modeling and machine learning feature selection, we found that the net charge in the antibodies and the amino acid composition in the variable region are key features which govern the viscosity behavior. We presented two predictive models for aggregation and viscosity based on machine learning trained models, facilitating early stage design.

602107 Development of Multi-Adjuvant Systems for Vaccines Against Infectious Diseases

Abstract:

The ability to elicit potent humoral and cellular immune responses is a key element required for the development of efficacious prophylactic and therapeutic vaccine formulations. For vaccines against intracellular pathogens, the induction of a balanced, long-lived, and potent $CD4^+$ and $CD8^+$ T cell activation and proliferation is critical, combined with robust antibody responses. One strategy is to activate innate immunity and proinflammatory responses through Toll-like receptor-independent cytosolic pathways. Importantly, engagement of these mechanisms has demonstrated its ability to provide protection against bacterial challenge, and several small molecules have been evaluated in preclinical and clinical studies as vaccine adjuvants against infectious diseases.

Small molecules can bind to cytoplasmic receptors triggering signaling cascades that activate innate immunity. Despite their potential as adjuvants, small molecules have some limitations that need to be overcome. For example, there may be low exposure in target tissues due to their size and charge, and are prone to enzymatic biodegradation; with some off-target toxicities. In order to address these challenges, the use of drug delivery systems is being explored.

In this work, three different polymeric nanoparticle platforms were evaluated for small molecule delivery and for their capability to act synergistically as co-adjuvants and elicit robust immune responses. In these studies, nanoparticles based on natural and synthetic polymers were fabricated using water/oil/water and oil/water emulsions. Particle characterization was performed using dynamic light scattering, scanning electron microscopy, and zeta potential measurements that confirmed the size, morphology, and charge of the resulting nanoparticles. Small molecule compounds were loaded into each of the platforms, and their loading and release efficiency were determined using liquid chromatography. Finally, these systems were evaluated using *in vivo* murine models following parenteral administration to assess their ability to induce cytokine secretion, antibody and cell-mediated immunity. The results obtained demonstrate the potential of polymeric particulate systems as dual delivery and co-adjuvant platforms for the development of small molecule-adjuvanted prophylactic and therapeutic vaccines.

603577 Development of a Harvest Small Scale Model to Simulate Continuous Centrifugation

Abstract:

Disk stack centrifugation is one of the main clarification steps used in large scale mammalian cell culture processes. During process development, pilot scale disc stack centrifuge is normally used to develop and characterize the harvest process for new cell culture processes due to the lack of smaller scale model for disc stack centrifugation. In order to increase speed of development, a small scale model (SSM) of a disk stack centrifuge is of high demand to enable harvest development in an earlier stage of process development. A small scale model can also improve throughput and reduce cost. In this study, we have developed a SSM for disc stack centrifugation using a capillary shear device combined with benchtop centrifugation. Capillary tube size, length, and flow rate were optimized to generate energy dissipation rates comparable to large scale disk stack centrifuges. Additionally, shear force from multiple centrifuges at different scales were standardized and compared. Mathematical correlations were established for energy dissipation, viable cell density, cell viability, and lactate dehydrogenase (LDH) activity. Our SSM offers an approach to understand how the separation process, such as centrifuge designs, sizes, and configurations affect filtration performance and product quality. This work can speed up harvest development during process scale-up in biopharmaceutical mammalian cell culture processes.

599615 Optimizing and Reducing Cleaning Production Loss in Pharma, Agrochemical and Specialty Chemical Companies Using a New Expert System.

Abstract:

Pharmaceutical, Biopharmaceutical, Agrochemical and Specialty chemical companies make many different products in their plants. This is true both for the active ingredient stages as well as the secondary of formulation processing. Cleaning and set up between products is critical and reduce productivity time up to 50% in some cases. The presentation will show the critical areas important in the modelling development in this expert system and how it will aid companies in scheduling their productions to reduce cleaning time and to reduce issues with cleaning.

599656 Revolutionize of Pharmaceutical Product Changeovers By Enhanced Cleaning with Microbubbles

Abstract:

Pharmaceutical, Biopharmaceutical, Agrochemical and Specialty chemical companies make many different products in their plants. This is true both for the active ingredient stages as well as the secondary of formulation processing. Cleaning and set up between products is critical and reduce productivity time up to 50% in some cases. Microbubbles have been shown to enhance many unit operations e.g. distillation. The benefits of microbubbles in cleaning will be presented along with the impact this will enable in regulatory/ registration processes.

603807 Achieve Process Intensification Using Simulated Moving Bed Technology

Abstract:

SMB (Simulated Moving Bed) technology has been successfully used in the petrochemical industry for almost half a century. Recently there has been developments in this technology in the pharmaceutical, nutraceutical, and biofermentation industry. The principles of process intensification if applied to existing processes can lead to development and redesign of existing processes to become more sustainable. Advantages of using SMB technology over other traditional unit operations will be discussed in this presentation.

This presentation will discuss a process and an adsorbent for the separation of ethanol and associated oxygenates from a dilute mixture of ethanol and associated oxygenates in water in the presence of organic compounds derived from a biofermentation process. After pretreatment, the separation is carried out in a simulated moving bed adsorption system employing a stationary phase adsorbent comprising fluorinated carbon or modified C18 silica gel selective for the adsorption of ethanol and associated oxygenates, such as 2,3-butanediol, with a mobile phase desorbent selected from the group consisting of methanol, ethanol, propanol, and methyl tertiary butyl ether.

An improvement to the above SMB process is also discussed in which increasing the number of raffinate streams alone or in combination with a hot regeneration zone within the SMB cycle can significantly reduce the capital and operating costs associated with the incorporation of the SMB process to produce ethanol and butanediol from biofermentation effluent. This process is an improvement over conventional distillation which is very energy intensive and not practical for dilute fermentation broths.

The process is useful for removing water from dilute aqueous mixtures of organic compounds comprising ethanol in dilute concentration in water and produced by fermentation, biomass extraction, biocatalytic, and enzymatic processes which are not economically recoverable by conventional distillation methods.

597551 Two-Zone Ligand-Assisted Displacement Chromatography for Producing High-Purity Rare Earth Elements with High Yield and High Productivity from Waste Magnets

Abstract:

Three rare earth elements (REEs), neodymium (Nd), praseodymium (Pr), and dysprosium (Dy), are essential ingredients of permanent magnets, used widely in electronics, motors, hybrid cars, generators, televisions, sensors, and windmills. Conventional methods for producing high-purity REEs employ two-phase liquid-liquid extraction methods, which require thousands of mixer-settler units in series or in parallel and generate large amounts of toxic wastes. In this study, a two-zone ligand-assisted displacement chromatography (LAD) system with a new zone-splitting method is developed for producing high-purity (>99%) Nd, Pr, and Dy with high yields (>99%) and high sorbent productivity from crude REE mixtures derived from waste magnets. The zone-splitting method based on selectivity-weighted composition factors enables a two-zone design to achieve two orders of magnitude higher productivity than that of a single column design. The design and simulation methods are based on first principles and intrinsic (or scale-independent) engineering parameters. They can be used to design processes for a wide range of feed compositions or production scales. The overall productivity of the two-zone LAD can exceed 100 kg REEs/m^3/day, which is 100 times higher than those of the conventional extraction methods. The LAD for the purification of the ternary mixture requires only three chromatography columns, a safe extractant, EDTA, and other environmentally friendly chemicals. Most of the chemicals can be recycled, generating little waste. This method has the potential for efficient and environmentally friendly purification of the REEs from waste magnets. The method may also help transform the current linear REE economy (from ores to pure REEs, to products, to landfills) to a circular and sustainable REEs economy.

595985 Parameter Estimation from Operation Data for Simulated Moving Bed Chromatography

Abstract:

Chemical products in daily lives, such as sugars, petrochemicals, and chiral compounds, are produced by chromatographic processes. Simulated moving bed (SMB) chromatography is one of the most widely used chromatographic processes. To design and operate an SMB process, a reliable mathematical model should be utilized without incurring excessive experimental costs. To this aim, one potential method is to utilize the data that contain product concentrations under different operating conditions; in many SMB processes in industries, such data are collected and stored in a large amount without effective use. Such data can potentially be used to improve the accuracy of the mathematical model obtained from single-column chromatographic tests.

In this study, based on the model correction method [1], we estimated the model parameters in the SMB model including those in a nonlinear isotherm. We obtained 35 datasets from a pilot plant study including product concentrations under different operating conditions. The SMB process employed the Advanced SMB (ASMB) operation developed by Organo Corporation [2]. As a case study, separation of fructose and glucose was considered. Our parameter estimation scheme with Tikhonov regularization found a parameter set from the data efficiently, and uncertainty of the estimated parameters was quantified as confidence intervals. Moreover, model validation was carried out using the experimental data that were not used in the estimation, which confirmed that our estimation improved the model prediction (Figure 1).

Reference: [1] S. Tie *et al.*, "Experimental evaluation of simulated moving bed reactor for transesterification reaction synthesis of glycol ether ester," Adsorption, vol. 25, no. 4, pp. 795–807, 2019. [2] Sato *et al.*, SEPARATION BY USE OF NEW ADVANCED-SMB PROCESS, 8th International Conference on Separation Science and Technology, Nagano, Japan 2008

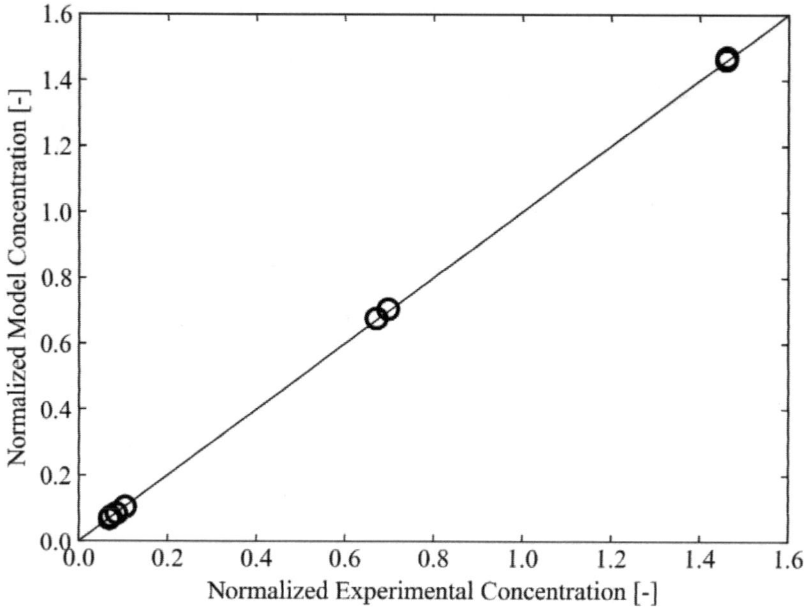

Figure 1. Model validation for normalized product concentrations by comparing predictions by the initial and the corrected model with experimental results.

605030 Continuous Separation of Multicomponent Aromatic/Aliphatic Mixtures By Simulated Moving Bed Adsorption: Combined Modeling and Experimental Approach

Abstract:

Simulated moving bed (SMB) chromatography has potential for high-efficiency separation of many valuable chemical mixtures. Much work in SMB-based separations has focused on binary mixtures, but considerably less attention has been devoted to the separation of multicomponent feeds. In this talk, we take a combined experimental and modeling approach to study the multicomponent separation of aromatics and aliphatics using a nano-porous silica adsorbent, which is of relevance in the separation of complex streams in petrochemical refineries. Our approach is based upon a more rigorous determination of adsorption, mass transfer, and SMB process model parameters based upon detailed experimental input (including column concentration profiles) from SMB experiments carried out in a 16-column "mini-plant" designed and constructed in-house at the Georgia Institute of Technology. Rigorous SMB process modeling with a simpler model mixture of components, along with SMB operation experiments with column-level composition sampling, allows us to demonstrate reliable prediction and validation of the effects of operational parameter changes on the compositional profiles and separation performance. We then generalize this approach by adding more components to the mixture, and obtain a very good agreement between model and experimental concentration profiles for a range of operating conditions and representative aliphatic and aromatic components in the feed. Hence, we demonstrate a robust multicomponent SMB process model with the capability to quantitatively predict the influence of key operating parameters such as extract, feed, desorbent and recycle flow rates, desorbent/feed ratio, and switch time on SMB separation results as well as column concentration profiles. Furthermore, conditions for clear separations of complex mixtures are also predicted and demonstrated.

602143 Flexible and Computationally Efficient Shortcut Distillation Column Models for Superstructure-Based Process Synthesis

Abstract:

Distillation is the most common unit process for separation in chemical/petrochemical facilities due to its capability to handle a wide range of feed flow rates/concentrations and to produce products with high purity[1]. Several methods have been proposed to estimate the energy requirement for a specific distillation task, while the Underwood equations[2] have been adopted in many works because of their simplicity. The Underwood equations enable the calculation of minimum vapor flow rates in the distillation column, which can be used to calculate the minimum energy requirement. Also, the Underwood equations can be extended to calculate the minimum energy of an entire distillation network to separate a zeotropic mixture[3]. Notably, the Underwood equations require a given set of components in the feed to calculate roots, which are essential to calculate the minimum vapor flow rates.

However, when we consider the synthesis of a system (e.g., separation network or combined reactor-separation network), component flow rates in the feed of a distillation system can vary, and some flow rates can be even zero, depending on decisions in other subsystems (e.g., reaction selections in upstream processes[4]). Variability in the feed, especially the presence of zero flow rates, complicates the use of the Underwood equations because it affects the number/location of their roots[5]. This limitation prevents us from adopting process synthesis approaches using superstructure-based optimization and thus finding novel solutions. Hence, it is desirable to develop a more flexible method that can handle this variability in the feed while successfully calculating the minimum energy for distillation.

Accordingly, in this work, we propose two novel reformulations of the Underwood equations, where the number/location of the roots are constrained depending on the presence of zero flow rates. Based on these reformulated Underwood equations, we propose two shortcut distillation column models which can be readily used for superstructure-based process synthesis. The proposed models can automatically identify adequate key components and corresponding energy requirement of a desired separation task while considering a broad range of types of splits including non-sharp/sloppy splits. Also, they can be readily utilized as submodules for the columns in the distillation network synthesis superstructure.

The key difference between the two models are 1) whether inactive roots are calculated and 2) how the recoveries of the key components are bounded. In the first model, only active roots (i.e., roots between the relative volatilities of the key components) are calculated depending on the key selection. The recoveries of the key components are bounded by constraints inspired by the Fenske equation, which can guarantee the physical feasibility of solutions with finite reflux and finite number of trays. In the second model, both active and inactive roots are calculated, and the inactive roots are used to enforce bounds on the recoveries of the key components, thereby guaranteeing that the solution is achievable with the minimum vapor flow rates. Also, based on the second model, we propose a shortcut distillation network model, which can be used to calculate an energy target for the separation of a zeotropic mixture without finding detailed distillation network configurations. This shortcut network model is particularly useful for preliminary process synthesis, where we want to estimate the energy requirement rather than to obtain detailed configuration of the network.

We close with a number of examples illustrating how the proposed models can be used to formulate and solve interesting and challenging process synthesis problems.

Reference

1. King CJ. *Separation Processes*. Courier Corporation; 2013.
2. Underwood AJ V. Fractional distillation of multicomponent mixtures. *Ind Eng Chem*. 1949;41(12):2844-2847.
3. Halvorsen IJ, Skogestad S. Minimum energy consumption in multicomponent distillation. 3. More than three products and generalized Petlyuk arrangements. *Ind Eng Chem Res*. 2003;42(3):616-629.
4. Ryu J, Kong L, de Lima AEP, Maravelias CT. A Generalized Superstructure-based Framework for Process Synthesis. *Comput Chem Eng*. 2019:106653.
5. Kong L, Maravelias CT. Generalized short-cut distillation column modeling for superstructure-based process synthesis. *AIChE J*. 2020;66(2):e16809.

604137 CFD Modelling of Thermal and Hydraulic Characteristics of a Rotating Packed Bed Reactor Used for CO2 Capture

Abstract:

As part of the nationally determined contributions (NDCs) defined within the United Nations Framework Convention on Climate Change (UNFCCC), an ambitious approach from each country is required to decrease the world's greenhouse gas emissions owing to their obligations towards the Paris Agreement. The prime share of these greenhouse gas emissions is coming from the power generation and industrial sector which alone encompasses 69.8% of the total gas emissions worldwide. In this context, pre and post-combustion carbon capture technologies have gained a lot of attention to mitigate climate changes along with meeting demands of the ever-rising power requirements around the globe with its 85% dependence on fossil fuel thermal power generation. Although research is being done to create greener power cycles, yet to meet the ambitious but essential goal set by the EU, carbon capture from the environment itself has become imperative. This novel idea is facing its own frontiers however, cost and size issues for example still need to be addressed. Rotating packed bed (RPB) reactor based on the process intensified (PI) technology could be a promising and cost-effective carbon capture technique. Experimental studies on the rotating packed bed reactors available in the literature provide vital information, nevertheless, constraints of cost and time confine the availability of enough experimental data required to understand complex flow patterns within RPB and to be used in the engineering design process to further improve its performance. High-end computational resources could be employed at this point to make up for the limited availability of experimental data. While CFD studies on the flow characteristics within RPB are available in the literature, investigations on the heat and mass transfer phenomena are rare. Since MEA solvents heat up during the carbon absorption process inside the RPB reactor evaluation of its heat transfer characteristics becomes a vital inquiry for its design as well as to address its intercooling requirements. Apart from thermal characteristics associated with RPB, hydraulic performance calculations bear equal significance as the pressure inside the RPB fluctuates incessantly subjected to varying flow area and centrifugal forces in the radial direction. The present investigation, therefore, focuses on the development of a CFD model so as to investigate the fluid flow and heat transfer characteristics inside the rotating packed bed reactor. The developed model is based on the volume of fluid (VOF) approach while its validation is accomplished using existing experimental data. Once model is validated, response surface method (RSM) is engaged to analyze the effect of control parameters associated with the RBP, i.e., liquid-gas flow rate ratio, inlet air-water temperature ratio, packing porosity, and rotational speed of rotor on the response parameters including heat transfer rates, liquid side Nusselt number, temperature drop, and pressure drops with its application to carbon capture. To implement the response surface methodology, the design of the experiment is carried out using a central composite design (CCD) technique while required data for RSM is generated using validated CFD model for various combinations of control parameters. Results suggest that the upsurge in the temperature of concentrated monoethanolamide (MEA) solvents is high enough to require intercooling. Moreover, granting all defined control parameters affect thermal and hydraulic characteristics associated with rotating packed bed, the impact of the inlet mass flow ratio and rotational speed is substantial and thus needs consideration in the design of RPBs. Finally, optimal ranges of the control parameters are reported based on the correlations developed using response surface methodology that will be helpful in further refining the design of rotary packed bed reactors.

602147 Distillation of Pyrophoric Materials UNDER High Vacuum

Abstract:

DISTILLATION OF PYROPHORIC MATERIALS UNDER HIGH VACUUM

Distillation is still the primary means of producing ultra-high purity raw materials for the microelectronic industries. While most microelectronic raw materials are distilled at atmospheric pressure, some materials such as dopants and other higher boiling point materials must be distilled at reduced pressures for either easier processing or to prevent thermal decomposition.

Many dopants must be distilled at extremely low pressures, which causes many problems with both product purity as well as significant safety concerns.

We will discuss process design as well as operating procedures with examples from both microelectronic materials as well as from the metal processing industry. Interesting aspects for system design and operations required for process safety will be discussed where these are closely related to high vacuum processing. These are especially important when dealing with materials that are air and/or water sensitive.

608599 Separation of Azeotropic Refrigerant Mixtures Using Extractive Distillation with Ionic Liquid Entrainers

Abstract:

Hydrofluorocarbons (HFCs) have been used as refrigerants globally since the 1990's and replaced chlorofluorocarbons (CFCs), which were linked to the depletion of the Earth's ozone layer. HFCs have zero ozone depletion potential (ODP), but some HFCs have a high Global Warming Potential (GWP). The next-generation of refrigerants, hydrofluoroolefins (HFOs), are being developed to replace HFCs in certain applications as a result of the Kigali Amendment to the Montreal Protocol. The Kigali Amendment became effective in January 2019 and limits the production and application of HFCs over the next two decades; therefore, the refrigerant industry is moving toward the production and marketing of HFOs and HFO/HFC refrigerant blends. This will require the recycling and repurposing of HFC mixtures, but many of the HFC mixtures are azeotropic or near azeotropic (i.e., R-410A, R-404A, R-407C, to name a few) making the separation of the components impossible using current methods.

Project EARTH (Environmentally Applied Research Towards Hydrofluorocarbons) aims to develop environmentally responsible methods to separate azeotropic HFC mixtures and recycle the pure component refrigerants. Extractive distillation is one method used to separate azeotropic mixtures with a similar boiling point. Ionic liquids (ILs) are being developed as novel entrainers for extractive distillation because of their negligible vapor pressure, thermal and chemical stability, and high and differential solubility with HFC refrigerants. Initially, Project EARTH will focus on the separation and recycling of a common commercial refrigerant mixture, R-410a, an azeotropic mixture containing 50 mass % R-32 (difluoromethane, CF_2H_2) and 50 mass % R-125 (pentafluoroethane, CHF_2CF_3). This presentation will provide results from ASPEN modeling and design and construction of the extractive distillation process for separating azeotropic HFC refrigerant mixtures.

610464 A Comparative Study of Protein a Membranes for the Rapid Purification of Monoclonal Antibodies

Abstract:

Chromatography is ubiquitous to therapeutic protein purification. When the protein is a monoclonal antibody (mAb), Protein A chromatography is used in the capture step to separate the product selectively from process impurities such as host cell proteins (HCPs), host cell DNA, viruses, charged variants, fragments, endotoxins and aggregates. Among the issues in resin-based Protein A chromatography is low throughput associated with flowrate-dependent performance. In particular, diffusion limitations require resin columns to be operated with long residence times to achieve satisfactory binding capacities.

Protein A membrane chromatography is a solution. Recent innovations in Protein A membrane chromatography have led to products with binding capacities on par with resin chromatography using residence times that are shorter by an order of magnitude or more. The objectives of this study were to provide the first preliminary comparison of all commercial Protein A membranes in terms of dynamic binding capacity, regeneration-reuse, and pressure-flow rate relationships. This presentation will report findings from a comparison of Purilogics Purexa™-A, Gore® Protein Capture Device, Cytiva HiTrap™ Fibro Prism A, and Sartorius Sartobind® A chromatography columns.

599542 Membrane Adsorbers for Medical Isotope Production

Abstract:

Radiopharmaceuticals offer promising new approaches for imaging and treatment for cancer patients; however, the current supply of these novel drugs is too low to support clinical trials. Copper-67 (Cu-67), can provide treatment (via beta decay) and imaging (via gamma rays). Cu-67 is purified using resin-packed columns, which are known to suffer from diffusion-limited transport. The practical implications of these limitations are (1) low column flow rates, (2) high elution volumes and (3) long purification times—all of which contribute to the bottleneck in the sustained supply of Cu-67. Due to the short half-life (2.58 days) of Cu-67, rapid purification is key for increasing the availability of the isotope for clinic trials and use.

A promising alternative to resin-packed columns are membrane adsorbers, which have seen recent success in the purification of biologics. This contribution describes our efforts to graft glycidyl methacrylate (GMA) from polyvinylidene fluoride (PVDF) membranes through activator generated by electron transfer, a type of, atom transfer radical polymerization (AGET ATRP). After grafting poly(GMA), diamine ligands of varying chain length (putrescine and ethylene diamine) are attached to the brushes through an epoxide ring opening reaction. Membranes are characterized by attenuated total reflection Fourier-Transform infrared spectroscopy throughout the synthesis process (grafting and ring-opening) to support functionalization. Membrane permeability was calculated from pure-water flux experiments in a dead-end filtration cell. The number of binding sites was quantified through acid-base titration and the binding capacity was calculated by modeling equilibrium batch adsorption data with the Langmuir isotherm. The results of this research are laying the groundwork for implementing membrane absorbers as a new separation material for medical isotope purification.

611028 Atomically Thin Graphene Membranes with Nanoscale Pores for Dialysis Based Separations

Abstract:

Atomically thin membranes from 2D materials offer unique opportunities for molecular separations with high permeance and high selectivity. Here, we report the fabrication of fully functional centimeter scale nanoporous (<2-3nm) atomically thin graphene membranes for molecular separation applications. Our work focuses on systematically addressing key challenges that have limited progress in field i.e. a) scalable nanopore formation methods including in-situ approaches, b) facile support casting, c) novel defect sealing methods and d) scalable manufacturing approaches. The synthesized nanoporous atomically thin membranes show higher selectivity and permeance (up to 100 times more) than state-of-the-art polymeric membranes and can be readily used for lab-scale dialysis.

Cheng P. et al. Nat. Nano. (submitted)

Kidambi P.R. et al. Adv. Mat. 2018

Prozorovska L. and Kidambi P.R. Adv. Mat. 2018

Kidambi P.R. et al. ACS App. Mat. & Int. 2017

Kidambi P.R. et al. Adv. Mat. 2017

Kidambi P.R. et al. Adv. Mat. 2017

Kidambi P.R. et al. Nanoscale 2017

607399 Effects of Flexibility and Ionic Strength on Membrane Filtration of Poly(dT_{60}) Single-Stranded DNA

Abstract:

Certain unique features inherent to membrane filtration technology offer several advantages for a variety of applications. Microfiltration (MF) and ultrafiltration (UF) membranes are commonly employed in the food processing, biotechnology, pharmaceutical, and water/wastewater treatment industries. Membrane filtration of flexible particles has sparked a renewed interest in DNA purification in pharmaceutical companies and filamentous virus filtration in water treatment plants. The ability of flexible particles to adopt conformations and enter membrane pores influences their filterability. Understanding the effects of mechanical flexibility on the rejection coefficient is crucial for designing and implementing of membrane separation. The main objective of this study is to obtain quantitative data on the transmission of single-stranded DNA (ssDNA) -which is designed for DNA therapeutics- through ultrafiltration membrane by considering the factors that govern the transmission of ssDNA through membrane pores. The effects of flexibility, transmembrane pressure (TMP), solution ionic strength, and membrane pore size on the rejection coefficient were investigated using track-etched polycarbonate ultrafiltration membrane in a stirred cell set-up. To characterize flexibility of DNA, Fluorescence Recovery After Photobleaching was integrated with Worm Like Chain polymer model to determine the persistence length. The results demonstrate that increasing TMP results in decreasing the rejection coefficient due to the ability of DNA to deform and pass through the pores easier in higher pressure. In addition, studying the effects of solution ionic strength shows the electrostatic interactions between the same charged DNA and membrane increase DNA rejection. Results generated from this study will be important for the optimal design of micro- and ultrafiltration systems for water and wastewater treatment and DNA purification in the biotechnological and pharmaceutical sectors.

611013 Antimicrobial Surfaces Applied in Dairy Membrane Modules

Abstract:

Downstream contamination of dairy processes by Listeria monocytogenes (LM) is a serious issue because of the associated mortality rate in immune-compromised individuals. Dairy membranes can be modified to include solid-phase antimicrobials to kill or mitigate growth of LM. The feed spacer mesh of spiral wound dairy membranes has been functionalized with Quaternary Amine Compounds (QACs), a family of broad spectrum antimicrobials that are effective against Gram-positive bacteria. The antimicrobial is covalently bound to the polypropylene surface to avoid leaching into the product stream. Growth curve studies with a surrogate have demonstrated the effectiveness of the surface, and well as the necessary robustness to withstand clean-in-place conditions that will be required when implemented in industry. We have also functionalized the surfaces with polymerized phenolic compounds and comparisons were made with QACs. Preliminary results will also be presented in milk and brine systems.

606111 High-Throughput Computational Screening of Metal-Organic Frameworks for Selective Adsorption of C_2H_6 over C_2H_4

Abstract:

C_2H_6/C_2H_4 separation is one of the most crucial processes in the petrochemical industry and currently practiced by energy-intensive distillation technology. As an alternative, adsorption-based separation is considered as technically feasible and economically viable, and there has been considerable interest to develop advanced adsorbents for C_2H_6/C_2H_4 separation. In this study, we computationally screen a large set (> 10000) of metal-organic frameworks (MOFs) to identify candidates for selective adsorption of C_2H_6 over C_2H_4. The quantitative relationships are established between adsorption performance metrics (C_2H_6 working capacity and C_2H_6/C_2H_4 selectivity) and structural descriptors (pore size and surface area), as well as isosteric heat. Top-performing 16 MOFs are identified with C_2H_6/C_2H_4 selectivity > 2.16 and C_2H_6 working capacity > 0.53 mol/kg, which are superior to the literature benchmarked MOF namely ZJU-120a. It is revealed that both pore size and pore shape are dominant factors to govern C_2H_6-selective adsorption. Moreover, the separation performance in MOF duplicates and geometrical analogues is elaborately analyzed and categorized. Finally, six strategies are proposed to boost C_2H_6/C_2H_4 separation by introducing catenation, aliphatic/aromatic linkers, pillar layers, framework flexibility, topological variants, and different metal nodes. These strategies provide bottom-up guidelines for the rational design of new MOFs for high-performance C_2H_6/C_2H_4 separation.

600270 Multi-Layer Connectivity-Based Atom Contribution (m-CBAC) Approach: Fast and Accurate Charge Assignments to MOFs for Adsorption Simulations

Abstract:

Metal-organic frameworks (MOFs) have shown great potential in a variety of energy-related applications such as gas separations and storage. With increasing interest in this emerging class of novel materials, molecular simulations can play a critical role in the efficient exploration of their adsorption properties and thus the MOF discovery. In these calculations, accurate charge assignments to the framework atoms are essential. To facilitate the large-scale computational simulation of MOFs, we develop a multi-layer connectivity-based atom contribution (m-CBAC) approach as an efficient, robust, and accurate approach for charge assignments. Different from the original connectivity-based atom contribution (CBAC) method, which uses 1st layer connectivity (i.e., directly bonded atoms) of a target atom, we expand on to incorporate multi-layer connectivity from the 0th layer up to the 2nd layer. This m-CBAC approach is developed with an automatic procedure of a supervised machine learning for classification, and it also allows further improvements along the time with more data to be included in the training set. In this study, an extensive set of ~2700 MOFs from the CoRE MOFs database with the density derived electrostatic and chemical (DDEC) charges are used to train the m-CBAC model. For the charge predictions, the m-CBAC approach assigns charges by successively searching from the highest-level layer (i.e., 2nd layer, presumably the most accurate) to the lowest-level connectivity (i.e., 0th layer, element type). With this approach, it becomes possible to assign charges for essentially all MOFs, with a high accuracy, and within seconds for a single structure. The m-CBAC approach is orders of magnitude computationally less expensive than quantum mechanical methods. The charges assigned using m-CBAC resemble the DDEC charges very well with a Pearson coefficient of as large as 0.988. The accurate charge assignments from m-CBAC are also reflected on their reliable predictions of CO_2 Henry coefficients in MOFs. Besides, a large-scale computation of the CO_2 Henry coefficients for ~12,000 MOFs in the recently released CoRE database has proved the capability of the m-CBAC approach. Overall, the m-CBAC approach can enable fast charge assignments for diverse MOFs with a good accuracy, and the software for m-CBAC charge assignments is made available along with this work.

602112 Incorporating Intrinsic Flexibility Effect into MOF Adsorption Properties Simulation at a Low Computational Cost

Abstract:

Computationally predicting the adsorption properties of Metal organic Frameworks (MOF) in a high throughput way has become a useful means of identifying promising materials for chemical separations. In almost all cases, high throughput simulations assume MOF structures as rigid for computational convenience. All MOFs, however, have intrinsic flexibility due to thermal vibration of atoms. Recent studies have suggested that this ubiquitous flexibility may have significant impacts on molecular adsorption in MOFs. Methods have been developed to predict the influence of this type of flexibility on MOF adsorption properties using molecular dynamics (MD) simulations, but these methods are computationally expensive and rely on the existence of reliable force fields.

We will describe a method of estimating the impact of intrinsic flexibility on MOF adsorption, which does not use any force fields for MOF framework. This approach uses experimentally available thermal vibration parameters to provide information about local flexibility. We show how this information can be used to reliably assess whether atomic vibrations make a significant impact on molecular adsorption in MOFs. Data from a variety of single component and binary mixtures adsorbed at dilute and non-dilute conditions forms our test sets. This method creates a useful way to establish the level of detail that is required for molecular modeling of adsorption in experimentally accessible MOFs to give quantitatively reliable results.

601234 Combining High-Throughput Molecular Simulations and Machine Learning to Optimize Adsorption Processes

Abstract:

High-throughput molecular simulations utilizing massively parallel high-performance computers allow now for the generation of an unprecedented amount of adsorption data. However, given the high dimensionality of state space (temperature, pressure, and composition) and/or the large number of porous materials governing the adsorption behavior, molecular simulation alone is not effective in finding optimal materials and conditions for adsorption processes. This talk will highlight the development of surrogate machine learning models that are trained on molecular simulation data and enable finding optimal material/condition combinations for gas storage or chemical separation applications: (a) desorptive drying for hydrogen-bonding solute/solvent mixtures in all-silica zeolites, (b) adsorption of BTEX mixtures onto zeolite nanosheets at membrane reactor conditions, and (c) gas storage in diverse porous materials.

596883 Computational Screening of Metal-Catecholate Functionalized Metal-Organic Frameworks for Room Temperature Hydrogen Storage

Abstract:

Hydrogen is a promising alternative to fossil fuels as it is non-toxic and carbon-free. However, due to its extremely low volumetric density, the storage and transport of hydrogen for practical applications remain a significant challenge, and high pressure and/or cryogenic temperature is required. Adsorption-based storage utilizing nanoporous adsorbents such as metal-organic frameworks (MOFs) can greatly reduce the storage pressure, but cryogenic or sub-ambient temperatures are required with current adsorbents, which limits the scope of applications. In this work, we aim to search for hydrogen-storage adsorbents that allow room temperature operation by looking at MOFs functionalized with metal catecholate groups, which have highly unsaturated open metal sites and thus greatly enhanced binding strength for hydrogen. We screened a dataset of 2736 zirconium-based MOFs that was constructed in a combinatorial fashion with wide varieties of topologies and linkers. By counting the possible sites that can be functionalized with metal catecholate groups, we were able to obtain the theoretical maximum hydrogen uptake in gravimetric and volumetric units for all of the MOFs. For the 100 MOFs with the highest theoretical uptakes, we built the functionalized structures on the computer and conducted grand canonical Monte Carlo (GCMC) simulations to predict the hydrogen uptake at adsorption (296 K, 100 bar) and desorption (296 K, 5 bar) conditions. The deliverable capacities were calculated by taking the difference of uptakes at those conditions, and we observed >5 wt% and >20 g/L deliverable capacities for some MOFs, which are very high for room temperature pressure swing adsorption cycles. Structure-activity relationships between the structural properties of MOFs and their hydrogen uptakes were also established.

611189 Pore Level Decomposition of Adsorption Isotherms in Metal Organic Frameworks

Abstract:

Metal Organic Frameworks (MOFs) demonstrate a variety of unique features due to their ability to selectively adsorb and retain guest molecules from gas and liquid phases. Adsorption properties are directly related to the specifics of the pore structure. A better understanding between the MOF structural parameters and adsorption properties is required for the guided design of novel MOFs and their modifications for advanced applications. MOFs are considered as nanoporous crystals processing regular 3-dimensional pore networks composed of pore compartments than differ in size, shape and chemistry. Here, we present a Monte Carlo (MC) simulation methodology for decomposition of the adsorption isotherm into the fingerprint isotherms, which reflect adsorption in individual pore compartments [1]. We consider two characteristic MOF structures, PCN-224 and ZIF-412, for which the distribution of adsorbate between the individual pore compartments was studied experimentally using adsorption crystallography based on *in-situ* XRD [2]. The pore network in PCN-224 represents a cubic network of alternating micropore channels and junctions of size 1.5 and 2.5 nm. The pore network in ZIF-412 has a complex micro-mesopore structure with three type, with pores of ca. 2.1, 2.5, and 3.9 nm. Employing MC simulation, we constructed the adsorption isotherms of three most common adsorbate probes (argon, carbon dioxide and nitrogen) in the ideal crystallographic models of PCN-224 and ZIF-412 and decomposed them into the fingerprint isotherms for the individual pore compartments. The results of simulation are found in qualitative agreement with the experimental data [2]. This work is motivated by the need of novel pore structure characterization methods, which account for the multiplicity of morphologies of different families of MOFs. Comparison of the reference fingerprint isotherms with the experimental adsorption isotherms allows to assess to what extend the structure of the practical sample differs from that of the ideal crystal due various types of defects, residual solvents, degree of hydration, pore blockage, etc.

This work is supported by the National Science Foundation (CBET grant no. 1834339)

1. Dantas, S.; Neimark, A. V., Coupling Structural and Adsorption Properties of Metal-Organic Frameworks: From Pore Size Distribution to Pore Type Distribution. *ACS Appl Mater Interfaces* **2020,** *12* (13), 15595-15605.
2. Cho, H. S.; Yang, J.; Gong, X.; Zhang, Y. B.; Momma, K.; Weckhuysen, B. M.; Deng, H.; Kang, J. K.; Yaghi, O. M.; Terasaki, O., Isotherms of individual pores by gas adsorption crystallography. *Nature Chemistry* **2019**.

602529 Modeling of Adsorption in Pristine and Defective Metal Organic Framework Materials

Abstract:

Metal-organic frameworks (MOFs) have attracted significant interest for adsorption due to their high degree of tailorability and large specific surface areas. The combination of tailorability and well-defined crystalline pores makes MOFs very promising candidates for highly selective adsorption.

Molecular-level modeling of adsorption and diffusion in MOFs almost always relies on models of MOFs that are defect-free (pristine). However, all real MOFs have defects, which affect adsorption by changing the types of pores within the MOFs. A fundamental understanding of how defects impact adsorption is important for identifying the limits of the performance of real material, developing improved design rules for new improved materials, and predicting and maximizing utilization of the material.

In this study, we focus on the study of how adsorption correlates with the number and types of defects. We initially consider UiO-66 with different levels of missing linker defects. The structures of the generated defective MOFs were optimized using periodic density functional theory with the CP2K simulation package. Adsorption isotherms were generated by carrying out grand canonical Monte Carlo (GCMC) simulations in Raspa. We investigated the effect of different adsorbate-adsorbent charge schemes by comparing isotherms with no framework charges and atomic charges calculated using DDEC6 and EQEQ methods. Different potentials for adsorbate were also tested for more accurate modelings.

595527 Purification of Monoclonal Antibodies Using a z^2 Membrane Chromatography (z^2MC) Module

Abstract:

Monoclonal antibodies (mAbs) comprise the largest category of biopharmaceuticals. Protein-A affinity chromatography is the standard method by which mAbs are purified. Protein-A resins are prohibitively expensive and the separation process involves mAb elution under acidic pH conditions, which could potentially destabilize the protein and promote its aggregation. Moreover, the protein-A ligand cannot distinguish between mAb charge variants. Also, aggregated forms of the mAb can bind to protein-A and therefore any aggregates present in the cell culture supernatant is co-eluted with the mAb. Hence, an alternative to protein-A based separation would be desirable. In this study, we examine mAb purification using two different types of chromatographic devices having equivalent volume: resin based (Capto S ImpAct) cation exchange chromatography column, and cation exchange z^2 membrane chromatography or z^2MC module, a novel separation device designed and developed within our group. The z^2MC device addressed some of the problems associated with the resin-based column. For example, the product obtained with the z^2MC device was purer than that obtained with resin based cation exchange chromatography and the separation process was significantly faster. In this presentation we highlight the advantages of using the z^2MC module for high-resolution protein separations.

605417 Understanding Membrane Performance during Combined Tangential Flow and Depth Filtration

Abstract:

Rapid advancement in upstream cell culture operations has led to a significant increase in product titers. However, this high-level of productivity is accompanied by a rather high cell density which places a significant burden on traditional downstream clarification and purification operations. This is especially true for the initial clarification steps. Depth filtration, centrifugation and tangential-flow filtration (TFF) are used to remove cells, cell debris and other impurities. Depth filters consist of a thick porous bed that can trap particles within the filter matrix. Screen type filters on the other hand largely reject particulate matter by surface filtration. In tangential flow filtration the feed flows parallel to the membrane surface while the filtrate flows perpendicular to the feed. This mode of operation tends to suppress cake growth. The BioOtimal MF-SL from Asahi KASEI represents a hybrid TFF-depth filter. The performance and especially fouling behavior of the BioOptimal filter is not well understood. Here a combined pore blockage and cake filtration model has been developed to describe the fouling behavior of this filter in normal flow mode whereas a resistance in series model has been used to describe the performance in TFF mode. Significant insights have been obtained into the fouling mechanism and optimal operation conditions of this filter.

609240 Membrane Selectivity for Protein Filtration Probed By Atomic Force Microscopy

Abstract:

Even though synthetic membranes have been around since the early 20th century, predicting their separation performance remains a substantial challenge. This work shows how atomic force microscopy (AFM) could be used to probe membrane selectivity for bio-separations. We selected 0.2 μm PVDF and PES membranes that were surface functionalized to render them hydrophilic, and that had relatively low and high affinity toward proteins, respectively. These two membranes, characterized by SEM, ATR-FTIR, Z-potential and contact angle, showed similar properties in terms of surface charge and hydrophilicity. Thus, these techniques were unable to predict different membrane selectivity toward biomolecules during filtration. Streptavidin (SA) was selected because of its stability in a wide range of operating conditions, and immobilized onto nano- and micro-particles for filtration experiments and atomic force microscopy measurements, respectively. To gain insight into the interaction forces between particles and membranes, the solution conditions (pH = 4.0, 5.0, 7.4, 9.0 and ionic concentration [salt] = 1, 10, 100 mM), and nanoparticle support material (polystyrene, silica and gold) were varied. AFM was used in force-volume mapping mode: 400 measurements, scanning 20 x 20 μm^2 surface area for each condition. Based on adhesion forces and protein pulling events, we identified stronger interactions between SA and PES compared with SA and PVDF. These data correlated with results from filtration experiments. This work demonstrates that AFM could be used to screen proteins and operating conditions in order to reduce fouling or increase protein capture.

611489 Multimodal Membrane Adsorbers for Clearance of Minute Virus of Mice

Abstract:

During purification, manufacturers of mAbs and other therapeutic proteins must demonstrate reduction of 10^3 -10^5 or more virus particles than is estimated in a single dose equivalent of the unprocessed bulk. The level of virus clearance is calculated by summing the clearance obtained from individual unit operations having different mechanisms of action (e.g. size exclusion and ion exchange). Depending on the antibody titer, estimates of the number of virus particles in a single dose equivalent could be as high as 10^{10}-10^{15} retrovirus-like particles per mL. In addition, removal of adventitious virus using a model parvovirus is required. In recent years, several membrane adsorbers with high binding capacity have evolved for the removal of contaminants for the polishing steps during downstream purification of biologics. In particular, multimodal membrane adsorbers with tunable pore sizes and ligand properties have generated tremendous interest due to their many possible applications. Here the removal of Minute Virus of Mice (MVM), host cell proteins (HCPs) and DNA using multimodal membrane adsorber Puri™-MQ was studied. The membrane performance was evaluated over a range of feed and elution conditions. The effects of pH and HCP concentration on MVM clearance were investigated. Comparison with other membranes at similar conditions demonstrates that multimodal membranes are highly effective for virus clearance.

611229 Thermodynamic Mechanism of Self-Heat Recuperative Distillation System

Abstract:

In the last several years the authors have been developing the self-heat recuperation technology for the heating and cooling thermal processes in which the heat of an effluent stream is recuperated and recycled for feed stream heating by the adiabatic compression of gas and/or vapor. Applying this technology to the distillation system, a drastic energy saving of more than 80% can be achieved.

In this study, the thermodynamic mechanism of self-heat recuperative distillation system is clarified in terms of exergy analysis by using module expression process flowsheet and temperature-entropy diagram.

595045 A Systematic Procedure for the Optimal Design of Operable and Flexible Reactive Distillation Systems

Abstract:

The chemical industries have recently been focusing on developing intensified operations (processes and/or units) in an effort to reduce environmental impact (e.g. energy consumption, waste treatment etc.) while maintaining or improving their competitiveness in the global market. Reactive distillation is one of the most established examples of Process Intensification (PI) and it combines reaction and separation into one single unit, offering potentially significant energy and/or capital savings as well as improvements in reaction selectivity and yield (Luyben and Yu, 2008 and Kiss, 2013). As the operating windows for reaction and separation may differ, and may only overlap for limited conditions, the design and operation of the process may be challenging (Harmsen, 2007). When also facing parameter uncertainties (e.g. in reaction kinetics) this challenge is amplified, and as a consequence, the implementation of reactive distillation in industry is slow. Although recent research has elucidated many aspects of reactive distillation, the extension of conventional distillation design techniques to reactive systems is still a challenge especially when the reactions is relative slow, thus requiring a significant liquid hold-up, which is normally minimized in conventional distillation equipment. Therefore, a systematic and rigorous method for the design of reactive distillation processes under uncertainty, including an evaluation of the associated controllability, has not yet been fully established (Wang et al., 2010).

Different optimisation strategies have been used for the design of reactive distillation systems. One of the most rigorous concepts is optimisation based on a process superstructure, a concept which was first introduced for the optimal design of simple distillation columns (Sargent and Gaminibandara 1976). This network representation method determines all the desired column design and operation specifications by solving a mathematically complex optimization problem. In this work, the concept of distillation column superstructures will be extended to reactive distillation systems in order to determine both structural and operational decisions including but not limited to: total number of stages, feed stage locations, existence of side draw streams, reactive zones etc. In addition to a single reactive distillation column, the synthesis of more complex reactive distillation processes, e.g. including a pre-reactor, a sequence of reactive/non-reactive distillation columns etc., will also be considered in order to investigate a large number of possible process alternatives.

The optimization problem considered is a Mixed-Integer Non-Linear Problem (MINLP), as the design variables are both continuous (e.g. reflux ratio) and integer (e.g. number of stages). Flowsheets with such a high level of complexity often suffer from initialisation issues as well as from the existence of a large number of local optima due to their non-convex nature. A demanding optimisation strategy is therefore employed to ensure that initialisation issues are tackled and to increase the possibility of locating the global optimum. The solution of the MINLP firstly determines the existence of the units and secondly, their design and operating parameters in case they exist. In this way, the optimal design of various process alternatives can be determined, thus allowing the selection of the most suitable process alternative based on the given objective function, in our work a cost function. The software used in this work is gPROMS ProcessBuilder (Process Systems Enterprise, 2020), selected due to its rigorous mathematical solvers and excellent optimization capabilities.

The optimisation strategy is demonstrated using a number of different chemical systems with varying difficulty in terms of the separation task (easy/difficult separations) and different kinetics (fast/slow reactions), and chemical equilibrium (towards products or reactants), in order to illustrate how the strategy can provide insight into which factor(s) (reaction, separation or their combination) is more dominant for the design and operation of the intensified process. The MINLP considered for each case aims to minimize the total annualised (capital and operational) cost of the process, subject to constraints such as required product purity.

As real plants are generally subject to disturbances, the controllability and operability of the process also need to be evaluated before settling on a process alternative. The control characteristics of the optimal steady state process found using the superstructure strategy are therefore initially investigated in the frequency- and time-domains by considering controllability indices and the performance of conventional control schemes under various disturbances which could potentially be introduced (e.g. feed flow rate disturbance), respectively. This investigation will indicate how operable the optimal design is, how reaction kinetics and separation parameters impact on the stability and controllability of a process and will establish whether any adjustments are needed in order to render it viable from a controllability perspective. The frequency domain analysis is performed using MATLAB (MATLAB 2020) and the time-domain analysis using gPROMS ProcessBuilder (Process Systems Enterprise, 2020).

The optimal design is obtained by the rigorous optimization of a mathematical model as described above. This model, however, depends on a number of model parameters, which are key for the performance of reactive distillation systems. A robust design should tolerate a certain level of parameter uncertainty otherwise it may be considered practically infeasible. For this reason, the final step of the methodology presented in this work is the investigation of the impact of uncertainty on the optimal design to determine exactly how sensitive the optimal process is to key model parameters and if needed, will revise the optimal design accordingly.

Based on a number of industrially-relevant case studies, this work will demonstrate how the optimal design, control performance and sensitivity to parameter uncertainty can be evaluated using a Process Systems Engineering approach for reactive distillation systems. Challenges around the formulation of the superstructure (flowsheet initialisation, software limitations etc.), as well as methods employed to overcome those, will be described. The importance of computational tools which can accommodate all steps of the investigation, and therefore guarantee consistency in the results supporting the development of systematic methodologies, will be highlighted.

In conclusion, this work will provide a systematic methodology for how to select the most suitable reactive distillation process design, based on various (simple and more complex) process configuration alternatives as expressed by the superstructure, that is operable as well as flexible enough to tolerate input uncertainty. A detailed insight into the economic and energy benefits (expressed by the optimal MINLP results) as well as into the potential operability/flexibility limitations (found using the controllability/uncertainty analysis) will be provided which may hopefully pave the way for an even greater acceptance of reactive distillation systems.

References

Harmsen, G. J. (2007). Reactive distillation: The front-runner of industrial process intensification: A full review of commercial applications, research, scale-up, design and operation, *Chem. Eng. Process.*, 46, 774–780.

[1]Kiss, A. A. (2013). *Advanced Distillation Technologies: Design, Control and Applications*, Wiley, Chichester.

Luyben, W. L. and Yu, C. C. (2008). *Reactive Distillation Design and Control*, Wiley, New Jersey.

Process Systems Enterprise, gPROMS, www.psenterprise.com/gproms, 2020.

Wang, S. J. et al. (2010). Plantwide Design of Ideal Reactive Distillation Processes with Thermal Coupling, *Ind. Eng. Chem. Res.*, 49, 3262-3274.

MATLAB (2020). 9.4.0.813654 (R2018a). Natick, Massachusetts, The MathWorks Inc.

Sargent, R. W. H. and K. Gaminibandara (1976). *Optimum Design of Plate Distillation Columns. Optimization in Action*;. London, Academic Press.

605909 A Comparison of Cryogenic Air Separation Processes Based on Exergy Analysis

Abstract:

Cryogenic distillation is the most widely used technique for industrial-scale production of oxygen and nitrogen from air. A two-column process is commonly employed. This work aims to compare seven (7) alternative two-column designs for the cryogenic air separation process based on exergy analysis.

Steady-state simulation models are developed using Aspen Plus®. The feed conditions (500 ton/h of ambient air with 50% relative humidity) and product purities (99% for both oxygen and nitrogen) are fixed for all models. Steady-state simulation results are used to calculate the exergy efficiencies of various equipment and their contribution to the total exergy destruction. It is observed that the compression section is the largest source of exergy destruction and accounts for ca. 30–45% of the total exergy destruction in the process. The low-pressure column (ca. 20–40%) and the multistream heat exchanger (ca. 10–20%) are the 2nd and the 3rd largest sources of exergy destruction, respectively. Less than 10% of the total exergy destruction occurs in the high-pressure column.

The total exergy destruction rate is ca. 32.6–38.4 MW for various process designs. A thermally-integrated Petlyuk-like arrangement of high- and low-pressure columns, where the air feed is split into 2 fractions, both of which are fed to the high-pressure column, and where both products are obtained from the low-pressure column, offers the lowest total exergy destruction.

605884 A Heat-Pump-Assisted Reactive Distillation Process for Methyl Lactate Hydrolysis

Abstract:

Reactive distillation (RD) is a prominent example of process intensification. From a thermodynamic perspective, continuous removal of product(s) from the reaction mixture allows the reaction equilibrium to shift in the forward direction, improving both the conversion and the selectivity of the process. From an economic perspective, combining reaction and separation in one physical equipment can significantly reduce both the capital investment and the operating expenses. However, because of the relatively low thermodynamic efficiency of conventional distillation superstructure, the conventional RD processes are still highly energy intensive. Many alternative designs have been reported in literature that can improve the overall process efficiency by reducing the thermodynamic losses in distillation columns.

In this study, we demonstrate the advantage of a heat-pump-assisted design over a conventional design for methyl lactate hydrolysis in an RD column. First, a rigorous simulation model is developed for the conventional RD column in Aspen Plus® and reference cost analysis is performed. Then, an alternative design based on mechanical vapor recompression (VRC) is developed and optimized for maximum energy savings. The optimized VRC system shows that the reboiler (that is, the hot utility) can be eliminated from the process. Overall, the VRC-based RD column offers 54–56% savings in direct operating/energy costs and, assuming a payback period of 3 years, 18–20% savings in the total annualized cost (TAC) relative to the conventional RD column.

596992 Diffusion Growth Mechanism of Five-Fold Twinned Ag Nanowire

Abstract:

Five-fold twinned nanowires can be produced with high aspect ratios from diverse face-centered cubic (fcc) metals. However, the mechanism of their anisotropic growth is still poorly understood. We develop a model combining deposition and surface diffusion to predict the diameters and the aspect ratios of Ag nanowires. We describe two aspects of nanowire growth, denoted as the seed process and the wire process. In the seed process, the deposition time is much longer than the surface diffusion time, so that the seed shape only depends on the difference between interfacet diffusion times. The structure of nanowire seeds is similar to Marks decahedra with {111} notches and {110} steps that accelerate interfacet diffusion leading to wire growth. The result exhibits that the decahedron seeds could lead to three kinds of products with different strucutures as nanowires, decahedron particles and giant decahedron seeds. We focus on the seeds growing to nanowires and introduce the wire process, which occurs subsequently and continues until the diffusion time is equal to the deposition time. We predict the aspect ratio by contrasting the diffusion rate from {100} to {111} to the deposition rate on {100}. The aspect ratios predicted in our simulations are in the experimental range.

597237 Volume Effects on Acetaminophen Nucleation through an Induction Time Approach

Abstract:

Confined crystallization is an approach to develop nano-crystalline active pharmaceutical ingredients which improves dissolution and bioavailability properties of a formulated drug. To elucidate on this, the effect of volume reduction to the nucleation of acetaminophen in ethanol was explored. Using a supersaturated solution of acetaminophen in ethanol at volumes of 1000 μL to 25 μL, induction times for the primary nucleation event were obtained experimentally. The probability distribution of induction times showed that as sample volume is reduced, the induction time increases exponentially. The induction time at the volume of 25 μL is obtained to be at 4214 minutes, which is at a 5400% increase from the observed induction time of 78 minutes when the volume is set at 1000 μL.These results demonstrate the effect of volume reduction to nucleation induction times. Confinement may relate to very low probabilities of a nucleation event.

600993 Amorphous Intermediate States during the Precipitation of Cerium Oxalate: Towards a New Model

Abstract:

The precipitation is a common step in industry which is relatively ill-understood. In the last decades, many efforts were made to make numerical simulations of this step to avoid using the trial and error method, especially for the precipitation of active compound like plutonium oxalate. Many parameters are required to handle these numerical simulations, including the nucleation rate (the number of particles which appears in the reaction media by unit of time and of volume).

The nucleation rate is usually rationalized according to the Classical Nucleation Theory (CNT). This theory predicts the nucleation rate, assuming that thermal fluctuations of the reaction media lead to the formation of some "clusters" of monomers with the same composition and (?) symmetry as the crystal. These clusters tend to immediately dissolve into the reaction media unless they reach a particular size, called the critical size, from which they can grow fast to a macroscopic size. The so-called nucleation rate is the rate at which clusters reach this critical size. However, the CNT is more and more questioned, as it has been evidenced for many systems, including cerium oxalate, that nucleation occurs through amorphous states which are currently ignored by the CNT and which lead to an significant underestimation of the nucleation rate, up to 400 orders of magnitude.

In this work, the nucleation of cerium oxalate, a non-radioactive surrogate for plutonium oxalate, has been studied using cryo-microscopy and small and wide x-ray scattering (SAXS and WAXS). Here we show that cerium oxalate has two possible non-classical nucleation mechanisms: at low concentrations, nucleation in presence of amorphous nanoparticles, and at higher concentration, nucleation involving both a transient dense liquid phase and amorphous nanoparticles.

601001 Lanthanide Precipitation from Spent Ni-MH Battery Leachates: Kinetic Study and Phase Identification

Abstract:

Securing access to critical metals required for high-performance technologies, particularly the light rare earth elements (REEs = La, Ce, Nd, Pr), has become a major challenge for import-dependent economies such as the EU. The recycling of spent nickel metal hydride (Ni-MH) batteries from hybrid electric vehicles (HEV) serves as an increasingly-attractive secondary source of REEs thanks to the recent development of hydrometallurgical processes.[1] In order to recover REEs, precipitation from pregnant leach solutions (PLS) in sulfate media, using Na_2SO_4 is often reported in the literature. However, little consideration is given as to whether and how sodium ions influence the precipitation efficiency and selectivity, and detailed phase characterization of the products is rarely reported. This work focuses on a better understanding of the precipitation phenomenon and mass balances in this complex system, by coupling pilot-scale experiments on industrially-sourced PLS, in-depth precipitated particles identification, and thermodynamic calculations.

The initial PLS used in this study originates from the pilot-scale leaching of industrial samples of spent Ni-MH battery powders in H_2SO_4 media, according to a process described in [1]. The metallic content of the battery waste leachate consists of 71%mol Ni, 10 %mol REEs and 19%mol of other metals like Co, Mn and Fe. In the view of metal recycling, it is of great interest to find a selective precipitation pathway to recover REEs from such a complex solution. For each precipitation experiment, 1.5 L of PLS was added to a double-jacketed batch reactor; solution temperature was regulated using a cryostat and pH was monitored. In a standard experiment, a 2.9 M Na_2SO_4 solution was maintained at 40 °C and added to the PLS at constant flow rate, in simple jet, using a membrane pump. Mechanical agitation was implemented with a 3-blade Teflon marine propeller at 400 rpm. The influence of three operating parameters on precipitation efficiency and kinetics was studied: temperature (25 – 60 °C), Na:REEs molar ratio (1:1 – 4:1, by increasing the total addition time of Na_2SO_4 solution at a constant flow rate) and Na_2SO_4 solution addition (all at once or slowly at 7 $g.min^{-1}$ implying an initial semi-batch mode). Suspension volumes were regularly sampled, filtered on 0.45 µm syringe filters, and solutions were analyzed by ICP-OES. Precipitation efficiency was calculated as the proportion of the REEs initially present in solution that have precipitated. After 1 h, the suspension was filtered on a P3 Büchner filter and rinsed with demineralized water at room temperature. The obtained crystals were dried in an oven at 80 °C overnight before in-depth characterization using multiple techniques: SEM-EDX, powder XRD, TGA, microwave digestion in aqua regia and ICP-OES.

Results compiled in Figure 1) a) illustrate the evolution of the proportions of Ni, Co, Mn, Fe, La, Ce, Nd and Pr remaining in solution when precipitation is carried out at 60°C, for a Na:REEs molar ratio of 4:1 and a Na_2SO_4 solution flow rate of 7 $g.min^{-1}$. The simultaneous drop of pH when REE concentrations start to decrease (Figure 1) b) demonstrates that pH monitoring serves as an indicator of precipitation initiation and therefore allows determining the induction period, which is of about 3 min in this case. Curves in Figure 1) a) flatten out after approximately 30 min, suggesting that equilibrium has been reached. After 1 h, the precipitation efficiencies of La, Ce, Nd and Pr reach 99.3 %, 100.0 %, 99.7 % and 87.6 %, respectively, corresponding to 70 ppm of La, 0 ppm of Ce, 4 ppm of Nd and 52 ppm of Pr remaining in solution. REE precipitation is highly selective in this configuration since 99 % of Ni, 99 % of Co, 98 % of Mn, and 99 % of Fe remain in solution for further recovery. The molar ratios of REEs in the precipitate during the semi-batch and batch modes are reported in Figure 1) c). These ratios remain relatively constant during the 1 h run, suggesting that the same phases form throughout the whole precipitation period.

Precipitation efficiencies and aqueous REE concentrations are summarized in Table 1 for all experimental conditions. This compilation highlights the influence of operating parameters. For a fixed Na:REEs molar ratio of 4:1, increasing the temperature from 25 to 60 °C greatly increases precipitation efficiency of all REEs and yields faster precipitation kinetics; for example, 74.4 % of La is precipitated at 25 °C while at 60 °C, 99.3 % is precipitated. Similarly, increasing the Na:REEs molar ratio from 1:1 to 4:1 at 60 °C, strongly increases the precipitation efficiencies of all REEs. Furthermore, adding the equivalent volume of Na_2SO_4 solution at once yields the same precipitation efficiencies than when the solution is added at a 7 $g.min^{-1}$ flow rate but greatly increases precipitation kinetics; however, rapid injection favors nucleation generating very small particles and consequently, after 1 h, filtration was more difficult (low filtration rates). It is interesting to note that, for a given temperature or for a given Na:REEs molar ratio, precipitation efficiencies of the REEs is always as: Ce > Nd > La > Pr.

Powder XRD analyses of the precipitates yield diffraction peaks that lie between those of pure $NaLa(SO_4)_2.H_2O$ and $NaCe(SO_4)_2.H_2O$ double sulfates. Crystal lattice parameters were calculated and exhibit intermediate values between those of pure double sulfates suggesting that precipitates are solid solutions of lanthanide-alkali double sulfates. Structures are assigned to the hexagonal crystal system[2], which can be observed on a representative SEM micrograph provided in Figure 2. The chemical composition of the solids was determined by TGA (number of water molecules n) and ICP-OES analysis (elemental content of S, Na, La, Ce, Nd and Pr). The resulting solid compositions are gathered in Table 2 as $Na_x(REE)_1(SO4)_y.nH_2O$, where x and y are the respective Na and S proportions relative to the total REEs content. The water content (n) is close to 1 for each solid, which proves the formation of monohydrate-type crystals. The mass and electrical (3+x–2y) balances yield that both Na and SO_4 contents are lower than the stoichiometric ratios ($x<1$ and $y<2$), and that the overall charge of such compounds is about +0.35 – +0.40. This charge balance inconsistency, which is currently investigated, is most likely due to the incorporation of additional ions in the structure. As indicated in Table 2, REE molar ratios in the mineralized precipitates are relatively equivalent for all experiments. Therefore, it is concluded that the same lanthanide-alkali solid solution forms regardless of the temperature or the Na:REEs molar ratio.

In parallel, equilibrium calculations were performed using the mixed-solvent electrolyte model implemented in the OLI software, which includes an accurate thermodynamic description of rare earth sulfates recently developed by Anderko et al.[3] Calculations were carried out with the various initial PLS conditions (pH, temperature, and ion concentrations). They foresee the precipitation of the compounds $NaLa(SO_4)_2.H_2O$, $NaCe(SO_4)_2.H_2O$, $NaNd(SO_4)_2.H_2O$ and $NaPr(SO_4)_2.H_2O$ in all conditions, which is in good agreement with XRD characterizations. The calculated REEs aqueous concentrations, detailed in Table 1, are also rather close to experimental measurements, which indicates that the precipitation reactions are close to thermodynamic equilibrium after 1 h. More specifically, REE aqueous concentrations strictly meet calculated concentrations at 60 °C and for a 4:1 Na:REEs molar ratio. At 60 °C, calculations show that the solubilities of all lanthanide-alkali double sulfates decrease with increasing Na^+ concentration, which explains why increasing the Na:REEs molar ratio improves precipitation efficiencies at constant pH. Moreover, for a Na:REEs molar ratio of 4:1, increasing the temperature decreases the double sulfates solubilities (apart from NaCe(SO4)2.H2O which shows the opposite trend). Therefore, based on both experimental results and thermodynamic trends, we conclude that the configuration (60 °C and 4:1 Na:REEs molar ratio) yields the best precipitation efficiencies for La, Ce, Nd and Pr (99.3 %, 100.0 %, 99.7 % and 87.6 %, respectively).

Key takeaways of this study are a straightforward approach to perform a grouped extraction of REEs contained in complex solutions originating from the leaching of industrial samples of spent Ni-MH batteries. Using a concentrated Na_2SO_4 solution, highly selective precipitation was obtained after 1 h at 60 °C and for a Na:REEs molar ratio of 4:1, whereby only 70 ppm La, 0 ppm Ce, 4 ppm Nd and 52 ppm Pr remain in solution and more than 98 % of the major elements do not co-precipitate. Thanks to multi-analytical characterization backed by thermodynamic calculations, the precipitated crystals were determined to be lanthanide-alkali double sulfate solid solutions of overall composition $Na_{0.79}La_{0.59}Ce_{0.20}Nd_{0.07}Pr_{0.14}(SO_4)_{1.72}.H_2O$. Results suggest that the same lanthanide-alkali solid solution forms regardless of the experimental conditions.

[1] M. Zielinski, L. Cassayre, P. Destrac, N. Coppey, G. Garin, B. Biscans, Leaching mechanisms of industrial powders of spent nickel metal hydride batteries in a pilot-scale reactor, ChemSusChem. 4 (2020) 616–628.

[2] A.C. Blackburn, R.E. Gerkin, Sodium lanthanum(III) sulfate monohydrate, NaLa(SO4)2.H2O, Acta Crystallogr. C. 50 (Pt 6) (1994) 833–838.

[3] G. Das, M.M. Lencka, A. Eslamimanesh, P. Wang, A. Anderko, R.E. Riman, A. Navrotsky, Rare earth sulfates in aqueous systems : Thermodynamic modeling of binary and multicomponent systems over wide concentration and temperature ranges, J. Chem. Thermodyn. 131 (2019) 49-79.

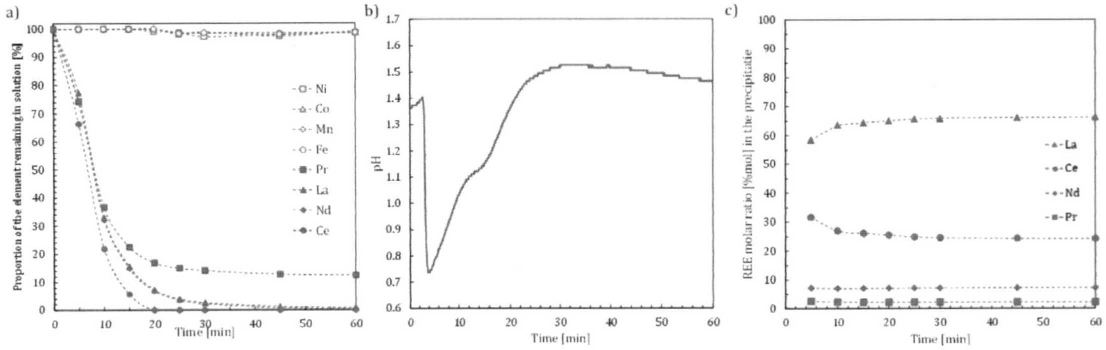

Figure 1 (a) Evolution of the proportions of Ni, Co, Mn, Fe, Pr, La, Nd and Ce remaining in solution [%] at 60 °C, for a Na:REEs molar ratio of 4:1 and a 7 g.min^{-1} Na_2SO_4 solution flow rate, (b) corresponding pH evolution and (c) evolution of the REE molar ratio [%mol] in the precipitate calculated from ICP analyses of the solution

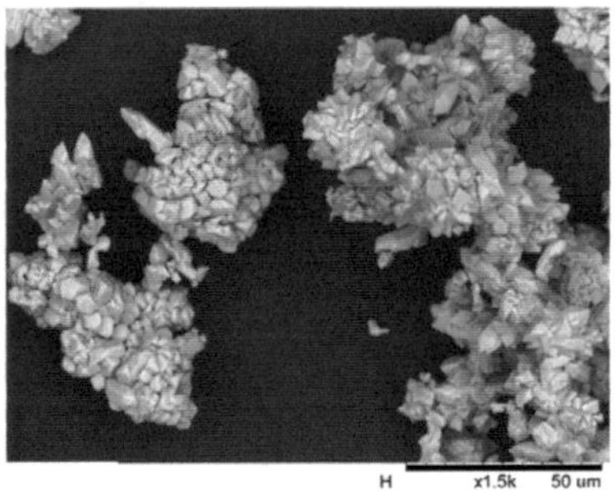

Figure 2 SEM micrograph of the precipitate obtained after 1 h of precipitation at 60 °C, a Na:REEs molar ratio of 4:1 and Na_2SO_4 solution flow rate of 7 g.min^{-1}.

Operating conditions	Influence of temperature			Influence of Na:REEs ratio			Influence of Na_2SO_4 flow rate			
Temperature [°C]	25	40	60	60	60	60	60	60	60	60
Na:REEs molar ratio	4:1	4:1	4:1	1:1	2:1	4:1	2:1	2:1	4:1	4:1
Na_2SO_4 flow rate	7 g.min^{-1}	7 g.min^{-1}	7 g.min^{-1}	7 g.min^{-1}	7 g.min^{-1}	7 g.min^{-1}	7 g.min^{-1}	At once	7 g.min^{-1}	At once
Precipitation efficiencies after 1 h of precipitation [%]										
Pr	69.3	84.3	87.6	69.1	80.0	87.6	80.0	82.7	87.6	14.4
La	74.4	92.9	99.3	71.7	92.4	99.3	92.4	93.0	99.3	1.2
Nd	77.5	94.4	99.7	74.1	93.5	99.7	93.5	93.7	99.7	1.2
Ce	87.5	100.0	100.0	78.7	95.6	100.0	95.6	94.4	100.0	1.9
Concentrations in solution after 1 h of precipitation [mol/L]										
Pr	0.0009	0.0005	0.0004	0.0011	0.0006	0.0004	0.0006	0.0005	0.0004	0.0004
La	0.0198	0.0056	0.0005	0.0224	0.0059	0.0005	0.0059	0.0058	0.0005	0.0010
Nd	0.0019	0.0005	0.0000	0.0024	0.0006	0.0000	0.0006	0.0006	0.0000	0.0001
Ce	0.0035	0.0000	0.0000	0.0059	0.0012	0.0000	0.0012	0.0015	0.0000	0.0005
Equilibrium concentrations according to OLI modelling [mol/L]										
Pr	0.0003	0.0003	0.0003	0.0034	0.0008	0.0003	0.0008	0.0008	0.0003	0.0003
La	0.0043	0.0016	0.0005	0.0069	0.0014	0.0005	0.0014	0.0014	0.0005	0.0005
Nd	0.0002	0.0001	0.0001	0.0011	0.0002	0.0001	0.0002	0.0002	0.0001	0.0001
Ce	0.0012	0.0021	0.0043	0.0263	0.0114	0.0043	0.0114	0.0114	0.0043	0.0043

Table 1 Precipitation efficiencies of La, Ce, Nd and Pr [%], corresponding concentrations [mol/L] remaining in solution after 1 h of precipitation and equilibrium concentrations [mol/L] as calculated by OLI modelling, with respect to operating conditions.

Parameter studied	Temperature [°C]	Na:REEs molar ratio	Na_2SO_4 flow rate	Precipitate composition after 1 h of precipitation
Influence of temperature	25	4:1	7 g.min^{-1}	$Na_{0.90}La_{0.58}Ce_{0.21}Nd_{0.07}Pr_{0.14}(SO_4)_{1.76}\cdot 1.1H_2O$
	40	4:1	7 g.min^{-1}	$Na_{0.87}La_{0.58}Ce_{0.21}Nd_{0.07}Pr_{0.14}(SO_4)_{1.74}\cdot 1.0H_2O$
	60	4:1	7 g.min^{-1}	$Na_{0.79}La_{0.59}Ce_{0.20}Nd_{0.07}Pr_{0.14}(SO_4)_{1.72}\cdot 1.0H_2O$
Influence of Na:REEs molar ratio	60	1:1	7 g.min^{-1}	$Na_{0.74}La_{0.58}Ce_{0.21}Nd_{0.07}Pr_{0.14}(SO_4)_{1.68}\cdot 1.0H_2O$
	60	2:1	7 g.min^{-1}	$Na_{0.78}La_{0.59}Ce_{0.20}Nd_{0.07}Pr_{0.14}(SO_4)_{1.70}\cdot 1.0H_2O$
	60	4:1	7 g.min^{-1}	$Na_{0.79}La_{0.59}Ce_{0.20}Nd_{0.07}Pr_{0.14}(SO_4)_{1.72}\cdot 1.0H_2O$
Influence of Na_2SO_4 flow rate	60	4:1	7 g.min^{-1}	$Na_{0.79}La_{0.59}Ce_{0.20}Nd_{0.07}Pr_{0.14}(SO_4)_{1.72}\cdot 1.0H_2O$
	60	4:1	At once	$Na_{0.78}La_{0.59}Ce_{0.20}Nd_{0.07}Pr_{0.14}(SO_4)_{1.70}\cdot 1.0H_2O$

Table 2 Precipitate composition after 1 h of precipitation with respect to operating conditions, deduced from microwave digestion of the precipitates, ICP-OES and TGA analyses.

604083 Molecular Modifiers Suppress Struvite Formation Via Unique Mechanisms

Abstract:

The crystallization of struvite ($MgNH_4PO_4 \cdot 6H_2O$) is relevant for nutrient (nitrogen and phosphorus) recovery in water purification and scale formation in pipelines, and it is a primary component of so-called infection stones arising from urinary tract infections. Research efforts have focused on understanding the effects of different parameters (e.g., supersaturation, pH, inhibitors) on crystallization of struvite; however, prior studies have only focused on the use of bulk techniques and have not provided mechanistic insights into these controls.

Herein, we present a comprehensive approach to evaluate struvite formation at both macroscopic and microscopic length scales in the absence and presence of various inhibitors. We conducted *in situ* kinetic measurements to estimate the extent of inhibition posed by each inhibitor on struvite formation. To this end, we employ a microfluidic platform to track the anisotropic growth rates of struvite crystals under various conditions, including comparisons of solutions with and without the presence of modifiers[1]. Furthermore, we elucidate the mechanism of inhibitory action by probing the surface dynamics in real time via *in situ* atomic force microscopy (AFM), which has proven to be a powerful technique to elucidate molecular level details of crystallization[2,3]. The results reveal unique mechanisms facilitating complete inhibition of growth and in select cases the unprecedented suppression of nucleation, which is not commonly observed for even the most potent modifiers of mineralization. Collectively, our findings identify highly efficient and commercially available modifiers that inhibit struvite formation. This study also uncovers a new class of inhibitory mechanisms that have potentially broader applicability to other minerals, thus establishing new paradigms for the prevention of commercial scale.

1. Kim, D., Olympiou, C., McCoy, C. P., Irwin, N. J., & Rimer, J. D. (2020). Time-Resolved Dynamics of Struvite Crystallization: Insights from the Macroscopic to Molecular Scale. *Chemistry–A European Journal*, *26*(16), 3555-3563.
2. Chung, J., Granja, I., Taylor, M. G., Mpourmpakis, G., Asplin, J. R., & Rimer, J. D. (2016). Molecular modifiers reveal a mechanism of pathological crystal growth inhibition. *Nature*, *536*(7617), 446-450.
3. Farmanesh, S., Ramamoorthy, S., Chung, J., Asplin, J. R., Karande, P., & Rimer, J. D. (2014). Specificity of growth inhibitors and their cooperative effects in calcium oxalate monohydrate crystallization. *Journal of the American Chemical Society*, *136*(1), 367-376.

610225 Solvent Structuring and Viscosity Effects on Thermodynamics and Kinetics of Crystallization

Abstract:

The structuring of the solvent at the crystal-solvent interface largely regulates the activation barrier for incorporation of solute into the crystal. Depending on the groups exposed on the crystal surface, the solvent at the interface may be ordered or disordered [1,2]. In general, the consequences of solvent structuring on the molecular level interactions comprising crystallization are poorly understood. Organic solvents, in which the bonds that could support a defined fluid structure are weak and varied present a particular challenge to understand the effect of solvent structuring on thermodynamics and kinetics of crystallization.

We explore the thermodynamic parameters of crystallization of a group of supramolecules known as porphyrins in organic solvents. Thermodynamic characteristics of crystallization were used as probes for the presence of solvent structuring which reveals that the interactions that govern the enthalpy and entropy of crystallization are dominated by the van der Waals contact between the solvent aliphatic tails and the non-polar groups of the etioporphyrin I (a free base porphyrin).

Time-resolved *in situ* atomic force microscopy monitoring of the step dynamics on (010) face of etioporphyrin I in two types of solvents, alcohols (octanol, butanol and hexanol) and DMSO, of varying aliphatic chain length and subsequent solvent viscosity demonstrates the impact of solvent structuring, solute-solvent interactions and solvent viscosity on the barrier of solute incorporation into kinks. The relation between step kinetic coefficient and solvent viscosity reveals that the crystallization of organic small molecules is not governed by diffusion limitation but rather by solute-solvent interaction at the kinks. The correlation between the step velocity and the solute concentration on the (010) face of etioporphyrin I for different solvents reveal that for solvents belonging to the same homologous series, solute incorporation scales with the solute diffusion coefficient. All-atom MD simulation for solvent structuring at the crystal interface emphasizes the importance of solute-solvent binding at the crystal interface.These emerging insights offer guidance for polymorph selection, chiral separations, structure design and numerous other open questions relevant to myriad crystallization systems in the pharmaceutical and chemical industries.

References

1. Vácha, R., Zangi, R., Engberts, J. B. F. N. & Jungwirth, P. Water Structuring and Hydroxide Ion Binding at the Interface between Water and Hydrophobic Walls of Varying Rigidity and van der Waals Interactions. *The Journal of Physical Chemistry C (2008)* **112**, 7689-7692.

2. Meister, K., Strazdaite, S., DeVries, A. L., Lotze, S., Olijve, L. L. C., Voets, I. K., & Bakker, H. J. Observation of ice-like water layers at an aqueous protein surface. *Proceedings of the National Academy of Sciences (2014)* **111**, 17732-17736.

611564 Identifying Crystal Morphologies to Optimize Growth Kinetics of Covalent Organic Frameworks – an Example of Cof-5

Abstract:

Covalent Organic Frameworks (COFs) have gained significant attention in the field of material science over last two decades [1]. 2D boronate-ester linked frameworks like COF-5 are porous, crystalline, organic polymers with low density, that have numerous applications such as gas storage and as membranes for various industrial purification applications. COF-5 exhibits high thermal stability, high surface area and permanent porosity, widening the scope of its applicability. Controlling crystal morphology is the most critical variable to maximize the applicability of these polymeric frameworks. Computing steady state and dynamic morphologies of COF-5 in a specific environment is essential for gaining comprehensive understanding of ways to control it. Face-specific growth rates have been previously utilized to understand these morphologies. The faces of these crystals are divided into three classes: flat, stepped and kinked faces, according to the Hartman-Perdok theory. This theory also states that layer by layer growth occurs on the F-face of the crystal lattice, which is the basis of studies conducted on COF-5 framework. Attachment energy calculations have been carried out along with kink sites distribution calculations to evaluate growth rates. The predicted rates have been used to understand the morphology of COF-5 crystals and the results are compared with experimental values, providing a deeper insight on optimizing process conditions for morphology control.

596311 Particle Remobilization in Filtration Membranes during Flow Interruption

Abstract:

Anomalously high levels of particles are often observed to pass through filtration membranes upon resumption of flow following interruption. However, it has been challenging to obtain direct information about the underlying mechanisms. Here, we employed a highly multiplexed single-particle tracking approach, which enabled the visualization of particle transport in microfiltration membranes under flow conditions that included stoppage intervals and therefore provided direct evidence for particle remobilization during flow stoppage. Our results suggest that during flow interruption, a sub-population of trapped/retained particles detach from their retention sites and may escape to less trapped areas through Brownian motion within the pore space. Particles that do not diffuse far enough during flow stoppage generally become trapped again upon the resumption of flow. As a result, the fraction of particles that are remobilized increases with stoppage time, since longer periods of Brownian motion permit diffusing particles to enter less confined regions of the membrane. This mechanistic information about particle transport remobilization will permit the design of filtration flow profiles that meet customized requirements for particle retention during complex filtration cycles.

604008 The Application of Tangential Flow Filtration (TFF) in Large-Scale Therapeutic Nanoparticle Processing

Abstract:

A key challenge in the "bench-to-bedside" translational research in nanomedicine is the scale-up synthesis and processing of nanomaterials to achieve precise control of the nano-properties. We have developed nanoparticle (NP) formulations encapsulating lumefantrine, which is a hydrophobic Biopharmaceutics Classification System class II drug and can be used as a cure for malaria. For scale-up processing, a reliable method to concentrate NP suspension is required to bridge the gap between a huge volume of dilute NP suspension produced and limited liquid-handling capacity of solidification equipment such as a spray dryer. A tangential flow filtration (TFF) system has been constructed to increase the NP concentration to greatly improves the NP solidification efficiency. Additional NP surface modification is needed to prevent NP aggregation. The shear rate and transmembrane pressure of NP flow within different types of filters are carefully controlled to prevent NP accumulation at the filter membrane. Next, scalable and continuous spray drying is applied to NP suspension of increased concentration to obtain dried powders with long-term storage stability. The dissolution kinetics of dried NP powders are shown to be significantly improved compared to that of crystalline lumefantrine in simulated fasted and fed intestine media. The technique we developed provides a promising route for scale-up processing of therapeutic NPs.

597153 Status of Modeling and Visual Observation of Mineral Scaling in High Temperature Aqueous Solutions

Abstract:

During petroleum and natural gas production, deposits of mineral scales throughout the production facilities cause not only production losses, but also equipment damage, and enhanced corrosion leading to serious safety concerns. While scaling at temperatures below 150 °C is relatively well studied, scaling at high-temperature, high-pressure (HTHP) conditions, such as those typically encountered in offshore ultradeep reservoirs, is a significant challenge and is poorly understood. A thermodynamic model needed for predicting Gibbs energies of aqueous species and, therefore, simulating scaling in HTHP conditions representative of ultradeep offshore oil and natural gas reservoirs is being developed. This presentation will provide the status of the model development along with experimental visualization of the scale formation observed in our laboratory.

602434 A Biophysical Outlook of Viral Inactivation with Arginine and Common Buffers in Biologics Manufacturing

Abstract:

A need for robust virus inactivation processes and environmentally friendly materials had prompted an exploration of innovative and alternate strategies in biologics manufacturing. A common detergent, Triton X-100, often relied on for inactivation of enveloped viruses, is on the edge of disbarment due to its detrimental effect on the environment. In this interest, two inactivation methods based on arginine and commonly used buffers were explored. Past studies have shown arginine to inactivate certain viruses under various conditions. However, the optimal conditions and mechanisms for fully inactivating all enveloped viruses by arginine is not known. For this reason, we tested a matrix of parameters such as pH, incubation time and buffer type to discern an optimal condition for inactivation of all enveloped viruses. Our results demonstrated that larger viruses - herpes simplex virus (HSV-1) and pseudorabies virus (SuHV-1) were effectively inactivated with > 4 LRV as compared to the smaller sized viruses - bovine viral diarrhea virus (BVDV) and equine arteritis virus (EAV). However, the optimal conditions resulted in different inactivation values between the similar herpes family viruses, HSV-1 and SuHV-1. Mechanistic understanding is being derived from the study of mechanical properties, virus-host interaction, and changes in the virus membrane. The second study explored the efficiency of commonly used buffers in the downstream train. Various buffers such as glycine, acetate, and citrate at pH 4 were explored to determine the inactivation efficacy of the above-mentioned viruses. A similar trend of inactivation based on virus size was observed for all the buffers. The efficiency of the buffers considering both the composition and incubation time was in the order of glycine < acetate ≤ citrate. Overall, this work directs towards the trend in phasing out Triton X-100 inactivation of enveloped viruses during biologics manufacturing and effects of commonly used buffers in the biologic and viral DSP.

611545 The Effects of Filtration Flux on the Clearance of Minute Virus of Mice during Viral Filtration

Abstract:

Viral safety is one of the major concerns in the production of mammalian cell and plasma derived biotherapeutic products. Demonstration of virus clearance is required by the regulatory agencies. Virus filtration is a size exclusion-based virus clearance method and is conducted industrially in dead-end mode. It is often operated in a batch mode under constant pressure. Membrane performance is shown to be affected by feed condition (pH, ionic strength and feed buffer components) as well as the operation condition (flux, pressure, time). During continuous bioprocessing, constant flux is a more desirable operation mode. Here the effects of flux operated under constant flux filtration mode and the solution condition on virus retention are systematically investigated for the filtration of monoclonal antibodies spiked with minute virus of mice (MVM) using three commercially available virus filters. Surprisingly, virus breakthrough was observed for two of the three filters when low constant flux operation was performed for a relatively longer periods of time. The interplay between virus loading on the filter, flux, membrane fouling and filtration time on virus retention will be discussed.

611537 Outside-in Hemodialysis: A Study of Toxin Clearance

Abstract:

For a patient with chronic or acute kidney failure, it is advantageous to perform hemodialysis continuously and for long durations in order to gradually remove excess fluids and improve toxin removal. Unfortunately, such long-term treatments are often hampered by the short lifetimes of the hollow-fiber dialyzers used in toxin removal, which arise due to blood clots formed in these fibers. The blood clots often block an entire fiber, so as more fibers are clogged, the surface area of the dialyzer and waste clearance are drastically lowered.

Labib and Dukhin (Novaflux Inc.) postulated that this clotting limitation can be overcome by a new approach that switches flow on the blood and dialysate side. In the outside-in filtration (OIF) configuration, dialysate flows inside the fibers and blood flows through the inter-fiber space. The researchers showed that OIF would allow prolonged hemodiafiltration with little dialyzer clogging.[1] With OIF, even if small inlet thrombi form that would block fibers in conventional intraluminal blood flow, these clots will have minimal effect on the blood flow and filtrate flux in OIF due to the three-dimensional system of interconnected hydrodynamic flow channels in the inter fiber space. However, very few studies have looked into the impact of OIF on toxin clearance and the level of mass transport of solutes with different sizes. The objective of this study is to experimentally assess the transport of different types of uremic toxins under both outside-in (OIF) and inside-out (standard) configurations. For small solutes, toxin clearance is relatively similar between the OIF configuration and the standard configuration, for both diffusive and convective transport. For larger toxins, however, we observed a noticeable difference in clearance level when OIF mode is used. We will discuss the physical mechanisms behind these observations, and the consequences of these observations for clinical applications.

[1] Dukhin SS, Labib M, et al. Outside-in hemofiltration for prolonged operation without clogging. Journal of Membrane Science 464 (2014).

600762 Study of the Spike (S) Glycoprotein from the Sars-Cov-2 As a Possible Source of Translocating Peptides of Biomedical Interest

Abstract:

Membranes are natural selective barriers necessary for the correct function of human cells as they control the entry of diverse molecules into the intracellular space. Although small and polar molecules such as amino acids and ions can come across the membrane through channels, larger macromolecules such as proteins and nucleic acids generally fail to do so. This represents a significant obstacle for efficient intracellular drug delivery, which is essential to increase the bioavailability of pharmacological agents for the treatment of numerous diseases[1,2]. Commonly, diverse methods have been studied to mediate the uptake of large molecules, including mechanical and electrical transfection techniques, but also viral carrier systems[2].

Some of these methods have been successfully tested *in vitro* but exhibit biosafety and cytotoxicity issues[3]. In consequence, there is an increasing interest in the development of novel delivery methods with low cytotoxicity but also high transduction efficiency[1]. An attractive alternative is the cell-penetrating peptides (CPPs), which can be obtained from different origins including signal peptides, viral proteins, or antimicrobial peptides[4]. CPPs are generally short peptides with lengths between 5-30 amino acids, positively charged or amphipathic, and rich in arginine and lysine[3]. Due to their ability to intermingle with the phospholipids of membrane bilayers, some CPPs are able to undergo translocation, and consequently can be considered as promising candidates for the delivery of biologically active molecules to cells.

The novel coronavirus SARS-CoV-2 has attracted significant attention over the past few months as it is responsible for the current global sanitary emergency where more than 1.3 M cases have been confirmed and over 70,000 people have died[5]. The Spike (S) glycoprotein has been thought to be responsible as the major determinant of the viral tropism towards human cells[6]. This protein has a 180 kDa molecular weight and is displayed at the viral surface as a trimer composed of two major domains[7]. The first one is the S1, which contains the receptor-binding domain (RBD) responsible for mediating the receptor binding (Angiotensin-converting enzyme 2). The second one is the S2, which allows the membrane fusion through the exposure of a fusion protein that is activated by proteolytic cleavage in a site upstream (S2') and proteolytically primed at the interface of the S1 and S2 domains. Transmission of the genetic material into the host cells has been attributed to proteases in priming, receptor binding, and some ionic interactions controlling the stability of the virus[7].

By recognizing the strong interaction between de spike (S) glycoprotein of SARS-CoV-2 and the angiotensin-converting enzyme 2 from the cellular membrane of the lung cells, here we aimed at finding motifs that could serve as possible sources of peptides capable of intermingling with membranes, and eventually with superior translocating potency. For this purpose, a prediction of the tertiary structure of the S protein from the SARS-CoV-2 was performed by homology, using Phyre Server, with the S protein from the bat coronavirus RaTG13. This was selected due to its closeness to the SARS-CoV-2 virus as it shares more than 93% of the identity in the S gene[8]. Simultaneously, the structure was also predicted *de novo* using the iTASSER Server to assure that predictions were robust enough for biophysical interaction studies. To determine the motifs with significant membrane activity (and potential translocation ability), a prediction of the possible formed transmembrane helices was carried out using TMHMM Server v. 2.0. Finally, the selected sequences were studied into detail via molecular dynamics (MD), using a mode membrane.

References:

4. **Derakhshankhah, and S. Jafari** "Cell penetrating peptides: A concise review with emphasis on biomedical applications" *Biomedicine & Pharmacotherapy*, 108, 1090–1096. (2018)
5. **W. Huang, and H.-J. Lee** "Cell-penetrating peptides for medical theranostics and targeted drug delivery" *Peptide Applications in Biomedicine, Biotechnology and Bioengineering*359–370. (2018)
6. **W. Huang, H.-J. Lee, L. M. Tolliver and R. S. Aronstam** "Delivery of Nucleic Acids and Nanomaterials by Cell-Penetrating Peptides: Opportunities and Challenges" *BioMed Research International*, 1–16. (2015)
7. **A. El-Baky, V. N. Uversky, and E. M. Redwan** "Virucidal activity of cell-penetrating peptides of viral origin"*Journal of Biomolecular Structure and Dynamics* 36, 1739–1746. (2017)
8. **H University and Medicine** "Coronavirus Resource Center" (2020)
9. **J.G. Hulswit, C.A.M. de Haan1 and B.-J. Bosch** "Coronavirus Spike Protein and Tropism Changes" Coronaviruses *Advances in Virus Research, 29–57.* (2016)
10. **A. Jaimes, N. M. André, J. K. Millet, and G. R. Whittaker** "Structural modeling of 2019-novel coronavirus (nCoV) spike protein reveals a proteolytically-sensitive activation loop as a distinguishing feature compared to SARS-CoV and related SARS-like coronaviruses" (2020)
11. **Lan, J. Ge, J. Yu, S. Shan, H. Zhou, S. Fan, Q. Zhang, X. Shi, Q. Wang, L. Zhang and X. Wang** "Structure of the SARS-CoV-2 spike receptor – binding domain bound to the ACE2 receptor" *Nature.* (2020)

608800 Combinatorial Discovery of Ligand-Fucntionalized Microbeads for Selective Enrichment of Methylated and Unmethylated DNA

Abstract:

DNA methylation patterns play a vital role in determining gene expression in different diseases including cancer. In addition, therapeutic plasmid DNA (pDNA) molecules, used in non-viral gene therapy, are frequently methylated by the enzymes present in bacterial strains employed in their bio-processing. Current approaches to isolate methylated plasmids and genomic DNA involve antibodies, which are mainly geared towards research-scale applications. Consequently, new approaches that can lead to the enrichment of methylated or unmethylated DNA from biological samples and bioprocessing media, in a rapid and cost-effective manner, can lead to transformative advances in these fields. We have developed aminoglycoside-derived microbeads, approximately 10μm in diameter, for binding DNA based on Coulomb interactions between the beads and DNA molecules. These parental microbeads were modified with different amino acids, lipids, and chemotherapeutic drugs, in order to obtain a diverse library of ligand-modified microbeads. The library of ligand-modified microbeads was screened for binding unmethylated and methylated pDNA (a model plasmid that encodes for luciferase protein) in order to identify lead candidates that showed preferential binding to methylated pDNA over the unmethylated analog or vice versa. Adsorption isotherms were determined for both sets of leads and the data were fit to a Langmuir isotherm model. Elution of adsorpbed DNA was investigated with different buffers and organic modifiers. Finally, quantitative structure activity relationship (QSAR) models were employed to obtain physicochemical insights into molecular interactions that govern ligand binding to methylated as well as unmethylated DNA. In conclusion, our results indicate that molecular engineering of a chemically diverse ligand-modified microbead platform can lead to rapid identification of novel ligands for enriching methylated or unmethylated DNA for applications in bio-manufacturing of gene therapy vectors and in the rapid detection of diseases, including in low-resource settings.

607410 Apohemoglobin Haptoglobin Complexes Attenuate the Hypertensive Response to Low Molecular Weight Polymerized Hemoglobin

Abstract:

Background: The presence of low molecular weight (MW) hemoglobin (Hb) species in previous generations of Hb-based oxygen (O_2) carriers (HBOCs) restricted their application as red blood cell (RBC) substitutes. Among the myriad of strategies to produce Hb, using glutaraldehyde crosslinking to form polymerized Hb (PolyHb) remains popular due to its low cost and excellent scalability. However, the commercially developed PolyHbs, HBOC-201® (Biopure Corp, Cambridge, MA, USA) and PolyHeme® (Northfield Laboratories Inc., Northfield, IL, USA), primarily contained fractions of material at or below 250 kDa in MW. These low MW PolyHb and unpolymerized Hb are able to extravasate into the interstitial space where they scavenge nitric oxide (NO), resulting in vasoconstriction, systemic hypertension and oxidative tissue injury. Recent strategies to mitigate hypertension have used adenosine and nitroglycerine to attenuate the hypertensive response of HBOC-201®. While effective at controlling systemic hypertension, administration of adenosine and nitroglycerine must be carefully controlled to prevent hypotension and cannot address HBOC extravasation into the tissue space. Instead, we propose that employing the naturally occurring mechanism of Hb detoxification is a potential strategy to mitigate systemice hypertension resulting from administration of low MW PolyHb. The naturally occurring Hb scavenging protein, haptoglobin (Hp), plays a pivotal role in detoxifying stroma free Hb in the blood. Recently, we determined that Hp preferentially binds to low MW PolyHb (MW < 256 kDa). By binding to Hp, the molecular diameter of low MW PolyhHb is effectively increased. This increase in size may reduce tissue extravasation and subsequent NO scavenging of PolyHb.

Methodology: In this study, we synthesized a low MW PolyHb. The hydrodynamic diameter, O_2 affinity, and size distribution were measured. The Hp used for this study was purified from human Cohn Fraction IV derived from pooled human plasma. The resulting Hp contained both Hp2-1 and Hp2-2 phenotypes. Hp binding to PolyHb was confirmed *in vitro* with stop-flow fluorescence spectrometry and size exclusion chromatography. Hp was stabilized by binding it to heme-free Hb (apoHb) in the form of an apoHb-Hp complex. Systemic parameters, including heart rate (HR) and mean arterial pressure (MAP), were used as indicators of reduced hypertension. Additionally, intravital microscopy was used to examine how co-administration of apoHb-Hp influences functional capillary density (FCD), vascular tone, and blood flow.

Results and Conclusions: Polymerization of Hb resulted in decreased O_2 affinity and increased MW. PolyHb contained a significant fraction of low MW species (< 250 kDa). Hp was able to bind low MW PolyHb. Hp binding resulted in a significant increase in the average size of PolyHb. In the apoHp-Hb group, there was negligible change in MAP, FCD, HR, and microhemodynamics compared to baseline conditions. When compared to the systemic and microcirculatory changes observed in the apoHb and saline groups, the relatively small changes in the apoHb-Hp group indicate that Hp-based species may serve as materials to counteract the pressor effects of low MW HBOCs that are capable of binding to Hp. Thus, the administration of a Hb scavenging apoHb-Hp solution maintained physiologic hemodynamics during the transfusion of PolyHb.

607825 Single Gas Multi-Normal Energy Distribution Models for Predicting Mixed Gas Adsorption Equilibria from Non-Uniform Heterogeneous Extended Langmuir and Hias Models

Abstract:

A new multi-normal energy distribution (MNED) model was developed for correlating single gas equilibrium adsorption isotherms with any number of NED functions. This new model utilizes the single process Langmuir model in the integrand of the adsorption integral equation, along with the MNED function. The parameters from the MNED model were then used to predict mixed gas adsorption equilibria of binary and ternary experimental data for both perfect positive and perfect negative correlations from two models that both utilize the cumulative MNED function obtained from the single gas correlations. The first model uses the extended Langmuir model in the integrand of the simplified adsorption integral equation, along with the cumulative MNED function, formulating a non-uniform heterogeneous extended Langmuir (NUHEL) model. The second model is the heterogeneous ideal adsorption solution (HIAS) theory. The experimental data used in this study was the single, binary and ternary adsorption equilibria for CO_2, H_2S and C_3H_8 on H-Modernite from Talu and Zwiebel [AIChE 32, 1263 (1986)]. The single gas correlations from the MNED model will be presented and contrasted against the dual and three process Langmuir (DPL and TPL) models, now referred to as the multi-process Langmuir (MPL) model. Then, the binary and ternary predictions from the NUHEL and HIAS models will be presented and contrasted against those from the MPL model. It suffices to state at this point that all three mixed gas models predict correctly that CO_2 and H_2S correlate in a perfect positive fashion, with both of these gases correlating in a perfect negative fashion with C_3H_8. The accuracy of the three mixed gas models, along with their ease of use, will be discussed.

611220 Deformation of Nanoporous Materials in the Process of Binary Adsorption: Methane Displacement By Carbon Dioxide from Microporous Carbons.

Abstract:

Phenomenon of adsorption deformation of nanoporous materials has recently attracted a lot of attention in adsorption and materials science communities. Various theoretical and molecular simulations approaches have been suggested to predict the stress and strain induced by single component adsorption. However, most practical applications, like adsorption separations or selective gas storage, involve multicomponent competitive adsorption. Here, we develop a thermodynamic method based on the notion of the adsorption stress to predict the deformation effects upon binary adsorption. As a practically important example, the process of displacement of methane by carbon dioxide from microporous carbons is considered. This process is the foundation of the secondary hydrocarbon recovery from shales and coalbeds associated with carbon dioxide sequestration.

A general equation is derived to relate the adsorption stress and the adsorption isotherm of multicomponent mixture. Further this theory is adapted for phenomenological models employed for multicomponent adsorption of microporous carbons, such as Langmuir, Dubinin, and IAST. The results are compared with available literature data and Monte Carlo simulations on model structures.

The proposed approach may have multifaceted applications in modeling behavior of hydrocarbon mixtures in micropores of geosorbents, gas separations and carbon capture on flexible adsorbents.

612287 On the Universality of Capillary Condensation and Adsorption Hysteresis Phenomena in Ordered and Crystalline Mesoporous Materials

Abstract:

Using adsorption experiments for argon over a wide range of temperatures from above the triple point (i.e., 87 K) down to 65 K, coupled with advanced molecular simulation studies, the underlying mechanism of capillary con- densation and hysteresis in a mesoporous metal-organic framework (MOF),MIL-101(Cr) is explored. The results obtained on the mesoporous MOFs are compared with the experimental and molecular simulation studies of argon adsorption and phase behavior in mesoporous molecular sieves (e.g., MCM-41 silica) of similar pore size and over the same temperature range (65–87 K). The studies provide insights into the nature of confinement effects on the adsorption and phase behavior of fluids in mesoporous MOFs. The obtained results suggest that certain features of adsorption, capillary condensation, and associated hysteresis phenomena in mesoporous MOFs and in mesoporous molecular sieves are governed by the same underlying physics, despite the striking differences in pore structure and chemical composition. Hence, the findings are of importance for advancing characterization meth- odologies based on gas adsorption, but also for predicting the behavior of mesoporous MOFs in various potential applications.

[1] K.A.Cychosz , M.Thommes*, L. Sarkisov*. Advanced Materials Interfaces, 2020, in press

596828 Prediction of Adsorption of CO_2 and Hydrocarbons in Zeolites: A Strong Test of Atomically Detailed Models

Abstract:

As an increasingly valuable complement to experiment, computational modeling approaches such as Grand Canonical Monte Carlo (GCMC) and Molecular Dynamics (MD) have been used to predict adsorption and diffusion properties ahead of experiment for guest molecules in porous materials like zeolites and interpret experimental observations at the molecular level. Successful implementation of GCMC and MD simulations requires reliable force fields (FFs) to describe the interactions of guest molecules with porous frameworks. Developing accurate and transferable FFs for adsorbed and diffused molecules is challenging.

Since first-principles quantum mechanical (QM) methods are capable of accurately predicting intermolecular interactions, deriving FFs from QM data without experimental input is a promising solution. We previously used this approach to develop a FF for adsorption of CO_2 in siliceous and cationic (Na and K) zeolites. This FF accurately predicts experimentally observed adsorption properties and shows good transferability across different zeolite topologies and Si/Al ratios. In this work, we extend this FF for CO_2 adsorption in ammonium ZSM-5, a zeolite reference material RM 8852, where good reproducibility and high reliability of the experimental isotherm data is available. We also independently performed experiments with RM 8852 by microcalorimetry. The results show that at low pressures good agreement was obtained between predictions and experiments. At high pressures, however, deviations were observed. We demonstrate that the charge-balancing cations in the experimental material are the predominant source of the discrepancy between simulation and experiment at high pressures; the experimental sample treatment causes deammoniation. In addition, accounting for a small amount of non-crystalline mesoporosity in this ZSM-5 brings predictions in much better agreement with experiments.

We also extend this FF to hydrocarbons. Most previous studies on development of FFs for hydrocarbon molecules in zeolite materials focus on prediction of adsorption properties. However, accurately reproducing adsorption data is not sufficient to guarantee the accuracy of predicting diffusivities. We demonstrate an approach to develop FFs that can accurately predict both adsorption and diffusion properties of hydrocarbons in siliceous zeolites.

602296 Conformational Changes during Adsorption and Chromatographic Separation of Bivalent Bi-Specific Antibodies - Biomolecular Perspectives

Abstract:

Proteins can exhibit conformational changes while adsorbed to the surface of resins used for chromatographic separations resulting in complex elution behaviors both for ion exchange chromatography and for hydrophobic interaction chromatography. Both unfolding with loss of secondary structure and reversible conformational changes have been reported. In either case, recovery and separation efficiency are impacted. Additionally, these phenomena complicate the development of mechanistic models since the classical, thermodynamics-based approaches for the description of protein surface interactions do not take such phenomena into account. In this talk, we present recent advances that we have made to gain a biomolecular understanding of the underlying phenomena and to model chromatographic elution of bivalent bi-specific antibodies whose multi-domain characteristics result in particularly complex elution behaviors.

598374 Enhanced Multiple Air Pollutant Abatement in the High-Gravity Rotating Packed Bed: Theoretical Approach Towards Practical Application

Abstract:

The application of a high-gravity rotating packed bed (HiGee RPB) for multiple air pollutants (i.e., SO_2, NO_x and particle matters (PM)) reduction from industrial flue gas emissions was developed and deployed. In this research work, a green process intensification by using the industrial alkaline wastes such as slags or fly ash leachate as the absorbent was developed, creating the synergetic removal of NO_x-SO_x-CO_2 and PM process to save the chemicals and energy consumption. High alkalinity solution and calcium ions contents in the solid leachate facilitate the accelerated carbonation and other acid gas. The combination of oxidation and absorption in the HiGee RPB can reduce the NO_x emission. The design criterial of HiGee RPB for acid gas pollutant including the determination of key operating factors such as liquid-to-gas-ratio and rotating speed as well as the mass transfer characteristics. The theoretical approach goes to investigate the mass transfer rate and enhancement factor caused by high-gravity field and chemical reaction. In addition, the PM collection performance using HiGee RPB is superior to the conventional wet scrubber. The process exhibits a high removal efficiency with a short residence time under ambient temperature and pressure. Several sub-pilot HiGee systems for air pollutants emission reduction were successfully deployed at the industries in Taiwan. The results showed that the application of the HiGee technology was a proven technology.

599579 Thermally Stable Electrospun Polyimide Fiber Membranes for Separation of Oil-in-Water Emulsions

Abstract:

One of the major drawbacks of polymeric membranes applied to the treatment of oily wastewater is that the membranes have relatively poor thermal stabilities and chemical resistance. Due to membrane fouling, frequent foulant removal processes becomes necessary, which reduce the lifetime of the polymeric membrane and incur additional economic and environmental costs. Therefore, there is a need to improve membrane stability and fouling resistance for this clean and relatively cheap technology to advance. In this work we report the use of electrospinning technology, which has been shown to produce membranes with improved robustness against fouling, to generate fibrous membranes made of P84 polyimide, a co-polymer known for its thermal stability and excellent chemical resistance. The P84 membranes were challenged by emulsions of dodecane and motor oil in water, stabilized by sodium dodecyl sulfate (SDS), in a constant pressure dead-end configuration. The resultant flux, oil rejection, sensitivity to fouling, and flux recovery were evaluated. When challenged by dodecane and motor oil, the P84 membranes exhibited moderate oil rejections of 86.6 ± 1.5 % and 88.0 ± 2.4 %, respectively, while subsequent foulant removal by heat treatment restored permeate fluxes to 108 ± 15 % and 93 ± 17 % of their original values. A comparison of P84 membranes to PVDF, PSF and PA6(3)T membranes shows that the P84 membranes have the best robustness against fouling, which is attributed to electrostatic repulsions and the oleophobic nature of the P84 membrane, as well as the ease with which such membranes can be regenerated. These results indicate that the electrospun P84 membrane is a promising candidate for microfiltration used in the treatment of water contaminated by emulsified oil.

595523 Experimental Research on the Structural Parameter of the Coupling Apparatus of Granular Bed with Cyclone

Abstract:

Abstract: Dust in the flue-gas turbine is a key bottleneck restricting the long-time operation of the fluid catalytic cracking units(FCC). Developing a high efficiency/low pressure drop cyclone is quite essential. Dust carried in the flue gas containing many kinds of metallic elements influences on the subsequent reaction performance and fluidization characteristics. Small particles(<3μm) accumulate on the pressure-side surface of rotors, while large particles(>20μm) induce erosion at the trailing edge of rotors. As the conventional dust removal equipment in FCC unit—cyclone, however, cannot separate particles less than 5μm. The design of this apparatus is based on the idea of combining different dust removal mechanisms including the centrifugal separation, the inertial collision and the interception. A gas-solid separator coupling cyclone and cross-flow particle bed was investigated. Reasonable coupling with different separation mechanisms makes up for the inherent disadvantages of the third cyclone and achieves the purpose of collaborative strengthening. The goal of this study is to evaluate the effects of the main structural parameters—swirling space ratio on the performance of fine dust; by analyzing the flow field distribution on the built-in granular bed surface, the separation mechanisms in different regions of the separator were studied. It was found that: (1) the pressure drop decreased with an increasing swirling space ratio as expected while the separator with large swirling space ratio has a desired operation stability; (2) when their total separation efficiency exceeded 99%, cyclone shell in the large swirling space ratio apparatus contributions more to the total separation efficiency which reduced the load of built-in particle bed effectively; (3) the variation of swirling space ratio strengthened the division characteristics between granular bed and cyclone shell, the separation efficiency of the particles less than 5μm improved significantly in the large swirling space ratio; (4) according to different separation mechanisms, the apparatus was divided into two regions along the axial direction, the cross-flow diffusion region and the entrainment effect region. Regionalization of granular bed provided theoretical basis for establishing the dust removal model for the coupling apparatus.

607359 Enzymatic Reactive Crystallization of Cephalexin Using Immobilized Penicillin G Acylase from *E. coli*

Abstract:

Beta-lactam antibiotics have been used extensively to treat bacterial infections for nearly a century. The two largest produced beta-lactam antibiotic classes, semi-synthetic penicillins (e.g. amoxicillin) and cephalosporins (e.g. cephalexin), constitute the largest amount of sales of antibiotics for the treatment of bacterial infections. Currently the industrial production of beta-lactam antibiotics is conducted largely via batchwise chemical synthetic pathways consisting of multiple reaction steps, use of protecting groups, harsh solvents, and very low (-40°C) temperatures. Using these methods, the chemical synthesis route produces a large amount of chemical waste with an E factor of roughly 50 and is highly energy intensive due to the operation of low temperature refrigeration equipment. Additionally, due to increases in labor and raw material costs, much of the production of beta-lactam antibiotics has moved to China and India [1]. Considering recent events regarding COVID-19, the disruption international supply lines could lead to widespread shortages of highly important antibiotics in Western countries [2]. Improving the economic viability and environmental sustainability of beta-lactam antibiotic production is paramount to globalizing the production of these important drugs to decrease the chance of shortages.

Recently, we have been working to develop a continuous enzymatic beta-lactam antibiotic production process to address many of the issues of the current batchwise chemical synthesis processes to produce cephalexin and amoxicillin. These drugs were selected because they are the two highest volume manufactured and prescribed antibiotics worldwide and have been included in the world health organization (WHO) list of essential medicines as key access antibiotics [3]. Next, the move from chemical synthesis to enzymatic synthesis decreases waste production and increases yield due to high enantioselectivity of enzymes and the ability to operate in a mild temperature aqueous environment. We have expressed, purified, and engineered penicillin G acylase from *E. coli* (EcPGA), an industrially relevant enzyme used to produce precursors for semi-synthetic beta-lactam antibiotics, to synthesize cephalexin and amoxicillin. Additionally, we have immobilized EcPGA on various solid supports via covalent attachment, allowing for the recycle of the biocatalyst and decreasing the overall cost of the catalyst for production. Next, the induction of crystallization of the API within the reaction vessel protects the desired product from hydrolysis by EcPGA into a minimally soluble byproduct. The *in-situ* crystallization of the API enables the downstream separation of the solid crystals from the mother liquor into a filtration, washing, and drying protocol.

By combining these elements, we have conducted several successful seeded batch reactive crystallization experiments to produce solid cephalexin crystals. We found that after three crystallization experiments of three hours each, the catalyst suffered a 20% decrease in specific activity. This decrease may be due to crystallization within the pores of the solid catalyst, as the catalyst experienced no measurable deactivation under reaction conditions without crystallization. We have also studied the effect of different sized support particles as well as different support materials (e.g. agarose, methacrylate, polypropylene) on the diffusive properties of cephalexin and its precursors as well as on-line stability of the immobilized biocatalyst. Additionally, during reactive crystallization, after approximately four hours, the scarcely soluble byproduct, phenylglycine, generated enough supersaturation to precipitate and contaminate the solid phase. The progress of cephalexin crystallization and growth as well as phenylglycine precipitation were detected via *in-situ* microscopy and focused beam reflectance measurement (FBRM). We are currently working to implement these same elements and use these experiments to inform the design and operation of continuous MSMPR. The shift from batch to continuous production will improve the capacity and decrease the overall footprint of the production process as well as decrease the amount of off-specification product due to more robust control during steady state operation. The implementation of process analytical technology (PAT) such as Fourier transform infrared spectroscopy (FTIR) will allow for the on-line monitoring of the liquid phase components while FBRM and *in-situ* microscopy will allow for the on-line monitoring of the solid phase components. Additionally, reverse phase HPLC will be used to confirm final product purity. With the information of the components' concentrations in both phases, we will study and develop a control strategy to maintain a "state-of-control" operation of the reactive crystallization process for the production of cephalexin and amoxicillin.

References:

[1] Elander, R. P. (2003). Industrial production of β-lactam antibiotics. *Applied Microbiology and Biotechnology*, *61*(5-6), 385–392. doi: 10.1007/s00253-003-1274-y

[2] Oehler, R. L., & Gompf, S. G. (2020). Shortcomings in the US Pharmaceutical Supply Chain. *Jama*. doi: 10.1001/jama.2020.1634

[3] WHO Model Lists of Essential Medicines. (n.d.). *The SAGE Encyclopedia of Pharmacology and Society*. doi: 10.4135/9781483349985.n433

607528 Recrystallization Development for Small Molecule API (Active Pharmaceutical Ingredient) Islatravir

Abstract:

Islatravir is a nucleoside reverse transcriptase inhibitor small molecule API (active pharmaceutical ingredient) targeted for HIV PrEP (pre-exposure prophylaxis). Islatravir has multiple crystal forms and a complex phase diagram. This talk will focus on the conversion from the monohydrate to an anhydrous form. To produce anhydrous API, we have designed a slurry conversion process, during which the less stable monohydrate form dissolves and a more stable anhydrous form crystallizes simultaneously. The solution composition also comes into play, as the process requires operating within a specific window of water-activity for effective form conversion. The slurry conversion process is also designed to proceed in a controlled manner to reduce crystallization of another kinetically favored phase. In addition, Merck has recently published a novel biocatalytic synthetic scheme for the synthesis of Islatravir[*]. This presentation will discuss differences in anhydrous crystallization performance (morphology, particle size, and milling performance) when starting with API from a traditional chemical synthetic route vs. the novel biocatalytic route. In this talk, we discuss the factors we can manipulate to design the most robust process, ultimately producing anhydrous API with consistent properties across various batches.

[*]Huffman et al. "Design of an in vitro biocatalytic cascade for the manufacture of islatravir." Science, 2019, 366 (6470), 1255-1259.

609555 Seeding Strategies for the Selective Crystallisation of Model Proteins

Abstract:

The recognition of the selective nature of biopharmaceutical drugs and the advances in biotechnology have led to over 246 approved products with cumulative revenues reaching $140 billion [1]. Upstream biomanufacture of proteins has advanced, such that the time and expense bottleneck in biopharmaceutics lies in the chromatographic purification steps [3-5]. Crystallisation is a more economic and energy efficient alternative purification step, producing crystalline proteins with higher purity and stability [3, 6]. However, this technology is yet to mature to manufacturing scale and robustness.

Here, we report our optimised crystallisation, additive and seeding strategies to crystallise model proteins and demonstrate selective crystallisation of target proteins from a binary mixture.

Mesoporous silica nanoparticles (NPs) are shown to be effective heterogenous nucleants for the selective crystallisation of lysozyme from a binary mixture of lysozyme and thaumatin. From hanging drop solution conditions which can crystallise both proteins, the 'hard template' SBA-15 NPs are found to not only promote the nucleation of lysozyme, but also suppress the crystallisation of thaumatin. These conditions are also effectively scaled up to the ml level. Further to this, the effectiveness of graphene oxides and DNA polymeric additives are also examined as seeds for protein crystallisation promotion.

Here, we report on the use of several templating strategies for the crystallisation of proteins demonstrating the feasibility for selective crystallisation of proteins for downstream separations.

ACKNOWLEDGEMENT

This study acknowledges by the SCoBiC project funded by the UK's EPSRC (EP/N015916/1).

References

1. Demain, A.L., REVIEWS: The business of biotechnology. Industrial Biotechnology, 2007. 3(3): p. 269-283.
2. Yang, H. and J.H. ter Horst, Crystal nucleation of small organic molecules, in New Perspectives on Mineral Nucleation and Growth. 2017, Springer. p. 317-337.
3. Basu, S.K., et al., Protein crystals for the delivery of biopharmaceuticals. Expert opinion on biological therapy, 2004. 4(3): p. 301-317.
4. Parambil, J.V., et al., Effects of Oscillatory Flow on the Nucleation and Crystallization of Insulin. Crystal Growth & Design, 2011. 11(10): p. 4353-4359.
5. Delmas, T., M.M. Roberts, and J.Y. Heng, Nucleation and crystallization of lysozyme: Role of substrate surface chemistry and topography. Journal of Adhesion Science and Technology, 2011. 25(4-5): p. 357-366.
6. Harrison, R.G., et al., Bioseparations science and engineering. 2015: Topics in Chemical Engineering.
7. Shah, U.V., et al., Heterogeneous nucleants for crystallogenesis and bioseparation. Current Opinion in Chemical Engineering, 2015. 8: p. 69-75.
8. Shah, U.V., D.R. Williams, and J.Y.Y. Heng, Selective Crystallization of Proteins Using Engineered Nanonucleants. Crystal Growth & Design, 2012. 12(3): p. 1362-1369.

609008 Developing a Scalable Templating-Platform for Enhancing Crystallisation of Bio-Active Proteins

Abstract:

Over the last decade, bio-active peptides have emerged as the next generation of bio therapeutics with a rapidly growing demand [1]. A lack of cost-effective and reliable processes for downstream purification is a main industrial challenge for meeting the growing demand of bio-active peptides [2]. Crystallisation of peptides is a novel strategy to debottleneck downstream purification, offering a more economical and efficient technique than expensive traditional techniques [3].

Due to their low solubility and large molecular size, one of the main challenges in peptide crystallisation is the lack of controllability of the nucleation. However, very little has been done in using templates for controlling nucleation to achieve a consistent and reproducible crystallisation process. Chen et al. recently reported that highly mesoporous inorganic surfaces, referred to as "hard templates", can remarkably induce significantly faster and more controlled nucleation, even at low supersaturation [4]. In our recent work, we have successfully demonstrated a "soft templating" strategy by using soft, organic molecules, such as amino acids, to enhance nucleation. In buffer solution, amino acids form various interactions with peptides, whereby electrostatic interactions have been identified as most important. Therefore, amino acids that exhibit strong electrically charged residual groups for the relevant buffer conditions were investigated. In this presentation, the impact of types, amount and surface charge of the amino acids on the crystallisation performance of insulin at various supersaturation will be reported. With the aim of providing a strategy suitable for high-throughput peptide crystallisation, the possibility of scaling this templating strategy up, from the microlitres to the tens of millilitres, will be evaluated.

Insulin was used as model peptide and comprises of two chains, A and B, with a combined 51 amino acids and an overall molecular weight of 5.8 kDa. The impact of amino acids on crystallisation is evaluated by means of measuring induction time, crystal yield, and size distribution as a function of time. Citric buffer with a pH between 5 and 7 was used with zinc sulphate as the precipitant. The impact of amino acids on crystallisation is evaluated by monitoring the peptide concentration with UV-vis spectroscopy every 20-30 minutes. Using this, the induction time and crystal yield are obtained. Furthermore, the pH is measured at the start and the end of the crystallisation. Images obtained with a microscope are analysed to evaluate the crystal sizes and shapes. The experiments were carried out in batch crystallisers with volumes up to 15 millilitres to demonstrate the scalability.

Our recent results show that amino acids reduce the induction time of insulin crystallisation by up to 40% when compared to the induction time with no amino acids. The faster decrease in insulin concentration when amino acids are added also implies an increase in the crystal growth rate. As amino acids accelerate the crystallisation process, a yield of 90% can be achieved in a shorter period of time in crystallisers with volumes of up to 15 millilitres. Preliminary investigations show that the type and amount of amino acid added have only a minor impact on solubility, leading to a slight increase. Instead of impacting the solubility significantly, the charged residual-groups of the added amino acids increase the pH significantly. This increase in pH leads to an increase in stability of the tertiary structure of insulin by altering its physio-chemical surface properties which result in a faster nucleation. As a result of the more controlled crystallisation with amino acids, a narrower crystal size distribution can be obtained while achieving the same mean crystal size (~20 μm) when compared to the mean crystal size obtained with no amino acids. The insulin crystals obtained with amino acids show a more uniform crystal size distribution with a smaller span for crystalliser volumes reaching from the scale of microliters to the tens of millilitres. The soft templating strategy could be successfully demonstrated for crystalliser volumes of up to 15 millilitres while identifying mixing as one of the main limiting factors of scaling-up.

It has been demonstrated that amino acids as organic templates have the potential of enhancing the controllability of insulin crystallisation even at low supersaturations and to the scales of tens of millilitres. These includes a significant reduction of residence time needed in the crystalliser to obtain a desired crystal yield by maintaining a high level of consistency. Although different polar amino acids were investigated, the magnitude of advancing crystallisation differs dramatically between them as well as the ratio of amino acids concentration to insulin supersaturation.

Bibliography

1. Lau, J.L. and M.K. Dunn, *Therapeutic peptides: Historical perspectives, current development trends, and future directions.* Bioorg Med Chem, 2018. 26(10): p. 2700-2707.
2. Yang, H., et al., *Development and Workflow of a Continuous Protein Crystallization Process: A Case of Lysozyme.* Crystal Growth & Design, 2019. 19(2): p. 983-991.
3. Roque, A.C.A., et al., *Anything but Conventional Chromatography Approaches in Bioseparation.* Biotechnol J, 2020.
4. Chen, W., et al., *High Protein-Loading Silica Template for Heterogeneous Protein Crystallization.* Crystal Growth & Design, 2019. 20(2): p. 866-873.

609313 Sensor Integrated Continuous Microfluidic Mixer for Protein Screening at Constant Supersaturation

Abstract:

Sensors are great assets to quick and sensitive detection inside miniaturized devices were the amount of sample is small. Connecting an integrated microfluidic system to a control unit increases the accuracy of the experiments due to the automated adjustments of parameters. One of the examples for the potential application of a sensor integrated microfluidic system is screening crystalline materials where there is a high demand for accurate measurement of the concentration within the crystallization zone. In our previous studies, we developed a continuous microfluidic mixer for screening crystalline materials at the constant supersaturation, where the homogeneous concentration was calculated using COMSOL Multiphysics. However, it was not possible to perform an in-situ measurement for the actual concentration to experimentally verify that. Here we have coupled the microfluidic mixer design with an electrochemical sensor to enable in-situ measurement of the actual concentration inside the mixing zone. Moreover, integration of a sensor within the continuous microfluidic platform we can control the imposed supersaturation by manipulation of the flowrates. To validate the sensor integrated platform experimentally, we have done the protein crystallization of lysozyme and NaCl. Products of this crystallization process can exhibit different morphologies according to the ratio of the starting solutions. Several different experimental conditions are explored to study the morphologies obtained at assigned initial concentration as well as the growth kinetics.

606008 Particle Size Control of a Drug Substance through Process Crystallization

Abstract:

The solid state characteristics of an Active Pharmaceutical Ingredient (API) have great impact on the Drug Product (DP) performances and its formulation development approach.

One Critical Quality Attribute (CQA) that has to be controlled in order to ensure the success of a given formulation strategy in terms of manufacturability is the API Particle Size Distribution (PSD). In this case study, the API (PSD) had to be controlled to produce a blend which was homogeneous and flowable once combined with the chosen excipients and, in parallel, to reduce the risk of segregation during the DP manufacturing process.

The initial drug substance batch available for the formulation prototype production and selection showed a D(v,0.9) around 350mm. A mechanical PSD reduction via FitzMill was required in order to tune the drug substance PSD to match the PSD of the chosen excipients to ensure proper blend flowability and homogeneity attributes as well as to mitigate the risk of segregation during the encapsulation. As the alternative, the control of the PSD via process crystallization was investigated. Ultimately an anti-solvent mediated seeded crystallization process was developed. This process was able to deliver drug substance in the desired PSD range and avoided the need for post isolation milling.

604742 How Does Membrane Microstructure Affect Performance?

Abstract:

The microstructure of membranes plays a significant role in controlling their overall efficiency, capacity and selectivity during filtration. Most commercial polymer membranes are fabricated using a phase inversion (PI) process that does not allow controlling the formation of the membrane microstructure. Hence, empirical pre- and post-fabrication methods are used to improve performance. The overall goal of this work is to replace PI with fabrication methods that rationally control the formation of the membrane microstructure and hence improve particle filtration performance. Our approach is to propose a microstructure design and then test that design computationally by passing a swarm of particles of different properties (size, surface forces) through the membrane microstructure to determine its selectivity and capacity. Here we release micron-size particles in a KCl solution flowing through the 3D crystal-like membrane structure. Using AFM in force measurement mode, zeta-potential, and surface tension measurements, the particles' physical and chemical properties were determined experimentally. The extended DLVO (xDLVO) theory, which includes van der Waals, electrostatic and Lewis acid-base interactions, was invoked to describe the interactions between the particles and the membrane pore walls. Combining these forces with fluid and particle drag, a comprehensive model was used to track particles through the microstructure. Discrete element method (DEM) was used with open-source software (Multiphase Flow with Interphase eXchanges, MFix) to describe the motion of the particles in the 3D microstructure. By modulating the details of the 3D structure (i.e with different pore size distribution), selectivity and capacity were obtained and further used for selecting an optimal microstructure with desired filtration performance. The novelty of this work lies in the combination of computational modeling incorporating measurement of molecular interactions to design and fabricate membranes with improved performance.

611429 Dissipative Particle Dynamics Simulation of PVDF Membrane Formation Via Non-Solvent Induced Phase Separation Process

Abstract:

Polymeric membranes produced by Non-solvent Induced Phase Separation (NIPS) have been extensively used across a wide array of microfiltration and ultrafiltration processes. However, despite their broad applications, current fabrication processes are largely empirical, determined largely by trial and error. Better insights into the factors affecting membrane morphology would be of great use. To that end, Dissipative Particle Dynamics (DPD), a mesoscale simulation tool, was used to simulate membrane formation process involving PVDF in DMF with H_2O as a nonsolvent. The DPD simulations were performed for a total of several million beads and over several microseconds. The interactions between these beads are determined according to Flory-Huggins thermodynamic theory and continuum transport equation. Significant insights have been obtained on the effects of polymer concentration and polymer molecular weight on the pore size, pore size distribution and membrane morphology.

596308 Effect of Trehalose on the Structure and Function of Aquaporin-1 Under Osmotic Stress from Molecular Simulation

Abstract:

Understanding how freezing-related stresses impact membrane behavior at the molecular level is key to developing next generation cryoprotectants and cryopreservation strategies for a number of applications, including organ banking and cold-tolerant agriculture. In this study, we used all-atom molecular dynamics to investigate how trehalose, a common cryoprotectant, interacts with an Aquaporin-1 tetramer embedded in a lipid bilayer. Specifically, we investigated this interaction under osmotic stress, a common cause of cell death in the presence of extracellular ice formation. To mimic osmotic stress, the system was modelled with two bilayers to form an extracellular and intracellular space where the concentration of ions and trehalose are easily varied. The effect of trehalose on the structure of the protein-membrane complex was then monitored, as well as its effect on the transport of water through the Aquaporin-1 complex.

611444 Predicting the Conductivity Increase By Volatile Salts Penetration in Direct Contact Membrane Distillation (DCMD) Permeate Water

Abstract:

Global freshwater crisis in recent years has forced worldwide demands for alternative water resources. As the major proportion of water body on the earth is salty water, desalination techniques have drawn significant attention. Accordingly, conventional pressure-driven desalination methods have been developed and optimized to provide solutions to salty water reuse. However, pressure-driven desalination methods have to get over the osmotic pressure contributed by water salinity, which limited its capability when treating high salinity water.

Direct Contact Membrane Distillation (DCMD) is an emerging thermal desalination process with high salt rejection. The only driving force of DCMD is the vapor pressure difference between feed and permeate caused by temperature difference, which grants the capability of treating high salinity water such as seawater and shale gas produced water. The low capital and operational cost of DCMD when integrated with waste heat and solar energy also leads to a promising commercialization prospect. In DCMD system, the hydrophobic membrane inhibits the permeation of liquid feed due to the high surface tension between hydrophobic membrane and aqueous phase, while allows the transport of water vapor through the membrane. This performance of DCMD will be limited by the volatile salts in the feed when volatile molecules pass the membrane in gas phase and contaminate the permeate water. The dissociation of volatile salts will then increase the permeate conductivity, deteriorating this significant evaluation index for permeate quality.

In this study, a conductivity model describing the transport of volatile salt for DCMD systems was developed according to data from bench-scale DCMD experiments for synthetic shale gas produced water with volatile contaminants (i.e. ammonium chloride, ammonium acetate, sodium acetate and phenol). It is shown that the existence of volatile salts in feed will cause the permeate conductivity increase immediately after the desalination process starts and the increase rate is determined by multiple factors of the contaminants (i.e. partial pressure, molecular weight, dissociation constant, etc.) as well as configuration parameters of DCMD system (i.e. temperature, flow rate, membrane pore size, etc.). Lab-scale DCMD experiments were also employed to verify the prediction, it is indicated that the volatile salts can penetrate the membrane and increase permeate conductivity without fouling the membrane, the permeate quality is compromised due to the invasion of these impurities and may fail the discharge standards. This study shows that DCMD treatment with certain polluted water may require pre-treatment to eliminate the impact of volatile salts and fulfill the discharge requirements.

611608 Modeling of Chemically-Specific Separations in Lyotropic Liquid Crystal Self-Assembled Nanoporous Membranes

Abstract:

The nanopores that form in cross-linked lyotropic liquid crystal (LLC) membranes are chemically complex and uniform in size which makes them promising for highly selective separations. We have developed an atomistic molecular model of an LLC membrane that is maximally consistent with experimental X-ray diffraction for the purpose of studying small molecule transport within it.

We have characterized transport of water, sodium ions and small polar solutes within the pores of our model. We find that the transport rate of a species is dependent not only on molecular size, but on chemical functionality as well, leading to anomalous diffusive behavior. Electrostatic interactions between the membrane and solutes provide an interesting diversity of mechanistic behaviors. In general, all solutes perform intermittent hops between lengthy periods of entrapment, and a range of different trapping mechanisms are responsible for this behavior. The particular frequency of each mechanism depends to large extent on the identity of the chemical species, meaning that such membranes have significant capability for separations.

We use our detailed mechanistic understanding to construct a range of stochastic models rooted in anomalous diffusion theory, which we can use to project out the long term behavior of the solutes and ultimately predict selectivity. Using the insights developed in this study, we can begin to think about how to redesign existing LLC membranes in order to perform solute-specific separations.

599872 Molecular Simulations and Machine Learning for Multicomponent Adsorption: BTEX Separation with Zeolite Membranes

Abstract:

Separation of *p*-xylene from mixtures of benzene, toluene, ethylbenzene, and xylenes (BTEX) is a critical industrial process, identified as one of "seven chemical separations to change the world" [1]. Membranes that permit selective transport of *p*-xylene, but not its isomers, *o*-xylene, *m*-xylene, and ethylbenzene, could dramatically improve the efficiency of *p*-xylene production. Zeolite membranes are stable under the high temperatures and pressures of the catalytic process; membranes synthesized with the MFI framework enable high selectivity and flux for *p*-xylene separation when tested at pressures < 1 bar and at temperatures up to 425 K [2]. However, at high temperatures and pressures, adsorption of aromatics into zeolite membranes saturates the pores, shutting down transport.

Process design of a *p*-xylene-selective membrane reactor requires knowledge of adsorption and transport through the membrane across a range of industrial process conditions. To address the question of multicomponent adsorption into MFI zeolite at elevated temperatures and pressures, we performed high-throughput Monte Carlo simulations in the Gibbs ensemble to predict BTEX adsorption across the expansive "process space" of potential operating conditions. The transferable potentials for phase equilibria (TraPPE) force field was extended to accurately predict vapor-liquid coexistence curves for alkyl-substituted aromatics. Then, unary and multicomponent adsorption isotherms and fluid phase properties were predicted across a multidimensional composition space, and from liquid, vapor, and supercritical phases. These simulations provide individual data points that are used to fit an artificial neural network, which provides a single, self-consistent, continuous, and differentiable hypersurface describing the properties of both the fluid and adsorbed phases across the full range of realistic process conditions.

[1] D.S. Sholl and R.P. Lively, Nature, 532, 2016, 436-437.
[2] M.Y. Jeon, et al. Nature, 543, 2017, 690-694.

597147 Stimuli-Responsive Carbon Nanotube Membranes for Chemical Warfare Protection

Abstract:

Stimuli responsive materials have many diverse uses in areas such as drug delivery, sensors, dynamic surfaces, and advanced textiles. For fabrics applications, smart protective garments that rapidly transport water vapor and autonomously block chemical threats are expected to enable effective personal protection without adding undue physiological stress to the user.[1] However, the incorporation of these seemingly incompatible properties into a single responsive system remains elusive. In this work, we demonstrate a bistable membrane that can rapidly, selectively, and reversibly transition from a highly breathable state in a safe environment to a chemically protective state when exposed to organophosphate threats such as sarin.[2] Autonomous response to chemical stimuli is achieved through the physical collapse of an ultrathin copolymer layer grafted to the membrane surface, which efficiently gates transport through membrane pores composed of single-walled carbon nanotubes (SWCNT). The adoption of nanometer-wide SWCNTs for ultrafast moisture conduction enables a simultaneous boost in size-sieving selectivity and water-vapor permeability by decreasing nanotube diameter, thereby overcoming the breathability/protection trade-off that limits conventional membrane materials. Moisture vapor transport rates that rival commercial breathable fabrics coupled with the ability to block both biological (by size sieving) and chemical threats make these membranes promising materials for next-generation protective garments for military, first responders, and medical personnel.

References

1. Bui, E. R. Meshot, S. Kim, J. Peña, P. W. Gibson, K. J. Wu, F. Fornasiero, *Adv. Mater.*, 28 (2016) 5871.
2. Li, C. Chen, E. R. Meshot, S. F. Buchsbaum, M. Herbert, R. Zhu, O. Kulikov, B McDonald, N. Bui, M. L. Jue, S. J. Park, C. Valdez, S. Hok, C. J. Doona, K. J. Wu, T. M. Swager, F. Fornasiero, *Adv. Funct. Mater*, in press (2020)

This work was performed under the auspices of the U.S. Department of Energy by Lawrence Livermore National Laboratory under Contract DE-AC52-07NA27344.

605467 Remote Performance Modulation of Ultrafiltration Membranes By Magnetically and Thermally Responsive Polymer Chains

Abstract:

Stimuli responsive ultrafiltration membranes that respond to both an external magnetic field and the local temperature have been developed. Surface initiated activator generated electron transfer (AGET) atom transfer radical polymerization (ATRP) has been used to graft poly(N-isopropylacrylamide) (PNIPAm) from the surface of 300 kDa regenerated cellulose ultrafiltration membranes. A superparamagnetic nanoparticle was attached to the end of the polymer chain. The polymerization initiator was selectively attached to the entire membrane surface, only the outer membrane surface or only the inner pore surface. The DI water flux as well as the flux and rejection of bovine serum albumin (BSA) was investigated in the absence and presence of a 20 and 1000 Hz oscillating magnetic field.

In an oscillating magnetic field, the tethered superparamagnetic nanoparticles can cause movement of the grafted PNIPAm chains or induce heating. A 20 Hz oscillating magnetic field maximizes movement of the grafted PNIPAm chains. For PNIPAM chains grafted from the outer membrane surface movement of the grafted polymer chains leads to break up of the concentration polarization boundary layer. This results in an increase in the observed rejection of partially rejected BSA. A 1000 Hz oscillating magnetic field leads to greater induced heating. PNIPAm displays a lower critical solution temperature (LCST) at 32 °C. Since the heat is generated at the interface with the PNIPAm chains heating leads to dehydration and collapse of the PNIPAm chains above their LCST. For PNIPAm grafted from the inner pore surface an increase in water and BSA flux is observed as well as a decrease in BSA rejection due to collapse of the chains. This work highlights the versatility of selectively grafting polymer chains containing a superparamagnetic nanoparticle from specific membrane locations. Depending on the frequency of the oscillating external magnetic field membrane properties may be tuned for a specific separation.

606526 Tuning Pore Size and Robustness of Membranes Formed By Scalable Self-Assembly of Random Copolymer Micelles

Abstract:

Membrane separations are energy-efficient, simple, and scalable. Yet, their broader use is limited by the separation capabilities of membranes prepared by conventional methods, typically confined to size-based separations. Self-assembly of functional polymeric materials is a powerful method for designing membranes capable of new separations, including the separation of organic compounds of similar size from each other. This can potentially be achieved by membranes that mimic biological pores such as porins, with pores that are only slightly larger than the target solute and functionalized with groups that selectively interact with one compound over another. Our group has developed a method for preparing such membranes by the spontaneous self-assembly of random amphiphilic copolymer in methanol into micelles that are then coated onto a support to form a selective layer of tightly packed micelles. These membranes, produced by a simple manufacturing process, have 1-4 nm pores lined with carboxylic acid groups that enable charge-based separation of organic solutes. Additionally, the carboxylic acid groups can be converted to a wide range of possible functional groups through coupling reactions. This can provide selectivity towards solutes by affinity. Because of the distinctive geometry and selectivity of these membranes, it is valuable to assess the versatility of this technology. The goal of this study was to tune the pore size and increase the robustness of these thin film composite membranes. The pore size was tuned by changing the composition of the copolymer and also by changing the casting solvent, while the robustness of the system was altered by different crosslinking procedures.

611777 Dual-Functional Nanofiltration Membranes Exhibit Multifaceted Ion Rejection and Antifouling Performance

Abstract:

The incorporation of charged functional groups is often used to facilitate the selective rejection of multivalent ions over monovalent ions by nanofiltration (NF) membranes. However, the incorporation of charged functionality can accelerate membrane fouling. Multifunctional membrane architectures that incorporate distinct layers of chemical functionality over their cross section provide a strategy for balancing the competing demands associated with making NF membranes that are both fouling-resistant and ion rejecting. Here, poly(trifluoroethyl methacrylate-*co*-oligo-(ethylene glycol) methyl ether methacrylate-*co*-(3-azido-2-hydroxypropyl methacrylate)) copolymer parent substrates, which can be modified using the copper(I)-catalyzed azide-alkyne cycloaddition (CuAAC) reaction, were exposed to a series of reactive solutions containing alkyne-terminated molecules to create dual-functional membranes. Using appropriately controlled exposure conditions, the CuAAC reaction propagates into the copolymer substrate as a reactive front. This phenomenon allows for the creation of layered domains of distinct anti-fouling zwitterionic and ion rejecting sulfonate functionality. The ion rejection and fouling propensity for a family of dual-functional membranes was examined. For membranes with a thin anti-fouling layer on top of an ion rejection layer, the rejection of 1 mM K_2SO_4, 87%, was comparable to the value for full charge control membranes, 90%. Moreover, when exposed to a fouling solution containing bovine serum albumin (BSA), these dual-functional membranes exhibited an 18% decline in normalized flux and recovered 99% of their flux upon rinsing. Comparatively, the full charge membranes exhibited a 44% decline in normalized flux and recovered 65% of their flux upon washing. As such, the results demonstrate that the controlled functionalization process reported here is capable of balancing anti-fouling and ion rejection capabilities. Furthermore, the versatile nature of the click chemistry mechanism at the center of this process offers a means by which to design and fabricate multifunctional membranes for numerous future applications.

608572 Spiral-Wound RO Elements Performance Tuning with Surface Nano-Structured Polyamide TFC Membranes

Abstract:

Membrane surface modification is a popular approach to mitigate membrane fouling and, in some cases, can be also utilized to tailor desalination performance of polyamide (PA) based thin film composite (TFC) reverse osmosis (RO)/nanofiltration (NF) membranes. However, scalability of membrane surface modification approaches that are suitable for producing commercial-scale spiral-wound elements is seldomly reported. To date, scalability of membrane surface modification has been only demonstrated for *in situ* surface modification of commercial spiral wound PA RO/NF membrane elements, accomplished via either redox-initiated graft polymerization of a hydrophilic monomer (e.g., glycidyl methacrylate (GMA)) or surface coating of a polydopamine layer. In contrast to the above, membrane surface nano-structuring (SNS) via atmospheric pressure plasma-induced graft polymerization (APPIGP) can be easily implemented in the downstream process of PA TFC flat sheet membrane manufacturing. Therefore, the SNS polyamide (SNS-PA) sheets can be produced prior to fabrication of spiral wound elements. Moreover, the previous study has shown that SNS of a PA TFC brackish water RO (BWRO) base membrane with a tethered poly(acrylic acid) (PAA) layer (SNS-PAA-PA) can achieve a seawater RO (SWRO) membrane performance. In the present study the scale-up potential of APPIGP was evaluated by synthesizing large SNS-PAA-PA flat sheet membranes (30" x 24") that are suitable for fabricating commercial scale spiral wound membrane elements (2.5" in outer diameter and 21" in length). The scaled-up synthesis of the SNS-PAA-PA membrane sheets was accomplished by utilizing an APP source mounted on an XYZ robot capable of scanning a large membrane sheet area, followed by graft polymerization of a hydrophilic monomer (i.e., acrylic acid) in a custom-made reactor designed for the large membrane sheet. The synthesized SNS-PAA-PA membrane sheets were then used to fabricate 2.5 x 21" spiral-wound RO elements. Compared to the commercial SWRO membrane element of the same size (Dow SW30-2521), the SNS-PAA-PA spiral-wound elements had up to ~2 times greater L_p and similar or lower B. Moreover, the results of a fouling study with model foulants (e.g., bovine serum albumin and alginic acid) indicate that the surface tethered PAA layers reduce membrane fouling and increase membrane cleaning effectiveness.

599454 Functionalized and Reactive Membranes for Water Applications

Abstract:

The integration of multifunctional materials in membrane pores for environmental and separation applications requires a high level of control of both the characteristics of the base polymeric support layer, as well as, its corresponding pore configurations. Membrane processes provide a highly flexible separation technique for water treatment, toxic materials degradation, and materials recovery. Microfiltration membranes are primarily used for the removal of bacteria and particles, but by appropriate pore functionalization with polymeric charge groups or integration of nanoparticles of these MF membranes (termed Functionalized Membranes) one can use these as high capacity metal capture membranes to catalysis and pH/temperature flux modulations with separation selectivity. The dependence of conformation properties of polyelectrolytes provides tunable separation and membrane flux control by pH and temperature-based stimuli responsive properties. Many current treatments for the reclamation of contaminated water sources are chemical-intensive, energy-intensive, and/or require post-treatment due to unwanted by-product formation. We demonstrate that through the integration of nanostructured materials within pore-functionalized synthetic membrane platforms (such as, PVDF membrane pores functionalized with poly-acrylic acid, and/or PNIPAM), we are able to conduct environmentally important reactions for toxic organic degradation and selective toxic material sorption from water without the addition of expensive or harmful chemicals. In addition, we have fabricated GO-PAA composite membranes by layer-by-layer assembly with an aim to integrate advanced sulfate radical-based oxidation reactions for efficient removal of chlorinated organics from water. The presentation will include: (a) pH and temperature responsive properties of bench-scale to full-scale pore functionalized membranes, (b) temperature modulated sorption/desorption of toxic PFOA by PVDF membrane pores functionalized with PNIPAM, and (c) Fe/Pd nanoparticle creation in PDVF membrane pore domain, and in GO-PAA composites for water and catalysis related applications. This research has been supported by NIH-NIEHS-SRP, NSF EPSCoR, and Chevron Corporation.

613144 Industrial Crystallization – the Intersection of Solid State Chemistry and Process Engineering

Abstract:

Crystallization is a complicated molecular recognition and self-assembly process that is completely managed by manipulating macroscopic process variables. Oftentimes, the crystallographic state of the resulting product is critial to its performance. The efficacy of products as diverse as pharmaceuticals and pesticides is often very sensitive to polymorph, for example. In addition, the particle size distribution depends on multiple process parameters in complicated, often nonlinear ways. As many crystallized products have become more advanced, with increasing demands on product performance, industrial crystallization has had to adapt. Pratitioners tend to come to the field from either process engineering or solid state chemistry. While it is possible to stay in one's own "lane" and contribute to the field, most successful practioners today are at least conversant with both sides of the chemistry/engineering "divide". Cross-training and collaboration often contribute greatly to successful crystalline product development and scale-up.

I come to industrial crystallization from process engineering, but have come to appreciate the necessity of understanding solid-state chemistry to adequately develop processes and products. I'll illustrate the interplay between process engineering and solid-state chemistry with several examples from several industries, and review where I think continued collaboration can lead to crystalline product control and improvement.

602066 Crystallization Optimization By Enzymatic Control of a Reactive Crystallization

Abstract:

Tunable reactions can be used to control precisely the generation of supersaturation in reactive crystallization systems. Enzymes, with high reaction specificity and predictable rate enhancement, enable fine control over a number of reactions, and have been demonstrated in several reactive crystallization systems. In this study, penicillin G acylase (PGA) is used to synthesize and simultaneously crystallize cephalexin, starting from two cephalexin precursors, in a controlled manner. It was recently shown that cephalexin, which typically forms high-aspect-ratio needle-shaped crystals, can form lower-aspect-ratio crystals when supersaturation is generated in a slow, controlled manner (Li et al., 2019). These lower-aspect-ratio crystals have many desirable properties including faster filtration (by an order of magnitude) and improved powder flowability and handling.

This study aims to show that enzyme control of cephalexin reactive crystallization can produce crystals with a more desirable aspect ratio while also providing the benefits of combined synthesis and separation (greater productivity, conversion). The mechanism of PGA synthesis of beta-lactam antibiotics such as cephalexin is known (McDonald et al. 2017); here an immobilized form of PGA is used to enable recycling of the enzyme. Use of process analytical technology (PAT) combined with a reaction-diffusion-crystallization model (Salami et al. 2020, *in preparation*) lends insight into the phenomena critical to successful implementation of this novel means of supersaturation control. *In situ* microscopy and focused beam reflectance measurement (FBRM) show that the amount of crystal surface area is important to both aspect ratio and secondary nucleation. The model suggests that the amount of enzyme, and unexpectedly the immobilized bead size, is critical in dictating the reaction timescale. In general, this study demonstrates how catalysts can lend control over the crystallization process, giving process designers an additional knob to turn in the synthesis of new routes to important pharmaceutical products.

Li, M., et al. (2019). "Optimizing the Aspect Ratio of Cephalexin in Reactive Crystallization by Controlling Supersaturation and Seeding Policy." Transactions of Tianjin University **25**(4): 348-356.

McDonald, M. A., et al. (2017). "Enzymatic reactive crystallization for improving ampicillin synthesis." Chemical Engineering Science **165**: 81-88.

Salami, H., et al. (2020). "Model development for enzymatic reactive crystallization of β-lactam antibiotics, a reaction-diffusion-crystallization modeling approach." Computers & Chemical Engineering **in preparation**

610666 Continuum of Solute Dimers Drives Distinct Growth Modes on Anisotropic Faces of Etioporphyrin

Abstract:

One of the basic tenets of the classical theories of crystal nucleation and growth is the Szilard postulate, which states that molecules from the supersaturated phase join a nucleus or a growing crystal individually. Over the years, numerous crystals have been found to nucleate by assembly of monomers and grow by sequential association of monomers to growth sites. Recent experiments in bio- and geomimetic environments have spotlighted a partial deviation from the Szilard rule, a bimodal distribution of growth species that arises from the assembly of ordered or amorphous structures in the solution, which then incorporate into the crystal as a whole. Here we report a complete violation of the Szilard rule. In octanol solutions of etioporphyrin, a planar molecule whose crystals have attractive electronic and optical properties, the solute populates a continuum of quasi-parallel dimers that differ from the conformation in the crystal by shift and orientation. Molecular absorption and luminescence spectroscopies and all-atom molecular dynamics (MD) simulations reveal that the stabilities of the dimers are within the thermal energy and enforce free conversion between the distinct structures. The monomer concentration in the solution in below 0.1 %. Despite the structural complexity of the solute, the velocity of the steps on the (010) faces of etioporphyrin crystals (measured using time resolved in situ AFM) increases linearly with the total solute concentration. The observed linearity indicates that the molecular reaction at the growth sites in monomolecular and the majority of solute dimers seamlessly adopt the crystallographic conformation. By contrast, step growth on the other major face, (001), scales with the square of the solute concentration. We eliminate three potential mechanisms that may manifest as super-linear growth: inaccurate solubility determination, escalating kink density at elevated supersaturation, and adsorption of step pinners on the crystal surface. The disqualification of the three scenarios enforces the conclusion that the quadratic kinetic law indicates a bimolecular process of incorporation. We demonstrate that in contrast to numerous other solution grown crystals, solute molecules directly incorporate in the kinks and do not reach the growth sites after first adsorbing on the terraces between steps. The paucity of solute monomers precludes a collision between two monomers to form a crystallographic dimer as the mechanism of the crystallization reaction. MD evaluation of solute interactions with kinks on the (001) face reveal a molecular trap selective for solute dimers located on the approach to all kinks, which necessitates a collision with a second dimer to ease the captured solute towards the kink. The results with etioporphyrin demonstrate the lack of correlation between the solute state and the incorporation kinetics driven by the solute. In a broader context, these observations illuminate the immense diversity of unexplored crystallization scenarios that pave the long road to crystallization control.

610307 Sterol Chemistry, but Not Lipid Domains, Modulates the Disruptive Effects of Nanomaterials on Membrane Models

Abstract:

Introduction: The interactions of nanoparticles with cell membranes have been increasingly studied due to the development of nanoparticle-based applications across a wide range of industries. Disruption of the membrane by nanoparticles can have deleterious effects on cell viability. Most studies have focused on the properties of the nanoparticles and less attention has been given to the role of the membrane structure in regulating the interactions. Depending on the lipid composition of the cell membrane, the membrane may exhibit heterogeneous phase segregation of lipids into disordered and ordered domains (lipid rafts). Formation of domains is influenced by the number and type of sterol present in the membrane, providing a facile method to modulate domain formation. The current study utilizes this method to examine the role of membrane domains in regulating the disruptive effects of silica nanoparticles on model membranes. However, in modulating the sterol composition of the model membrane, it is also possible to observe the impact of individual sterol chemistry on particle-membrane interactions. We are particularly interested in how sterol hydroxyl group orientation, sterol ring double bond placement, and R group identity impact the membrane.

Materials and Methods: Unilamellar vesicles 100 nm in diameter were synthesized with equimolar concentrations of sphingomyelin (SM), dipalmitoylphosphotidylcholine (DOPC), and a sterol. Various sterols were chosen to either promote or hinder domain formation in vesicles. Vesicles containing either cholesterol, ergosterol, coprostanol, or androstanol were synthesized. It was expected that vesicles containing cholesterol or ergosterol would exhibit ordered domain formation and vesicles containing coprostanol or androstanol would have few or no ordered domains. Plain silica nanoparticles with diameter 39.0 ± 2.9 nm and zeta potential = -8.22 ± 0.41 mV were used to study nanoparticle-membrane interactions. Membrane models were exposed to nanoparticles at a particle concentration of 0.01 g/L. Fluorescence anisotropy of diphenylhexatriene (DPH) and GUV imaging using rhodamine-PE allowed for quantification of domain formation in vesicles with different sterols. Following this, the degree of vesicle disruption upon nanoparticle exposure was quantified by incorporating the self-quenching fluorescent dye carboxyfluorescein into the vesicles and measuring its leakage over time.

Results and Discussion: DPH anisotropy indicated that vesicles containing cholesterol or ergosterol have higher lipid order compared to vesicles containing coprostanol or androstanol, evidenced by greater anisotropy values for cholesterol (0.20 ± 0.02) and ergosterol (0.21 ± 0.02) than for coprostanol (0.15 ± 0.01) and androstanol (0.13 ± 0.01) vesicles at 25 °C. GUV imaging with confocal microscopy complemented anisotropy results and showed distinct segregated ordered domains in vesicles containing either cholesterol and ergosterol and no domains in vesicles containing coprostanol. Nanoparticle leakage assays yielded no discernable relationship between physical membrane order or membrane disruption by particles. Leakage for vesicles containing cholesterol was $20\% \pm 6\%$ while for ergosterol it was $37\% \pm 1\%$ (measurements taken at 25 °C). For coprostanol vesicles at the same temperature, leakage was $11\% \pm 2.5\%$ and for androstenol leakage was $75\% \pm 9\%$. Currently, we are investigating the impact of sterol chemistry (i.e. presence and location of double bonds, stereochemistry, and presence of R groups) on the nanoparticle membrane interaction to determine if this has a more dominant impact on interactions as opposed to presence of physical domains. We intend to expand or vesicle characterization using FRET analysis and to perform nanoparticle binding assays using FRET and GUV imaging to determine nanoparticle localization at the membrane.

Conclusions: Altering the sterol composition of lipid membrane models alters the membrane structure and affects assembly of ordered/disordered domains. However, there does not seem to be a predictable relationship between the degree of ordered domain formation in the vesicle and the nanoparticle interaction. We are investigating the role of sterol chemistry in modulating nanoparticle-membrane interactions.

607755 Simulations of the Bilayer-Mediated Assembly of Cationic Nanoparticles on Phase-Separated Lipid Bilayers

Abstract:

Understanding interactions between functionalized gold nanoparticles (NPs) and lipid bilayers is essential for their use in biomedical applications. Experiments and continuum-scale simulations have suggested that small (~5 nm diameter) cationic NPs aggregate when adsorbed to the surface of zwitterionic lipid bilayers despite being stable in aqueous solution, suggesting that bilayer-mediated interactions facilitate assembly. Determining the origin of these effects and the impact of other bilayer features, such as lipid phase separation, on the interfacial assembly of NPs would enable the design of NPs for applications that require NP assembly, such as photothermal therapy.

In this work, we use chemically specific coarse-grained molecular dynamics simulations to show that cationic NP adsorption to lipid bilayers is driven by a balance between favorable NP-lipid interactions and the unfavorable mechanical deformation of the bilayer to create local curvature. The minimization of induced curvature near topological features, such as boundaries between stiff and flexible phase-separated bilayer domains, enhances NP adsorption. Since these driving forces do not depend on solvent-mediated interactions, we next parameterize an implicit solvent model to model larger length-scale phenomena involving multiple adsorbed NPs. We measure the bilayer-mediated interactions between two adsorbed NPs by comparing the aggregation free energy on a bilayer to that in solution. These simulations indicate that the bilayer contributes an attractive interaction between the nanoparticles associated with the reduction in induced curvature. We further show that that the association of 3 NPs is less favorable than the aggregation of 2 NPs and leads to linear aggregates. Finally, we find that NPs align along phase boundaries and prefer dimerization as opposed to larger structures. These simulation results indicate that bilayer curvature mediates interactions between adsorbed NPs and favors alignment near phase boundaries. Future investigation will include the influence of size, shape, and surface chemistry on the self-assembly of adsorbed NPs.

606005 Fabrication and Characterization of Free-Standing Asymmetric Membranes

Abstract:

All biological cell membranes maintain an electric transmembrane potential of around 100 mV, due in part to an asymmetric distribution of charged phospholipids across the membrane. This asymmetry is crucial to cell health and physiological processes such as intracell signaling, receptor-mediated endocytosis, and membrane protein function. Experimental artificial membrane systems incorporate essential cell membrane structures, such as the phospholipid bilayer, in a controllable manner where specific properties and processes can be isolated and examined. Methods to fabricate asymmetric model membranes are in their infancy, and many do not allow for quantitative control over solvent inclusion, lipid composition, and/or the ability to dynamically change and monitor conditions on either side of the membrane, which is necessary to characterize membrane physicochemical properties. To address this experimental gap, a platform was developed to fabricate and characterize planar, free-standing, asymmetric membranes with control over the lipid composition and solvent. First, a thin film balance was used to form a freestanding membrane by adsorbing aqueous phase lipids to an oil-water interface and subsequently thinning the oil to form a bilayer. This lipid-in-aqueous approach resulted in the formation of membranes with similar thickness and compressibility as previous lipid-in-oil methods, based on membranes composed of zwitterionic DOPC and anionic DOPG. By controlling the lipid composition in the aqueous compartments on either side of the oil within the thin film balance, asymmetric membranes were formed. This asymmetry was qualitatively demonstrated using a fluorescence quenching assay and quantitatively characterized through voltage-dependent capacitance measurements. Stable asymmetric membranes with DOPC on one side and DOPC/DOPG mixtures on the other were created with transmembrane potentials ranging from 10 to 25 mV. Increasing membrane charge asymmetry increases the offset voltage and also increases the stiffness of the membrane. These initial successes demonstrate a viable pathway to quantitatively characterize asymmetric bilayers that can be extended to accommodate more complex membrane processes in the future.

602283 Perfluoroalkyl Substances (PFASs) Partition into Bacterial Cell Membranes and Cause Unexpected Lipid Ordering

Abstract:

Perfluoroalkyl substances (PFASs), commonly found for example in non-stick coatings and waterproof materials, are persistent environmental pollutants that do not degrade naturally within the environment. Referred to as "forever chemicals", PFASs bioaccumulate across the food chain, binding strongly to proteins and partitioning into fatty regions such as cellular (lipid bilayer) membranes. Specific to membranes, PFASs are known to accumulate in and disorder model membranes comprised of single component, often zwitterionic, lipids. Our goal is to extend PFAS partitioning studies to live, intact bacterial membranes and bacterial membrane extracts to determine how complex lipid mixtures, including anionic lipids that are expected to repel the anionic headgroup of PFASs, respond to PFAS accumulation. *Alcanivorax borkumensis*, an alkane-degrading marine bacterium, was used as a model organism. A range of carboxylate and sulfonate PFASs were examined at concentrations ranging from nanomolar to micromolar to determine the role of perfluoroalkyl chain length (C_4-C_{10}) and headgroup (COO^- and SO_3^-) on *A. borkumensis* growth and *in situ* membrane ordering. This includes the legacy C_8 compounds perfluorooctanoic acid (PFOA) and perfluorooctanesulfonic acid (PFOS). All PFASs lead to unexpected lipid ordering of the membrane based on fluorescence anisotropy measurements, with longer chain PFASs leading to the greatest ordering. Cell growth also unexpectedly increased in the presence of PFASs, with higher PFAS concentrations leading to greater cell growth. Monolayer studies using *A. borkumensis* membrane extracts are being conducted to determine why lipid ordering, and possible lipid condensation, is observed for all PFASs examined despite the varying degrees of surface activity.

602539 Glycocalyx Osmolarity Modulates Macromolecular Binding to Cell Surfaces

Abstract:

The plasma membrane of mammalian cells is now understood to be a highly complex surface. Mass spectrometry, microscopy, and bioinformatics have revealed great diversity in size, composition, and post-translational modifications of cell surface glycoproteins and glycolipids. While physical crowding of the glycocalyx has been shown to induce membrane deformation, a clear understanding of the physical factors that contribute to crowding-associated phenomena remain unknown. In this talk, we demonstrate that 2D crowding of the cell surface glycocalyx directly modulates the binding affinity of soluble macromolecules to the plasma membrane. We introduce a set of molecular probes and use their binding isotherm to make a direct calculation of the in-plane osmotic pressure resulting from surface crowding. Importantly, we show that the osmotic pressure reduces the equilibrium binding affinity of an IgG antibody (among other macromolecules) to the surface, by up to a factor of ~7x, meaning that affinity is context dependent. Surprisingly, we find that negative charges contributed by sialic acid glycans decrease cell surface osmotic pressure far more than molecular weight or number density, accounting for ~50% of the reduction in affinity of IgG antibody on red blood cell membranes. We combine statistical mechanics theory, molecular simulations, and in-vitro reconstitution to highlight the impact of cell surface crowding on macromolecular binding and other cell behaviors mediated through the plasma membrane.

[Takatori S and Son S contributed equally to this work.]

602852 Role of Lipid Headgroups on Lateral Organization and Phase Transition Dynamics in Binary Mixture of Phospholipids and Cholesterol: What Can We Learn from Lipid Model Systems?

Abstract:

Biological membranes are highly complex structures comprising of a wide variety of phospholipids with differing headgroups, hydrocarbon chain lengths, and degrees of unsaturation, which distinguishably interact with cholesterol. The nature of lipid-cholesterol interactions, therefore, has been a long-standing interest not only to biochemists and biophysicists but also to clinicians, as it can provide important insight on physiologic and pathologic cellular pathways.

In this presentation, we describe the study performed to better understand the lateral interactions of two different types of phospholipids, i.e. 1,2-dipalmitoyl-sn-glycero-3-phosphocholine (DPPC) and 1,2-dipalmitoyl-sn-glycero-3-phosphoethanolamine (DPPE) mixed with cholesterol in the membrane plane. The morphological heterogeneity originated from liquid-liquid immiscibility and corresponding phase transition behaviors were systematically studied in using a fluorescence microscopy coupled Langmuir trough as a function of composition and lateral pressure. To further evaluate the phase dynamics associated with monolayer compressibility, dilatational rheological responses of corresponding binary mixtures were also determined using the pendant drop tensiometer.

We anticipate that our results will provide a critical link as to how subtle changes in lipid compositions and surrounding conditions can alter membrane dynamics and structures, which, in turn, affect biological performance and functionality in cell membranes at large including myelin sheath of which defects and/or structural disruption can lead to neurodegenerative diseases such as Multiple Sclerosis.

601184 Encapsulation and Phase Behavior of Lipid/Copolymer/Protein Hybrid Biomembranes

Abstract:

Entrapment of biomembranes in mesoporous metal oxide gels has proven to be a challenge, as current and previous techniques utilize liposomes as biological membrane hosts. The instability of liposomes in mesoporous gels is attributed to their size and altered environment upon entrapment within the nanometer scale pores (5-50 nm). We have overcome these barriers by entrapping nanometer-scale lipid/protein/copolymer membrane assemblies such as styrene-maleic acid copolymer-stabilized lipid nanodiscs. We have investigated the phase behavior of the lipids in addition to the structure, localization, and environmental polarity of the involved proteins mainly by spectroscopy methods. We then compare the thermal stability of gel-entrapped bacteriorhodopsin in purple membrane or lipid nanodiscs and find differences related to oligomeric states of the protein. Next, will be discussed recent studies of hybrid biomembranes combining block copolymers and phospholipids. The chemical diversity afforded by inclusion of block copolymers is particularly relevant to the formation of stealth liposomes or stimuli-responsive membranes (e.g. pH- or thermo-responsive systems), which can aid in the control of drug retention and release. However, the phase behavior of these hybrid biomembranes has not been methodically explored. To this end, we have studied small (~100 nm) vesicles, copolymer-stabilized nanodiscs, and giant vesicles with membranes made of polybutadiene-*block*-polyethylene oxide (PBd-PEO) and dipalmitoylphosphatidylcholine (DPPC) using fluorescence spectroscopy and by fluorescence microscopy.

599662 Mechanistic Studies on Daptomycin-Induced Phase-Transitions on MODEL Lipid Membranes: Effect on Membrane Permeability

Abstract:

Objective: For infections caused by multidrug resistant Gram-positive pathogens the cyclic-lipopeptide antibiotic Daptomycin is considered a potent treatment option. Its bactericidal activity is attributed to a calcium- and PG lipid-dependent disruption in the membrane potential resulting in the release of bacterial contents. Formation of oligomeric-Daptomycin membrane pores has been proposed as an action mechanism. In this study, we propose a different mechanism on model lipid membranes.

Methods: We studied the interactions of Daptomycin with model lipid bilayer vesicles with symmetric and/or asymmetric lipid leaflet compositions. We characterized the release kinetics of fluorescent probes encapsulated in vesicles while systematically varying lipid headgroups and acyl-tails.

Results: Our findings suggest that the increase in membrane permeability by Daptomycin is due to the formation of transient, leaky lipid interfaces that span the bilayer. These interfaces originate right on the boundaries of lateral, phase-separated, well-packed lipid domains induced by Daptomycin and, importantly, are registered across the bilayer. The following two-step process is proposed. In the first step, the binding of Daptomycin on vesicles results in formation of well-packed single-(outer)leaflet lipid-domains via its association with the PG-headgroups only. In this process, a well-packed outer lipid-leaflet domain occurs independent of the extent of unsaturation of the tails of PG-lipids that partition in the Daptomycin-induced domain. In turn, for this initial phase-transition to result in increase of the membrane permeability, the outer lipid leaflet domains need to induce cross-registered phase-transitions of also well-packed lipids in the inner leaflet. This second phase-separated domain occurs only when the tails of the lipids in the inner leaflet are saturated independent of their headgroups. The extent of release via the interfacial domain boundaries depends on the different acyl-tail lengths of lipids across the lateral inner-leaflet-lipid-interfaces.

Conclusions: These results support the hypothesis that Daptomycin increases bilayer membrane permeability via the formation of domains with transiently leaky-interfaces.

598004 Single-Molecule Studies of Cadherin Binding Interactions in Biomimetic Lipid Bilayer Models

Abstract:

Cadherin transmembrane proteins are responsible for cell adhesion in a number of tissues and therefore, modulate a variety of biological processes, such as tissue morphogenesis, cell motility, force transduction, and macromolecular transport across tissue boundaries. Furthermore, deficiencies in cadherin-mediated adhesion are correlated with several diseases, such as a number of skin diseases and cancers. Cell-cell adhesion is facilitated via adherens junctions, consisting of adhesive, trans-interactions, and lateral, cis-interactions, but the molecular role and interplay of these interactions is a topic of discussion. Here we used single-molecule microscopy to show the drastic, double cooperativity, of cis- and trans-interactions and how this can facilitate large junction formation using a biomimetic lipid bilayer cell adhesion model. Notably, the addition of cis-binding capability resulted in a 25-fold increase in trans-binding lifetimes between epithelial-cadherin extracellular domains. The nature of this cooperativity appears to be primarily due to allostery, as opposed to avidity. It has been thought the primary role of cis-interactions is to provide molecular ordering within junctions, increasing their size, but junctions still may form in the absence of cis-interactions. Furthermore, cis-interactions have been predicted to cause an elevated cadherin surface density in the adhesive zone between cells, potentially increasing adhesive binding strength though cluster avidity. We have directly observed strong cooperativity between cis- and trans-interactions, providing vital insight into the molecular mechanism of junction formation, and therefore, cadherin-mediated cell adhesion.

598421 Investigating the Role of Calcium on Coronavirus Fusion with Host Cells

Abstract:

Coronaviruses (CoVs) are a diverse family of enveloped viruses responsible for causing respiratory and/or enteric diseases across a wide range of species, causing widespread outbreaks in the human population. The expansive animal host range of CoVs can be attributed to their ability to adapt to diverse cellular environments (*i.e.* pH, ions, proteases, receptors), and exploit different entry pathways to mediate viral-host cell membrane fusion and infect cells. Ions have arisen as an important factor in the viral membrane fusion mechanism. Within the CoV family, our team has found that calcium availability leads to increased virus infection of cells. Previously, we observed that calcium availability also promotes insertion of SARS FP into the host cell bilayer and subsequent membrane lipid ordering, pointing to the possible role of calcium in interacting directly with the FP during CoV membrane fusion activity.

Due to the highly conserved nature of the FP between SARS- and MERS-CoV, we chose to investigate the specific binding pockets of Ca^{2+} within SARS- and MERS-CoV. We used site-directed mutagenesis and infectivity assays to pinpoint specific residues that lead to changes in infectivity when calcium is present or not. We identified potential calcium-binding residues by substituting the charged residues (*i.e.* aspartic acid, glutamic acid) in the FP with non-charged amino residue, alanine. We compared the infectivity of mutant and wild-typed CoV VLPs under calcium-rich or poor environments using available FDA-approved calcium blocking drugs. We further show that this enhancement could be attributed to Ca^{2+} increasing SARS- and MERS-CoV FP fusion-relevant membrane ordering. Although both SARS-CoV and MERS-CoV FP interact with Ca^{2+}, our data suggests differences in their interactions. SARS-CoV FP can bind two Ca^{2+}, whereas MERS-CoV FP can only bind one Ca^{2+}. Overall, these results demonstrate that Ca^{2+} have specific interactions with CoV FP.

607865 Stochastic Simulations of Motion of Solute Molecules during the Process of Crystallization Yields the Rate Constants for the Processes of Crystallization

Abstract:

Theoretical and computational understanding of crystallization processes can significantly replace the trial-and-error methods in place for crystallization at different pharmaceutical, chemical and fertilizer industries. However, the self-assembly of molecules during the process of crystallization spans multiple timescales which makes full computational modeling of crystallization process difficult. Recently, with the help of molecular simulations, it was shown that dynamics of solvation shell determines the crystal structure and growth rates. However, more appropriate computational approach is required for calculation of rate constants for nucleation and growth of crystals. In this study, stochastic simulations of glutamic acid molecules were performed. The molecular features of glutamic acid were approximated using the already established models to retain its molecular feature while reducing the computational time. Such approximation allows simulating higher number of glutamic acid molecules in a simulation box for longer times. Findings of this study yield in rate constants for glutamic acid crystallization which reasonably agree with experimental observations. The approach also shows potential to be applied to other crystallization systems.

604531 Irreversible Inhibition of Barite Crystallization: A Unique Mechanism for Treating Scale Formation

Abstract:

The mineralization of sparingly soluble compounds in process equipment may result in corrosion, equipment damage, and cause reduction or loss of finished product for oil and gas production, wastewater treatment, and manufacturing processes. Conventional treatments rely on the use of caustic solutions that pose a negative environmental impact. Designing effective biodegradable chemical treatments to reduce or eliminate scale formation requires an understanding of the molecular-scale interactions of chemical additives in supersaturated environments at the crystal interface. Here we implement a cooperative approach to design and test environmentally friendly chemical treatments for scale formation using a combination of surface science techniques and microfluidic technology. Crystallization can be suppressed in highly supersaturated solutions using dilute quantities of modifiers. These studies focus on barium sulfate (barite), a highly insoluble and common scaling mineral. In this presentation we will describe the effect of organic acids on barite crystallization in bulk assays under flow and in quiescent conditions. Using *in situ* scanning probe microscopy, we explore the interfacial interactions between the modifiers and different surfaces of barite over a range of supersaturated conditions to elucidate the role of these inhibitors during crystal growth. Our findings identify unique crystal-modifier interactions that suppress barite crystallization via an irreversible inhibition mechanism. Collectively, these studies provide new fundamental understanding of crystal growth modifier action towards the design of improved scale treatments.

611669 Multi-Step Crystallization of Self-Organized Spiral Eutectics

Abstract:

A method for the solidification of metallic alloys involving spiral self-organization is presented as a new strategy for producing large-area chiral patterns with emergent structural and optical properties, with attention to the underlying mechanism and dynamics. This study reports the discovery of a new growth mode for metastable, two-phase spiral patterns from a liquid metal. Crystallization proceeds via a non-classical, two-step pathway consisting of the initial formation of a polytetrahedral seed crystal, followed by ordering of two solid phases that nucleate heterogeneously on the seed and grow in a strongly coupled fashion. Crystallographic defects within the seed provide a template for spiral self-organization. These observations demonstrate the ubiquity of defect-mediated growth in multi-phase materials and establish a pathway toward bottom-up synthesis of chiral materials with an inter-phase spacing comparable to the wavelength of infrared light. Given that liquids often possess polytetrahedral short-range order, our results are applicable to many systems undergoing multi-step crystallization.

600638 Modeling Impurity-Mediated Crystal Growth

Abstract:

Crystallization is commonly used in industrial processes to convert solute molecules dissolved in solvent to a structured solid state. Pharmaceutical companies often crystallize APIs in the form of organic molecules to selectively formulate specific crystal habits for optimal bioperformance. Crystal engineering is also of importance for developing catalysts with tailored surfaces to maximize active sites. Furthermore, tuning crystallization is desirable for varying electrical and optical properties in the field of electronic materials and for altering the impact sensitivity of energetic materials. Given the ubiquity of crystal growth in industrial processes, there is substantial demand for predictive and mechanistic modelling of crystallization. Crystallization of organic molecules is well understood for ideal systems (i.e., Kossel crystals with a single centrosymmetric growth unit). There is interest in studying crystal systems in which non-idealities are introduced, as these are more representative of realistic conditions. One such non-ideality involves the presence of impurities or 'imposter molecules.' There are numerous notable examples of crystal growth systems in which the presence of impurities significantly influences crystallization processes. Examples include "antifreeze" proteins and antimalarial medications. While these systems are prevalent and of importance in many fields, there is currently no comprehensive model that captures the full effect of impurities on crystal growth processes.

The goal of this research is to investigate the effect of impurities on the crystal growth process and to develop theoretical models for the mechanisms by which impurities influence crystal growth and hence affect crystal morphology and size. Impurities affect growth kinetics at the scale of kink attachment and detachment events, which are too fine to examine experimentally in real time. Thus, we use simulations to study the proposed mechanisms for growth inhibition. We employ Kinetic Monte Carlo (KMC) to simulate the time evolution of centrosymmetric organic crystal growth. Various mechanisms have been proposed to explain the growth-inhibiting effect of impurities, such as Cabrera and Vermilyea's *step-pinning* model and Doherty and Sizemore's *spiral-pinning* models. Model equations have been developed in accordance with these mechanisms to quantify the effect of inhibition on the crystal face. We use KMC simulations to examine the limits of these equations under different growth regimes and to elucidate the form of a more general model. The simulations serve to quantify the degree of growth inhibition of a simple crystal growth system populated with impurities. The results suggest a monotonic decrease of the step velocity with increasing impurity concentration, though the form of this dependence is based on the *nature of the impurity distribution*. Furthermore, the simulations suggest a critical impurity coverage at which crystal growth is halted completely.

611583 Monte Carlo Simulations of Hydrothermal Metal Organic Framework (MOF) Crystal Growth

Abstract:

Metal organic frameworks (MOFs) are crystalline materials which are well-known due to their applications in separations, catalysis, and chemical capture, among others, brought forth by their large surface areas and high porosity. These materials are structurally versatile as they are composed of metal cluster and organic ligand pairs, which can be strategically switched to make MOFs with certain functionalities. A subset of MOFs with a "rod-packing" motif ("rod-MOFs) are promising as breakthrough materials for size-selective, passive chemical separation processes due to their fine unidirectional channels. The main challenge for these applications, however, rests on the optimization of the crystallization process of these MOFs to make them larger than their conventional micron scales and with negligible intergrowths.

Crystallization through the hydrothermal method is a common process used to synthesize certain classes of Metal organic frameworks (MOFs), including MIL-53, which falls under the rod-MOF classification. Though the process is ubiquitous to the formation of a variety of crystals, control of the dynamics of the growth process itself remains a challenge as the closed nature of the batch system typically used for these reactions is reminiscent of a "black box". A wealth of literature exists on the exploration of crystal growth dynamics through Monte Carlo (MC) and Molecular Dynamics (MD) simulations, as well as powerful models which incorporate both. However, these models have seldomly been applied to investigate the crystal growth kinetics of MOFs, and more specifically, the MOF crystallization phenomena for special cases where the inputs to the system are changing periodically.

Herein we apply our understanding of the physics of crystallization, statistics and well established kinetic Monte Carlo algorithms (kMC) based on the solid-on-solid (SOS) model and Metropolis MC to develop a deeper insight on MOF crystallization phenomena under hydrothermal conditions.

609413 The Role of Water in the Crystallisation and Nucleation of Glycine Homopeptides

Abstract:

The combination of amino acids into peptides and proteins, through peptide bonds, are the building blocks of life on earth. The last decade has seen a significant increase in the use of these materials in the treatment of chronic and metabolic diseases (e.g. cancer, obesity and diabetes).[1-2] However, the general peptides product process still has severe problems, like low solubility, proteolytic degradation and physiochemical instability.[3] Peptide crystallisation, as a good alternative, can help us get the peptide structure and get ideal physicochemical products. Water as a most popular solvent used in peptide and protein crystallisation will not only support the bioactivity of these materials, but also can stabilise the more complex molecular conformations that arise from long chain biomolecules, through satisfying the multiple hydrogen bonding sites of these materials.[4-5] Solving how the water molecule effect the peptides crystal structure can help us to uncover the details of the peptide conformation and waters role in their crystallisation, giving insight into their roles in the crystallisation of highly hydrated protein structures.

Glycine is an important and well-studied amino acid, being abundant in proteins and utilized as a pharmaceutical excipient. Its simple hydrogen side chain makes this an ideal case study candidate for biopharmaceutical research.[6] Here we investigate the crystallisation of pure and hydrated forms of mono-, di- and tri-glycine. However, only mono- and tri-glycine are known to crystallise in hydrated forms. The crystallisation of mono, di- and tri-glycine from aqueous solutions was compared in this research. The single crystal structure of triglycine dihydrate (TGDH) form was solved for the first time, which can help us to analysis the interaction between water and peptide molecules. For Glycine homopeptides, crystal forms of up to five glycine residues in a peptide chain are known, and the antiparallel β sheet structure was found is the most stable crystalline form for triglycine anhydrate. However, from the crystal structure of triglycine dihydrate, the backbone of it trends to bend in water to find a more stable form.

The activation free energy (Gc) of nucleation was calculated based on induction time measurement at different supersaturation and different temperature. Compare with mono- and di-glycine, triglycine has the greatest decrease in Gc at 283.15K when the form is dihydrate. Molecular modelling was also used to investigate the delicate balance of conformation and packing forces, as a function of chain length.

In summary, we solved the first TGDH crystal structure and explored the water role in the hydration conformation of peptides. The modeling work help us to find several competing effects which are driving TGDH. It also gives protein crystallisation a new insight about the interaction between water and protein, this discovery has an important implication for pharmaceutical engineering and biology research.

Reference

1. Cicero, A. F. G.; Fogacci, F.; Colletti, A., Potential role of bioactive peptides in prevention and treatment of chronic diseases: a narrative review. *Br J Pharmacol* **2017,** *174* (11), 1378-1394.
2. Marqus, S.; Pirogova, E.; Piva, T. J., Evaluation of the use of therapeutic peptides for cancer

treatment. *J Biomed Sci* **2017,** *24* (1), 21.
3. Otvos, L., Jr.; Wade, J. D., Current challenges in peptide-based drug discovery. *Front Chem* **2014,** *2*, 62.
4. Scoppola, E.; Sodo, A.; McLain, S. E.; Ricci, M. A.; Bruni, F., Water-peptide site-specific interactions: a structural study on the hydration of glutathione. *Biophys J* **2014,** *106* (8), 1701-9.
5. Wyttenbach, T.; Liu, D.; Bowers, M. T., Hydration of small peptides. *International Journal of Mass Spectrometry* **2005,** *240* (3), 221-232.
6. Koleva, B. B.; Kolev, T. M.; Spiteller, M., Structural and spectroscopic analysis of hydrogensquarates of glycine-containing tripeptides. *Biopolymers* **2006,** *83* (5), 498-507.

595426 N-Nitrosodimethylamine (NDMA) Formation during Advanced Oxidation Process (AOP) Treatment Following Reverse Osmosis for Potable Water Reuse

Abstract:

A 0.7 ng/L concentration of N-nitrosodimethylamine (NDMA) in drinking water is associated with a 10^{-6} lifetime excess cancer risk. NDMA is a particularly challenging compound for utilities purifying municipal wastewater effluents, where sewage contains higher concentrations of NDMA compared to conventional water sources, and NDMA can form from precursors when chloramines are applied upstream of microfiltration and reverse osmosis (RO) within advanced treatment. In order to reduce NDMA concentrations to below the 10 ng/L California Notification Level, UV-based advanced oxidation processes (AOP), reliant on radical production, such as UV/hydrogen peroxide (UV/H_2O_2), and, more recently, UV/free chlorine (UV/HOCl) are applied to RO permeates. While UV treatment effectively degrades NDMA, the potential for simultaneous NDMA formation during UV-based AOP treatment to partially offset this degradation has not been evaluated.

The objective of this laboratory-scale study is to evaluate the potential for NDMA formation during UV/hydrogen peroxide (UV/H_2O_2) and UV/free chlorine (UV/HOCl) AOP treatment. Interest in UV/HOCl processes has recently increased, as radical production from HOCl photolysis can be higher than from H_2O_2 photolysis resulting in more effective contaminant degradation. However, previous research suggests that the potential for greater NDMA formation during UV/HOCl AOP treatment. In this study, we demonstrate that dark reactions associated with residual chloramines in RO permeates increase NDMA formation during UV treatment. Addition of HOCl to RO permeates for UV/HOCl AOP treatment results in rapid NDMA formation from dark reactions with chloramine and UV-mediated breakpoint chlorination reactions. While these NDMA formation pathways and NDMA photodegradation occur simultaneously, our study shows that NDMA concentrations after UV/HOCl treatment can be double those post UV/H_2O_2 treatment. A mechanism for formation will be proposed. Given our results, in order to meet NDMA California Notification Levels with UV/HOCl treatment, higher UV fluence is required, and must be considered alongside higher radical production prior to implementation.

610968 Assessment of Enhanced Advanced Oxidation Processes for Treatment of Chlorinated Solvents

Abstract:

This paper addresses the process interactions resulting from combinative advanced oxidation processes, involving Fenton's reagent, ultraviolet light, titanium dioxide, ozone (O_3), hydrogen peroxide (H_2O_2), sonication or acoustic cavitation, among others. A comprehensive review of the technical literature has been performed, and the data ascertained have been critically analyzed and evaluated to determine whether the oxidative effects are additive, synergistic, or antagonistic. Predictions based on the individual unit operations are made and compared against the behavior of the combined unit operations. The data reported in this paper focus primarily on treatment of petroleum hydrocarbons (e.g., benzene, toluene, ethylbenzene, and xylene) and chlorinated solvents. The kinetic data included in this presentation are obtained both from the technical literature and from research investigations of the authors. The results of this study are assessed and discussed in this paper. The first-order kinetic rate constants are compared with other researchers.

595922 Plasma-Assisted Degradation of Perfluorobutane Sulfonate (PFBS)

Abstract:

This study investigates plasma-assisted degradation of perfluorobutane sulfonate (PFBS), a compound very few advanced oxidative and reductive technologies are able to degrade. PFBS, a member of a group of per- and polyfluoroalkyl substances (PFAS), is used as a replacement of perfluorooctane sulfonate (PFOS) in many consumer products such as nonstick pans, wire coating and firefighting foams. Recent studies have shown that PFBS is as persistent in the environment and bio-accumulative in people as its long-chain counterparts and human exposure to it can lead to serious health effects such as cancer, liver damage and immune system and thyroid changes. Here, the degradation of 10 mg/L of PFBS was investigated using a point-ring plasma reacor electrode configuration in which plasma is formed in argon gas and is contacting the surface of water containing dissolved PFBS. Experiments were conducted with and without cetyltrimethylammonium bromide (CTAB), a surfactant, whose purpose was to interact with and concentrate PFBS at the plasma-liquid interface, the region of chemical reactivity. No removal of PFBS was observed without the surfactant. With CTAB, ~99% of PFBS was removed from the bulk liquid and concentrated at the interface. At the interface, 67% of the 99% concentrate was degraded which resulted in 43% defluorination and 42% desulfonation. The dosage of the surfactant also played a role in the degradation. With step-wise CTAB dosing, defluorination rate increased markedly from 43% to 67%, and sulfate production from 42% to 68%. Experiments conducted with different surfactant types revealed that only cationic surfactants such as CTAB affect PFBS degradation, mainly through electrostatic attraction. Electrical discharge plasma appears to be a viable technology for the degradation of non-surfactants, in particular short-chain PFAS.

597753 Graphene Quantum Dots Based ZnO Nanomaterials for Degradation of Pollutant and Microbial Deactivation in Wastewater Treatment Plant.

Abstract:

Advanced oxidation processes (AOPs) offer enhanced contaminants destruction and low by-product and secondary pollution problems compared to conventional treatments. Photocatalysis is the mostly used form of AOPs. The photocatalyst can inactivate microorganisms and degrade organic materials by producing reactive oxygen species (ROS) upon photon irradiation. ZnO is an effective photocatalyst. It is characterized by a large bandgap energy, cost-effectiveness, biocompatibility and a high electron mobility. However, the efficiency of ZnO is reduced due to electron-hole (e^-/h^+) pairs rate recombination and inability to harvest a large part of the sunlight. In this work, different shapes of ZnO nanomaterials nanoflakes (ZnO-NFs) and zinc oxide nanorods (ZnO-NRs) were covalently linked to graphene oxide quantum dots (GQDs). They were linked to reduce the bandgap and decrease the e^-/h^+ pairs rate of recombination of the photocatalysts. The photocatalytic and antimicrobial behaviour of the graphene quantum dots based ZnO nanomaterials (ZnO-NFs@GQDs and ZnO-NRs@GQDs) were investigated. Several microscopic, spectroscopic and thermal analytic techniques were employed to confirm the successful preparation of the nanomaterials. Optical analysis showed that the calculated bandgap decreased from 2.98 eV to 2.61 eV and from 3.00 eV to 2.79 eV for the linked ZnO-NFs@GQDs and ZnO-NRs@GQDs, respectively. Data obtained from photolumiscence (PL) showed the same trend with a larger decrease in PL obtained for ZnO-NFs@GQDs indicating a large decrease in the e^-/h^+ pairs rate of recombination, thus a better photocatalyst. The photocatalytic experimental results revealed that the dyes were absorbed at 573 nm. A noticeable red-shifting of the dyes peak was evident with increase in time. The maximum absorption peak of the dyes at 573 nm decreased significantly with time for ZnONFs@GQDs than for ZnO-NRs@GQDs. The reduction in absorbance in the visible region indicated the decolourization through deprotonation (cleavage of the N−N bond) and desulfonation (ring opening). However, ZnO-NRs@GQDs showed greater efficiency in inhibiting bacterial growth for *E. coli* and *S. aureus* strains under the experimental conditions. The antimicrobial efficiency of the nanocomposites was in the order of ZnO-NFs < ZnO-NFs@GQDs < ZnO-NRs < ZnO-NRs@GQDs with greater efficiency being against *S. aureus*. The findings in this study imply that graphene quantum dots based ZnO nanomaterials can be used for the removal of dye pollutants and bacterial deactivation in wastewater treatment plant.

608957 Increasing Sodium Percarbonate Fenton like Process By Glutamic Acid Assisted Zero-Valent Iron for the Degradation of Methylene Blue

Abstract:

Zero valent iron (ZVI) catalyzed sodium percarbonate (SPC) Fenton like process assisted by glutamic acid (GA) for the degradation of methylene blue (MB) was evaluated in this study. Almost complete degradation of MB was observed within 40 min under optimum conditions when the molar ratios of ZVI/GA/SPC were 1/3/4 and the mineralization efficiency achieved was 42% within 3.5 hours. The detection of reactive oxygen species (ROSs) elucidated the significant generation of hydroxyl radicals and their key role in the degradation of MB. The experimental results and density functional calculations were found in good agreement thus, proving the more generation of ROSs in the ZVI/GA/SPC system for the efficient degradation of MB. The major intermediate products were identified and pathway mechanism was proposed based on mass spectrum analysis. Matrix effect of water constituents were examined to elucidate the efficiency of ZVI/GA/SPC system. In conclusion, actual wastewater results demonstrated significant potential of ZVI/GA/SPC system for the future aspirations.

Keywords

Sodium percarbonate, Fenton like process, Reactive oxygen species, Glutamic acid, DFT simulation.

594973 Ionizing Radiation-Induced Catalytic Degradation of Antibiotics in Water Matrices Using Fe/C Nanomaterials Derived from Iron-Based MOFs

Abstract:

The serious threat of antimicrobial resistance to human health has aroused the development of potential technologies to abate antibiotics from aqueous solutions. High-energy ionizing radiation, using electron beam or gamma rays, is considered to be more economic and efficient on a large-scale application than other AOPs such as UV radiation. H_2O molecules absorb radiation energy and water radiolysis occurs to generate the species involving the powerful oxidant ·OH and reductant e_{aq}^- and H·. ionizing irradiation has the advantages of good penetration in water matrices, insensitivity to suspended particles, no residuals produced and operation at room temperature. The public safety concern about radiation and high cost of investment and operation are the major factors to limit its application.

Recently, the iron-based metal organic frameworks (MOFs) have increasingly aroused great research attention in catalyzing the Fenton-like reaction and photocatalytic reaction to degrade organic pollutants in aqueous solution. MOFs are the hybrid porous crystalline materials constructed of the central metal /clusters coordinated to organic ligands, which have the large surface area and porosity, tunable topology and flexible structure. The Fe-based MOFs-derived Fe/C nanomaterials (DMOFs) are fabricated by using Fe-based MOFs as precursors through thermal pyrolysis under controlled temperatures and inert atmosphere. DMOFs remained the high specific surface area and porosity of MOFs and exhibited high catalytic activity and stability.

In present work, we proposed a novel Fe/C nanomaterial fabricated using Fe-based MOFs as precursors through thermal pyrolysis to catalyze gamma irradiation-induced degradation of antibiotics, cephalosporin C (CEP-C) and sulfamethazine (SMT) in aqueous solution. The adsorption kinetics of the two antibiotics on DMOFs were firstly investigated. The performances of gamma/DMOFs system was evaluated in terms of the antibiotics removal and mineralization extent in comparison to that using gamma irradiation alone. The degradation intermediates of antibiotics were identified and the physicochemical properties of DMOFs before and after irradiation was assessed. This study will provide a new insight about an efficient technique to decompose antibiotics in water matrices by catalytic ionizing radiation.

Results showed that DMOFs have the regular octahedrons structure of MOFs with high porosity. It consists of element C, Fe and O and Fe^0 with a fraction of Fe_3O_4 and Fe_2O_3 were identified. DMOFs addition was more efficient to enhance the degradation of antibiotics having a higher adsorption capacity like SMT. The degradation rate of CEP-C and SMT raised by 1.3 times and 1.8 times, and the TOC reduction at 1.0 kGy reached 42% and 51%, respectively by gamma/DMOFs treatment, compared to TOC removal of 20.2% (CEP-C) and 4.5% (SMT) by gamma irradiation alone. This was attributed to the enhanced generation of ·OH catalyzed by DMOFs during gamma irradiation. Overall, the degradation rate of both CEP-C and SMT was higher at acidic conditions and declined at alkaline pH of 9.0. It is noted that the adsorption capacity of both antibiotics declined with increasing the solution pH, which might have negative effect on antibiotic degradation. Regarding to the iron leached, it was much higher (up to 0.9-1.2 mg/L) at pH 4.0, while it maintained at low levels (0-0.2

mg/L) when the solution pH was higher than 5.2. Taken into account of the higher antibiotic decomposition rate and the minor iron leaching, pH value of 5.7 ± 0.5 was suggested for practical application.

The antibiotic molecules were destroyed, leading to the formation of intermediates of the N-S bond cleavage for SMT, C-N bond cleavage and β-lactam ring and dihydrothiazine ring opening for CEP-C. DMOFs exhibited good stability and recyclability as exposed to gamma irradiation. The crystal structure, functional groups and magnetism of DMOFs changed slightly after gamma irradiation, which made it possible to be recycled and reused. DMOFs were promising to enhance the degradation of antibiotics during gamma irradiation.

611104 Electrically-Heatable Graphene Aerogels As Nanoparticle Supports in High-Pressure CO_2 Capture

Abstract:

Hydrogen (H_2) is a key feedstock in the chemical industry and refineries worldwide, mainly used to produce ammonia and methanol, and upgrade heavier crude oils. Additionally, H_2 can be used in electrochemical cells or in combustion engines with the advantage of eliminating pollutant emissions by end users. Therefore, development of efficient, low carbon routes to produce H_2 is essential to meet its current and growing demand.[1,2] The so-called sorption-enhanced H_2 processes (SE-H_2) are one of the most promising and immediate intensification strategies to boost the amount of H_2 produced with efficient in situ carbon capture and storage (CCS), especially in steam methane reforming (SMR) based routes.[3] However, the selection of suitable materials and the relatively low maturity of the existing SE-H_2 configurations are hindering their commercial deployment.

Hydrotalcite-derived nanoparticles (HTs) are the most important group of CO_2 adsorbents for SE-H_2 since they show their highest CO_2 sorption performance in the temperature (473-773 K) and pressure range of interest (1-30 bar). Under these conditions, CO_2 physisorbents (e.g. zeolites, activated carbons) have low capacities and selectivities, while other potential CO_2 chemisorbents (lithium zirconates and calcium oxides) exhibit slow adsorption kinetics or require high energy to be regenerated. Despite the positive sorption characteristics of HTs for SE-H_2, their overall multicycle performance in terms of capacity, stability and energy efficiency needs to be further improved before they can be successfully used commercially. In this regard, it has been reported that significant enhancement in the CO_2 adsorption performance of HTs can be obtained by dispersing them on high surface area materials such as carbons.[4]

In this work we address the main challenges of SE-H_2 by supporting Mg-Al HT nanoparticles on 3D-reduced graphene oxide (rGO) aerogels, which are particularly suitable carbon structures due to the matching morphology and surface charge between the HT and the graphene support. For this purpose, unsupported Mg-Al HTs are synthesised via co-precipitation. To prepare HT/rGO supported aerogels, a well-dispersed GO suspension and a pre-fabricated wet-paste of HT are mixed. The aerogels are obtained by freeze drying and reduced with 5 % $H_2(N_2)$ at 1273 K. Physicochemical characterisation techniques such as BET, TGA, XRD, TEM, SEM, XPS and state-of-the-art equipment including an Intelligent Gravimetric Analyser (IGA), are used to gain fundamental insights into adsorptive performance under realistic temperature and pressure conditions for SE-H_2. Full details of the experimental methodology used are given elsewhere.[5]

The results of the work show that hydrotalcite-derived Mg-Al nanoparticles are successfully supported on rGO aerogels at very high HT loadings (~80 wt %), which is needed for the design of compact and efficient sorption units. The HT/rGO aerogel hybrids exhibit significantly higher CO_2 uptake compared to the unsupported HTs, showing an enhancement of ~230% at 573 K and PCO_2 = 8 bar. Temperature-pressure swing (573-673 K, 10 mbar-10 bar) tests show that the initial capacity of the pure HT drops by 20-30% upon cycling while the uptake of the aerogel hybrids is reduced only by 1-5%. The substantial benefits obtained by the presence of the rGO aerogel in HTs are attributed to electrostatic compatibility, excellent electrical conductivity, and enhanced mesoporosity and particle dispersion. The novel HT/rGO aerogel hybrids markedly outperforms other HT-carbon supported systems (e.g. GO, rGO, carbon nanotubes and nanofibers), and other type of relevant materials (e.g. activated carbons, MOFs, metal oxides and zeolites) in terms of capacity and stability. Under the operating cyclic conditions used, the CO_2 isotherms of the adsorbents can be modelled by the Freundlich equation, which is characteristic of heterogeneous surfaces.

In this contribution, we provide the first data on the CO_2 adsorption performance of carbon supported Mg-Al HTs at both elevated temperature and partial pressures of CO_2 relevant for industrial SE-H_2. In addition, we present for the first time the use of electrically-conducting rGO aerogels as very efficient supports to enhance the CO_2 adsorption properties of HTs. Research efforts are currently being devoted to extend this ground-breaking study by assessing the performance of the novel HT/rGO hybrids with other metal combinations such as Cu-Al and Ni-Al that have shown to be highly active and stable for the most relevant catalytic reactions involved in SE-H_2. Emphasis is given to further investigate the mechanical properties and thermal stability of the materials for their use in fixed bed and fluidised bed reactors. In addition, the regeneration of the hybrids by temperature-pressure swing adsorption is complemented by Joule-heating analyses. The use of the Toth and the linear driving force models is explored as a source of useful information for the prediction of SE-H_2 units. The feasibility of using this type of hybrids in other relevant H_2-CCS intensified processes such as adsorptive desulfurization and chemical looping combustion (CLC) is also under investigation.

Acknowledgements

This work is supported by the ELEGANCY, EU H2020 project and the Imperial College London Research Fellowship Grant. The authors are grateful to Milo Shaffer and David Chadwick for discussions.

References

1. IEAGHG report, Techno-economic evaluation of SMR based standalone hydrogen plant with CCS, (2017).
2. Valdsund, K. Jordal and R. Anantharaman, "Hydrogen production with CO_2 capture," Int. J. Hydrog. Energy, 41, pp. 4969-4992 (2016).
3. Iruretagoyena, K. Hellgardt, D. Chadwick, "Towards autothermal hydrogen production by sorption-enhanced water gas shift and methanol reforming: A thermodynamic analysis", Int. J. Hydrogen Energy, 43, pp. 4211-4222 (2018).
4. Iruretagoyena, Supported layered double hydroxides for sorption enhanced hydrogen production, Springer Theses, Switzerland, (2016).
5. Menzel, D. Iruretagoyena et al., "Electrically-Heatable Graphene Aerogels as Nanoparticle Supports in Adsorptive Desulfurisation and High-Pressure CO_2 Capture", *submitted.*

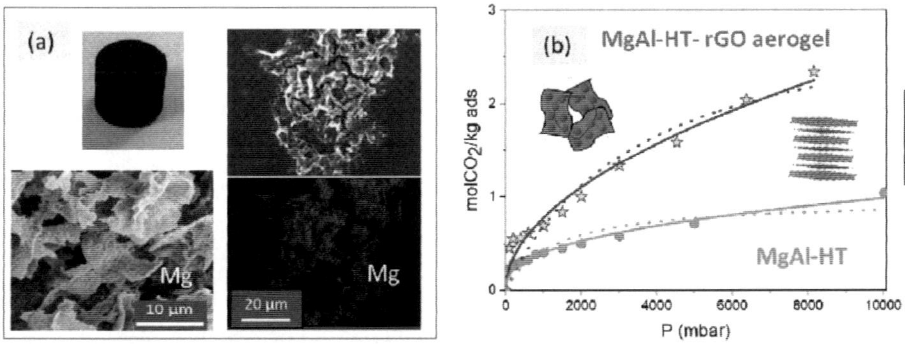

(c)

Adsorbents	Normalized CO_2 capacity	
	1st cycle	6th cycle
Mg-Al-HT	1	0.82
Mg-Al-HT -rGO aerogel	1	1

Representative figure. (a) Digital image of aerogel monolith, SEM and EDX mapping of HT/rGO. (b) High pressure CO_2 adsorption isotherms of unsupported HT and HT/rGO aerogels. (c) Normalised cycle capacity upon TSA-PSA cycling.

608974 Multiscale Screening of Adsorbents for Post Combustion Capture: Effect of the Vacuum Swing Adsorption Process Configuration on the Adsorbent Performance

Abstract:

Carbon capture and storage (CCS) has been identified as the near-term solution for mitigating climate change effects with the simultaneous use of fossil fuels for the production of energy. Bulk of global CO_2 emissions come from power plants burning fossil fuels. The current benchmark technology for post combustion CO_2 capture is absorption using amines. Adsorption process has been explored as an alternative technology for capturing and concentrating CO_2 from these large point sources. In the case of an adsorption process, the selected adsorbent should possess a very high capacity for CO_2 and high CO_2 selectivity. The benchmark material for adsorption-based carbon capture is Zeolite 13X and several published literature studies are available on this material [1-6]. Recent developments include new materials such as metal organic frameworks (MOFs), covalent organic frameworks (COFs) and Zeolite imidazolate frameworks (ZIFs). Many these new materials are computationally screened using molecular simulations and ranked based on metrics such as CO_2 capacity CO_2/N_2 selectivity etc, despite the limited availability of accurate force fields. However, these metrics were not reliable and the optimal behaviour of a material in a real process will also depend on the specifics of process configurations. Therefore, a reliable way to rank adsorbents is to study their performance in an adsorption process cycle[4,5]. The current approach to rank adsorbents on a process scale is to perform detailed process optimization while fixing the cycle configuration. However, in order to obtain the true potential, each adsorbent must be mapped to an appropriate cyclic process and very few studies have done a detailed cycle synthesis to obtain the best cycle for a given material[3,6].

In this work, we have employed two different cycles namely a 4-step vacuum swing adsorption (VSA) cycle with light product pressurization and a 6-step VSA cycle with reflux and light product pressurization, described in earlier publications[4,5]. These cycles were used to evaluate the performance of four adsorbents namely zeolite 13X, CPO-27-Ni, HKUST-1 and Silicalite. The information on adsorption equilibrium and kinetics were obtained from lab scale volumetric experiments as well as molecular simulations. Detailed optimization of the cycles was carried out using non-dominated sorting genetic algorithm (NSGA-II) in MATLAB to arrive at operating conditions that satisfy 95% CO_2 purity and 90% CO_2 recovery with minimum energy consumption and maximum productivity. The performance of the different VSA process was evaluated using both the actual experimental data and predictions from molecular simulations. Here we demonstrate the differences in performance with respect to isotherms obtained from experiments and molecular simulations as well as the cycle configuration.

Key words: Adsorption, carbon capture and storage, vacuum swing adsorption, metal organic frameworks

1. Krishnamurthy, S., et al., *CO_2 capture from dry flue gas by vacuum swing adsorption: A pilot plant study.* AIChE J, 2014. **60**(5): p. 1830-1842.
2. Li, G., et al., *Competition of CO_2/H_2O in adsorption-based CO_2 capture.* Energy Procedia, 2009. **1**(1): p. 1123-1130.

3. Haghpanah, R., et al., *Cycle synthesis and optimization of a VSA process for postcombustion CO_2 capture.* AIChE J, 2013. **59**(12): p. 4735-4748.
4. Rajagopalan, A. K.; Avila, A. M.; Rajendran, A., Do adsorbent screening metrics predict process performance? A process optimisation-based study for post-combustion capture of CO_2. *Int J Greenh Gas Con* **2016,** 46, 76-85
5. Khurana, M. and S. Farooq, *Simulation and optimization of a 6-step dual-reflux VSA cycle for post-combustion CO_2 capture.* Chem Eng Sci, 2016. **152**: p. 507-515.
6. Reynolds, S. P.; Mehrotra, A.; Ebner, A. D.; Ritter, J. A., Heavy reflux PSA cycles for CO_2 recovery from flue gas: Part I. Performance evaluation. *Adsorption* **2008,** 14, (2), 399-413.

601815 Multi-Technique Study of Oxidative Degradation of Supported Poly(ethylenimine) sorbents for CO_2 capture

Abstract:

The increase in CO_2 concentration in the air has been a growing subject of attention for scientists in the past few decades. By holding on to the current policies and regulations, it is anticipated that the CO_2 level in air will be more than 450 ppm by 2035, leading to a minimum increase of 2 °C in average atmospheric temperature. [1] Therefore, technologies that lead to a reduction of the CO_2 concentration in the air, such as direct adsorptive CO_2 capture from air (Direct Air Capture, DAC), are expected to play an important role in the coming years.

DAC sorbents can be prepared by loading amines into different mesoporous inorganic supports such as alumina, silica, and carbon. Typically, primary and secondary amine groups react with CO_2 to form carbamates and bicarbonates. However, one of the key challenges involved with employing amine sorbents is their sensitivity to deactivation by NOx and SOx, amine leaching, and simultaneous exposure to heat and oxygen. Previous studies have shown that primary and secondary amines can readily oxidize to imines and carbonyl/imide species at the elevated temperatures used for sorbent regeneration if significant oxygen is also present.[2] Considering the significant concentration of oxygen in air, understanding the chemistry and kinetics of this oxidative degradation process plays a crucial role in designing long-lifetime sorbents for direct-air-capture.

In this research, the oxidative degradation of a prototypical class 1 sorbent [poly(ethylenimine) impregnated into commercial mesoporous γ-Al_2O_3] at different operational conditions (temperature, water content, oxygen content, and oxygen exposure time has been analyzed. A suite of analytical techniques including calorimetry, thermogravimetric, elemental, and IR analysis, has been used to measure the rate of oxidation in situ. The results are then used to understand the kinetics and thermodynamics of the oxidative degradation reactions. The trend observed in the obtained heat of reaction from calorimetry studies, carbonyl peak intensity from IR, and CO_2 loss from thermogravimetric analysis are in agreement with each other. Using the calorimetry and elemental analysis data, a rate expression has been developed, and an overall activation energy is calculated for the concerted suite of oxidation reactions.

References:

1. Anwar, M. N.; Fayyaz, A.; Sohail, N. F.; Khokhar, M. F.; Baqar, M.; Yasar, A.; Rasool, K.; Nazir, A.; Raja, M. U. F.; Rehan, M.; Aghbashlo, M.; Tabatabaei, M.; Nizami, A. S., CO_2 utilization: Turning greenhouse gas into fuels and valuable products. *Journal of Environmental Management* **2020,** *260*, 110059, DOI: 10.1016/j.jenvman.2019.110059.
2. Ahmadalinezhad, A.; Sayari, A., Oxidative degradation of silica-supported polyethylenimine for CO_2 adsorption: insights into the nature of deactivated species. *Physical Chemistry Chemical Physics* **2014,** *16*, 1529-1535, DOI: 10.1039/C3CP53928H.

595580 Comprehensive Assessment of the Stability of Amine-Grafted Mesoporous Silica during Cyclic CO_2 Capture Using Steam Regeneration

Abstract:

The hydrothermal stability of triamine-grafted, large-pore SBA-15 CO_2 adsorbents was studied using steam stripping. Following two 3-h cycles of steam regeneration, lower CO_2 uptakes, lower CO_2/N ratios, and slower adsorption kinetics were observed relative to fresh samples, particularly at the lowest adsorption temperature (25 ^oC). CO_2 adsorption measurements for a selected sample exposed to 48 h of steam stripping depicted that after the initial loss during the first exposure to steam (3 to 6 h), the adsorptive properties stabilized. For higher adsorption temperatures (i.e., 50 and 75 ^oC), however, all adsorptive properties remained almost unchanged after steaming, indicating the significance of diffusional limitations. TGA and FT-IR spectroscopy results for grafted samples before and after steam stripping showed no amine leaching and no change in the chemical nature of the amine groups, respectively. Also, a 6-cycle CO_2 adsorption/desorption experiment under dry conditions showed no thermal degradation. However, N_2 adsorption measurement at 77 K showed significant reductions in the BET surface area of the grafted samples following steaming. It is proposed that exposure to steam restructured the grafted materials, causing mass transfer resistance. It is inferred that triamine-grafted, large-pore SBA-15 adsorbents are potential candidates for CO_2 capture at relatively high temperatures (50 to 75 ^oC; e.g. flue gas) combined with steam regeneration.

610532 Integrating Process Design and Materials Genomics for Energy Efficient Adsorption-Based Capture Technologies

Abstract:

Separations use as much as 15% of our energy consumption; the future of our planet may depend on the efficiency at which we capture CO_2, currently one of the boldest separation research areas.[1]Recent developments in synthetic chemistry allow us to synthesize millions of different nanoporous crystalline materials. For instance, by combining a metal node with an organic linker, one can synthesize a nearly infinite number of metal organic frameworks (MOFs).[2]However, designing and constructing tailor-made materials for specific applications is one of the key research challenges that still remains to be addressed. In other words, a systematic approach on how to select from such a large number of materials, the ones that are optimal for a given carbon-capture process, is still lacking.

In this work, we integrate recent developments in Material Science and Materials Genomics,[3]in which we generate in silico millions of candidate materials for CO_2capture, with process design and techno-economic and environmental analyses. The lack of such integration has been identified as one of the key bottlenecks that limit the prospect of novel materials for carbon capture technologies to reach the market. This study is carried out within the framework of the PrISMa project (http://www.act-ccs.eu/prisma), which is an international effort that aims to accelerate the transition of energy and industrial sectors to a low-carbon economyby developinga technology platform to tailor-make cost-efficient carbon capture solutionsfor a range of different CO_2sources and CO_2use/destinations.Our aimed technology platform will allow us to tailor-make carbon capture solutions in terms of an effective carbon price (ECP) that makes a particular process economically viable. To achieve that, we integrate materials and process design, supported by computational materials genomic approaches, machine-learning, high-throughput material synthesis and characterization, device-performance testing, value chain analysis and systematic process design.

Through this integration, we aim to change the paradigm on how novel materials are developed for chemical engineering applications. Some recent results[4]from the PrISMa collaboration involved the design of a metal organic framework that can capture CO_2in wet flue gasses (see Figure 1). As a similar approach can be developed for other separations, we expect the impact of this work in terms of the potential to decrease the time to market for novel materials to go beyond carbon capture applications.

Acknowledgements

The PrISMa Project (No 299659) is funded through the ACT programme (Accelerating CCS Technologies, Horizon2020 Project No 294766). Financial contributions made from: Department for Business, Energy & Industrial Strategy (BEIS) together with extra funding from NERC and EPSRC Research Councils, United Kingdom; The Research Council of Norway, (RCN), Norway; Swiss Federal Office of Energy (SFOE), Switzerland; and US-Department of Energy (US-DOE), USA, are gratefully acknowledged. Additional financial support from TOTAL and Equinor is also gratefully acknowledged.

References

[1] B. Smit and S. Garcia, *Carbon capture and storage: making fossil fuels great again?*Europhys. News **51**(2), 20 (2020) http://dx.doi.org/10.1051/epn/2020203

[2] H. Furukawa, K. E. Cordova, M. O'Keeffe, and O. M. Yaghi, *The Chemistry and Applications of Metal-Organic Frameworks*Science **341**(6149), 974 (2013) http://dx.doi.org/10.1126/Science.1230444

[3] P. G. Boyd, Y. J. Lee, and B. Smit, *Computational development of the nanoporous materials genome*Nat Rev Mater **2**(8), 17037 (2017) http://dx.doi.org/10.1038/natrevmats.2017.37

[4] P. G. Boyd, A. Chidambaram, E. Garcia-Diez, C. P. Ireland, T. D. Daff, R. Bounds, A. Gladysiak, P. Schouwink, S. M. Moosavi, M. M. Maroto-Valer, J. A. Reimer, J. A. R. Navarro, T. K. Woo, S. Garcia, K. C. Stylianou, and B. Smit, *Data-driven design of metal-organic frameworks for wet flue gas CO2 capture*Nature **576**(7786), 253 (2019) http://dx.doi.org/10.1038/s41586-019-1798-7

Figure 1: A molecular representation of the carbon capture material Al-PMOF (picture by Mohamad Moosavi and Kevin Jablonka; iRASPA software was used for visualisation)

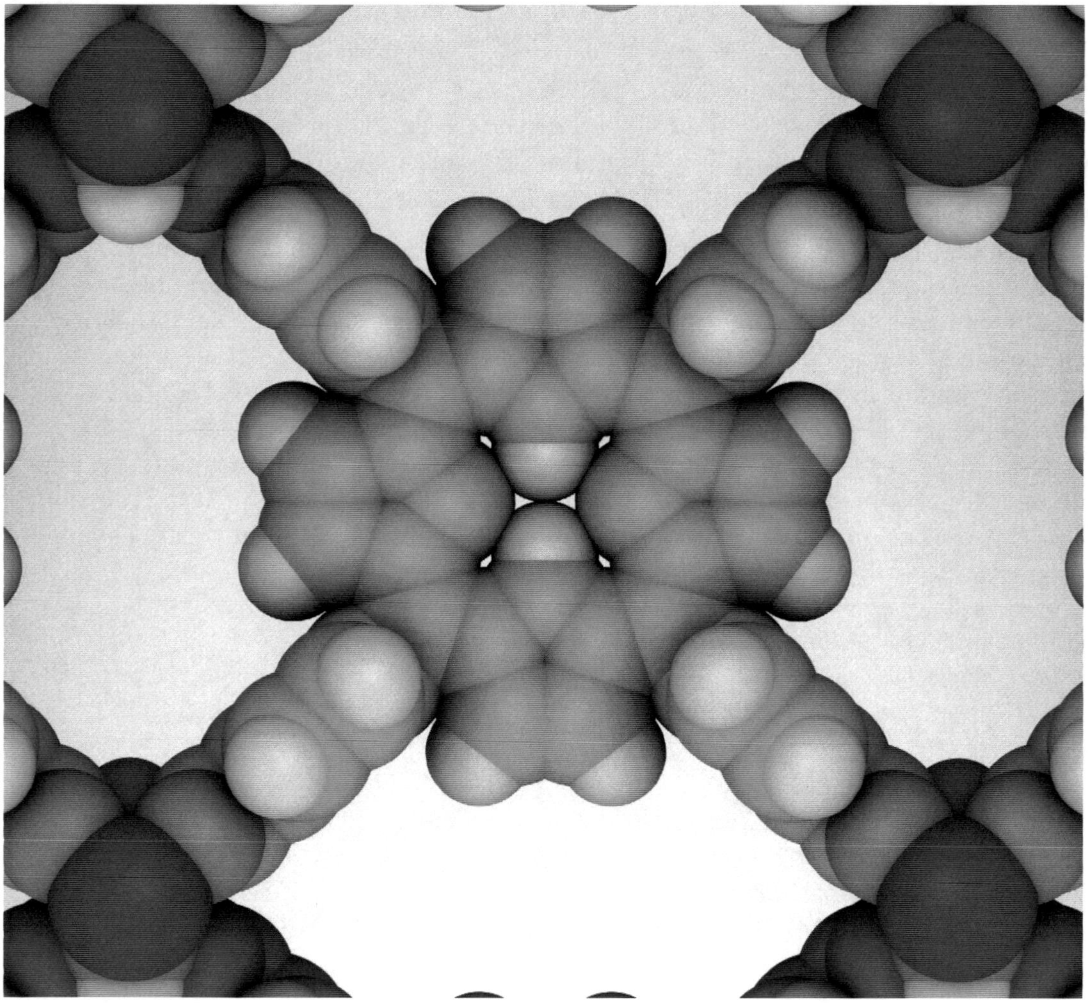

612123 Technoeconomic Assessment of Optimized Adsorption Processes for Post-Combustion CO_2 Capture in Hydrogen Plants

Abstract:

Hydrogen is a clean fuel and plays a key role towards building a low-carbon sustainable energy society. While there are renewable sources to produce hydrogen, most of the global hydrogen production relies on fossil fuels like natural gas. To enable low-carbon emission hydrogen production, hydrogen production with CO_2 capture and storage remains an interim plausible solution. The most common route for hydrogen production involves steam methane reforming (SMR) technology. In SMR-based hydrogen plants, one of the main CO_2 emission sources continues to be SMR furnace flue gas.

Vacuum swing adsorption (VSA) technology has been widely considered for post-combustion CO_2 capture. Given very few technoeconomic studies exist for adsorptive-based CO_2 capture, detailed process design based on economics remains a key challenge. In order to fully understand the potential of VSA technology to enable low-carbon hydrogen production, this study provides a technoeconomic assessment of VSA for post-combustion CO_2 capture from SMR furnace flue gas. To this end, an integrated technoeconomic optimization methodology which takes into account process models, scale-up design, vacuum pump performance and comprehensive costing model is developed. Several operating conditions are thoroughly optimized for minimum CO_2 capture cost using genetic algorithm-based optimization with requirements of 95% CO_2 purity and 90% CO_2 recovery. Since the choice of adsorbent is critical for process design, commercial zeolite 13X and metal organic frameworks (MOFs) are assessed for their overall technoeconomic performances and compared against a benchmark solvent case. In addition, the impact of several parameters like vacuum pump efficiency, adsorbent cost, adsorption columns, etc. on the capital and operating costs are also considered. Further, many studies use multiobjective optimization of minimization of energy consumption and maximization of productivity to approximate the optimal cost performance. It is shown that the cost performance approximated through energy-productivity optimization may not reflect the true optimal cost performance obtained by considering cost minimization. At the meeting, the effect of choosing appropriate objective functions and the importance of comprehensive costing model to evaluate adsorption-based processes will be discussed.

Keywords: post-combustion CO_2 capture; adsorption; H_2 plant; process optimization; technoeconomic analysis; metal-organic frameworks; zeolite.

Acknowledgements

This publication has been produced with support from the NCCS Centre, performed under the Norwegian research program Centres for Environment-friendly Energy Research (FME). The authors acknowledge the following partners for their funding contributions to the NCCS Centre: Aker Solutions, ANSALDO Energia, CoorsTek Membrane Sciences, Gassco, KROHNE, Larvik Shipping, Norcem, Norwegian Oil and Gas, Quad Geometrics, Statoil, TOTAL, and the Research Council of Norway (257579/E20). Funding from Canada First Research Excellence Fund through University of Alberta Future Energy Systems is acknowledged.

595579 Long-Term Heel Transformation during Cyclic Adsorption-Desorption of Volatile Organic Compounds in a Full-Scale Adsorber-Desorber

Abstract:

This study elucidates the different mechanisms contributing to heel formation during cyclic adsorption-desorption of volatile organic compounds (VOCs) from a painting process in a full-scale adsorber-desorber. Two batches of beaded activated carbon (BAC), one subjected to normal operation and the other inadvertently exposed to oxygen during desorption, were investigated. Both batches displayed sharp increases in apparent density during early cycles, which indicate high heel formation due to occupation/blockage of high-energy adsorption sites. Thermogravimetric analysis and gas chromatography-mass spectrometry results identified non-desorbed adsorbates, adsorbate reaction by-products and char from adsorbate decomposition. The main heel formation mechanisms include physisorption and chemisorption followed by thermal oxidation, pyrolysis and eventually char formation. We hypothesize that non-ideal heat and purge gas distribution across the desorber is the reason for the accumulation of non-desorbed species and heel formation, shortening the lifetime of the adsorbent. In addition to pyrolysis reactions that partially contributed to heel formation in the absence of oxygen, accumulation of thermal oxidation products also contributed to the deterioration of adsorbent performance in the presence of oxygen during desorption. Moreover, extended exposure of the non-desorbed physisorbed and chemisorbed heel species to repetitive desorption cycles ultimately transformed them to permanent (non-desorbable) heel via polymerization and char formation. Therefore, to reduce heel buildup and maximize long-term adsorber performance, it is recommended to (i) ensure effective desorption conditions to prevent or minimize the accumulation of non-desorbed physisorbed species, and (ii) maintain sufficiently low oxygen levels in the desorption purge gas to minimize the detrimental effects of any species that do accumulate.

607811 New 100 Hz Volumetric Frequency Response System: Design, Operation and Analysis

Abstract:

Frequency response methods have been developed as an alternative technique for the determination of mass transfer mechanisms controlling the transport of gases into and out of microporous adsorbent materials. Recent results in the literature are showing that this technique seems to be the best for discriminating the controlling mechanism. It is also possible to reveal two or more different controlling mechanisms that may dominate in different frequency ranges. This presentation will focus on the design and operation of a new 100 Hz volumetric frequency response (VFR) system, and the analysis of the resulting amplitude and phase lag data.

Like the previous 10 Hz system constructed at USC, the apparatus consists of two sections, the gas handling/adsorption section and the driving section. The first section is comprised of the tubing for gas dosing and vacuuming into the adsorption chamber, differential and absolute pressure sensors, and all the necessary isolation valves. The second section includes the metal bellows for volume perturbation, motor, eccentric cam, position indicator, and electrical and data acquisition systems. It operates like the previous one using LabView hardware and software, but up to much higher frequencies.

With respect to analysis, the effects of and corrections for three undesirable physical phenomena observed at intermediate (0.001-10 Hz), high (> 1 Hz) and very high (> 20 Hz) frequencies will be shown and discussed. The first of these three phenomena is well described in the literature and is associated with heat that is generated due to gas compression in sections of the working volume that do not have the ability to remove heat. The other two phenomena are probably due to local pressure drop and mechanical vibration. The ability of the instrument to measure amplitude and phase lag data up to about 90 Hz will be demonstrated for BPL activated carbon in both pelleted and structured forms.

SATURATION LOADINGS ON ZEOLITES VARY SIGNIFICANTLY WITH REDUCED TEMPERATURE IN THE RANGE 0.60 TO 3.5

Dana M. Abouelnasr, Georgia Institute of Technology, Atlanta, GA and Kevin F. Loughlin, Retired, Kingston, Ontario

Biography

Dana Abouelnasr, Senior Lecturer in the Coulter Dept of Biomedical Engineering, Georgia Institute of Technology. Formerly Professor of Chemical Engineering, American University of Sharjah, UAE. Kevin F. Loughlin, Retired Professor of Chemical Engineering, American University of Sharjah, UAE.

Abstract

Saturation loadings on zeolites 4A, 5A and 13X are reported for 56 species over a reduced temperature range 0.6 to 3.5. The data is culled from the literature for the past 64 years from isotherms having a slope of dlnq/dlnp tending towards 1. The subcritical region is modeled using a theoretical equation employing a Rackett liquid density model and crystallographic information of the zeolites; the supercritical region is modelled by an empirical linear loading correlation for individual species. The saturation loading results for both regions are consistent with the proposed models.

Purpose/Introduction

Starting in 2009, the purpose of our studies has been to quantify saturation loadings on three zeolites: 4A, 5A, and 13X from data posted in the literature and to model the resulting findings. The data posted in the literature for 56 species over the past 64 years is taken from tabulated Tables or by digitizing the relevant Figures; saturation loadings are determined from log-log plots for isotherms exhibiting a slope of $dlog\, q/dlog\, p$ approaching zero. Loadings are normalized to account for binder content. For each adsorbent/adsorbate pair, isotherms across studies are compared for consistency. For each saturated isotherm, the final loading and reduced temperature are recorded. To compare across adsorbent/adsorbate pairs, q_{max} is normalized by $q_{max,c}$ (defined later in this paper) and plotted against T_r. This procedure is independent of isotherm models.

Our objective is to provide a model for q_{max} and thereby reduce the number of parameters required for isotherm modeling. For example, the Langmuir isotherm model parameters are reduced from two (K_H and q_{max}) to one (K_H). The model can be generalized to other adsorbents.

Experimental Design/Method

Our observations indicate that T_{CA}, the temperature at which adsorbed species transition to supercritical behavior, is often lower than T_C, the critical temperature from vapor-liquid equilibria. The saturation loadings for subcritical temperatures can be theoretically modeled for microporous zeolites from first principles for zeolite crystals assuming 100 % accessibility of the large-cage void for the adsorbate and from the Rackett model for the sorbate density (1):

$$q_{max}\left(\frac{g}{100\ g\ Z}\right) = 100\frac{\varepsilon_Z}{\rho_Z}\left(\frac{P_c\ MW}{R\ T_c}\right) Z_{RA}^{-\{1+(1-T_r)^{0.2857}\}} \tag{1}$$

where ε_Z is the zeolite void fraction (large-cage only), ρ_Z is the zeolite crystallographic density, and Z_{RA} is a particular constant for the modified Rackett equation. Values for Z_{RA} are given by Spencer and Danner (1).

This equation can be rearranged to:

$$\frac{q_{max}}{q_{max,c'}} = Z_{RA}^{-\{(1-T_r)^{0.2857}\}} \tag{2}$$

where

$$q_{max,c}\left(\frac{g}{100\ g\ Z}\right) = 100\frac{\varepsilon_Z}{\rho_Z}\left(\frac{P_c\ MW}{R\ T_c\ Z_{RA}}\right) \tag{3}$$

The physical interpretation of $q_{max,c}$ is the saturation loading at the critical temperature, as predicted by the model. A plot of Equation 1 is presented in Figure 1 for diatomic molecules on 5A zeolite. The plots illustrate a monotonic decrease from a high loading at a low reduced temperature (in this Figure, we have started at $T_r = 0.3$) to a low loading at $T_r = 1$; the difference in q_{max} between the two reduced temperatures is a factor of approximately 3.

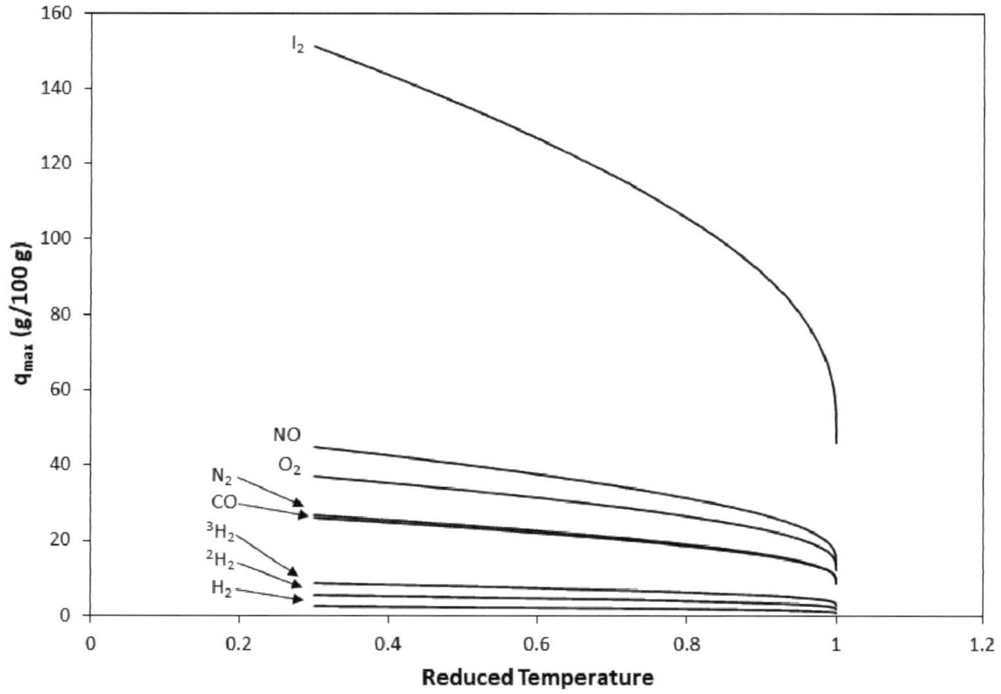

Figure 1 Plot of the subcritical model of saturation loadings for diatomic species on 5A zeolite. Only those species for which saturation loadings have been collected on 5A Zeolite from the literature are presented.

Supercritical data appear relatively linear when plotted as q_{max} versus T_r; consequently, a simple linear model for the supercritical range can be derived, and the intersection of this line with the subcritical model gives T_{CAR}, the critical adsorbate reduced temperature. The intersection for T_{CAR} is generally below $T_r = 1$, corresponding to the critical VLE temperature. This result agrees with de Boer's text on "Dynamical Adsorption" where he has shown that T_{CAR} is significantly less than $T_r = 1$ (de Boer called this phenomenon the two-dimensional critical temperature) (2).

Major Findings/Results

Nitrogen isotherms on 5A zeolite are presented in Figure 2 as a representative set of isotherms. Figure 2 is a log-log plot of loading versus reduced pressure for 45 studies in the literature. The label for each isotherm is the T_r at which it was taken. The references for all the studies are provided in the paper by Abouelnasr and Loughlin (3). Isotherms with a slope of $dlog\ q/dlog\ p$ tending to zero are used to find q_{max}; 22 isotherms satisfy this criterion, and so 22 values for saturation loadings have been identified. In addition, isotherms with $dlog\ q/dlog\ p$ having a linear slope of 1 in the low concentration region are used to determine Henry constants. This has been done for nitrogen on 13X zeolite (4). We will also plan to perform this task for the all the data collected for every species on the 4A, 5A and 13X zeolite.

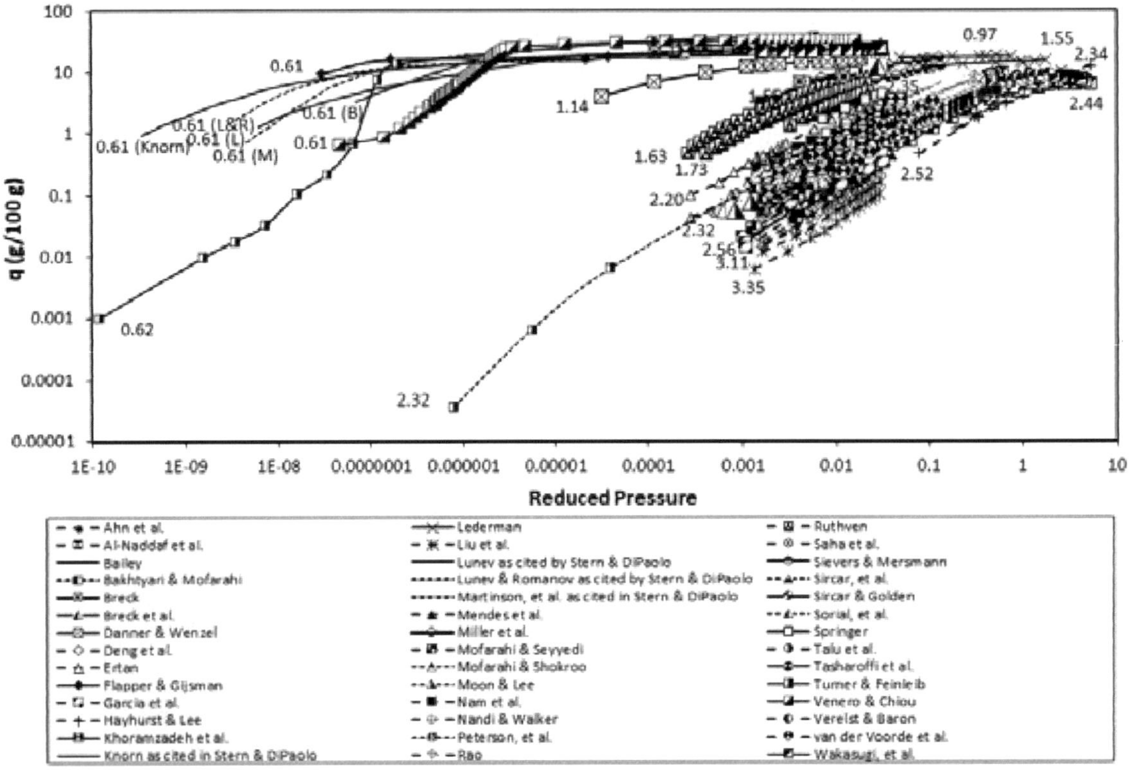

Figure 2. Plot of all nitrogen isotherms on 5A zeolite taken from the literature

The major findings are summarized in Figures 3 to 6. The $subT_{CAR}$ data for n-alkanes (C_1-C_8) on zeolites 5A and 13X are plotted in Figure 3 along with the model curves for Equation 1 for the largest and smallest Z_{RA} for these alkanes (and so they indicate the range for all pertinent curves). The data fit the model reasonably well; some of the low points below the C_1 model may be attributed to the difficulty in measuring q_{max} in glass apparata.

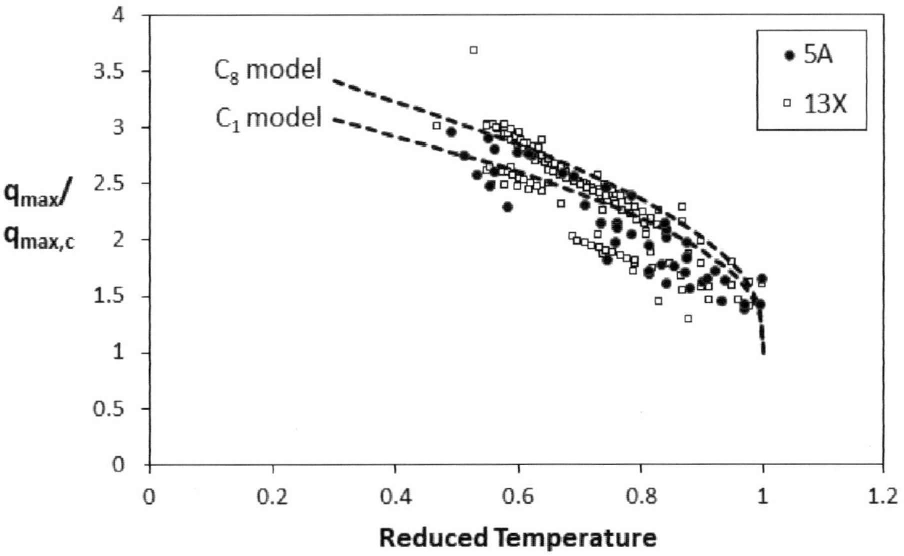

Figure 3. All subT_{CAR} data for n-C_1 through n-C_8 alkanes on 5A and 13X zeolite. The dashed lines represent the proposed model for CH_4 and C_8H_{18}, as they represent the range of values of Z_{RA} (0.289 and 0.257, respectively) and so the curves represent the range for all species in the Figure.

A similar plot is illustrated in Figure 4 for all the SubT_{CAR} data (50 species) excluding n-alkanes larger than n-C_8 on 4A, 5A and 13X zeolites; the observations from this Figure are similar to those from Figure 3. The two dashed lines are model curves for the largest and smallest Z_{RA} values.

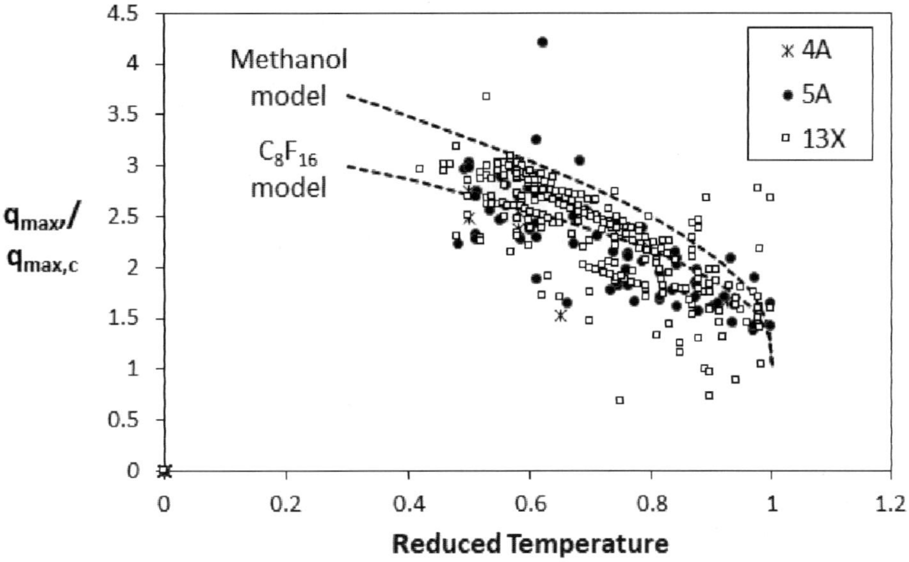

Figure 4. The subT_{CAR} plot for 56 species on 4A, 5A and 13X zeolite. These represent all species for which data have been collected, except for n-alkanes larger than n-C_8. The dashed lines represent the proposed model for C_8F_{16} and methanol, as they represent the range of values of Z_{RA} (0.296 and 0.236, respectively) and so the curves represent the range for all species in the Figure.

The superT_{CAR} plot for inorganic diatomic species on zeolites 4A, 5A and 13X is shown in Figure 5 along with the model curves for Equation 1 for the subT_{CAR} region for the largest and smallest Z_{RA} for these species (deuterium, with a Z_{RA} of 0.312 and CO, with a Z_{RA} of 0.290) and the linear regression fit for all data in Figure 5. The data for individual species has very high linear regression coefficients (> 0.95). The spread observed is due to the many species and zeolites portrayed; however, no definitive trend was observed between zeolites or species. The intersection of the linear line with the subT_{CAR} curve provides T_{CAR} and is mainly around 0.6 to 0.8. A further interesting point is that the ratio $q_{max}/q_{max,c}$ ranges from 2.75 to 0.5 indicating a significant change of density or molecular volume with reduced temperature.

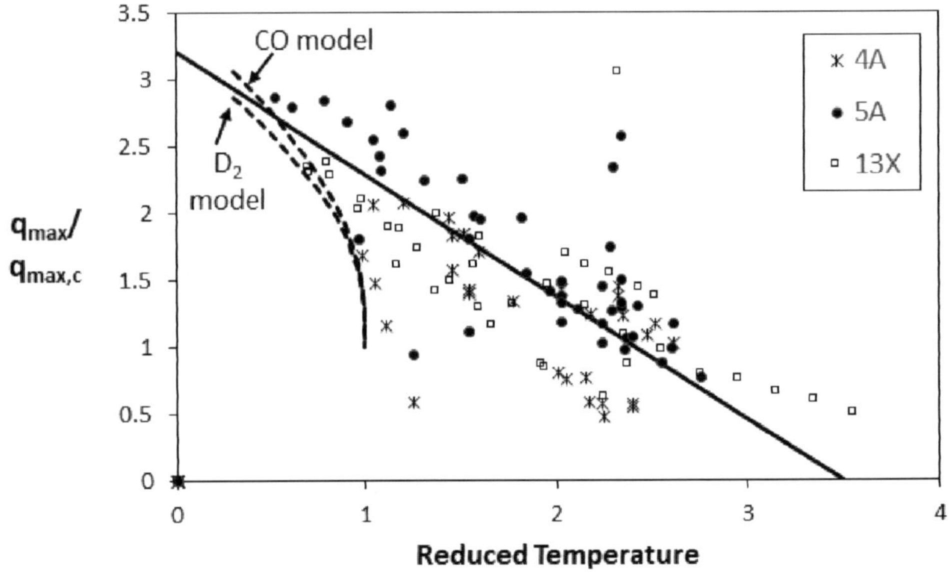

Figure 5. The superT_{CAR} plot for diatomic species on 4A, 5A and 13X zeolite. These represent all diatomic species for which superT_{CAR} data have been collected. The dashed lines represent the proposed T_{CAR} model for deuterium and CO, as they represent the range of values of Z_{RA} (0.312 and 0.290, respectively). These curves are presented only for reference, as the data are all superT_{CAR}. The solid line represents the linear regression of all the superT_{CAR} data on the Figure.

A superT_{CAR} plot for 50 species on 4A, 5A and 13X zeolites is illustrated in Figure 6; this represents all the species for which superT_{CAR} data have been collected, excluding n-alkanes larger than n-C_8H_{18} (3,5-11). The observations from Figure 6 are similar in all respects to the diatomic plot in Figure 5. In total 572 data points are included in Figure 6.

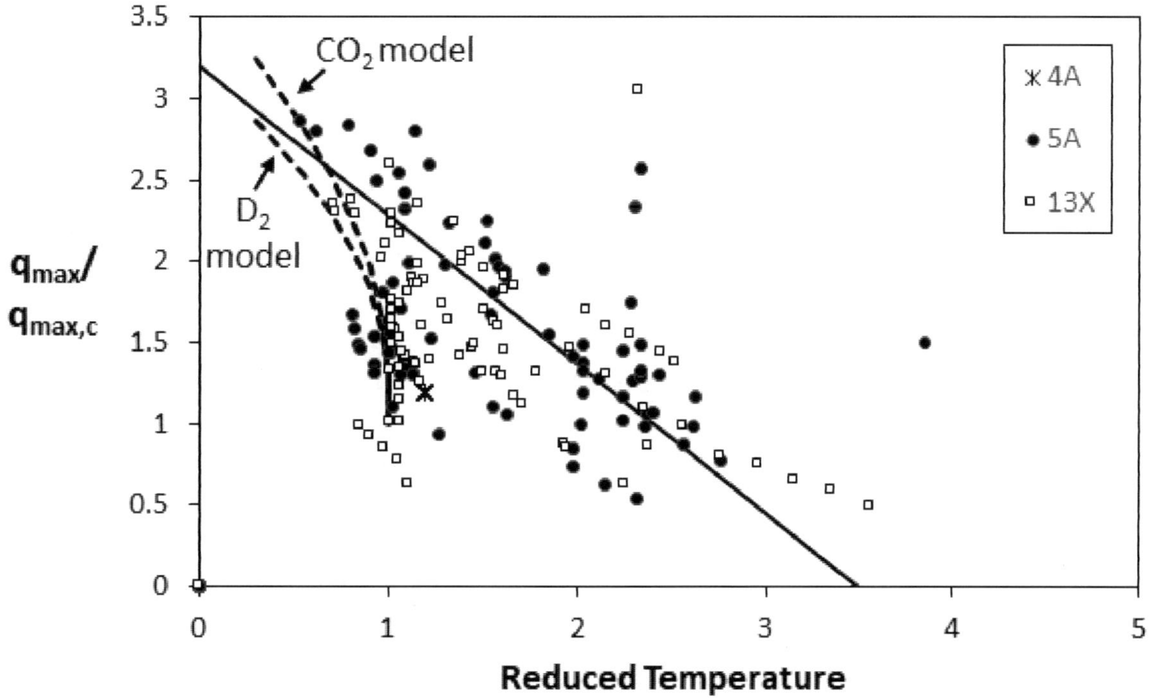

Figure 6. The superT_{CAR} plot for 56 species on 4A,5A and 13X zeolites. These represent all species for which superT_{CAR} data have been collected, excluding n-alkanes larger than n-C_8H_{16}. The dashed lines represent the proposed subT_{CAR} model for deuterium and CO2, as they represent the range of values of Z_{RA} (0.312 and 0.272, respectively). These curves are presented only for reference, as the data are all superT_{CAR}. The solid line represents the linear regression of all the superT_{CAR} data on the Figure.

Conclusions

The data indicate that q_{max} is not constant but varies significantly over a large T_r range reducing by quite a bit. This implies that that the sorbate density decreases with T_r and the mean molecular volume increases by the same factor. For example, the area occupied by a molecule of N_2 would be significantly different at the temperature at which surface analysis is made, 77K (Tr = 0.61) than at the temperature at which a catalyst might be used for chemisorption, 700K (Tr = 5.56) We will continue to study this subject and the Henry constant area to portray this data.

Keywords

Adsorption saturation loadings; 4A, 5A and 13X zeolites; inorganics, alkanes, alkenes, aromatics and fluorocarbons.

References

1. Spencer, C., & Danner, R. (1972). Improved Equation for Prediction of Saturated Liquid Density. *J. Chem. Eng. Data, 17*(2), 236-240.
2. De Boer, J. H., (1968), The Dynamical Character of Adsorption, Oxford University Press.
3. Abouelnasr, D. M., & Loughlin, K. F. (2020). Saturation Loadings on 5A Zeolite above and below the Critical Conditions: Inorganic Diatomic Species Data Evaluation and Modeling. *in preparation*.
4. Loughlin, K. F., & Abouelnasr, D. M. (2020). Evaluation of Henry Law constants for Nitrogen on NaX/13X zeolite. *in preparation*.
5. Abouelnasr, D., L. K., & Al Mousa, A. (2017). Saturation loadings on 13X (faujasite) zeolite above and below the critical conditions. Part III: Inorganic monatomic and diatomic species data evaluation and modeling. *Adsorption, 23*, 945-961.
6. Al Mousa, A., Abouelnasr, D., & Loughlin, K. (2015). Saturation Loadings on 13X zeolite; Alkane Hydrocarbons Data Evaluation and Modelling. *Adsorption, 21*(4), 307-320.
7. Al Mousa, A., Abouelnasr, D., & Loughlin, K. F. (2015). Saturation Loadings on 13X (Faujasite) above and below the critical conditions: Unsaturated and Cyclic Hydrocarbons, Data Evaluation and Modelling. *Adsorption, 21*(4), 321-332.
8. Dirar, Q. H., & Loughlin, K. F. (2013). Intrinsic adsorption properties of CO2 on 5A and 13X zeolite. *Adsorption, 19*(6), 1149-1163.
9. Loughlin, K. F., & Abouelnasr, D. M. (2009). Sorbate Densities on 5A zeolite above and below the critical conditions: n alkane data evaluation and modeling. *Adsorption, 15*, 521-533.
10. Loughlin, K. F., Abouelnasr, D. M., & Al Mousa, A. (2018). Saturation loadings on 13X (Faujasite) zeolite above and below the critical conditions. Part IV: inorganic multi-atomic species, halocarbons and oxygenated hydrocarbons data evaluation and modeling. *Adsorption, 24*(1), 81-94.
11. Agarwal, R. K. and Schwartz, J. A., (1988), High Pressure Adsorption of Gases on Activated Carbon by Potential theory, Carbon, 26(6), 873-887.

594214 Synthesis and Performance Appraisal of Palladium Nanoparticles Supported on Waste Macadamia Nutshell Biomass for Chromium(VI) Detoxification

Abstract:

Though South Africa boasts the largest production of chromium (Cr), the downside of this is environmental pollution, particularly by highly toxic Cr(VI), a well-known carcinogen. Its status as a global leader in the production of Macadamia nuts and palladium (Pd) affords the addition of Macadamia nutshells (MNS), a waste material in nut production, and palladium, a well-established catalyst, to the country's pool of resources available for use in the mitigation of Cr(VI) pollution. In this work, Pd nanoparticles (PdNPs) were grown on the surfaces of polyethyleneimine(PEI)-grafted MNS granules by the impregnation tetrachloropalladate ions (PdCl4 2-) and subsequent reduction with hydrazine (Figure 1). The Pd functionalized biomass and its precursors were characterized by Fourier transform infrared (FTIR) spectroscopy, scanning electron microscopy (SEM), and X-ray diffraction (XRD) analysis. SEM revealed the successive formation of layers of the polymeric scaffolds and PdNPs while FTIR suggested straining of the nanoparticle anchoring functional groups and XRD indicated the formation of Pd with a crystallite size of 4.8 nm. The catalytic efficacy of the supported PdNPs in the reduction of Cr(VI) to less toxic Cr(III) in synthetic wastewater was tested. The reusability of the nanomaterial also evaluated. The supported nanoparticles effectively catalysed the reduction of Cr(VI) and appeared to be a promising catalytic material for the treatment of Cr(VI) contaminated aqueous wastes such as electroplating wastewater using formic acid, a clean reagent that yields environmentally benign carbon dioxide and hydrogen gas as by-products.

595808 An Assessment of IAST for Predicting Mixed Gas Adsorption in MOFs

Abstract:

The ability to understand and design separation systems for complex mixtures under various realistic conditions will be a turning point for the separations community and is a key to transforming separation science. Adsorptive separation is an established technology and currently an active research area within separation science. However, to date its study has relied heavily on the Ideal Adsorbed Solution Theory (IAST) to predict mixture adsorption from pure component adsorption experiments while neglecting gas mixture adsorption experiments, especially for mixtures of more than two components. It has been noted that IAST is particularly limited in its ability to predict gas mixture adsorption in metal-organic frameworks (MOFs) for mixtures containing adsorbates of disparate sizes or sphericities, and especially for adsorbates of different polarities. Improving the ability to predict when mixture adsorption will deviate from IAST predictions is key to designing separation systems for complex mixtures. This knowledge gap in the separations field has persisted in large part due to the technically challenging nature of mixture adsorption measurement. Developing a better understanding of IAST requires measurement of mixture adsorption. This work is a study of how binary and ternary mixture adsorption on the MOFs UiO-66 and HKUST-1 deviates from IAST predictions for mixtures of CO_2, CH_4, and C_4H_{10}. In this work, a home-built, automated volumetric instrument is used to measure multicomponent gas adsorption.

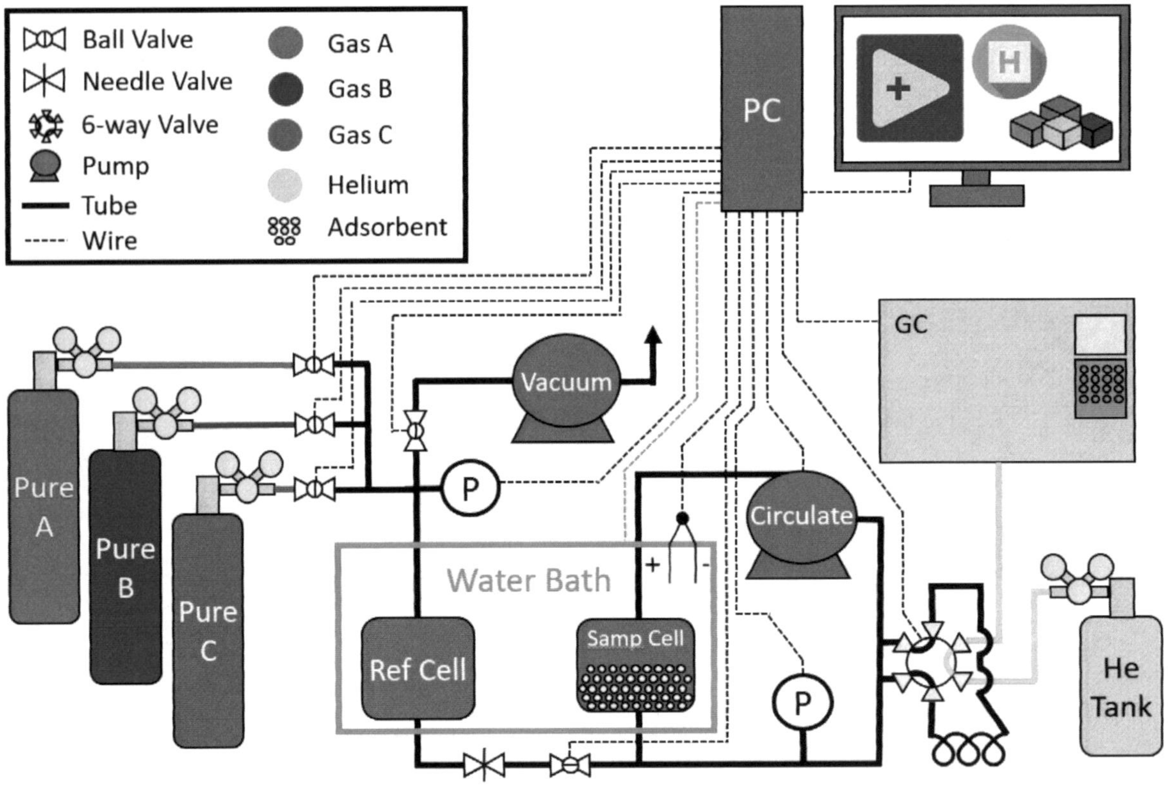

Figure 1. Simple schematic of the Multicomponent Gas Adsorption System (MC GAS) instrument used in this work to conduct binary and ternary adsorption measurements.

602599 Crystal Morphology Modification in a Nucleation Dominant Crystallization: A Tool from Traditional Chemistry

Abstract:

Crystal morphology engineering is an important component of process development in the pharmaceutical industry, since the crystal shape or habit directly influences downstream processing, formulation, and the physical properties of the API. Most commonly the desired habits of crystals are prepared through the modification of crystallization process parameters such as temperature, the crystallization solvent, and/or the addition of a co-solvent or an additive. However, in some cases, these methods do not successfully deliver the desired morphology. Here we present the use of Soxhlet extraction conditions as a valuable tool for morphology modification in a nucleation dominant crystallization process. In this case study, the process to produce a cocrystal between Vertex compound A and coformer X resulted in a fibrous morphology in every attempt under various conditions, leading to mixing and filtration issues and low yields. The use of the Soxhlet extraction method provided very slow mixing of the two compounds coupled with temperature cycling close to the solvent's boiling point to generate a needle morphology, which provided better filtration times, processability and yields, enabling a path forward for further API development.

599684 Process Development to Avoid Meta-Stable Liquid-Liquid Phase Separation during Crystallization of API

Abstract:

Crystallization is a critical unit operation in the production of small molecule active pharmaceutical ingredients (APIs), as it often determines the form and powder properties of the product. Therefore, crystallization process development focuses on identifying conditions that are controlled and scalable. In general, liquid-liquid phase separation or oiling out of product are undesirable as they may lead to uncontrolled nucleation and growth of particles that impacts critical quality attributes, such as particle size and level of impurities. Therefore, understanding liquid-liquid phase equilibria and solubility of the compound is critical in designing crystallization processes.

This presentation will describe how we experimentally investigated the liquid-liquid phase separation that occurred during crystallization of a small molecule API in a ternary solvent system of tetrahydrofuran, acetone and water. Although these solvents are typically miscible, the presence of the API resulted in liquid-liquid phase separation at the seed point under some conditions. In-line process analytical tools such as ParticleViewer and FBRM were incorporated to detect and understand the phenomena. A detailed phase diagram was generated for the crystallization design space as a function of solvent composition, concentration of API in solution and temperature. Along with solubility data, the phase diagram was used to design a process that avoids liquid-liquid phase separation throughout the entire crystallization. In addition, we will describe the impact of wet-milling during seed age that was implemented to ensure that full desupersaturation was achieved, which was critical in avoiding phase separation.

596447 Decoupling Nucleation and Growth for Continuous Crystallization in Microfluidics

Abstract:

1. **Introduction**

Microfluidics provide several advantages such as high efficiency, easy control and safety over conventional, and mostly batch reaction systems[1]. Therefore it can be seen as a replacement for batch crystallization as one of the most important separation and purification process especially in pharmaceutical industries[2,3].

The aim of this project is the design of a modular and decoupled microfluidic system consisting of separate nucleation and growth sections. The strategy persued in this project is using the two- phase interface in microreactors as a heterogeneous nucleation site. In this regard micro- bubbles are introduced in the system to control the location of the nucleation site and particle aggregates and to limit wall interactions. The velocity of the bubbles determines the residence time and using a gas phase enables a straight forward separation, see Figure 1.

2. **Experimental study**

2.1. Nucleation section

For designing the nucleation section, in the first step, the generation of micro-bubbles (i.e. nitrogen as an inert gas) in capillaries is studied. The total bubble surface area per unit volume of the reactor is related to the bubble diameter and the number of bubbles present in the reactor volume. Bubble diameter, bubble velocity and bubble generation frequency were extracted from the images captured by the high-speed camera using an inhouse-generated Matlab script. Using this information, the bubble surface area available as a heterogeneous nucleation site is further calculated and the effect of gas and liquid flowrate is investigated, see Figure 2.

This is followed by a systematic study of nucleation of an organic compound (i.e. cooling crystallization of paracetamol) using the developed system employing microbubbles. The particle size distribution is measured at the outlet of the channel subjected to the microbubble residence time and the nucleation and growth kinetics are extracted.

The continuous crystallization experiments provided proof that the micro-bubbles act as a heterogeneous nucleation site. An enhanced nucleation rate compared to the operation without bubbles was observed, and the presence of micro-bubbles results in the formation of more crystals, which indicates that nucleation is faster than without bubbles, i.e. the metastable zone width is reduced, see Figure 3. Furthermore, quantifying the crystal yield showed an increase by more than a factor of 2 in the presence of micro-bubbles.

2.2. Growth section

Further on a growth section is added to the nucleation section which is kept at the constant temperature equal to the outlet of nucleation section to allow formed nuclei to grow. Sample is collected at the outlet of the growth section for further analysis of particle size distribution, see Figure 4.

4. Conclusions

A decoupled crystallization set-up consisting of different nucleation and growth sections is designed using microreactors. Nucleation is enhanced using microbubbles acting as heterogeneous nucleation sites providing lower surface energy interface for the nucleation to start. Further a growth part is added to allow the formed nuclei grow within the constant temperature growth section.

In case of the continuous crystallization of paracetamol the results allow the conclusion that the low-energy heterogeneous surface introduced by the bubbles promotes crystal nucleation. In addition, adding micro-bubbles enables the continuous crystallization for long operating times, as the bubbles act as carriers for the formed crystals and effectively prevent wall deposition and thus clogging.

References

[1] Gunther, K. F. Jensen, Multiphase microfluidics: from flow characteristics to chemical and materials synthesis, Lab Chip , 2006, 6, 1487–1503

[2] S. Flowers, R.L. Hartman, Particle Handling Techniques in Microchemical Processes, Challenges 2012, 3, 194-211

[3] Wu, S. Kuhn, Strategies for solids handling in microreactors. Chim Oggi. 2014; 32(3):62-66.

[4] Fatemi, Z. Dong, T. Van Gerven, S. Kuhn, Microbubbles as Heterogeneous Nucleation Sites for Crystallization in Continuous Microfluidic Devices. Langmuir (2018). doi:10.1021/acs.langmuir.8b03183

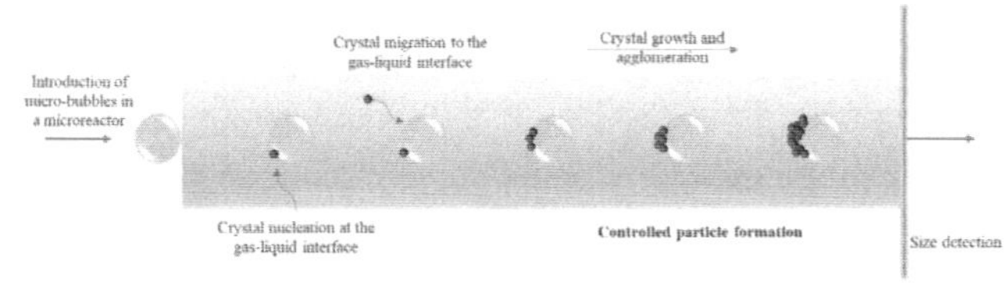

Figure 1. Microbubbles used as a heterogeneous nucleation site [4]

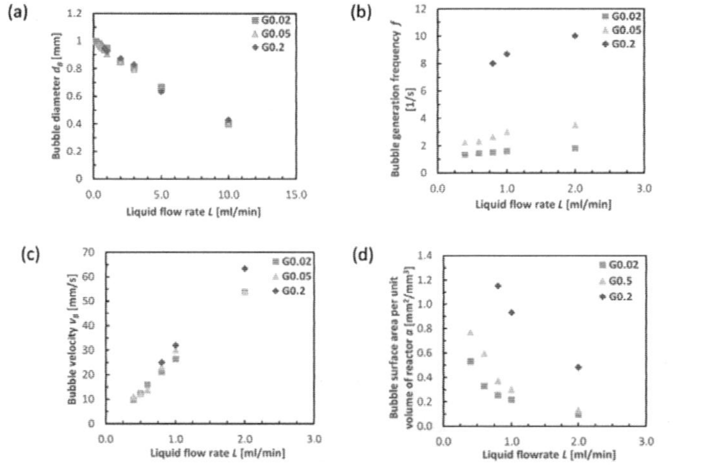

Figure 2. Effect of gas and liquid flow rates on (a) size of the microbubbles; (b) frequency of microbubble generation; (c) microbubble velocity; and (d) microbubble surface area per unit volume of the reactor for [4]

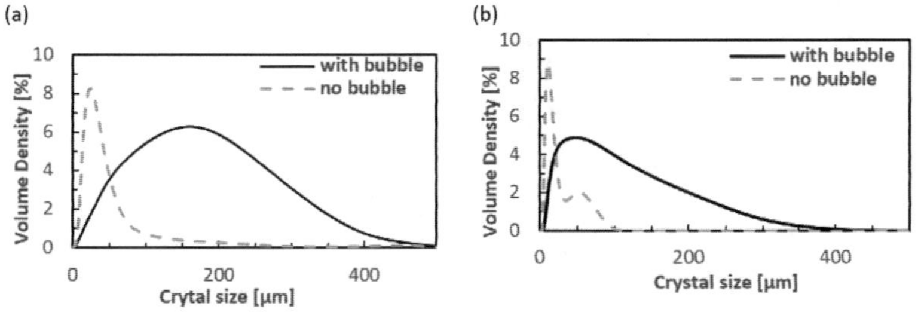

Figure 3- Effect of micro-bubbles on the crystal size distribution for a growth time of 1 minute and a supersaturation of S=1.81 and the gas flow rate G=0.05 ml/min and liquid flow rate of: (a) L=0.5 ml/min, (b) L=1 ml/min

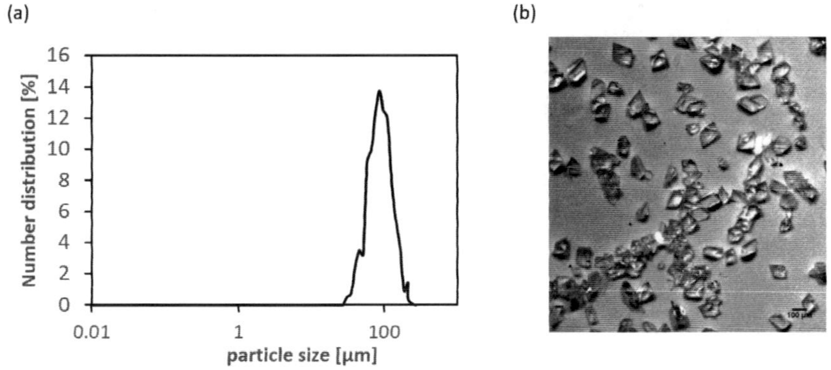

Figure 4- (a) particle size distribution, (b) microscope images of paracetamol crystals for growth section with the length of 1.4m at T=25°C with a supersaturation of S=1.81 and the gas and liquid flowrates 0.05 and 1 ml/min, respectively

597227 Effects of Volume Reduction on Glycine Nucleation

Abstract:

Crystallization in confinement is a technique used both as an approach for nanocrystal development and as way to gain access to unstable crystal polymorphs. To elucidate this, the effect of decreasing volumes in the nucleation of glycine in water was explored. Using a supersaturated aqueous solution of glycine at decreasing volumes from 1000 μL to 25 μL, induction times for primary nucleation were obtained experimentally. Analysis of probability distribution of β-glycine induction times showed that as sample volume decreases, induction time increases. Image analysis revealed that β-glycine formation serves as the primary nucleation event; however, primary nucleation shifts toward γ-glycine nucleation at volume sizes below 100 μL. These results demonstrate that feasibility of accessing metastable and unstable crystal polymorphs with the aid of confinement. However, confinement also increases induction times which relates to very low probabilities of a nucleation event.

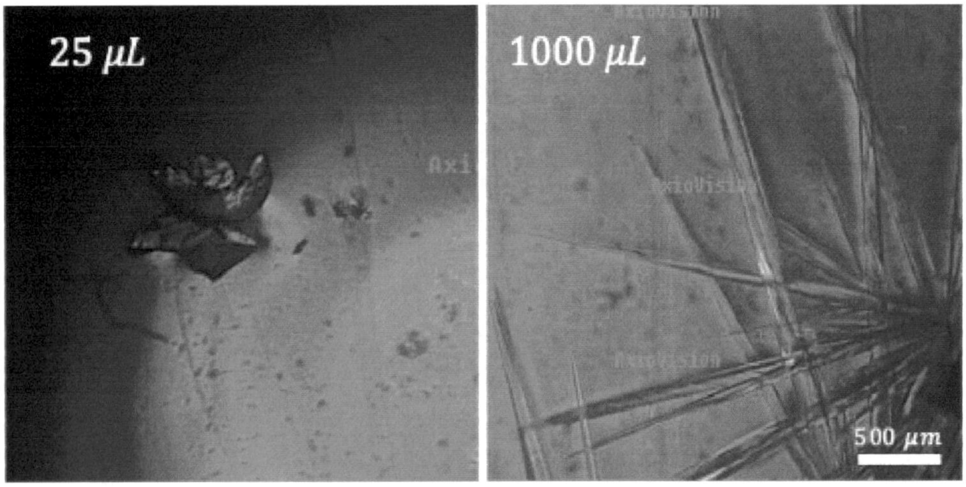

597873 Formation of Solid Solutions As a Key Impurity Retention Mechanism in Solution Crystallization in the Presence of Structurally Similar Impurities

Abstract:

The main purpose of crystallization is to purify the product by removing undesired impurities. In pharmaceutical industry, crystallizations at different steps in the synthesis are relied upon to control the final quality of the active pharmaceutical ingredient (API). During these operations, the main challenge often stems from dealing with structurally similar impurities, which tend to be difficult to remove. Presented herein is an investigation into how this type of challenging impurities are purged mechanistically in the crystallization, using pharmaceutical case studies. Through these examples, it is shown that formation of substitutional solid solutions is the underlying mechanism for impurity retention. Impurities molecularly replace the product molecules in the crystal lattice during the crystallization. This lattice incorporation entrapment occurs unevenly during the crystallization resulting in a gradient in impurity levels throughout the material. A mathematic framework combined with a previously reported SLIP test is used to elucidate this varying impurity entrapment through a so-called *material impurity distribution* (MID). The impurity levels are seen in many cases to change by an order of 20 across the material. The effect of the entrapment can also be observed in the physical properties of the material, e.g. by a significant reduction in crystallinity, increase in solubility and earlier onset melting event.

609881 Design of a Continuous Crystallization for a Drug Substance Using a Dynamically Mixed Reactor

Abstract:

The pharmaceutical industry has predominantly relied on batch processes for manufacturing of drug substance as well as drug products. However, longer campaign times, labor-intensive nature of batch manufacturing and batch to batch variations in the product quality pose significant drawbacks. In view of this, the Pharma industry is turning towards continuous manufacturing, going beyond flow chemistry to continuous crystallization and downstream operations for the synthesis of drug substances. Continuous crystallization offers many advantages such as better control over the solid form and particle size, and the potential for an end-to-end continuous manufacturing process.

The current work describes the design of a continuous crystallization process for a drug substance in order to integrate with upstream and downstream continuous processes. It was not only important to achieve the targeted solid state form (trihydrate, Form I) but also ensure that the filtration rate could be increased to enable integration with the downstream continuous filter and minimize the cleaning frequency. The presented case study discusses in detail the approach, starting from lab-scale batch experimentation to development of a continuous crystallization process at lab followed by scale-up design for a capacity of multi-tons per month.

As a first step, fundamental solubility data was collected followed by identification of critical parameters impacting nucleation and transformation between the various forms. It was observed that the solvent (alcohol) to antisolvent (water) ratio, seed loading and temperature had a significant impact on the solid form. A batch crystallization process was first established by combining antisolvent and cooling crystallization to consistently generate the desired Form I, while ensuring the absence of other solid forms.

The translation of the batch process to continuous was initially attempted in MSMPRs in series but proved unsuccessful as an undesired form started to appear over time despite seeding. It was hypothesized that several more MSMPRs would be required to achieve the desired form but this would occupy a huge footprint and increase hold-up at scale. While plug flow reactors would be suitable, the risk of fouling is high, particularly when flow rates are reduced to achieve sufficient residence time. Scouting for an appropriate technology lead to evaluation of dynamically mixed reactors from AM Technologies®, which were able to handle a slurry and provide a mixing profile close to plug flow decoupled from the residence time. At lab scale, the Coflore® Agitated Cell Reactor (ACR), which is a bench-top mechanical agitated reactor comprising ten small cells, was able to aid in controlled crystallization of the desired Form 1 with well defined crystals and a consistent particle size distribution. Since additional reactors and residence time were required to gradually cool the slurry from 65°C to 25°C, the ACR was combined with three CSTRs in series (forming a mixed suspension mixed process set up). The trials were executed by combining a stream of alcohol containing the dissolved solute (at 65°C) and another stream of hot water containing seeds (at 65°C) in the ACR. The slurry exiting from the reactor was continuously transferred into the CSTRs, where it was subjected to stage-wise cooling in each CSTR. The residence times were designed to achieve maximum crystallization in the ACR and minimize secondary nucleation in the subsequent CSTRs.

After establishing the success and robustness of the continuous crystallization process at lab-scale, the process was scaled up in a 100L, production-scale version of the dynamic mixing reactor - Coflore®

Rotating Tubular Reactor (RTR). Optimization of the process parameters revealed that maintenance of higher temperatures (~65 °C), , enhanced mass transfer at the point of nucleation along with the presence of Form 1 seeds, and higher ratios of anti-solvent were critical for achieving the desired polymorph while low mixing shear aided in crystal growth. The growth of these fine, needle-shaped crystals was essential to achieving a sufficiently high filtration rate for the downstream continuous separation of solids. The RTR, which operates as a ten-stage, actively mixed, self-baffled reactor, provided high wall shear with radial mixing while maintaining a low mixing shear, thus enabling crystal growth and improving the filtration rate while minimizing fouling.

In summary, the work presents key aspects of design and scale-up of a continuous crystallization for a drug substance with regard to both process and technology selection. One of the main advantages of this continuous process was consistency in the solid form, particle size distribution resulting in improved filtration of the slurry, and ability to integrate with upstream and downstream continuous processes.

Internal communication number: IPDO IPM-00637

594302 Organic Solvent Nanofiltration in Pharmaceutical Manufacturing – Membrane-Based Peptide Synthesizer

Abstract:

Separation processes in pharmaceutical industry often involve the isolation or removal of high added value products, such as active pharmaceutical ingredients (APIs), from organic solvents. Organic solvent nanofiltration (OSN) technology has since gained interest from the industry as it allows separations to be carried out directly in organic solvents by simply applying a pressure gradient across a membrane. In this work, OSN is applied to the synthesis of one class of APIs - peptides. The industry is currently dominated by solid-phase synthesis since it overcomes the problem of inefficient intermediate isolation encountered in liquid-phase synthesis. Although this allows rapid and automated synthesis to be carried out, solid-phase synthesis faces other challenges such as low crude purity due to poor mass transfer and scalability.

To fill in the gap, we have developed a membrane-based peptide synthesiser to synthesise peptides in liquid phase and utilise OSN for intermediate isolation. This retains all liquid-phase advantages, while benefiting from facile separations to facilitate solid-phase-like rapid synthesis. The following synthesis cycle was devised for the synthesiser: coupling of amino acid (AA) to the growing peptide intermediates, inactivation and deprotection of excess AA, and diafiltration to remove all reaction debris. At the end of the cycle, growing peptide intermediates are retained by the OSN membrane and ready for next coupling.

To deliver rapid synthesis, repeated syntheses and intermediate isolations need to take place continuously inside the synthesiser without any liquid transfers in between. Under this condition, OSN membrane being the key element of the synthesiser must withstand harsh chemical environments and possess superior long-term stability. Crosslinked Polybenzimidazole (PBI) membrane was chosen for its high chemical stability and mechanical strength. To increase the selectivity between peptide intermediates and reaction debris, the membrane surface was modified with polymer brushes to alter the surface hydrophilicity-hydrophobicity balance. This classs of PBI membrane was proven to have excellent performances which remained constant after many synthesis cycles. To enhance the diafiltration efficiency, the membrane-based peptide synthesiser was configured into a 2-stage cascade system. Continuous synthesis cycles took place in the 1st stage, with 2nd stage purposed to recycle partial flow back to the 1st stage.

Our progress to date has validated the membrane-based peptide synthesiser through the synthesis of model peptides (5 and 10 AA) with crude purities that are higher than, or at least comparable to, peptides synthesised via solid-phase. Currently, we are working on the development of full process automation and the synthesis of drug candidates eg. Octreotide (8 AA). We envisage that this membrane-based synthesiser will become a new standard for pharmaceuticals manufacturing.

594925 Complex Aromatic Hydrocarbon Mixture Osro Separation Using Defect-Engineered Torlon® Hollow Fiber Membranes

Abstract:

The fractionation of complex liquid hydrocarbon mixtures is an important emerging area of membrane science. Polymeric asymmetric hollow fiber membranes can enable this concept, especially if size and number of defects in the membrane skin layer can be precisely engineered. Here, we fabricated various "defect-engineered" hollow fiber membranes by modifying the composition of the volatile solvents of the polymeric solution as well as spinning parameters to improve the organic solvent reverse osmosis (OSRO) separation performance of complex aromatic hydrocarbon mixtures. The quality of the membranes was probed using both gas and liquid permeation tests, which provided the data for a useful fundamental analysis that estimates the pore size and surface porosity of the asymmetric hollow fiber membranes. That information, combined with organic solvent permeation data, provides an engineering platform to estimate the relative permeation rates of organic molecules within a complex mixture. For instance, we will assess the ability of combined transport modality permeation models (i.e., sorption-diffusion + hydraulic flow) to predict the membrane's ability to fractionate complex mixtures of aromatic hydrocarbons. Useful approximations and apparent deficiencies in this transport "framework" will be discussed. Although we have employed hydrocarbon mixture fractionation to illustrate our approach, we believe the membrane-based fractionation concept to be highly relevant across a range of existing and emerging separation challenges, including biorefinery separations and downstream separations in futuristic "e-refineries"

603876 SET-LRP Modification of Cross-Linked Polyimide Membranes for Organic Solvent Nanofiltration

Abstract:

Abstract

A novel surface modification technique for polyimide membranes their characterization and organic solvent filtration performance is described. The resultant brush layer and polyimide support are characterized and tested for organic solvent filtration. The process involves treating crosslinked polyimide membranes with surface initiators followed by Single Electron Transfer- Living Radical polymerization (SET-LRP) with alkyl methacrylates. Proof of grafting is established using ATR-FTIR, XPS and FESEM. The degree of grafting and brush height and hence selectivity and permeability are found to be a function of monomer type and polymerization time. The filtration performance of these composite membranes was measured using the permeability of 10 organic solvents and the selectivity and fouling of an organic dye (Rose Bengal) in ethanol. These studies provide a basis for designing polymeric membranes comprising brush selective layers with defined microstructure not possible with the widely popular phase inversion process.

596345 Fundamental Origin of Flux Non-Linearity in Organic Solvent Reverse Osmosis and Organic Solvent Nanofiltration

Abstract:

Organic solvent reverse osmosis (OSRO) and nanofiltration (OSN) are attracting interest due to compactness, modularity and superior energy efficiency. However, while recent research efforts mostly focused on the synthesis and fabrication of solvent resistant composite polymer membranes, fundamental understanding of chemical and physical aspects that govern solvent and solute transport in OSRO-OSN membranes remains entirely unexplored. The transport mechanism itself is poorly understood: some researchers have hypothesized a solution-diffusion mechanism, others a pore-flow mechanism. Finally, others have considered a combination of the previous two mechanisms. A peculiar OSRO-OSN feature is the flux non-linearity vs. Dp, that is, a negative departure of flux from linearity is observed starting from Dp = 10 atm, which has been the subject of a long standing debate in the literature. Despite membrane compaction has been invoked to explain this phenomenon, this hypothesis has no quantitative support.

In this study, we critically discuss the hypothesis of membrane compaction. To demonstrate that the molecular origin of flux non-linearity is purely thermodynamic, we propose, for the first time, a thermodynamic-diffusion framework which describes solvent transport in OSRO-OSN membranes in terms of the concentration gradient produced by the applied pressure across the membrane. The model predictions agree very well with experimental data. Solvent diffusion coefficient in the membrane increases with increasing Dp, which further confirms that flux decline is not related to membrane compaction. The developed framework allows to quantify both frame of reference and non-ideal thermodynamic effects on solvent diffusion coefficients in OSN membranes.

This study demonstrates that the solution-diffusion model provides a satisfactory description of small molecule transport in OSN membranes, without the need to resort to pore-flow or more complicated transport models.

599234 Multicomponent Liquid Hydrocarbon Transport in Polymeric Membranes

Abstract:

The separation of multicomponent, complex liquid mixtures, which are mixtures without a clear singular solvent, is an important, emerging area of membrane science. Transport models for polymeric materials have typically been applied to systems with three or fewer species; however, many critically important separations challenges involve complex mixtures containing dozens if not thousands of compounds. Here, we experimentally examine the ability of Maxwell-Stefan models to accurately predict complex mixture permeation in polymeric membranes based only on single component sorption and diffusion data. We specifically investigate the separation of a complex mixture of hydrocarbons covering a range of classes such as alkanes, cycloalkanes, alkyl aromatics, and polyaromatics. We observe strong sorption-based selectivity for aromatic compounds while diffusion-based selectivity dominates for aliphatic molecules. We utilize these experiments to test simplifying hypotheses regarding the estimation of sorption and diffusion coefficients of any arbitrary hydrocarbon in our polymer material. These simplifying hypotheses potential enable extension of our predictive capabilities to any N-component mixture of hydrocarbons, of which there are many industrially-relevant streams.

601731 Computational Evaluation of Carriers in Facilitated Transport Membranes for Post-Combustion Carbon Capture

Abstract:

Incorporating molecular amines as mobile carriers in facilitated transport membranes (FTMs) has been demonstrated to significantly enhance the CO_2 permeance and CO_2/N_2 selectivity of the membrane for CO_2 capture from flue gas. In this study, by employing computational techniques such as the density functional theory, molecular dynamics studies, and Monte Carlo simulations, the role of mobile carriers is systematically studied at the molecular level from the perspectives of the amine–CO_2 reaction chemistry, diffusivities of carriers and gases, and N_2 sorption. The water uptake of FTMs with and without mobile carriers is also experimentally quantified. The introduction of mobile carriers is shown to enhance the diffusivities of CO_2 reaction products and the water uptake compared to FTMs without mobile carriers. The choice of mobile carriers is also demonstrated to influence the separation performance. Computationally, certain carriers are shown to possess a faster reaction kinetics and a higher CO_2 loading capacity; these carriers also demonstrate a better CO_2 permeance experimentally. The good consistency between the computational and experimental observations validates the computational models employed and insights generated in this study. The outcomes of this work shed light on the future design and selection of carrier structures, and the adopted computational approaches can be employed to discover promising mobile carrier candidates.

601067 A Modified Free Volume Model to Designing High-Performance Polymers for Gas Separation

Abstract:

Understanding the structure and property relationship is critical to designing high-performance membrane materials for gas separation, particularly for materials with balanced permeability and selectivity above or near the Robeson's upper bound. Herein, the free volume model is adapted to describe the upper bounds for the separation of several gas pairs, such as O_2/N_2, He/CH_4, CO_2/CH_4, and H_2/CO_2. The results are more or less consistent with the conventional activation energy model reported in the literature. More importantly, this model can be used to predict the structure of polymers with targeted separation properties on demand. I will demonstrate that for the separation of specific gas pairs, polymers with specific free volumes can be identified to achieve the best performance. The model can also satisfactorily predict the effect of temperatures on separation properties, which may be needed for industrial H_2/CO_2, CO_2/CH_4, H_2/C_3H_8, and O_2/N_2 separations. Such an understanding may provide an effective route for the high throughput design of membrane materials through data mining or machine learning.

602370 Global Optimization of Multicomponent Membrane Cascade

Abstract:

Membranes have been playing an important role in chemical industry. Especially in the past decades, substantial advancements have been achieved on membrane materials which makes membrane one of the most promising separation technologies in the future. However, a single membrane permeator is still economically infeasible for some real-life applications, especially those requiring high recovery and purities of the products. This limitation could potentially be overcome using a membrane cascade, in which multiple membrane stages are connected to perform the separation. However, determining the optimal membrane cascade configuration and the optimal operating condition is challenging due to the complexity of the cascade structure and the membrane model itself. The prior literature that reports global optimal solutions assumes simpler mixing models such as well-mixing[1], which our simulations show deviates significantly in its predictions from experimental results. The research that has used more accurate crossflow model has either failed to derive global optimal solutions or only applies to the binary case[2].

In the current work, a global optimization algorithm has been developed to determine the optimal membrane cascade in terms of energy consumption. Given the feed composition, permselectivity of each of the components, product purity, and recovery requirement, the optimal membrane cascade and the optimal operating condition can be easily determined. The algorithm is tested over several real-life applications such as NGL recovery, H_2 recovery from hydrocarbon mixture, and acid gas removal from shale gas to evaluate the performance of the algorithm.

Reference

1. Alicia Aliaga-Vicente, Jose A Caballero, and Maria J Fernandez-Torres. Synthesis and optimization of membrane cascade for gas separation via mixed-integer nonlinear programming. AIChE Journal, 63(6):1989–2006, 2017.
2. Qi and M.A. Henson. Optimization-based design of spiral-wound membrane systems for CO_2/CH_4 separations. Separation and purification technology, 13(3):209–225, 1998.

602408 On the Numerical Solution of Rigorous Membrane Module Models Involving Nonideal Multicomponent Mixtures

Abstract:

On the Numerical Solution of Rigorous Membrane Module Models Involving Nonideal Multicomponent Mixtures

Dylan Weber[1], Chau-Chyun Chen[2], Joseph Scott[1†]

1 Department of Chemical and Biomolecular Engineering, Georgia Institute of Technology, Atlanta, GA 30332, USA

2 Department of Chemical Engineering, Texas Tech University, Lubbock, TX, 79409, USA.

Corresponding author; Email: joseph.scott@chbe.gatech.edu

Abstract

In this presentation, we consider the problem of efficiently and reliably solving rigorous models of membrane processes involving nonideal multicomponent mixtures. Numerous applications of membranes involve nonideal multicomponent mixtures, including water purification, carbon capture, hydrogen separation, olefin/paraffin separation, and benzene derivative concentration. Moreover, these applications use many different membrane materials that can interact with such mixtures to produce complex coupled transport phenomena that cannot be captured with simple constant permeability models (e.g., zeolites, carbon nanotubes, ionic polymers, mixed-matrix MOF, graphene-oxide, and specialty polymers).

Membrane process models can be viewed as having two distinct parts: (i) a local flux model, which relates the driving force at an arbitrary point on the membrane surface to the transmembrane flux of each component at that point, and (ii) a global module model, which governs the driving force at every point on the membrane surface as a function of flowrates, module geometry, flow configuration, etc. For nonideal multicomponent mixtures with coupled transport, evaluating the local flux model at a single point on the membrane already requires the solution of nonlinear two-point boundary-value problem in differential-algebraic equations (DAEs). We presented an efficient and reliable shooting algorithm for solving this problem based on a general Maxwell-Stefan (MS) coupled transport model in a recent contribution to the 2020 AIChE Spring Meeting. However, embedding such a local flux model within a global model of an entire membrane module leads to a considerably more difficult problem. Industrially, membrane process modules can take the form of hollow fiber, spiral wound, flat plate, tubular, or plate-and-frame units. Moreover, each of these module types can be operated with various flow configurations, including co-current, counter-current, or crossflow patterns. Finally, there is also significant interest in membrane modules with chemical reactions occurring on the feed side or within the membrane itself. As a result, accurate models of complete membrane modules often take the form of a multidimensional boundary value problems in nonlinear partial differential algebraic equations (PDAEs), often with stiffness along multiple dimensions.

This problem of solving membrane module models numerically has been researched extensively over the past half century. Existing methods include full discretization approaches that solve all of the governing equation simultaneously at a set of nodal points, as well as a variety of shooting-type algorithms applicable under various simplifying assumptions. However, nearly all existing methods for simulating full membrane modules with nontrivial flow configurations assume a very simple form for the embedded local flux model. Typically, the local flux is computed using a constant permeability linear driving force model, with slightly more complex nonlinear algebraic expressions used in some cases. On the other hand, prior works using detailed local flux models that can capture coupled multicomponent transport using the Maxwell-Stefan framework have done so only under the assumption of well-mixed volumes on the feed and permeate sides. This reduces the system to a single local flux calculation, but can significantly over-estimate performance metrics compared to practical industrial units. To date, an effective numerical solution procedure for models including both detailed MS local flux models and realistic module configurations has not been proposed in the literature.

In this talk, we will present our recent progress towards an efficient and reliable numerical method for solving full membrane module models for nonideal multicomponent mixtures with detailed MS-based local flux calculations and realistic flow configurations. Numerical experiments comparing our tailored membrane algorithms to general purpose full discretization and shooting methods will be presented. Finally, we will discuss the applicability of key simplifying assumptions that can significantly reduce computational complexity in certain cases.

602026 An Industry Wish List for Molecular Modeling and Process Simulation of Membrane Separation Processes

Abstract:

(Note: This abstract is submitted for presentation at a virtual conference. The authors are unable to travel during the year of a global pandemic.)

The chemicals industry strives to produce high purity products while minimizing inputs such as feedstocks and energy. Separations are a key component to reaching these goals. Many separations processes require phase transitions that are energy intensive. Membranes present an enticing separation technique that avoids such transitions. However, engineers require a high degree of confidence in a unit operation before it will be used in a production facility. Molecular modeling and process simulations can play a key role in accelerating membrane adoption. This presentation reviews some of the opportunities where liquid separating membranes could be used and discusses the derisking process that occurs before adoption into a chemical facility.

603169 New Developments in Polybenzimidazole (PBI) Membranes for Electrochemical Devices

Abstract:

Polybenzimidazole (PBI) polymers have been extensively studied for high temperature PEM fuel cells operating at 100-200°C. As part of our ongoing work in developing and using the sol-gel PPA process, we have been exploring the effects of chemical structure on membrane properties, including issues related to membrane stability and durability. Membranes produced from this process showed the ability to maintain high levels of phosphoric acid (PA) and high proton conductivities while simultaneously exhibiting low levels of PA loss during operation in many simulated duty cycles. The knowledge base created by a broad search of different PBI backbone chemistries and variations in molecular architecture (random and block copolymers, crosslinking, etc.) originally directed at high temperature fuel cells, is now being used to design PBI polymers for other electrochemical devices such as hydrogen pumps, electrolyzers, and flow batteries. In many cases, the different operating conditions of the distinct devices advocate for a specifically designed polymer structure tailored for its environment. In these devices, the general polybenzimidazole backbone structure provides outstanding chemical and thermal stability, high ionic conductivities when doped with acids, and resistance to oxidatively challenging environments. In this talk, we will review some of the emerging devices and describe a new process for producing PBI membranes with high proton conductivities, improved mechanical properties, and excellent performance in multiple devices.

613189 Proton Exchange Membranes for Low Relative Humidity Fuel Cell Operation Based on Poly(Phenylene Sulfonic Acid)

Abstract:

Novel, composite, fuel cell membranes have been fabricated from a non-fluorinated polymer using a pore filling technique. The membranes consist of a mechanically stabilizing skeleton from an electrospun poly(phenylene sulfone) (PPSU) fiber mat and a thermally crosslinkable poly(phenylene sulfonic acid) (cPPSA) proton conducting ionomer that fills the interfiber voids. The work builds on two prior studies carried out in the labs of Prof. M. Litt and Prof. P.N. Pintauro demonstrating: (1) Exceptionally high proton conductivity, even at very low RH (down to 20%RH), of cPPSA solution cast films, and (2) Excellent durability of electrospun perfluorosulfonic acid (PFSA)/PPSU composite membranes, exceeding that of the pristine PFSA. Unfortunately, the high proton conductivity of cPPSA films, especially at low relative humidity, was offset by their significant brittleness, which made handling of those materials, including fabrication of membrane-electrode-assemblies, extremely difficult. The proposed pore-filling strategy should alleviate the brittleness problem leading to a robust, low RH-conductive PEM that is a true competitor to PFSA membranes and suitable for commercial applications.

Experiments focused on optimizing the cPPSA/PPSU ratio to achieve the desired balance of tensile strength and proton conductivity, and on minimizing membrane thickness to obtain area specific resistance (ASR) lower than 0.02 S/cm in the 40-90% RH range at 80°C. cPPSA copolymer was synthesized using Ullmann coupling copolymerization of 4,4′-dibromobiphenyl 3,3′-disulfonic acid with 1,4-dibromobenzene-2,5-disulfonic, followed by grafting certain fraction of backbone sulfonic acid groups with biphenyl linker [1]. The PPSU fiber mat was electrospun from NMP/acetone solution. Pore-filling was carried out by pouring a solution of cPPSA in methanol over the framed mat, followed by heating at 70°C to evaporate solvent. The cPPSA was crosslinked by an additional heating step, in a vacuum oven at 210°C for 1-2 hours. The resultant membranes had excellent proton conductivity, 5-10 times greater than that of Nafion^{Ò} 211, in the 40-90% RH range at 80°C. Details of the membrane fabrication and characterization work will be presented in this talk.

[1] M. Litt and R. Wycisk, Poly(Arylenesulfonic Acids) with Frozen-In Free Volume as Hydrogen Fuel Cell Membrane Materials, Polymer Reviews, 55, 307 (2015).

606616 Membranes for Artificial Photosynthesis: A Tradeoff in Ion Transport and Product Crossover

Abstract:

The requirements of membranes utilized in artificial photosynthesis devices are multi-faceted and potentially incongruent. Artificial photosynthesis devices produce carbon-neutral liquid fuels (e.g., ethanol) from abundant resources, namely CO_2 and water, using energy harnessed from the sun. Polymer electrolyte membranes are incorporated into these devices to (1) mitigate the crossover of liquid fuel products that leads to product loss when the product is re-oxidized at the anode, (2) promote the transport of charge carriers between the anode and cathode, reducing the ohmic resistance, and (3) manage the difference in hydration states between the liquid-fed anode and gaseous-fed cathode, wherein dehydration at the cathode can lead to increases in ohmic resistance while flooding can lead to product distributions that disfavor liquid fuel products. It is important to consider these requirements together as many membranes demonstrate tradeoffs with respect to application requirements (e.g., permeability-selectivity tradeoff in gas separation and reverse osmosis). While ex situ studies have correlated polymer electrolyte membrane structure (i.e., free volume fraction) to either ionic conductivity or organic (neutral) solute permeation, few have correlated these two transport phenomena, as is relevant for artificial photosynthesis devices. Furthermore, most artificial photosynthesis studies have utilized commercial membranes, limiting correlation of device performance to explicit membrane physicochemical properties.

Herein, we utilize a polymer electrolyte membrane platform to explore the structure-transport relationships associated with the requirements for membranes utilized in artificial photosynthesis devices. Through systematic variation of the degree of functionalization of an imidazolium-functionalized poly(phenylene oxide) material platform, we fabricated membranes of varied water uptake and microstructure. By modifying the membrane's microstructure, we adjusted the ethanol (a commonly reported CO_2 reduction fuel product) permeation and bicarbonate (a common charge carrier) conductivity in the membranes. Membranes with a greater free volume fraction exhibited a higher bicarbonate conductivity and a higher ethanol permeability than those with a low free volume fraction. The trend of increasing ethanol permeability with increasing bicarbonate conductivity observed in this family of materials (Figure 1) presents a tradeoff in the membrane properties desirable for artificial photosynthesis (i.e., high charge carrier transport and low CO_2 product crossover). While optimal membrane properties remain undefined due to the diversity of device types and operational characteristics (e.g., current density) reported, systematic control of the membrane physicochemical properties presented herein enables the study of a variety artificial photosynthesis devices operating over a range of conditions. Understanding the structure-transport relationships of the membrane could inform rational design of membranes for artificial photosynthesis and other electrochemical applications.

Figure 1. The trade-off between ionic conductivity and ethanol permeability in the imidazolium-functionalized poly(phenylene oxide) membrane platform.

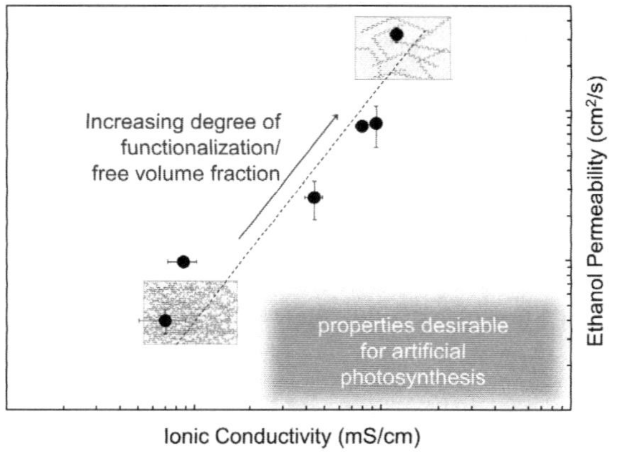

606771 High Performance Anion Exchange Membranes Based on Block Co-Polymers for Energy Conversion Applications

Abstract:

Anion exchange membranes (AEMs) have attracted increasing interest for various energy conversion applications, including fuel cells, electrolyzers, redox flow batteries, and reverse electrodialysis, in addition to water purification applications. Their advantages include having improved oxygen reduction reaction kinetics under alkaline conditions which can allow for the use of non-noble metal catalysts. Unlike proton exchange membranes (PEMs), AEM-based devices have the ability to utilize, or generate, more complex hydrocarbon fuels. Hydrocarbon based AEMs are a potentially less expensive and more environmentally preferable alternative to perfluorinated membranes, which have been essential to mitigate the aggressive oxidizing conditions present in PEM fuel cells and electrolyzers. AEMs can be constructed entirely from hydrocarbons, allowing the additional functionality of block copolymers to be accessed. For the breadth of electrochemical applications mentioned above to be implemented, there are several membrane properties that are required: high ionic conductivity, mechanical integrity under humidified or wet conditions, and chemical stability under potentially severe alkaline conditions dependent on the application. Simultaneously achieving these properties is still a major challenge in the field of AEMs, despite significant developments in backbone chemistry and cation stability.

One of the main hurdles is balancing the ion exchange capacity (IEC) against water uptake and swelling of the material. A higher IEC will contribute to having a higher ionic conductivity at the cost of increasing the hydration and swelling of the material, which can ultimately lead to mechanical failure. Many of the leading AEMs with high hydroxide ion conductivity (96 – 140 mS cm^{-1} at 80 °C in liquid water) suffer from excessive water uptake, > 100%, and this can be especially detrimental in liquid applications. Ongoing research has focused on identifying alkali stable polymer backbones and cation groups, since hydroxide ions can attack ionic groups along with the polymer backbone itself through multiple degradations pathways. Numerous cation chemistries have been explored, namely quaternary ammonium groups, of which trimethylammonium has been the most widely investigated. However, trimethylammonium functionalized polymers have been shown to have poor stability at high pH, and have low or no ionic conductivity at elevated temperatures. Marino and Kreuer studied 26 different quaternary ammonium groups at high temperature and pH and found that quaternized aliphatic-heterocycles with 5 and 6 membered rings were much more stable than the traditional trimethylammonium ions, since they were less susceptible to both nucleophilic attack and substitution.

Equally important is the chemistry of the polymer backbone, where it is desirable to have a hydrophobic backbone to diminish water uptake and membrane swelling and minimize, or eliminate, the presence of polar groups that can act as sites of attack by nucleophilic hydroxide anions. For these reasons, polyethylene is a great candidate for an AEM polymer backbone as it is stable under alkaline conditions, is hydrophobic, does not swell in aqueous solutions, is semi-crystalline, and has excellent mechanical properties. Block copolymers have been extensively investigated for use as backbones for AEMs because their structure induces a phase separated morphology, and well-interconnected hydrophilic domains have been shown to enhance water and ion transport through the membrane.

We have developed an ABA triblock copolymer of polychloromethylstyrene-*b*-polycyclooctene-*b*-polychloromethylstyrene (PCMS-*b*-PCOE-*b*-PCMS) with a scalable and efficient synthesis that was optimized for high ionic conductivity. Based on this previous work, conductivity was shown to increase with increasing lamellar d-spacing. Here we take a PCMS-*b*-PCOE-*b*-PCMS triblock precursor and modify it by hydrogenating the polycyclooctene (PCOE) midblock to polyethylene (PE). The ratio of PCMS:PCOE was kept similar to the previous work to ensure good mechanical integrity. Based on the work of Mahanthappa et al., a polydisperse midblock and narrow dispersity outer blocks were synthesized to provide a larger window for nanoscale phase separation. The polyethylene-based triblock copolymer was processed by hot pressing the polymer in a slurry of *p*-xylene into a film. Hot pressing was performed at two different temperatures, and the elevated temperature lightly crosslinked the PCMS blocks through scission of benzyl chlorides. The films were subsequently quaternized with either trimethylamine to yield benzyltrimethylammonium (TMA) or methylpiperidine to yield benzylmethylpiperidinium (MPRD) cations. The incorporation of PE in the midblock produced anion exchange membranes with enhanced hydroxide and chloride conductivity, in addition to excellent water uptake and interesting mechanical properties.

Based on similar chemistry we have also synthesized ABA polymers with isoprene mid blocks that may be hydrogenated to give amorphous poly(methylbutylene) PMB. In fact we can extend this chemistry for both PE and PMB to give ABABA or BABAB pentablock materials. This additional level of control enables us to totally tune the mechanical properties of these materials for a variety of demanding applications.

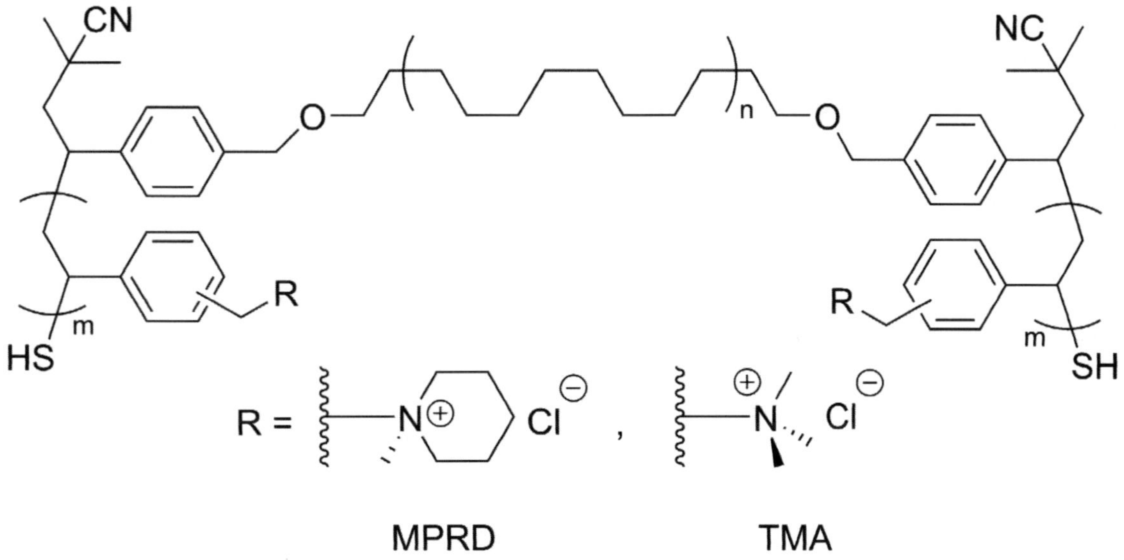

602179 Anion Exchange Membranes with Maximized Combination of Durability and Conductivity

Abstract:

Anion exchange membranes (AEMs) are recently being heavily investigated due to their potential use in electrochemical devices, such as fuel cells and electrolyzers, in order to promote the societal adoption of renewable-energy infrastructure. Due to the intermittent nature of renewable energies, it is envisioned that excess renewable energy from solar, wind, etc. could be stored in the chemical bond form of hydrogen, which then could be utilized to produce electricity via fuel cells when demand exceeds supply, or to power other applications, such as vehicles. To be used in electrochemical devices for economically-viable renewable energy storage and usage, AEMs must have good mechanical durability and alkaline stability as well as have high ion conductivity. To achieve all three necessary properties simultaneously, we have developed a new class of AEMs by following three design criteria. First, the polymer backbone consists of all carbon-carbon linkages to ensure good alkaline stability. Second, quaternized ammonium head-groups are tethered to the polymer backbone via a flexible alkyl chain to promote high alkaline stability and high anionic conductivity. Lastly, crosslinking is used to enhance the mechanical and dimensional stability under hydrated operational conditions. By following the aforementioned strategies, crosslinked polyphenylene AEMs were synthesized and demonstrated excellent alkaline stability (negligible IEC loss after submersion in 1M $NaOH_{(aq)}$ at 80 °C for 1,000 h), superior mechanical properties (stress $\geq$ 40 MPa, strain $\geq$ 140 %), excellent OH^- conductivity ($\geq$ 130 mS/cm) and promising electrolyzer performance (< 1.8 V, 500 mA/cm^2, 45 μm).

599975 CO_2 Utilization for CO Production Via Fuel Cell Enabled By CO_2-Selective Membrane

Abstract:

This presentation describes a method of CO_2 utilization for CO production via a fuel cell system that is enabled by a CO_2-selective membrane. We have synthesized a highly CO_2-selective facilitated transport membrane, showing an extraordinarily high CO_2/CO selectivity of greater than 1200 at 100 – 130°C that separates a CO_2 and CO mixture to give high-purity CO_2 on the permeate side and high-purity CO on the retentate side. The CO_2-selective membrane contained polyallylamine as the fixed-site carrier and the potassium 2-aminoisobutyrate as the mobile carrier to facilitate the CO_2 transport.

An embodiment of the CO_2 utilization is for the simultaneous production of CO and electricity using natural gas, i.e., methane (CH_4) feeding to the anode of a solid oxide fuel cell (SOFC). In this case, the CO_2 feed enters the cathode of the SOFC, where the CO_2 is converted to CO. As the conversion is not 100%, the cathode exhaust is typically a mixture of CO_2 and CO. This mixture enters a CO_2-selective membrane module, which separates the mixture to a high-purity CO product on the retentate side and a high-purity CO_2 stream on the vacuum permeate side. The methane enters the anode of the SOFC, which is oxidized into CO_2 while generating electrons to produce electricity via the external circuit (load). The CO_2 stream coming out from the anode may contain some CO due to incomplete oxidation of methane. The mixture produced at the anode may be sent to the same CO_2-selective membrane module or another one for separation. The separated CO_2 product may be combine with the CO_2 feed to go into the cathode. The methane feed may be converted to ethylene. Instead of the methane feed, the feed to the anode is paraffin, e.g., ethane or propane, for conversion to olefin, e.g., ethylene or propylene, or CO for conversion to CO_2 that is captured by the membrane.

608279 CoNi Alloy Nanoparticles Embedded in Metal-Organic Framework-Derived Carbon for the Highly Efficient Separation of Xenon and Krypton via a Charge-Transfer Effect

Abstract:

Separation of xenon from krypton is of industrial interests due to their wide applications in medical imaging, spacecraft propellants, insulation, commercial lighting and lasers. There is an urgent demand to develop an alternative separation technology to replace capital and energy intensive cryogenic distillation, which is the most popular in separation of Xe/Kr. Adsorption separation based on task-specific adsorbents is significant and promising. Herein, we propose a new strategy for Xe/Kr separation using MOF derived Ni/C and $Ni_{0.5}Co_{0.5}/C$ adsorbents via orbital hybridization. Gallate-based MOFs were used as precursors to prepare a series of carbon adsorbents by activating agent-free pyrolysis. Adsorption tests and breakthrough experiments were used to evaluate the performances of these carbon adsorbents. The optimal carbon adsorbent, $Ni_{0.5}Co_{0.5}/C$-700, exhibits relatively high Xe uptake (1.42 mmol/g) at ambient conditions. Moreover, the IAST selectivity of Xe/Kr at ambient conditions and Henry's selectivity of Xe/Kr for $Ni_{0.5}Co_{0.5}/C$-700 are high (24.1 and 20.5, respectively), surpassing all carbon adsorbents and most top-performing MOFs reported to date. The density functional theory calculations reveal that the great affinity with Xe benefits from the presence of Ni and Co nanoparticles in carbon adsorbents. 5pz orbital of Xe has closer energy level to metal orbitals than that of Kr, providing greater mixing and stronger interaction. Facile preparation process, high adsorption capacity, record-high separation selectivity and high stability ensure this adsorbent a promising candidate in industrial use. And the proposed new strategy provides new ideas for the adsorption separation of inert gases in the future.

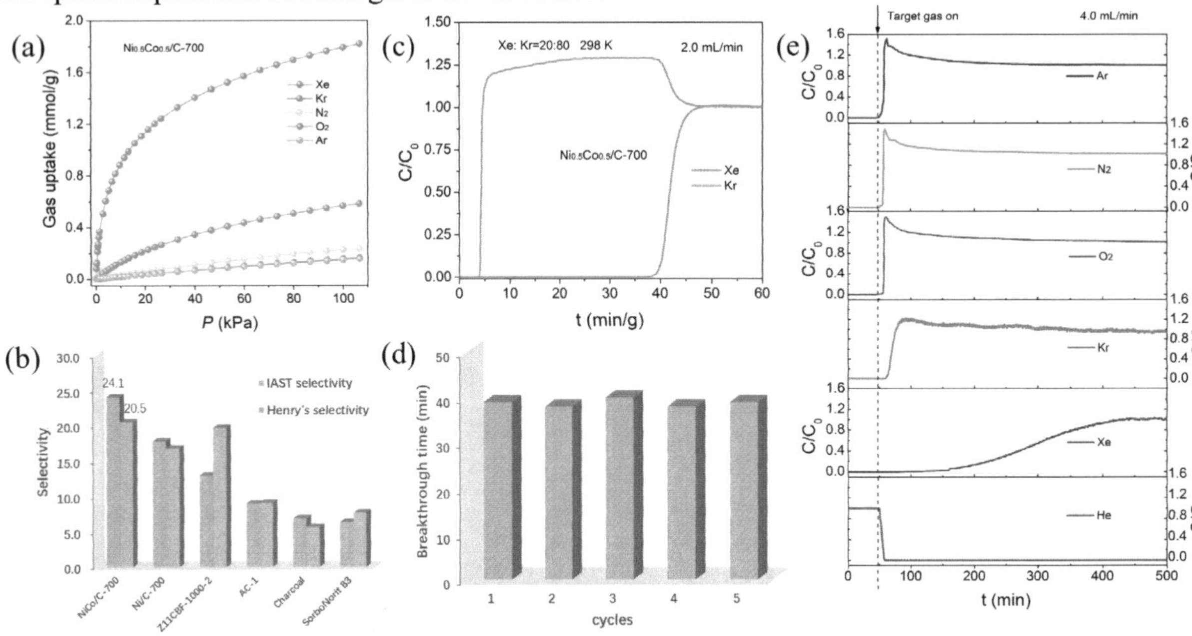

Figure 2. (a) Adsorption isotherms of $Ni_{0.5}Co_{0.5}/C$-700 at 298 K; (b)Performance comparison among reported carbon adsorbents; (c,e) Experimental breakthrough curves of $Ni_{0.5}Co_{0.5}/C$-700 and (d) Cycling performance of $Ni_{0.5}Co_{0.5}/C$-700.

596663 Porous Boron Nitride - Moving up the Scale for Use in Industrial Molecular Separations

Abstract:

Industrial separation processes account for 10-15% of global energy consumption[1]. The energy cost of large-scale gas and liquid separation processes like distillation could be significantly reduced by moving towards adsorption processes, i.e. using porous materials to separate molecules based on size and/or chemistry. An example of an inorganic adsorbent is porous boron nitride (BN): this material exhibits high surface area, high porosity and benefits from greater oxidative and thermal stability than common carbonaceous adsorbents[2]. In this work, we are exploring the potential of porous BN as an efficient and recyclable adsorbent for molecular separations at industrial scale.

Recently, our group developed a new method to produce porous BN with enhanced surface area, tuneable porosity[3] and promising liquid and gas separations performance. To further advance the use of porous BN, we must investigate its chemical stability. In fact, preliminary tests in our group pointed to the instability of the material in presence of water[4]. Considering this aspect, we functionalised the surface of porous BN via silanisation to enhance hydrophobicity and resistance to moisture. The samples were characterised before and after functionalisation using FTIR, XRD, XPS, TGA and SEM. The changes in hydrophobicity were monitored using water vapour sorption and contact angle measurements. In addition, the porosity of the material was probed before and after exposure to moisture and liquid water with nitrogen sorption. The results point to the efficiency of the approach to produce moisture-resistant BN-based adsorbents.

In parallel, we also investigated the formation mechanism of porous BN via the comprehensive characterisation of intermediates at different synthesis temperatures using a combination of analytical and spectroscopic tools, including X-ray absorption spectroscopy. The results of this study not only provide valuable insight on how to tune better the porosity of the material, but also on how to improve the hydrolytic stability discussed above. Therefore, this research paves the way for scaling up the synthesis of porous BN towards industrial applications.

1. Sholl, D. S.; Lively, R. P., Seven chemical separations to change the world. *Nature* **2016,** *532* (7600), 435-437.

2. Jiang, X.-F.; Weng, Q.; Wang, X.-B.; Li, X.; Zhang, J.; Golberg, D.; Bando, Y., Recent Progress on Fabrications and Applications of Boron Nitride Nanomaterials: A Review. *Journal of Materials Science & Technology* **2015,** *31* (6), 589-598.

3. Marchesini, S.; McGilvery, C. M.; Bailey, J.; Petit, C., Template-Free Synthesis of Highly Porous Boron Nitride: Insights into Pore Network Design and Impact on Gas Sorption. *ACS Nano* **2017,** *11* (10), 10003-10011.

4. Shankar, R.; Marchesini, S.; Petit, C., Enhanced Hydrolytic Stability of Porous Boron Nitride via the Control of Crystallinity, Porosity, and Chemical Composition. *The Journal of Physical Chemistry C* **2019,** *123* (7), 4282-4290.

605312 ZIF-14 Membrane Supported on ZnO Nanorods for Gas Permeation and Separation

Abstract:

Zeolitic imidazolate frameworks (ZIFs) area subclass of metal organic frameworks (MOFs) consisting of tetrahedral transition metal ions (zinc/cobalt) bridged by imidazole-type linkers. ZIF-14 is one of the promising ZIFs for gas separation. It features a ANA type zeolitic structure with Zn^{2+} metal nodes and 2-ethylimidazole (2-EtIm) linkers. The pore aperture of ca. 2.2 Å is ideal for molecular sieving for small gas molecules such as H_2. However, it is difficult to prepare a pinhole-free continuous ZIF-14 membrane by direct *in-situ* growth method since the low heterogeneous nucleation rates on ceramic substrates.

Here, we have designed an interfacial layer of aligned ZnO nanorods (NRs) for inducing the nucleation and growth of ZIF-14 thin films on porous α-Al_2O_3 tube substrate. The ZnO NR seed layer significantly reduces the nucleation barrier for ZIF-14 due to the dissolution and surface reaction of ZnO with the 2-EtIm linker in the hydrothermal synthesis. Highly crystalline, pinhole-free ZIF-14 membranes obtained in this work exhibit high performance for gas separation. The H_2 permeance reaches 1.7×10^{-8} $mol\cdot m^{-2}s^{-1}Pa^{-1}$, while the ideal separation factors for H_2/CO_2, H_2/N_2 and H_2/CH_4 are 20.2, 27.1 and 30.2, respectively. The high selectivity combined with its excellent hydrothermal stability indicates the reported ZIF-14 membrane as a promising candidate for gas separation. In addition, this preparation strategy has high reproducibility scalability and also is possible to extend the same principle to the preparation of other ZIF and MOF membranes.

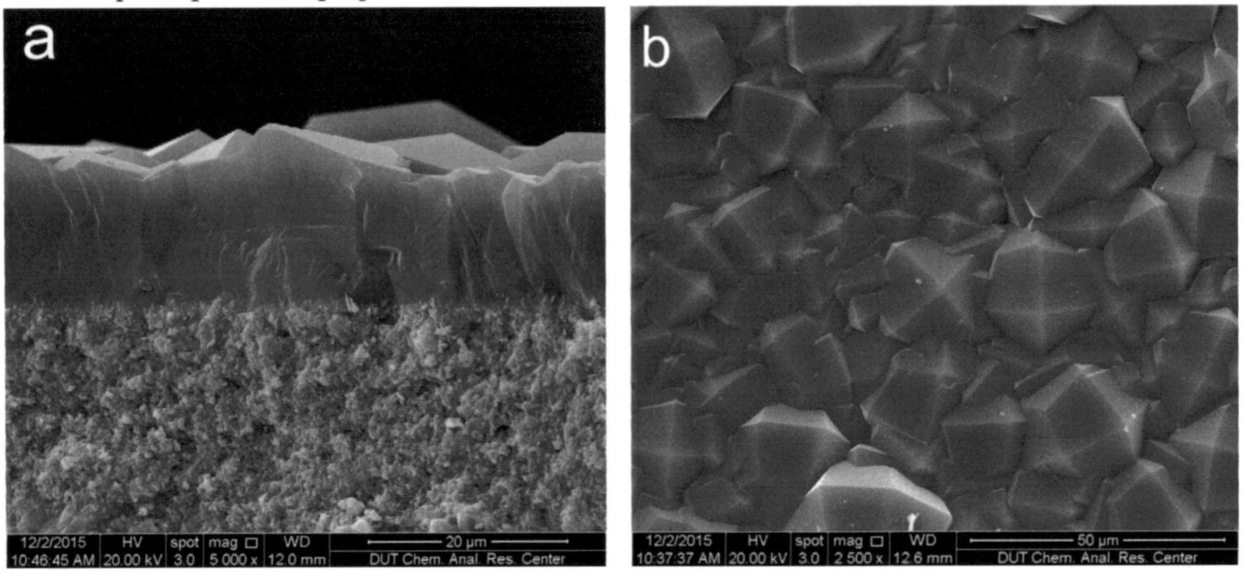

597881 Calcium Alginate Thin Films Derived from *Sargassum Natans* for the Selective Adsorption of Cd^{2+}, Cu^{2+}, and Pb^{2+} Ions

Abstract:

Calcium alginate is biopolymer, well known for its ion exchange characteristics and has often been exploited for heavy metal sequestration using spherical beads with high surface area per unit volumes. Thin-films, however, also have large surfaces for mass transfer due to their effective thicknesses (<100 µm) and thus allow for enhanced mass transfer properties. Furthermore, calcium alginate thin films have not been utilized or fabricated for heavy metal sorption, and thus presents the novel contribution in this research. Hence, this study aims to investigate the adsorption capacities of Pb^{2+}, Cu^{2+} and Cd^{2+}ions for calcium alginate thin films. Adsorption characteristics was investigated using Sigma-Aldrich, Protanal, Manugel and sodium alginate extracted from native *Sargassum Natans.* The Langmuir isotherm successfully modeled the adsorption phenomena of Pb^{2+} and Cu^{2+} with a maximum monolayer adsorption capacities of 4701 to 8475 mg/L for Pb^{2+} and 4292 to 6494 mg/L for Cu^{2+}. While the Freundlich isotherm indicated that Cd^{2+} binds heterogeneously with the alginate surface. Furthermore, it was found that process was predominantly governed by ion exchange whereby bidentate chelating coordination resulted in the respective ion binding affinity. Kinetic studies into adsorption behavior showed that the heavy metals followed the pseudo-second-order model with a maximum adsorption capacity of 1819 to 4489 mg/L for Pb^{2+}, 1805 to 3136 mg/L for Cu^{2+} and 1556 to 1781 mg/L for Cd^{2+}. Also, NMR characterization revealed that an increase in G-block monomer units followed an increase in adsorption capacity and kinetic rate constant. Insights into, the selectivity of the different ions revealed that the metal affinity for calcium alginate was as follows, $Pb^{2+} > Cu^{2+} > Cd^{2+}$. The results of this study successfully shows the potential of calcium alginate thin-films as effective adsorbents for heavy metal ions.

605086 Novel Porous Ternary Mg-Al-Ti Oxide Composite Nano-Adsorbents with Extremely High Arsenic Removal Capacity

Abstract:

Arsenic pollution of groundwater is a menacing issue in various parts of India as well as the entire world due to its toxicity and carcinogenicity. Removal of heavy metals from water by adsorption is one of the most effective approaches for its cost-effectiveness, ease of operation, high efficiency, process simplicity, and easy regeneration of materials. Porous materials are efficient adsorbents as they offer a large surface area for fast mass transfer and more active adsorption sites which result in high adsorption capacity. In this work, porous ternary Mg-Al-Ti oxide composites were synthesized by evaporation induced self-assembly using tri-block copolymer as soft-template. Composites with various elemental compositions were calcined at 400°C and 900°C to study performance and morphological differences. The 400°C calcined Aluminium rich composites showed high As(V) uptake efficiency of 265.6 mg/g, Magnesium-rich composites showed extremely high uptake of 1215 mg/g, and Titanium rich composites showed high uptake of 329 mg/g at pH~8 and 0.5g/L dose. Nanoparticles were characterized by XRD, SEM-EDX, BET surface area, N_2 adsorption isotherm. The BET surface areas were found to be 151.3 m^2 /g and 74.5 m^2 /g for 900°C calcinated Aluminum and Magnesium-rich samples respectively. SEM image and hysteresis loop in N_2 adsorption isotherm proved the porous nature. Detailed characterization and experiments were done to understand the adsorption mechanism and regeneration ability of the composites. The adsorption capacity of Mg-rich composite is significantly higher than the previously reported adsorbents. Due to their porous nature and extremely high adsorption capacity, the composites are promising candidates for arsenic removal through adsorptive and membrane assisted processes.

610841 A Fundamental Rietveld Refinement and DFT Study on the Sulfur Adsorption Sites of CuCe-Y Zeolite

Abstract:

Ion-exchanged Y zeolites have proven to be effective adsorbents for adsorptive desulfurization (ADS) of thiophenic compounds from hydrocarbon fuel. Compared with conventional hydrodesulfurization (HDS), ADS can operate at ambient conditions, while producing sulfur-free fuels. HDS requires extremely intense energy conditions to break down refractory sulfur compounds, but ADS can mitigate this challenge by utilizing mesoporous zeolites. Mesoporosity grants sulfur with high kinetic diameter access to internal active sites, thereby eliminating diffusion limitations. Recently, the combination of transition metals, such as Cu, and rare-earths, such as Ce, have shown promising synergy in increasing both the capacity and selectivity for sulfur.[1–3] In fact, promising ADS results using CuCeY zeolites have been well-established in the literature.[3–5]
This particular study focuses on the fundamental investigation of sulfur adsorption at the molecular level. Specifically, Rietveld refinement was performed on an x-ray diffraction (XRD) zeolite pattern to determine structural parameters, atomic composition and location of cations in CuCeY. Figure 1 shows the observed, calculated and differential XRD patterns of CuCeY. The difference curve indicates minimal discrepancy between the calculated and observed XRD patterns, leading to a relatively low residual error of $R_{wp} = 0.09$ and goodness of fit (GOF) = 3.35. Table 1 shows information regarding the final structure of CuCeY, which indicates that there are approximately 8 Ce^{4+} cations in the sodalite cage and 9 Cu^{2+} in the supercage, consistent with our XRF results. Using a zeolite model based on the results of the refinement, density function theory (DFT) studies were carried out to investigate the energetics of metal-sulfur interaction. Figure 2 shows the two-layer ONIOM cluster that was used to study the adsorption mechanism of various sulfur compounds (e.g. thiophene) on CuCeY. The computational results show strong adsorption of sulfur compounds on CuCeY and indicate selectivity towards sulfur over aromatic molecules. Subsequently, natural bond orbital (NBO) analysis was carried out on the high-theory QM layer. Both DFT and NBO analyses show significant charge transfer between the outer-shell orbitals of the cation and the sulfur molecule, as a result of σ-donation from the π-bond of the sulfur compound to the unoccupied orbitals of Cu, and the simultaneous backdonation of electron density from the filled 3d orbitals of Cu to the antibonding π* orbitals of the adsorbate. The combination of fundamental studies on CuCeY using Rietveld refinement and DFT to support experimental findings has been helpful in elucidating the role of bimetals on the adsorption of sulfur. These techniques will help guide the synthesis of more effective ion-exchanged zeolites as novel sulfur adsorbents.

References

1. Hernández-Maldonado, A. J. & Yang, R. T. Desulfurization of Diesel Fuels by Adsorption via π-Complexation with Vapor-Phase Exchanged Cu(I)−Y Zeolites. J. Am. Chem. Soc. 126, 992–993 (2004).
2. Velu, S., Ma, X. & Song, C. Selective Adsorption for Removing Sulfur from Jet Fuel over Zeolite-Based Adsorbents. Ind. Eng. Chem. Res. 42, 5293–5304 (2003).
3. Lee, K. X., Tsilomelekis, G. & Valla, J. A. Removal of benzothiophene and dibenzothiophene from hydrocarbon fuels using CuCe mesoporous Y zeolites in the presence of aromatics. Appl. Catal. B Environ. 234, 130–142 (2018).
4. Lee, K. X. & Valla, J. A. Investigation of metal-exchanged mesoporous Y zeolites for the adsorptive desulfurization of liquid fuels. Appl. Catal. B Environ. 201, 359–369 (2017).

5. Lee, K. X., Wang, H., Karakalos, S., Tsilomelekis, G. & Valla, J. A. Adsorptive Desulfurization of 4,6-Dimethyldibenzothiophene on Bimetallic Mesoporous Y Zeolites: Effects of Cu and Ce Composition and Configuration. Ind. Eng. Chem. Res. 58, 18301–18312 (2019).

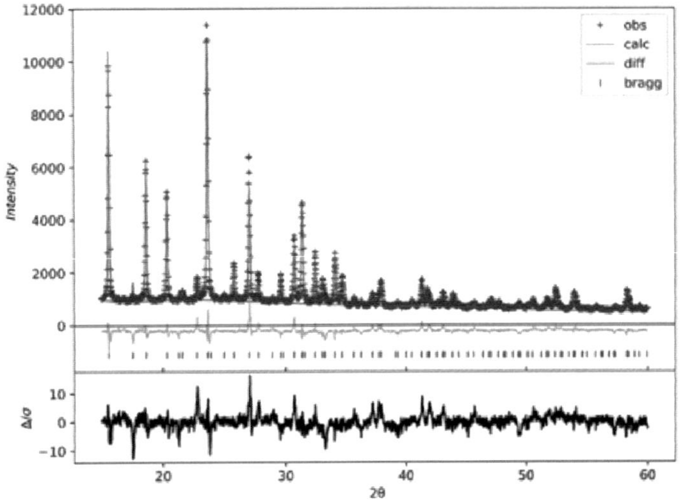

Figure 1: XRD pattern of CuCeY refined by the Rietveld method.

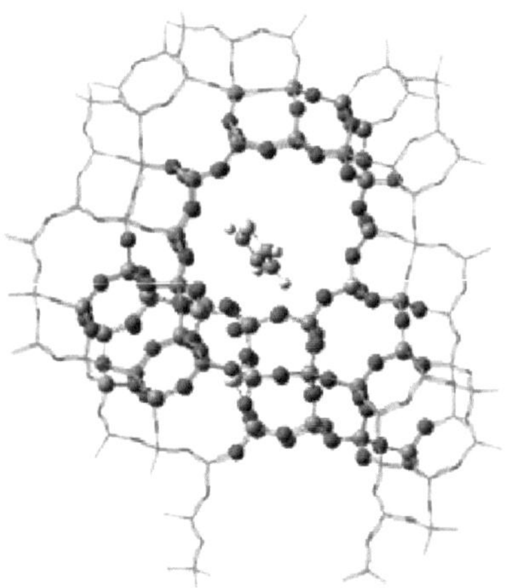

Figure 2: 172T ONIOM cluster model of CuCeY.

Table 1: Final structure model of CuCeY

Atom	x	y	z	U_{iso}	frac	multi	# atoms/uc
Si	-0.05392	0.12810	0.03691	0.00959	0.7212	192i	138.4704
Al	-0.05392	0.12810	0.03691	0.00959	0.2788	192i	53.5296
O1	0	-0.10533	0.10533	0.01573	1	96h	96
O2	0.00100	0.00100	0.13937	0.01112	1	96g	96
O3	0.07359	0.07359	-0.03661	0.01467	1	96g	96
O4	0.07076	0.07076	0.32845	0.00570	1	96g	96
Na (I)	0.01575	0.01575	0.01575	0.01000	0.1836	32e	5.8752
Ce (I')	0.07228	0.07228	0.07228	0.01271	0.1504	32e	4.8128
Ce (II')	0.16481	0.16481	0.16481	0.01742	0.0990	32e	3.168
Cu (II)	0.29070	0.29070	0.29070	0.09394	0.2907	32e	9.3024

601733 Self-Assembling Random Zwitterionic Terpolymers As Novel Fouling- and Chlorine-Resistant Nanofiltration Membrane Selective Layers

Abstract:

Nanofiltration (NF) is widely used for removal of multivalent ions and larger small molecules from water, in applications that range from water softening to water and wastewater treatment. Currently, most commercial NF membranes feature cross-linked polyamide selective layers that are prone to fouling and highly sensitive to chlorine, a common disinfectant used to manage biofouling. Zwitterions, functional groups with equal numbers of positive and negative charges, strongly resist fouling. Previously, our group has developed high flux, fouling resistant, size-selective membranes utilizing the self-assembly of random zwitterionic amphiphilic copolymers (rZACs) to form a network of zwitterionic effective "nanochannels" that allow the permeation of water and smaller solutes. These membranes are exceptionally fouling resistant, showing little to no flux decline during the filtration of a wide range foulants and complete flux recovery with a water rinse. They have effective pore sizes around ~1.5 nm, and exhibit limited salt rejection. Here, we aim to expand the separation capabilities of zwitterion-containing copolymer selective layers to enhance charged solute rejection and impart anion selectivity. For this purpose, we synthesized a charged zwitterionic amphiphilic copolymer (CZAC) comprised of zwitterionic, hydrophobic, and carboxylic acid-containing repeat units: poly(trifluoroethyl methacrylyate-*r*-sulfobetaine methacrylate-*r*-methacrylic acid) (PTFEMA-*r*-SBMA-*r*-MAA). The presence of charged groups, through dissociated MAA repeat units segregated into the zwitterionic nanochannels, leads to selective rejection of salts and charged organic compounds. The salt rejection values and water permeances of these membranes are comparable with commercial NF membranes. Yet, unlike typical NF membranes, CZAC membranes are highly resistant to fouling and exposure to chlorine. Furthermore, the selectivity of the membrane can potentially be further tuned by the functionalization of carboxylic acid groups on MAA repeat units to access additional separations. These features demonstrate the promise and versatility of this new polymer family for a wide range of membrane applications.

607200 Conductive Antifouling Membranes: Preserving Nanoscale Features in Polymers during Laser-Induced Graphene Formation

Abstract:

Membrane-based processes are widely accepted as one of the most promising technologies for water treatment, due to their high energy efficiency and relatively low overall cost compared to other treatment methods. However, membrane-based water treatment requires many additional pre- and post-treatment steps to remove contaminants and maintain membrane effectiveness. In particular, the treatment and prevention of fouling constitutes a large fraction of typical operational costs for membrane separations. One potential approach to combat fouling is to design conductive coatings that can prevent the attachment and growth of biofoulants, both electrostatically and via electrochemical generation of reactive oxygen species. Despite their potential, these conductive membrane coatings are often expensive, requiring additional chemicals and non-scalable methods to produce, e.g. carbon nanotube mats or other graphitic coatings deposited by vacuum filtration.

In this work, we explore the use of laser-induced graphene (LIG) for the creation of conductive ultrafiltration membranes. Porous polyethersulfone (PES) membranes are first treated with alumina using a process called sequential infiltration synthesis (SIS) before being irradiated with an infrared laser. We show that this alumina treatment, which can be scalably performed using roll-to-roll processing, can localize LIG formation to the surface of the membrane, preventing the buried, un-lased areas of PES from melting and losing their porosity during the lasing process. This allows the top-most layer of the PES to be a conductive coating that can be used to electrochemically degrade or remove contaminants (e.g. fouling mitigation). The formation of LIG is verified by scanning electron microscopy and Raman spectroscopy. The conductive layer is also shown to possesses low sheet resistances comparable to carbon nanotube mats, which is important for reducing power consumption in devices. Insight into the mechanism behind the improved stability to melting provided by SIS is provided by thermogravimetric analysis, differential scanning calorimetry, and Fourier-transform infrared spectroscopy. These insights are used to discuss the potential application of this approach to stabilize other materials during thermal cycling, or to create conductive coatings on other polymers.

596797 Phosphorene- Based Antifouling Membranes: Synthesis, Fabrication and Applications

Abstract:

Nanofiltration membranes (NF) have been largely developed and commercialized over the past decade because they show promise for the separation of neutral and charged solutes in aqueous solutions, offer low operation pressure and high flux. However, during nanofiltration, rejected materials can often accumulate on the membrane surface and/or within membrane pores to form a fouling layer. Phosphorene is a single-layer, two-dimensional (2D), exfoliated form of black phosphorus (BP). Phosphorene exhibits interesting and useful features, including its anisotropic electric conductance and optical responses which distinguishes it from other isotropic 2D materials such as graphene, molybdenum, tungsten chalcogenides. Phosphorene exhibits high absorbance, being a direct and narrow band gap semiconductor, and could effectively harvest low energy photons during photocatalysis. The band gap of phosphorene can be tuned sufficiently for photon absorption in the ultraviolet, visible light and the near-infrared region of the solar spectrum. Specifically, relevant to the field of membrane science, the band gap of phosphorene provides it with potential photocatalytic properties, which could be explored in making reactive membranes that can self-clean. Furthermore, phosphorene's ambi-polar transport characteristics (where both positively and negatively charged carriers conduct current under certain voltage bias conditions provides it with anti-microbial capabilities since applying electrical potential on conductive surfaces can hinder the development of biofilm. The overarching goals of this project are to first prove that phosphorene's electrical conductivity is anti-microbial, and it also possesses photocatalytic properties to destroy organic compounds; then, to develop novel nanocomposite membranes holistically from the initial investigation of phosphorene to testing. To this end, for the first time, membranes have been embedded with phosphorene. Membrane modification was verified by the presence of phosphorus on membranes, along with changes in surface charge, average pore size, and hydrophobicity. After modification, phosphorene-modified membranes were used to filter methylene blue (MB) under intermittent ultraviolet light irradiation. Phosphorene-modified and unmodified membranes displayed similar rejection of MB; however, after reverse-flow filtration was performed to mimic pure water cleaning, the average recovered flux of phosphorene-modified membranes was four times higher than that of unmodified membranes. Furthermore, coverage of MB on phosphorene membranes after reverse-flow filtration was four times lower than that of unmodified membranes, which supports the hypothesis that phosphorene membranes operated under intermittent ultraviolet irradiation can become self-cleaning. The antimicrobial potential of phosphorene will be tested using *Serratia marcescens*. The conductivity of phosphorene will be characterized to determine its antibacterial properties. This study will investigate the specific role of electric current in bacterial detachment and inactivation when a constant current is applied through phosphorene. The use of phosphorene to prevent biofilm formation has not been shown and will be examined in this study.

609285 Monitoring Protein Fouling on Patterned Membranes By Light Microscopy and Simulation

Abstract:

The goal of this project is to develop the basic science needed to design new fouling-resistant membranes. We present findings on the relationship between membrane surface patterning and protein fouling that support two hypotheses (i) geometric patterns will reduce membrane fouling compared to the control, and (ii) computational models will accelerate the discovery of new membrane surface morphologies and methods to improve membrane fouling resistance.

Experimental efforts combined flux decline measurements with confocal laser scanning microscopy (CLSM) to study the fouling of as-received and patterned polyethersulfone membranes during ultrafiltration of bovine serum albumin (BSA, model foulant). Patterned membranes were prepared by embossing ordered micropatterns on the surface. Both membrane types were labelled with a fluorophore, 5-DTAF. BSA labelled with Alexa Fluor 488 was used to visualize the fouling profiles with CLSM, which provided three-dimensional images of membrane surface patterns and co-localized protein foulant. In parallel, numerical simulations were used to compare fouling patterns with visual analysis of the CLSM images and to explore the full potential of the micropatterning. We utilized the customized simulator SUMs (Stanford University Membrane solver) within the OpenFOAM framework. The solver uses a finite volume method to study the dynamic couplings among flow, solute transport, and surface fouling. From the numerical study, we found that the micropatterning acts as a fouling-focusing mechanism. When the flow alternates, the foulant shifts its preferential accumulation zone and self-cleans the previously fouled region. This innovative design provides a possibility to consecutively alternate flow direction during operation to periodically clean zones of the membrane surface. The overall approach is expected to aid in the design of new membranes with tailored surface structures that prevent the irreversible deposition of foulants in prone-to-foul regions.

601745 Thiol-Affinity Immobilization of Casein-Coated Silver Nanoparticles on Polymeric Membranes for Biofouling Control

Abstract:

Introduction

Silver nanoparticles (AgNPs) have been widely studied for the control of biofouling on polymeric membranes due to their antimicrobial properties. However, nanoparticle leaching has posed a significant impediment against their widespread use. In this study, casein-coated AgNPs were chemically attached on CA membranes to minimize the leaching of particles.

Methods

AgNPs were attached through a procedure where the epoxide groups of polyGMA were open and attached with CYS, followed by the addition of thiol groups (from CYS) attached with AgNPs. The polymer complexes were then added to the membrane dope solution to minimize the number of post-synthesis steps. The membranes were cast via the non-solvent induced phase separation (NIPS) method, and then characterized and tested.

Results

The addition of polyGMA-CYS-AgNPs changed the cross-sectional structures by producing irregular structures and generating more small channels. XPS results verified the presence of sulfur and silver on the membranes cast with CYS and AgNPs, respectively. Dead-end and crossflow leaching studies showed little leaching of AgNPs. During BSA filtration studies, membranes containing AgNPs (M2 and M3) showed higher flux values as compared to the baseline CA membranes (M1), likely due to the presence of the NPs acting as pore formers. Lastly, membranes incorporated with AgNPs showed significant suppression of *Serratia marcescens*, as compared to membranes without AgNPs, indicated that the AgNPs membranes effectively suppressed the growth of *S. marcescens*.

Discussion

This study provided a one-step method to chemically incorporate AgNPs in the dope solution, and then cast CA/AgNPs membranes, which is favored to scale up. A quantitative method of testing antimicrobial performance of larger piece of membranes was proposed and proved to be effective.

610949 Dynamics of Nonliving Organic Matter in Membrane Fouling

Abstract:

Natural nonliving organic matter (NNOM) is well known as a foulant for many types of water purification membranes. Shearing environments, pH and ionic strength modify the self-assembled forms or aggregates of this material, affecting partitioning and dynamics. The aggregates form and disperse reversibly in the presence of shear, and hydrogen bonding has been suggested as a driving force for assembly. The argument that H-bonding energetics result only from weak attractions (<4–5 kcal mol−1]) is no longer chemically complete, and in this study we explore the presence of strong hydrogen bonding in shear-induced breakup kinetics for NNOM aggregates. Increased ionic strength was found to slow the aggregation kinetics of the physical hydrogel. Two phases of NNOM hydrogels were found to coexist, however, which we term here metachemical and physical hydrogels. One implication for membranes is that breakup on the retentate side of the membrane can lead to transport and reassembly on the permeate side, with downstream issues associated with disinfection byproducts.

600307 Covalently Grafting Graphene Oxide Onto UF Membranes to Improve Antifouling Properties

Abstract:

Graphene oxide (GO) is negatively charged and has good antifouling properties. However, its good dispersibility in water makes it challenging for membrane surface modification to improve antifouling properties. Herein, we demonstrate a versatile approach to covalently graft GO onto UF membrane surfaces with different hydrophilicities in aqueous solutions at room temperature. Specifically, we first functionalize the polysulfone (PSf) membrane surface with polydopamine (PDA), a bio-adhesive that can deposit on membranes with a variety of surface characteristics. Second, the GO with carboxyl acid groups on the edges can be activated by 1-ethyl-3-(3-dimethylaminopropyl) carbodiimide (EDC) and N-hydroxysuccinimide (NHS). Finally, the activated GO can be covalently grafted onto the PDA layer through the reaction of the carboxyl groups and amines in the PDA. The GO was prepared using the modified Hummers' method and then etched using H_2O_2 to reduce the size and maximize the grafting efficiency. The permeation and antifouling properties of the GO modified membranes are thoroughly evaluated in a constant flux cross-flow system. GO grafting improves the surface hydrophilicity without decreasing the water permeance. The GO-modification decreases the fouling rate by 63% and increases the critical flux by 20% when challenged using 1 g/L sodium alginate in the constant flux cross-flow system. The modified membranes demonstrate stability for continuous 48-h operation and interval cleanings using NaOH solutions. Given the simplicity of the GO grafting and the versatility of the PDA for various surfaces, our approach can also be used to post-modify modules for industrial use.

604490 Digital Design Via Mechanistic Modelling of a Continuous Crystallization Process with Periodic Wet Milling

Abstract:

Introduction

The attainable regions for critical quality attributes, such as Particle Size Distribution (PSD) and impurity levels, in the manufacture of Solid Oral Dosage Forms (SODF) can be highly dependent upon the type and operation of the crystallization process. In this work, we outline a step-wise workflow consisting of model validation, model-based technology transfer and process optimisation employed for the digital design of a continuous cascade cooling crystallization and wet milling process for manufacturing an Active Pharmaceutical Ingredient (API). Following this workflow, the mechanistic model was first validated using batch crystallization data and subsequently applied to describe the continuous crystallization process and with continuous and periodic wet milling, and explore the impact of varying process parameters, including duration of milling, rotor speed and frequency of milling operation during the crystallization process.

The step-wise approach taken for model validation and its subsequent application was as follows:

1. Model validation of crystallization kinetics for the pure API solid phase from batch de-supersaturation experimental runs.
2. Application of the validated crystallization kinetics to describe the continuous crystallization of pure API in a three-stage cascade process, as shown in Figure 1 below (wet milling was not considered at this stage).
3. Model validation of breakage kinetic parameters using batch data from a wet milling unit operated in a recycle with a batch crystallizer. A rotor-stator wet mill was utilised for this process, with a range of rotor frequencies and milling head generator configurations probed to assess the impact on the PSD over time and to fit the breakage kinetic parameters. Subsequent refinement of the breakage kinetic parameters utilizing continuous crystallization experiments, where intermittent/periodic wet milling was employed using an immersion wet mill in the first stage crystallizer.
4. Scenario analysis of the continuous crystallization and periodic wet milling process for this API, looking at the impact of process parameters on the Critical Quality Attributes (CQAs) of the product, including product PSD and dimer phase formation (impurity).

Conclusions

The main conclusions of the work include the following:

- The continuous crystallization process was unable to achieve the desired product PSD without wet milling, due to very low levels of secondary nucleation in the system. Optimal PSD quantiles (d10, d50 and d90) predicted for the product were significantly higher than the target PSD quantiles required to achieve the desired product performance for this compound.
- Addition of a wet milling step in a recycle with the continuous crystallization significantly increased the lower end of the attainable region in terms of PSD quantiles and impurity level,

allowing the process to comfortably achieve the desired range for product PSD quantiles.
- Mechanistic modelling approaches can be utilised to significantly reduce the development timelines, material consumption and efficiency of continuous API crystallization production processes, in particular cases where periodic operation of some units, in this case the wet mill, are employed.

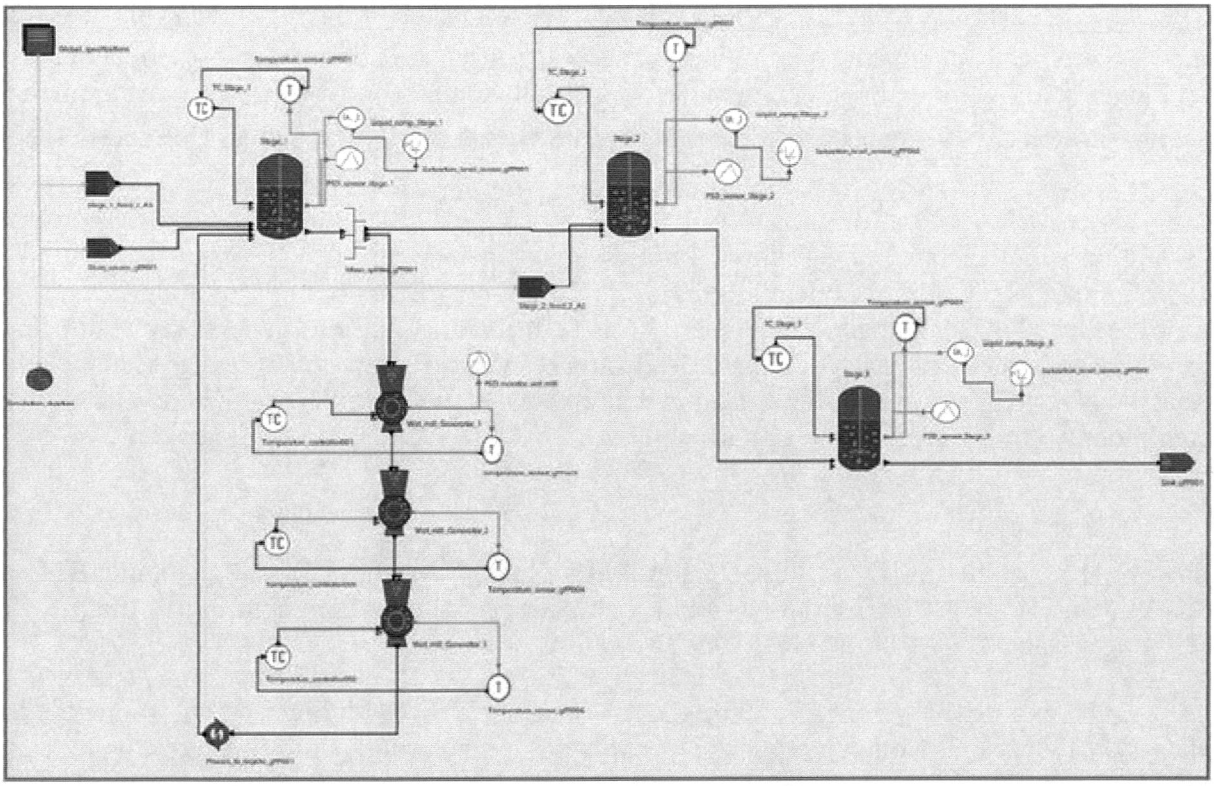

Figure 1: Simulation flowsheet used model for optimisation of continuous crystallization and wet milling process.

606179 Continuous Generation of Glycine Crystals with Pure Form and Narrow Size Distribution in Slug Flow

Abstract:

In pharmaceutical industries, continuous manufacturing of medicine has the potential to reduce capital, time and increase efficiency[1]. Continuous crystallization as the bridge between synthesis and formulation[2], the process and quality control is vital such as aggregation of crystals can affect the following blending and coating. Segmented/slug flow crystallization is one of the effective continuous process demonstrated to allow enhanced control of organic crystal properties such as size and shape[3–6].

This presentation will elaborate in the following topics[7]: (1) generate pure-form uniform size crystals in slug flow rapidly; (2) explore the operation region (meaning no secondary nucleation nor aggregation) in slug flow crystallizer. The crystal forms at different conditions generated by both slug flow crystallizer and batch crystallizer are discussed and compared. Experimental results validated the proposed crystallizer reduced production time and equipment cost while increased the operational region significantly compared to batch crystallization.

Reference

(1) Schaber, S. D.; Gerogiorgis, D. I.; Ramachandran, R.; Evans, J. M. B.; Barton, P. I.; Trout, B. L. Economic Analysis of Integrated Continuous and Batch Pharmaceutical Manufacturing: A Case Study. *Ind. Eng. Chem. Res.* 2011, *50* (17), 10083–10092.

(2) Lee, S. L.; O'Connor, T. F.; Yang, X.; Cruz, C. N.; Chatterjee, S.; Madurawe, R. D.; Moore, C. M. V.; Yu, L. X.; Woodcock, J. Modernizing Pharmaceutical Manufacturing: From Batch to Continuous Production. *J. Pharm. Innov.* 2015, *10* (3), 191–199.

(3) Jiang, M.; Zhu, Z.; Jimenez, E.; Papageorgiou, C. D.; Waetzig, J.; Hardy, A.; Langston, M.; Braatz, R. D. Continuous-Flow Tubular Crystallization in Slugs Spontaneously Induced by Hydrodynamics. *Cryst. Growth Des.* 2014, *14* (2), 851–860.

(4) Eder, R. J. P.; Schrank, S.; Besenhard, M. O.; Roblegg, E.; Gruber-Woelfler, H.; Khinast, J. G. Continuous Sonocrystallization of Acetylsalicylic Acid (ASA): Control of Crystal Size. *Cryst. Growth Des.* 2012, *12* (10), 4733–4738.

(5) Jiang, M.; Papageorgiou, C. D.; Waetzig, J.; Hardy, A.; Langston, M.; Braatz, R. D. Indirect Ultrasonication in Continuous Slug-Flow Crystallization. *Cryst. Growth Des.* 2015, *15* (5), 2486–2492.

(6) Lu, J.; Litster, J. D.; Nagy, Z. K. Nucleation Studies of Active Pharmaceutical Ingredients in an Air-Segmented Microfluidic Drop-Based Crystallizer. *Cryst. Growth Des.* 2015, *15* (8), 3645–3651.

(7) Mou, M.; Jiang, M. Fast Continuous Non-Seeded Cooling Crystallization of Glycine in Slug Flow: Pure α-Form Crystals with Narrow Size Distribution. *J. Pharm. Innov.* 2020, 22–24.

609118 Continuous Crystallisation Oscillatory Flow Platform for Protein Purification and Bio-Separation

Abstract:

Crystallization of proteins could be a potentially cost effective and robust method to purify protein molecules in the pursuit of the next generation of protein based biopharmaceutic therapies. However, the step from traditional small-scale high throughput screening methods, to crystallizers which can produce high yields of pure crystalline protein material, is still in its infancy.

Here, we present a platform for continuous oscillatory flow crystallization (COFC) (Figure 1) and its application for the crystallization of lysozyme[1,2]. A workflow is developed from µl screening experiments, to scaled up batch oscillatory flow crystallization (BOFC) and COFC experiments.

The induction time for lysozyme is found to be inversely proportional to the initial protein concentration, with the same relationship found between the induction time and the frequency and amplitude of the oscillation. However, it is also observed that the decrease in induction time comes at a cost to the crystal size and quality. In turn, the high shear environment from increased amplitude and frequency of the COFC can also damage the fragile protein crystals, resulting in the formation of more fine particles and an undesirable size distribution. This led to further investigation of slug flow mixing in the continuous crystalliser, where the crystal size and shapes are compared to the particles produced from the osciallatory flow reactor. Further work around optimization of the crystallization parameters and crystallizer geometry using the gCrystal platform to find the optimum scale up conditions are also discussed. The use of seeding and potential templating materials is also covered with a view to expanding the use of this crystallizer to less well characterized protein materials.

It is demonstrated that purification of proteins can be achieved by crystallization in a continuous mode. This route offers the possibility to revolutionise downstream separations of biopharmaceuticals

References

[1] H.Y. Yang, W.Q. Chen, P. Peczulis, J.Y.Y. Heng (2019): **Development and Workflow of a Continuous Protein Crystallization Process: A Case of Lysozyme**, *Cryst. Grow. Des*; 19, 983-991.

[2] H.Y. Yang, P. Peczulis, P. Inguva, X.Y. Li, J.Y.Y. Heng (2018): **Continous Protein Crystallisation Platform and Process: Case of Lysozyme**, *Chem. Eng. Res. Des.* 136, 529-53

Figure 1: Schematic of the continuous oscillatory flow crystallisation platform

609250 Temperature Control Approaches for Cooling Crystallization Using a Continuous Microfluidic Mixer

Abstract:

Cooling crystallization is a popular crystallization technique in pharmaceutical and fine chemical industries. The driving force for this process is the temperature gradient that consequently results in increasing the supersaturation ratio. However, it is very challenging to minimize the gradient range and maintain a constant supersaturation without sacrificing the mixing performance. In this study we have we have developed two techniques to enable temperature gradient within the crystallization zone using the previously developed continuous microfluidic mixer. For the first method, we have designed a cooling bath around the circular zone which has separated inlet and outlet. The cold stream is circulated around the zone using a peristaltic pump while the hot streams enter from the four main inlets of the device. Due to the temperature difference between the hot and cold streams, the solution within the crystallization zone will attain an equilibrium temperature lower than the hot streams. In the second method, saturated hot and cold streams will enter the circular zone and mixed to impose supersaturation inside the mixing zone. Both systems are modeled using COMSOL Multi physics and equilibrium temperature and imposed supersaturation are calculated respectively. These microfluidic devices are then 3D-printed using the Formlabs 3D-printer for experimental verification. Using these two systems we have performed cooling crystallization of L-Glutamic acid and L-Histidine using water as solvent. Here we have compared eight different supersaturation for each compound and measured the face specific growth rate followed by morphology and size distribution of the crystals. Comparative analysis has also been performed using a batch microtiter plate.

602048 “Filling the Gap” in Continuous Pharmaceutical Technologies: A Novel Continuous Filtration Drying System

Abstract:

In recent years, the world’s top pharmaceutical companies have started major research programs to switch from classical batch to continuous manufacturing of their active drug substances. There is a significant gap in technology for the continuous isolation. Alconbury Weston Ltd (AWL), a leading supplier of continuous crystallization and filtration equipment, has designed an innovative continuous carousel filter dryer (CCFD) which filters, washes and dries solids in small aliquots simultaneously in a continuous manner.

The core technology is based on well-established Nutsche filtration, processing thin cakes in a fully automated system. A state-of-the-art vision system can be used to track the cake and liquid heights during filtration and washing, improving product purity, yields and quality attributes. The CCFD is a complete plant which replaces a traditional batch production facility and fits into a walk-in fume cupboard.

The presentation will describe the challenges and issues observed in batch filtration systems and demonstrate how these challenges can be addressed using continuous filtration. The talk will demonstrate the collaborative efforts required for the development of a novel continuous processing technology through to a commercial product. AWL has worked with Strathclyde University and CMAC (Glasgow, UK) and Purdue University (Indiana, US) to investigate the advantages of continuous filtration. The presentation will include case studies carried out at both academic centres of excellence.

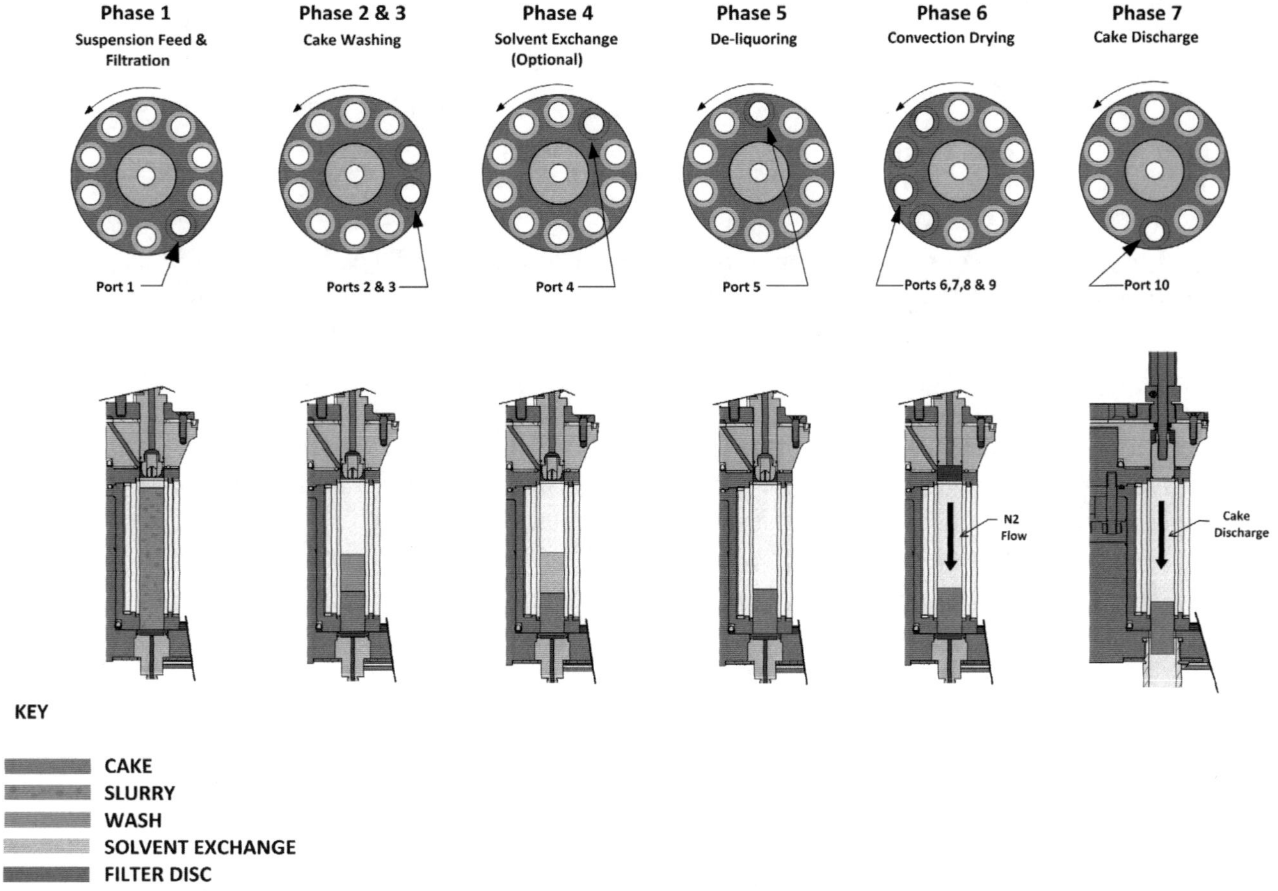

608687 Integrated Continuous Crystallization and Spray Drying of Insulin

- Download/View Extended Abstract

Abstract:

Insulin injection is commonly used for the treatment of diabetes. Pulmonary delivery through inhalation of dry insulin powders is commercially also available and offers a more convenient route compared to injection.[1] The manufacture of such powders is challenged by the need to produce crystalline particles with an aerodynamic diameter in the narrow range of 1 – 5 µm. Particles outside the optimal size range will not deposit well into the lungs. Amorphous particles produced from spray drying an insulin solution are limited by their fast dissolution,[2] which leads to a therapeutic effect of only several hours. Furthermore, in general, crystalline particles have a better stability and aerosolization performance and different drug release characteristics compared to amorphous particles.[3,4] Therefore, the manufacture of a dry powder with insulin crystals is of interest to potentially improve pulmonary delivery of insulin. The crystallization of insulin has been studied in various crystallizers.[5–7] However, those processes generally do not deliver crystals in the optimal size range for pulmonary delivery. Therefore, there is a need for novel crystallization processes that can produce insulin crystals with tailored size for the application of dry powder inhalation more efficiently.

The objective of this work is to develop and characterize a novel continuous crystallization process for insulin that is tailored for dry powder inhalation by integrating a segmented-flow crystallizer with spray drying for rapid solvent removal. Continuous crystallization has received growing interest in the fine-chemical and pharmaceutical industries due to typical advantages such as easier process control and a more consistent product quality compared to batch crystallization.[8] The benefits of integrating continuous crystallization with spray drying for the manufacture of dry powders for inhalation has been demonstrated by earlier work from our group for different small-molecule active pharmaceutical ingredients.[9] The present work is, to the best of our knowledge, the first demonstration of such process for a therapeutic protein.

The process consisted of a tubular crystallizer and a spray dryer. A calibrated two-channel peristaltic pump was used to introduce the feed and the precipitant solution into the crystallizer at the same flow rate through a T-mixer. A gas flow controller was used to introduce nitrogen with controlled flow rate via another T-mixer to create a segmented flow, which offers the benefit of a narrow residence time distribution.[10,11] The supersaturation was generated by both cooling and a change in pH. The tubular crystallizer, which consisted of two connected pieces of tubing of 14m long each, was immersed in a thermostatic bath to allow temperature control. At the outlet of the crystallizer, a buffer vessel was used for separation of nitrogen and to provide a steady flow to the spray dryer using another peristaltic pump. Finally, the dry powder was collected in a powder collector located below a cyclone. Initially, the crystallizer was operated stand-alone to identify optimal crystallization conditions. Subsequently, optimal crystallization conditions were applied for the integrated process. The morphology of the particles and their approximate size were characterized with optical microscopy and scanning electron microscopy (SEM). The mass median aerodynamic diameter (MMAD) and fine particle fraction (FPF) fraction was measured with a Next Generation Impactor (NGI).

The recovery of insulin from stand-alone crystallization experiments was in excess of 90% when using a combined flow rate of 1.2 ml/min and an insulin concentration of 2.4 mg/ml. A lower recovery was obtained at higher combined flow rates, and thus shorter residence times, in the range of 1.8 ml/min to 2.4 ml/min. A constant recovery could be approached in a short period of about 10 to 15 minutes when running at a high initial concentration in the range of 2.0 mg/ml to 2.4 mg/ml. Such fast start-up time is consistent with plug-flow conditions, which is an inherent benefit of a segmented-flow crystallizer compared to a mixed-suspension mixed-product-removal crystallizer. The produced particles appeared crystalline from SEM and were close to the desired size range for pulmonary drug delivery. The particles produced with the integrated process under optimal conditions were characterized by a MMAD in the range of 2 μm to 6 μm and an FPF of 20 - 40%, which indicated a good potential for pulmonary drug delivery. In conclusion, the demonstrated benefits of the novel process include a short total residence time, high recovery from crystallization, and suitable product quality attributes for pulmonary drug delivery of insulin without any added excipients.

1. References

1. Easa, N., Alany, R. G., Carew, M. & Vangala, A. A review of non-invasive insulin delivery systems for diabetes therapy in clinical trials over the past decade. *Drug Discovery Today* vol. 24 440–451 (2019).
2. Bailey, M. M., Gorman, E. M., Munson, E. J. & Berkland, C. Pure insulin nanoparticle agglomerates for pulmonary delivery. *Langmuir* **24**, 13614–13620 (2008).
3. Shenoy, B., Wang, Y., Shan, W. & Margolin, A. L. Stability of crystalline proteins. *Biotechnol. Bioeng.* **73**, 358–369 (2001).
4. Basu, S. K., Govardhan, C. P., Jung, C. W. & Margolin, A. L. Protein crystals for the delivery of biopharmaceuticals. *Expert Opin. Biol. Ther.* **4**, 301–317 (2004).
5. Hirata, G. A. M., Bernardo, A. & Miranda, E. A. Crystallization of porcine insulin with carbon dioxide as acidifying agent. *Powder Technol.* **197**, 54–57 (2010).
6. Parambil, J. V., Schaepertoens, M., Williams, D. R. & Heng, J. Y. Y. Effects of oscillatory flow on the nucleation and crystallization of insulin. *Cryst. Growth Des.* **11**, 4353–4359 (2011).
7. Chen, F. *et al.* Crystallization of bovine insulin on a flow-free droplet-based platform. in *AIP Conference Proceedings* vol. 1820 030003 (American Institute of Physics Inc., 2017).
8. Wood, B., Girard, K. P., Polster, C. S. & Croker, D. M. Progress to Date in the Design and Operation of Continuous Crystallization Processes for Pharmaceutical Applications. *Org. Process Res. Dev.* **23**, 122–144 (2019).
9. Hadiwinoto, G. D. *et al.* Integrated Continuous Plug-Flow Crystallization and Spray Drying of Pharmaceuticals for Dry Powder Inhalation. *Ind. Eng. Chem. Res.* **58**, 16843–16857 (2019).
10. Neugebauer, P. & Khinast, J. G. Continuous Crystallization of Proteins in a Tubular Plug-Flow Crystallizer. *Cryst. Growth Des.* **15**, 1089–1095 (2015).
11. Jiang, M. *et al.* Continuous-flow tubular crystallization in slugs spontaneously induced by hydrodynamics. *Cryst. Growth Des.* **14**, 851–860 (2014).

602621 Hybrid Membrane-PSA System for Efficient Oxygen Generation for Transportable, Modular Gasification Systems

Abstract:

Modular gasification systems need new low capital and operating cost approaches for the separation of O_2 from air to produce high-purity O_2 (>97%). Production of O_2 with large-scale cryogenic distillation is costly and not scalable to small- to commercial-scale systems. The feasibility of concentrating O_2 from air using a combination of targeted mixed matrix membranes (MMMs) with a process-intensified, process-integrated pressure swing adsorption (PSA) system was experimentally determined. Pre-concentrating O_2 with MMMs allows the integrated PSA system to generate high-purity O_2 with smaller PSA units, faster cycle times, and reduced power requirements to minimize the overall costs. High-capacity O_2 sorbent materials, commercial adsorbents, and O_2-selective adsorbents such as cerium-exchange zeolite, titanosilicates, and cerium-exchanged zeolite X with strontium were tested and characterized Adsorbents were characterized by BET analysis, energy-dispersive X-ray (EDX) spectroscopy, X-ray power diffraction (XRD), and equilibrium adsorption isotherm measurements. Porous asymmetric MMMs with high selectivity and permeability were synthesized using glassy polyimides, polyether sulphone (PES), and polyvinyl acetate and then modified with a thin layer of a zeolite or silicate nanomaterial. The nanomaterials were combined with a high permeability glassy polymer such as Poly (1-trimethylsilyl-1-propyne) (PTMSP) to form an asymmetric membrane onto a porous substrate (e,g, PES). A range of inorganic fillers (e.g., clays, zeolites, aluminosilicates) were used to evaluate selectivity and permeability. The effects of the membrane thickness, porosity, inorganic material loading, and operating conditions were determined. Lastly, a pilot-scale PSA system was operated using an O_2-enriched stream representative of the membrane-based concentrator to demonstrate the combined membrane-PSA approach. The incoming, O_2-enriched air was varied to determine the effects of O_2 concentration, temperature, pressure, and other process parameters on the total power and energy for operation, separation efficiency, and effluent gas O_2 concentration.

599985 Analysis of Membrane Processes for CO_2 Removal and H_2 Reuse for Solid Oxide Fuel Cells

Abstract:

In order to enhance fuel utilization in solid oxide fuel cells (SOFCs), part of the anode exhaust may be recycled to join the anode inlet stream. However, the CO_2 present in the exhaust would dilute the fuel concentration and lower the efficiency of the SOFCs. Few papers have considered removing CO_2 from the recycle stream, and none have reported on using a membrane process for this task. A CO_2-selective facilitated transport membrane was developed with fluoride- and hydroxide-containing species for the CO_2/H_2 separation, and the membrane exhibited a good stability, showing a CO_2 permeance of 108 GPU and a CO_2/H_2 selectivity of 106 at 120°C. Moreover, the membrane was scaled up to 21″ in width, which had consistent transport results as the lab-scale samples.

Herein, a process analysis was done on using the developed membrane for an SOFC system. First, due to its wide availability, air was used as the sweep gas. Furthermore, a vacuum was proposed for the permeate side to further enhance the driving force and allow for CO_2 capture. Also, as oxygen is absent, amine-containing membranes with better transport performances could be used for the vacuum process at 120°C. At 99% H_2 recovery, the air sweep process was estimated to remove 44.7% of CO_2 from the recycled anode exhaust, whereas the vacuum process could boost the removal to 76.9%. A less stringent H_2 recovery requirement could yield a higher CO_2 removal. For instance, a CO_2 removal of over 90% could be achieved when the H_2 recovery was relaxed to 96.7% for the vacuum operation. In addition, a high-level techno-economic analysis was conducted, without including the H_2 reused value. The CO_2 removal costs corresponding to the air sweep and vacuum processes were \$62.9/tonne and \$53.3/tonne, respectively, even with the added cost of the vacuum pump for the amine membrane process. The cost is even lower at \$35.4/tonne if the captured CO_2 is utilized for enhanced oil recovery.

601968 Conceptual Modeling and Design for Membrane-Based Hybrid CO_2 Capture Processes in Cement Industry

Abstract:

As the new climate system concluded in the Paris Agreement of the United Nations is implemented in 2021, it is expected that countries around the world will establish various climate change policies to achieve the greenhouse gas emissions target. In particular, to reduce the emission of CO_2, which is a major cause of climate change, research on CCS technology that directly captures CO_2 emitted from industrial sites such as power generation, cement, and steel petroleum has been actively conducted. This study focuses how to effectively integrate membrane-based hybrid systems for CO_2 capture process in cement process. For the CO_2 capture process, stand-alone systems based on membrane and cryogenic technology are analyzed, which are then compared with hybrid systems. A case study is carried out in hybrid systems to integrate two technologies and evaluate the effect of hybrid option. Then, the technical feasibility of the hybrid capture process is assessed at the level of conceptual design.

Acknowledgement

This research was supported by the Korea Carbon Capture & Sequestration R&D Center (KCRC) through the National Research Foundation of Korea (NRF) funded by the Ministry of Science, ICT and Future Planning (2014M1A8A1049338) and the Korea government(MSIT)(No. 2019R1A2C2002263

612016 A Unified Look at Uptake and Transport in Ion Exchange Materials for Rfbs

Abstract:

Ion exchange membranes are frequently used in redox flow batteries. However, there are few general guidelines on how to select membranes. Uptake and transport of ions and the factors underlying them are critical determinants of thebehavior of conducting membranes. Substantial work was carried out on issues such as partitioning of ions into membranes in the early studies of Nafion. The heavy focus on fuel cell applications narrowed the scope of many studies to an emphasis on proton conduction. However, with the increased emphasis on redox flow batteries and other, similar electrochemical devices, there is a need to build a stronger base of studies of ions other than protons or hydroxide in various membrane situations. Furthermore, the operating conditions of the RFB typically entails exposure of membranes to concentrated electrolytes. We will report recent studies of the uptake, conduction and transport of ions in various membranes that are exposed to both aqueous and non-aqueous environments. In this presentation, we will describe the separation of factors determining permeation through ion exchange membranes and the molecular basis of these parameters. By comparing these behaviors across different membrane types, we will begin to systematize our understanding of the interactions leading to observed phenomena. In addition, we will outline our recent work in developing a description of the factors influencing uptake of ins in to membranes.

Acknowledgements

I would like to gratefully acknowledge the current support of this work by the Office of Naval Research and the U.S. Department of Energy, Office of Electricity Delivery and Energy Reliability (Dr. Imre Gyuk).

611901 Performance of Commercial Polymer Electrolyte Membranes in Solar Fuels Devices

Abstract:

In the search for clean sources of energy, solar fuel devices (photoelectrochemical carbon dioxide reduction) are promising technologies for the production of renewable liquid transportation fuels. In a membrane electrode assembly type device, consisting of two electrodes separated by a hydrated polymer electrolyte membrane, the membrane must allow ion transport so that current can flow in the device, while inhibiting transport of CO_2 reduction fuel products between electrodes to minimize parasitic re-oxidation that contributes to device efficiency compromises. A prototype solar fuels device consisting of a copper gas diffusion cathode, iridium anode, and a Selemion AMV anion exchange membrane was assembled and used throughout this study. The stability of cell operation at varying potential was evaluated through cyclic voltammetry. Chronoamperometry was used to test the ability of the membrane to sustain a stable current as well as to measure the current density that the device could support. Membrane degradation was investigated by measuring the overall resistance of the prototype device using energy dispersive spectroscopy and potentio-electrochemical impedance spectroscopy before and after CO_2 reduction. Similar studies on other commercial membranes will be reported. By studying the capability of various commercial membranes to maintain a high and stable current density, further enhancement of solar fuels device operation may be achieved.

597956 Understanding the Thermodynamics and Kinetics for Water-Splitting in Micropatterned Bipolar Membranes

Abstract:

The palette of applications for bipolar membranes (BPMs) has expanded in recent years to numerous electrochemical energy and conversion applications including fuel cells, electrolyzers, and photoelectrochemical cells. BPMs have historically been deployed in electrodialysis setups for mineral acid and base production. The need to have disparate pH environments in electrochemical cells, and prevent species crossover, have motivated researchers to examine BPMs as an electrolyte separator. BPMs have the unique capability to split water into protons and hydroxide ions charge carriers in addition to conducting those ions in opposite directions to maintain current flow in the electrochemical setup. A BPM consists of a cation exchange membrane (CEM) appended to an anion exchange membrane (AEM) that have intimate contact and feature a water dissociation catalyst between the individual membrane layers.

Most materials related research for BPMs has focused on water-dissociation catalysts. There are few reports that investigate the importance of high-quality bipolar junction interfaces for improving water-splitting in BPMs. Although it has been shown that intimate contact between the oppositely charged layers at the interface improves BPM performance, no model or scaling relationships exists that relate bipolar junction interfacial area to the on-set potential for water-splitting and the current density to water-splitting. Our research has aimed to address this knowledge gap by preparing BPMs with systematically varied interfacial area values using soft lithography. Polarization experiments with the new, micropatterned interface BPMs reveal a 250 mV reduction in the on-set potential when increasing the interfacial area by 2.2x. With respect to current density, the same increase in interfacial area resulted in a 15% increase in current density at 2 V (i.e., marginal improvement). This talk explains these observations by examining the change in local electric field using Poisson-Boltzmann equation and limitations in enhancing reaction kinetics due to mass transfer. Finally, the soft-lithography approach devised was successful for making BPMs with different chemistries ranging from perfluorinated AEMs and CEMs to alkaline stable, ether-free poly(arylene) hydrocarbon AEMs. These polymer chemistries are more robust for fuel cell and electrolysis applications.

This work was supported from NSF Award #: 1703307.

602494 Development of Polymer Electrolyte Membrane and Ionomer Binder Materials for Electrochemical Energy Conversion Technology

Abstract:

Anion exchange membranes (AEMs) based on hydroxide-conducting polymers are a key component for anion-based electrochemical energy technology such as fuel cells, electrolyzers, and advanced batteries. Although these alkaline electrochemical applications offer a promising alternative to acidic proton exchange membrane electrochemical devices, access to alkaline-stable and high-performing polymer electrolyte materials has remained elusive until now. Despite vigorous research of AEM polymer design, examples of high-performance polymers with good alkaline stability at an elevated temperature are uncommon. Traditional aromatic polymers used in AEM applications contain a heteroatomic backbone linkage which is prone to degradation via nucleophilic attack by hydroxide ion.

In this presentation, I will highlight recent progress at the Bae group of Rensselaer Polytechnic Institute in the development of advanced hydroxide-conducting polymers for applications in AEMs and electrode ionomers. We have developed new synthetic methods that produce polymer design made of all C−C bond backbone and a flexible chain-tethered quaternary ammonium group and that provide an effective solution to the problem of alkaline stability. The advantage of good solvent processability and convenient scalability of the reaction process has generated considerable interest of these polymers, and they are considered leading candidates for commercial standard AEM. AEM fuel cell and electrolyzer tests of some of the developed polymer membranes showed excellent performance, suggesting that this new class of HCPs opens a new avenue to electrochemical devices with real-world applications.

602402 CO_2-Selective Membrane for Enhancing H_2 Utilization in Solid Oxide Fuel Cells

Abstract:

In order to utilize the H_2 remaining in the anode exhaust of a solid oxide fuel cell (SOFC), a CO_2-selective membrane was developed to remove CO_2 from the anode exhaust and recycle the concentrated H_2 back to the SOFC. As the cooled exhaust still has an average temperature of 120°C, it is challenging for polymeric membranes to maintain both high CO_2/H_2 selectivity and CO_2 permeance. Previously, facilitated transport membranes were developed using tetramethylammonium hydroxide (TMAOH) as the mobile carrier to selectively remove CO_2. However, TMAOH has been reported to degrade via a nucleophilic substitution mechanism, and a possible higher temperature of the anode exhaust in practice may exacerbate the degradation.

Herein, more stable fluoride- and hydroxide-containing species were investigated for their potentials of replacing TMAOH. The thermal stabilities of the quaternaryammonium-containing compounds were characterized by TGA. Moreover, quantitative results from NMR showed that tetramethylammonium fluoride (TMAF) was 14 times more stable than TMAOH at 130°C. Next, membranes with the optimal composition containing TMAF were synthesized, and their transport performances at different temperatures and feed CO_2 partial pressures were measured, respectively. The membrane showed a CO_2 permeance of 108 GPU and a CO_2/H_2 selectivity of 106 at 120°C, and it was stable for 120 h. The TMAF-containing membrane also displayed an improved stability of about 4 times better than the TMAOH-containing membrane at 130°C. To date, scale-up of 21″-wide membranes with a total length of 720 ft has been demonstrated, and the membranes exhibited consistent transport results as the lab-scale samples. Additionally, a techno-economic analysis showed that, when the H_2 recovery of the membrane process was set at 99%, the CO_2 removal cost was calculated to be $62.9/tonne. Lastly, a CO_2 removal of 76.9% could be obtained when the air sweep was replaced by vacuum on the permeate side.

613190 Bipolar Membranes with a High Interfacial Area 3D Electrospun Junction

Abstract:

Bipolar membranes (BPMs), typically laminated layers of anion-exchange and cation-exchange polymers, have the unique capability of splitting water into H^+ and OH^- at a potential near 0.83 V. Such membranes are used in electrodialysis membrane separation processes, water electrolyzers, CO_2 electrolysis cells, and self-humidifying fuel cells. We report here on recent developments regarding BPMs with a high interfacial area, 3D nanofiber water splitting junction.

A junction layer was made by the simultaneous electrospinning of anion-exchange and cation-exchange polymers, with the addition of catalyst nanoparticles to facilitate water splitting. Solvent vapor exposure and hot-pressing closed all interfiber pores, to create a dense film of interpenetrating and interlocking nanofibers of positively and negatively charged polymers. Dense films of solution cast anion and cation exchange polymers were then hot-pressed onto the junction layer to create a tri-layer BPM. BPMs were made with a junction layer of 12-15 mm and a total membrane thickness of 50-75 mm. Membranes were prepared with sulfonated poly(ether ether ketone) or perfluorosulfonic acid as the cation-exchange polymer and either quaternized poly(phenylene oxide) or AEMION^{ä} (an imidazolium-based polymer sold by Ionomr Innovations, Inc.) as the anion-exchange polymer. $Al(OH)_3$ or graphene oxide particles were used as the junction layer catalyst. Membranes were evaluated in an H-cell (steady-state current-voltage data collection) and for constant current operation in a flow cell with an aqueous Na_2SO_4 electrolyte. The 3D junction membranes performed exceptionally well: (i) the water dissociation potential was very low and (ii) the extended bipolar reaction zone for water splitting allowed for high current density operation (up to 1 A/cm^2) with no evidence of membrane degradation.

In this talk, the effect of membrane composition (the type/amount of anion and cation exchange polymers) and structure (the thickness of the layers) on short-term and long-term membrane performance will be discussed.

610886 Enhanced Selective Adsorption Desulfurization on CO_2 and Steam Treated Activated Carbons: Equilibria and Kinetics

Abstract:

Selective adsorption removal of sulfur (SARS) is one of the most promising alternative strategies to ultra-deep hydrodesulfurization (HDS) for lowering the sulfur content of liquid hydrocarbon fuels and meet stringent environmental regulations. This process employs solid adsorbents and, unlike HDS, it operates at less severe temperatures and pressures, without the costly hydrogen consumption.[1,2] Activated carbons (ACs) show great potential for SARS but require improvements in uptake and selectivity.

The current work examines comprehensively the influence of modifying a commercial-grade activated carbon (AC) by CO_2 and steam treatment for the desulfurization of dibenzothiophene (DBT) and 4,6-dimethyldibenzothiophene (4,6-DMDBT), which are sterically hindered sulfur compounds that are particularly difficult to remove from hydrotreated gas oils by conventional HDS methods. An untreated AC and a charcoal Norit carbon (CN) are used for comparative purposes. The materials are characterised using physicochemical techniques such as N_2 physisorption, HRTEM, SEM, XRD, ATR-FTIR, ICP and TGA analyses. The equilibria and kinetics of the materials are examined using both batch and fixed-bed flow measurements, respectively, at low adsorption temperatures (303 to 353 K) and ambient pressure. Full details of the experimental techniques used are given elsewhere.[3]

The results show that the steam and CO_2 treated ACs exhibited significantly higher sulfur uptakes than the untreated AC and CN samples due to an increased surface area and microporosity. The volume of micropores governs capacities mainly by physical p-p interactions. Surface chemistry effects were found to be relatively small. The steam treated AC appears to be especially effective to remove sulphur showing a remarkable uptake (~24 mg S•$gads^{-1}$), due to its smaller molecular size which favours its diffusion and thus leads to a better developed microporosity. The modified ACs showed similar capacities to both DBT and the sterically hindered 4,6-DMDBT molecules. Compared to other adsorbents, the modified ACs exhibited relatively high adsorption capacities.[1,2,3] The adsorption sites of all the samples assessed are characterized as heterogeneous due to the better fit to the Freundlich model. The kinetic breakthrough profiles are adequately described by the linear driving force (LDF) model. The findings of this work provide very useful information for the design of commercial adsorption units. Research efforts are being devoted to detailed selectivity and stability analyses and to the use of more rigorous adsorption model approaches.

Acknowledgements

This work is supported by the ELEGANCY, EU H2020 project and the Imperial College London Research Fellowship Grant.

References

[1] D. Iruretagoyena, R. Montesano, Selective sulphur removal from liquid fuels using nanostructured adsorbents, in: T.A. Saleh (Ed.), Nanotechnology in oil and gas industries: Principles and applications, Springer International Publishing, Cham, 133-150, 2018.

[2] R. Menzel, D. Iruretagoyena, Y. Wang, S. M. Bawaked, M. Mokhtar, S. A. Al-Thabaiti, S. N. Basahel, M. S. P. Shaffer, Graphene oxide/mixed metal oxide hybrid materials for enhanced adsorption desulfurization of liquid hydrocarbon fuels, Fuel, 531-536 (181), 2016.

[3] D. Iruretagoyena, K. Bikane, N. Sunny, H. Lu, S. G. Kazarian, D. Chadwick, R. Pini, N. Shah, Enhanced selective adsorption desulfurization on CO_2 and steam treated activated carbons: Equilibria and kinetics, Chem. Eng. Journal, 122356 (379), 2020.

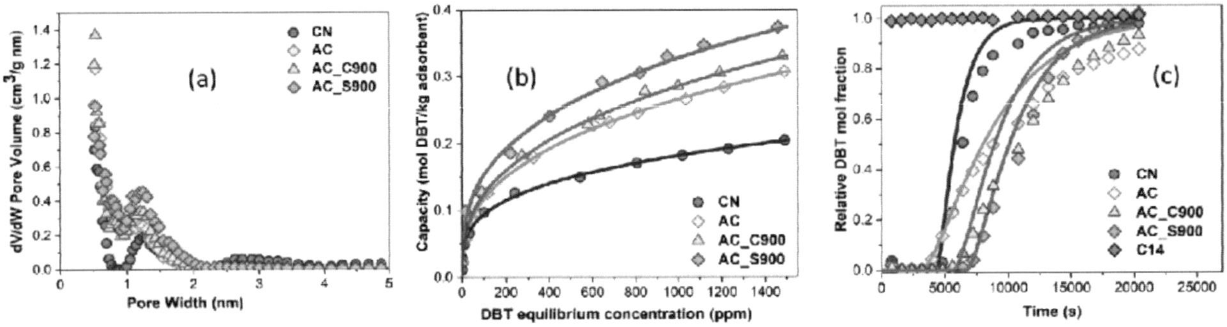

Representative figure. (a) Pore width distribution of CN (Norit carbon), AC (activated carbon), AC_C900 (CO_2 treated AC) and AC_S900 (steam treated AC). (b) Equilibrium isotherms, Freundlich fitting and (c) Breakthrough profiles, LDF fitting. Data shown for DBT (obtained using mixtures of equal concentration of DBT and 4,6-DMDBT).

597038 Self-Supported Branched Poly(ethyleneimine) Scaffolds for CO_2 Capture

Abstract:

Structured contactors in the form of monoliths for CO_2 capture not only combat the high energy consumption and corrosion rate observed in traditionally used aqueous amines, but also offer a low pressure drop in comparison to powder adsorbents during operation.[1] Developments so far in monoliths include attachment of functionalized nanoporous materials such as metal organic frameworks (MOFs), zeolites, carbons or inorganic/organic composites containing polymeric amines.[2] While amine materials have an advantage in terms of better sorption capacities in the presence of humidity, the need for a support that itself does not contribute directly to CO_2 capture remains non-ideal.[3,4] This work therefore focuses on utilizing widely used poly (ethyleneimine) (PEI) to synthesize wholly organic monoliths without any additional support, potentially reducing the sensible heat load in the thermal regeneration cycle. High molecular weight branched poly(ethyleneimine) was crosslinked with poly(ethylene glycol diglycidyl ether) in aqueous media at different cryogenic temperatures using an ice templating method.[5] The reaction temperature was varied to change the ice crystal structure in the templating process and thereby the nature of the pores in the scaffold. The relative ratio of PEI and crosslinker was varied and it was found that scaffolds formed at -196 °C containing 0.1 equivalent of crosslinker relative to the amines in PEI showed an optimal CO_2 adsorption capacity of 5.5 mmol/g after 12 h at 25 °C in a 10% CO_2 feed stream with 65% RH. Functionalization of PEI before crosslinking and its effect on the morphology of the scaffold as well as the CO_2 adsorption capacity has also been studied. It was found that a hydrophobic capping group resulted in dense polymer domains and using water miscible additives provides additional scope for positively affecting the morphology and thereby improving the CO_2 adsorption capacity using more dilute CO_2 feeds.

References:

(1) Rezaei, F.; Webley, P. Optimum Structured Adsorbents for Gas Separation Processes. *Chem. Eng. Sci.* **2009**, *64* (24), 5182–5191.

(2) Murdock, C. R.; Didas, S. A.; Jones, C. W. Direct Capture of CO_2 from Ambient Air. *Chem. Rev.* **2016**, *116*, 11840-11876.

(3) Darunte, L. A.; Terada, Y.; Murdock, C. R.; Walton, K. S.; Sholl, D. S.; Jones, C. W. Monolith-Supported Amine-Functionalized Mg_2(Dobpdc) Adsorbents for CO_2 Capture. *ACS Appl. Mater. Interfaces* **2017**, *9* (20), 17042–17050.

(4) Shi, Y.; Liu, Q.; He, Y. CO_2 Capture using Solid Sorbents. *Handbook of Climate Change Mitigation and Adaptation* **2016**, 1-56.

(5) Yoo, C. J.; Narayanan, P.; Jones, C. W. Self-Supported Branched Poly(Ethyleneimine) Materials for CO_2 Adsorption from Simulated Flue Gas. *J. Mater. Chem. A* **2019**, *7* (33), 19513–19521.

600400 Selective Recovery of Ammonia Nitrogen from Wastewaters with Transition Metal-Loaded Polymeric Cation Exchange Adsorbents

Abstract:

The presence of nitrogen compounds, such as ammonia and ammonium (total ammonia nitrogen, or TAN), in wastewater is concerning due to their contribution to eutrophication, which has a wide range of negative environmental effects. Typically, TAN is removed from wastewater using microbial nitrification-denitrification that converts TAN into nitrogen gas (N_2). This process reverses the Haber-Bosch process, which consumes roughly 1% of global energy to convert N_2 into ammonia. A more sustainable and economical approach would be to separate TAN from wastewater directly and use it for value-added products. Ion exchange is a well-established technology for removing ions from water, but due to low ion selectivity, resulting eluent streams currently require additional costly steps to further separate ions before value-added products can be made.

Therefore, we have modified commercial polymeric cation-exchange adsorbents by simple ion-exchange for selective ammonia removal and recovery. Since metal cations like Cu^{2+} and Zn^{2+} are known to form inner-sphere complexes with ammonia, we hypothesized that metal ligand exchange would drive selectivity towards ammonia in the presence of other common wastewater cations like Na^+, K^+, Mg^{2+}, and Ca^{2+}. However, because some anions (e.g. phosphate) and organics can also form complexes, we utilized the Donnan-Exclusion Effect of cation-exchange resins to repel them.

We loaded each metal into three resins with three different functional groups: acrylic acid (AA), iminodiacetic acid (IDA), and bis-picolylamine (BPLA). The goal was for metal to occupy all functional sites and limit ion exchange with nontarget cations, since heavy metal cations bind strongly to these groups. We then performed batch adsorption experiments at pH 9.5, based on recent reports of pH 9-10 as optimal for metal-ammine complexation. A pH of 9.5 is also similar to the pH of hydrolyzed urine, one of the most concentrated and ubiquitous ammonia-rich waste streams. Finally, we used batch sulfuric acid treatment to regenerate resins and recover ammonia ligands as ammonium sulfate.

Compared to commercial resins, metal-ligand exchange adsorbents exhibited higher ammonia removal capacity (8 meq/g) and recovery selectivity (N/K^+ concentration factor of 5) in binary equimolar solutions of TAN and K^+. During acid regeneration, we found an optimal pH range that caused under 0.7% transition metal elution through H^+ ion-exchange while recovering 60-70% of adsorbed ammonia. However, in urine-level cation solutions, divalent cation exchange increased transition metal elution and reduced ammonia adsorption. Considering optimal ammonia concentrations (200-300 meq/L) and pH (9-10) for metal-ligand exchange, we identified hydrolyzed urine as a promising candidate for selective TAN recovery. Ultimately, metal-ligand exchange adsorbents can advance nitrogen-selective separations from wastewaters.

604138 Sorbent Development for Microwave-Accelerated Regeneration of a Slurry (MARS) Process for Carbon Capture

Abstract:

Solvent regeneration is the most energy- and cost-intensive operation in aqueous amine-based capture processes. It often requires a large amount of superheated steam to raise the bulk temperature typically above 120 °C for reactive regeneration. This limitation leads to slower and inefficient capture/release stages that also hinder large-scale utilization.

In this work, a slurry of basic immobilized amine sorbent (BIAS) and polydimethylsiloxane (PDMS) oil has been developed for a CO_2 adsorption/desorption process aided by the Microwave (MW)-Accelerated Regeneration of a Slurry (MARS). Using non-aqueous silicone oil can minimize water use and also reduce the energy penalty for slurry regeneration because silicone oil has a much lower heat capacity but a higher boiling point than water. Previously, studies indicated that selective MW heating of CO_2-adsorbed BIAS can increase the CO_2 desorption flux by 10 times in comparison with thermal heating and may consume much less energy for CO_2 desorption, even at process temperatures of 65-85°C. In order to enhance the energy efficiency, the sorbent was modified by adding trace amount of MW-favored nano-materials to promote selective absorption of MW energy. Improved CO_2 desorption kinetics will be reported.

599058 Simple Preparation Method of K_2CO_3-Promoted Hydrotalcite-like Compounds for NO_x Adsorption at Low Temperature

Abstract:

Lean-burn engines operate at higher air-to-fuel ratios than stoichiometric ratio to reduce the fuel consumption and CO_2 emissions. Vehicle pollutants including carbon monoxide (CO), hydrocarbons, and nitrogen oxides (NO_x) can be simultaneously removed by three-way catalysts (TWC), however, the catalytic activity for NO_x reduction significantly decreases in the lean-burn condition that the oxygen concentration increases and exhaust gases temperature decreases. Among various NO_x removal systems replacing TWC, lean NO_x trap (LNT) that is based on the adsorption concept is considered a promising technology. In this system, NO is oxidized, and then stored on the adsorbent during lean condition. For successful application of LNT, it is necessary to develop NOx adsorbents with high adsorption performance.

Since the NO_x storage ability is strongly related to the basicity of storage materials, the incorporation of alkali metal is a powerful strategy for improving NO_x adsorption performance. In this study, we prepared K_2CO_3-promoted hydrotalcite-like compounds for NO_x adsorbent in a facile method. While the conventional method includes a step for synthesizing hydrotalcite-like compounds and additional step for introducing K_2CO_3, these two procedures could be achieved simultaneously through hydrothermal treatment. The NO_x adsorbents developed in this study presented excellent performance compared with previous works, and the role of K_2CO_3 on NOx adsorption was investigated based on the X-ray photoelectron spectroscopy (XPS) and in-situ diffuse reflectance infrared Fourier transform spectroscopy (DRIFTS).

606941 Neural Networks-Based Self-Driven Control Strategies for a Pharmaceutical Crystallization Process

Abstract:

In the pharmaceutical manufacturing, crystallization is the most applied downstream process to recover high-purity pharmaceuticals and fine chemicals. However, establishing control over this unit operation is still a major bottleneck particularly in the industry. The main reason is the nonlinear and stochastic dynamics of the crystallization process that challenge the control strategies and actions. In addition to this, in the absence of any online or inline measurements, the nonlinear relationship between the process outcome (e.g. crystal properties) and process inputs is difficult to understand and to establish. Fortunately, the latest advances in Process Analytical Technology (PAT) tools enable to monitor the critical solution- and crystal-state properties in real-time. This eases the understanding of the crystallization dynamics and helps to establish a correlation between the operating inputs and the critical properties of the crystal product. Thus different and advanced feedback control strategies can be applied based on the information obtained from PAT tools in order to establish some degree of control over crystal size, polymorphic form and morphology. Despite the significant progress, the control of industrial crystallization process is still an outstanding issue. One reason behind might be attributed to the fact that the state of the art crystallization control strategies fall into one of two restrictive categories: model-free (direct nucleation control, supersaturation control) and model-based. While model-free methods have the advantage of being simple and widely applied, some efforts are needed for the determination of a robust setpoint or robust sensor calibration that should be also valid in the presence of the disturbances in the system. On the other hand, model-based control strategies suffer from the complexity and the accuracy of the implemented model as well as the speed of solution time in order to achieve a reliable real-time optimization. However, a fast and successful transition between different scales of the crystallization operation in the pharmaceutical industry requires a simple but reliable control strategy. Therefore, differently from two categories, a third category can be attributed to the data-driven (based on process data) control strategies. Data-driven approaches have proven to be useful particularly for the complex processes that are troublesome and costly to develop a knowledge-driven model in a timely manner for the optimization and control of industrial processes. Therefore, in this work we present a data-driven control strategy based on the neural networks (e.g. radial basis functions and deep neural networks) for a pharmaceutical batch cooling crystallization process. The proposed data-driven control utilizes both real-time and offline the experimental data measured with different PAT tools (*e.g.* FBRM, FT-IR, image analysis) to optimize the cooling profile in order to achieve the desired crystal-state property profile throughout the process. A model pharmaceutical compound (e.g. ibuprofen) is selected as a case study and the robustness of the proposed control strategy is shown based on the comprehensive experiments in the presence of several process disturbances (initial supersaturation, impeller speed, the impurity of the solvent and seed size). The proposed neural-networks based control is a promising strategy that is easy to implement, fully-automated, and relies on relatively limited data for training. Hence, it is expected to contribute to quick process development and control, especially in the absence of comprehensive process understanding and historical data in pharmaceutical industries.

Acknowledgment

We would like to thank the Danish Council for Independent Research (DFF) for financing the project with grant ID: DFF-6111600077B.

606184 Modeling of Nucleation, Growth, and Dissolution Kinetics of Paracetamol from Ethanol Solution for Unseeded Batch Cooling Crystallization with Temperature Cycling Strategy

Abstract:

Crystallization is widely used for industrial separations and purifications. Batch crystallization is often operated in a cooling mode, with the temperature gradually lowered to increase supersaturation and induce nucleation and growth. However, a temperature-cycling strategy, having both cooling and heating phases, can drive the system in new directions to achieve larger crystals or altered shape. Population balance models are often generated to describe crystal growth, but few models exist that describe the effects of both cooling and heating. Even less common is the comprehensive modeling of nucleation, growth, dissolution, and disappearance of crystals. However, it is this disappearance of small crystals during heating that enables the generation of larger crystals during subsequent cooling phases.

In this presentation, a one-dimensional population balance model is presented for unseeded cooling crystallization, with temperature-cycling strategies to control the mean crystal size and the crystal size distribution. This model can predict the path of crystallizing paracetamol from ethanol solutions based on the characteristics of primary nucleation, secondary nucleation, growth, dissolution, and disappearance of crystals. To collect solute concentration data, attenuated total reflectance-Fourier transform infrared spectroscopy (ATR-FTIR) was used, and to collect the crystal size distribution data, sieve analysis was carried out. The model was developed from two experimental and parameter-estimation steps, for dissolution and crystallization, to provide independent estimates of the model parameters. To solve the partial differential equation, we applied the conservation element/solution element scheme for the discretization of the spatial domain. The population balance model employed boundary conditions to describe the nucleation and the disappearance of crystals.

Dissolution experiments were carried out in isothermal conditions with different temperatures and different sizes of seed crystals. Unseeded crystallization experiments were performed beginning with a clear solution, with temperature cycling to control crystal size. The dissolution and crystallization models achieved good agreement with the experimentally measured volume density distribution of crystals as well as the concentration and supersaturation trends. These results demonstrate that the model can be used to inform future open- and closed-loop control strategies.

598779 Modeling, Simulation and Design Strategies to Efficiently Apply Preferential Crystallization in a Continuous Process

Abstract:

Preferential crystallization (PC) is a strong technique capable to resolve mixtures of enantiomers that crystallize as a conglomerate forming system [1, 2]. It involves preferentially crystallizing a target enantiomer by adding pure seed crystals into a supersaturated solution at racemic composition. The non-target enantiomer crystallization is inhibited kinetically for a certain period, which is expressed in this study as a stop time (t_{stop}). After this period, crystallization of the counter enantiomer takes place and the purity of the solid phase drops. Due to unavoidable spontaneous nucleation of the non-target enantiomer, the classical approach of a single batch process has limitations. Therefore, it is desirable to delay the counter enantiomer nucleation and to benefit from a longer operation time window to harvest more pure product. One of the methods to achieve this is based on using racemization in the liquid phase.

Population Balance Models (PBMs) are a powerful tool to describe crystallization processes [3]. Due to the consideration of many details, the resulting set of equations is rather difficult to solve and expensive when applied for simulation and optimization. In this work, a shortcut model (SCM) is introduced, which is based on the assumptions of equally sized spherical particles and a single lumped kinetic mechanism [4]. The analysis considers as a case study of the enantiomers of DL-asparagine (DL-Asn) in water crystallizing as a monohydrate under used conditions. An evaluation of Key Performance Indicators (KPI) for the process (i.e. productivity and yield) is performed using the shortcut models. This model is found to be capable to describe main features of batch PC with and without racemization using only three differential equations, which are formulated from the mass balances of the preferred enantiomer in the liquid and solid phases (Figure 1). A systematic theoretical study of design strategies of PC in classical single batch crystallization with and without racemization in one-pot and also in two connected pots is performed using the shortcut model. Finally, a possible extension of the shortcut model to simulate preferential crystallization carried out in continuous mode and for resolving racemic compound forming systems will be discussed in the presentation.

1. Jacques (1994): Enantiomers, Racemates and Resolutions. Krieger Publishing Company Malabar, FL.
2. Coquerel (2007): Preferential Crystallization. Top Curr Chem 269, 1-51
3. Ramakrishna (2000): Population Balances. Theory and Applications to Particulate Systems in Engineering. Academic Press, London
4. T. Carneiro, S. Bhandari, E. Temmel, H. Lorenz, A. Seidel-Morgenstern (2019): Shortcut Model for Describing Isothermal Batch Preferential Crystallization of Conglomerates and Estimating the Productivity. Cryst. Growth Des. 19, 9, 5189-5203

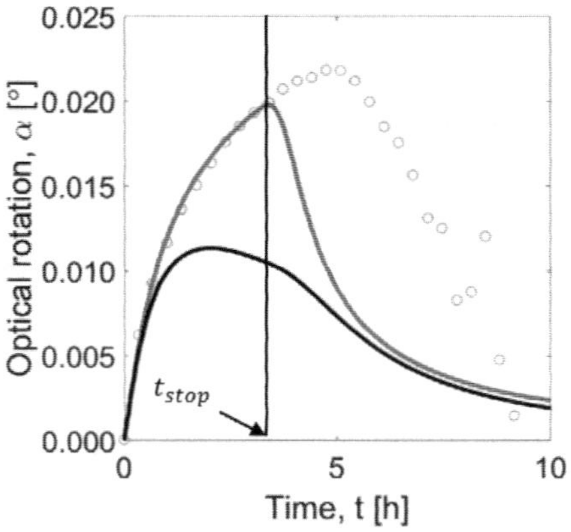

Figure 1: Comparison between SCM simulations and experiments of PC (DL-Asn/water). Red circles: Experimental profile of optical rotation as described in [4]. Black and red lines: predicted optical rotation with and without racemization, respectively

604894 Simultaneous Solvent and Process Optimization of Continuous Reactive Crystallization Processes Involving Electrolytes and Recycling Using the Electrolyte PC-SAFT

Abstract:

Solution crystallization is one of the most commonly used separation techniques in the pharmaceutical industry. Solvents, which are commonly present in crystallizing systems, have a significant impact on the attainable crystal yield, product quality, waste generation, energy duty, and safety of the process. The large number of different solvent types and their mixtures that are routinely used in the pharmaceutical industry makes solvent selection a challenging task, particularly when an optimal trade-off among different process-related performance criteria (i.e., product quality, yield, waste generation) needs to be made. Traditional solvent selection approaches based on heuristics are generally not sufficient for optimization due to the strongly interdependent nature of these criteria. As a result, the development of computer-aided solvent selection approaches for pharmaceutical crystallization processes has attracted substantial research interest. Early work mainly focused on solvent design for stand-alone crystallization [1]. However, solvent selection approaches for integrated operations (i.e., crystallization embedded within a process including downstream separations) are important when minimizing energy duty and waste generation. In particular, solvent recycling can be useful for reducing waste. Furthermore, the current trend towards continuous manufacturing in the pharmaceutical industry has made recycling more convenient [2]. We have developed an optimization-based thermodynamic approach for simultaneous solvent and process design for anti-solvent crystallization integrated with solvent recycling using the statistical associating fluid theory equation of state (PC-SAFT EoS) [3] in our earlier work [4]–[6].

In addition to downstream solvent separation and recycling, upstream reactions for the synthesis of active pharmaceutical ingredients (APIs) may also need to be considered in an integrated approach, which has received less attention. Furthermore, crystallization of APIs can sometimes be combined with a reaction in the same environment to achieve process intensification. However, thermodynamic modeling of such reactive crystallization can be complicated due to the presence of a reaction equilibrium, various types of chemical species and their intermolecular interactions, and the potential to form multiple solid phases (some of which are undesirable). Thus, the development of novel optimization-based solvent selection and design approaches for processes involving reactive crystallization and downstream separation and recycling is needed. Recycling is likely more important for reactive crystallization processes compared to anti-solvent crystallization processes, as reactants need to be considered for recycling in addition to solvents. Our recent work on an integrated approach for a reactive crystallization process of an API with solvent recycling [7] using the PC-SAFT EoS showed the high potential of such an approach. However, API synthesis reactions likely involve electrolytes, and they have an important impact on the phase and reaction equilibria in the process and, therefore, may affect the optimal solvent choice and process conditions, which has not been addressed yet.

The objective of this work is to develop an optimization-based simultaneous solvent and process design framework suitable for the continuous reactive crystallization processes involving electrolytes and downstream separation and recycling of reactants and solvents. The process configuration consists of a reactor/crystallizer, where the reaction and the product crystal formation occur simultaneously, a filter, and a flash separator, where the mother liquor is concentrated and partially recycled back to the crystallizer. The solvent type to be identified acts both as the reaction and the crystallization medium, thus affecting the extent of the reaction and the final crystal yield as well as the ease of recycling. Furthermore, the optimal choice of solvent type is influenced by the process operating conditions, and those two sub-problems (i.e., solvent selection and optimizing operating conditions) cannot be decoupled effectively [8]. Therefore, mathematical models and effective optimization approaches are needed for optimizing both of these sub-problems simultaneously. The development of a process model to relate process operating conditions with solvent choice is a crucial step. In the case of reactive crystallization, the reaction/crystallization medium is usually a complex mixture consisting of multiple solutes with the potential to form different solid phases. It is also well possible that some of those solutes are electrolytes, which calls for a thermodynamic model that can reliably capture all the different types of interactions in the medium. In this work, the PC-SAFT EoS [3] is used as the basis of a unified thermodynamic model to predict the reaction, solid-liquid and vapor-liquid equilibria in the process. The PC-SAFT EoS has been successfully applied in modeling the phase equilibria of a large number of pure substances and mixtures [3], [9], as well as complex mixtures consisting of API molecules commonly encountered in pharmaceutical processes [10]. Furthermore, the electrolyte PC-SAFT (ePC-SAFT) EoS can model electrolyte systems [11]. Therefore, we have embedded the ePC-SAFT EoS in the process model and optimization framework to capture the impact of electrolytes. The remainder of the process model is constructed from conservation balances and additional constraints to ensure the practical and physical feasibility of the process.

The resulting mathematical model is a mixed-integer nonlinear programming (MINLP) problem, which is computationally prohibitive for a model of a complicated continuous pharmaceutical process involving multiple unit operations, reaction and phase equilibria, different types of species, and recycling. Therefore, a continuous mapping method [8] is adopted to convert the original MINLP problem into a nonlinear programming (NLP) problem to make the original problem computationally tractable. The approach is successfully demonstrated for a case study involving reactive crystallization of dalfampridine. Electrolytes have a substantial impact on the equilibria of this system. An economic objective function involving the minimization of the total production cost of the API is employed, where the solvent type and process operating conditions are free variables. A solvent is represented by six pure component PC-SAFT parameters. These parameters are treated as continuous variables, and the resulting optimal values are mapped onto a solvent database from our previous work [4] to identify a real solvent. The reactor/crystallizer medium dielectric constant, which depends on the solvent type and the operating conditions, is an important property for the fugacity calculations of electrolytes [11]. For the optimization framework, the medium dielectric constant needs to be determined in a predictive manner. Therefore, a correlation is established to predict the dielectric constant of the medium using the pure component PC-SAFT parameters and process operating conditions.

The case study demonstrates the sensitivity of the optimal solution with respect to the trade-offs for economic performance, such as the extent of the reaction, crystal yield, recycle flow rates, and energy duty. Furthermore, the case demonstrates the importance of taking all those factors into account simultaneously. The proposed approach reveals a more competitive process design compared to a conventional dalfampridine synthesis case from the literature [12]. The inclusion of electrolytes in the thermodynamic model enables more realistic values for the equilibrium constants and the prediction of possible undesired salt precipitation. The latter can be avoided by imposing constraints.

This work was financially supported by the Research Grant Council of Hong Kong under Grant No. 16214418.

References

[1] A. T. Karunanithi, L. E. K. Achenie, and R. Gani, "A computer-aided molecular design framework for crystallization solvent design," *Chem. Eng. Sci.*, vol. 61, no. 4, pp. 1247–1260, Feb. 2006.

[2] S. Mascia *et al.*, "End-to-End Continuous Manufacturing of Pharmaceuticals: Integrated Synthesis, Purification, and Final Dosage Formation," *Angew. Chemie Int. Ed.*, vol. 52, no. 47, pp. 12359–12363, Nov. 2013.

[3] J. Gross and G. Sadowski, "Perturbed-Chain SAFT: An Equation of State Based on a Perturbation Theory for Chain Molecules," *Ind. Eng. Chem. Res.*, vol. 40, no. 4, pp. 1244–1260, 2001.

[4] J. Wang and R. Lakerveld, "Integrated solvent and process design for continuous crystallization and solvent recycling using PC-SAFT," *AIChE J.*, vol. 64, no. 4, pp. 1205–1216, Apr. 2018.

[5] J. Wang and R. Lakerveld, "Integrated Solvent and Process Optimization Using PC-SAFT for Continuous Crystallization with Energy-intensive Solvent Separation for Recycling," *Comput. Aided Chem. Eng.*, vol. 44, pp. 1051–1056, Jan. 2018.

[6] J. Wang, L. Zhu, and R. Lakerveld, "A Hybrid Framework for Simultaneous Process and Solvent Optimization of Continuous Anti-Solvent Crystallization with Distillation for Solvent Recycling," *Processes*, vol. 8, no. 1, p. 63, Jan. 2020.

[7] N. P. Mendis, J. Wang, and R. Lakerveld, "A Thermodynamic Approach for Simultaneous Solvent and Process Design of Continuous Reactive Crystallization with Recycling," in *30th European Symposium on Computer Aided Chemical Engineering*, 2020 (Accepted).

[8] A. Bardow, K. Steur, and J. Gross, "Continuous-Molecular Targeting for Integrated Solvent and Process Design," *Ind. Eng. Chem. Res.*, vol. 49, no. 6, pp. 2834–2840, Mar. 2010.

[9] M. Kleiner and G. Sadowski, "Modeling of Polar Systems Using PCP-SAFT: An Approach to Account for Induced-Association Interactions," *J. Phys. Chem.*, vol. 111, no. 43, pp. 15544–15553, 2007.

[10] F. Ruether and G. Sadowski, "Modeling the Solubility of Pharmaceuticals in Pure Solvents and Solvent Mixtures for Drug Process Design," *J. Pharm. Sci.*, vol. 98, no. 11, pp. 4205–4215, Nov. 2009.

[11] C. Held, T. Reschke, S. Mohammad, A. Luza, and G. Sadowski, "ePC-SAFT revised," *Chem. Eng. Res. Des.*, vol. 92, no. 12, pp. 2884–2897, Dec. 2014.

[12] D. S. Rane and M. M. Sharma, "New strategies for the Hofmann reaction," *J. Chem. Technol. Biotechnol.*, vol. 59, no. 3, pp. 271–277, Mar. 1994.

600035 Comparison of a One Dimensional and Two Dimensional Population Balance Models for the Crystallization Process of a Needle-Shaped API

Abstract:

Crystallization is the most widely used method for separating an active pharmaceutical ingredient (API) from a reaction product mixture. It functions to control purity, crystal form, particle morphology, and particle size distribution (PSD). The application of population balance modeling (PBM) to crystallization process development has been growing as a result of improvements in process analytical tools (PAT) and numerical software packages. One such improvement has been the ability to account for high aspect ratios particle morphologies, such needles or plates, by treating the major and minor particle axis as independent, and assigning them independent growth and nucleation kinetics. Many APIs crystallize with needle morphology, and modeling these particles as spheres, as is the case with one-dimensional PBM, fails to accurately predict PSD. The advantage of two-dimensional population balance modeling for a needle-shaped API is highlighted here by comparing the one-dimensional model that has been developed for an antisolvent crystallization with the two-dimensional model. While the one dimensional model is useful in optimizing the crystallization process to enhance desaturation kinetics, the two-dimensional model can be utilized to optimize process conditions for minimizing particle aspect ratio. The API utilized for this work had extremely slow desupersaturation, and even after aging the seed bed for more than twenty four hours, the solubility concentration was not achieved. The 1D PBM revealed that the crystallization had a high energy of activation, indicating that the growth kinetics were rate limited by the surface attachment of API molecules, and not diffusion limited. As such, agitation rate was not predicted to effect the desupersaturation, and this was corroborated by experimental work. The model directed the lab scale crystallization development to alter the composition of the mother liquor at the seed point, increasing the relative supersaturation at seeding, which indeed resulted in faster desupersaturation kinetics. However, the 1D model was not able to inform on process conditions for generating lower aspect ratio particles. Given that needles are a challenging morphology for many downstream processing operations, reducing the particle aspect ratio can be beneficial for enhancing proccessability. This capability requires two dimensional PBM.

607574 Morphological Population Balance Modeling of a Pharmaceutical Compound for Size and Shape Prediction for in-Silico Experimental Design

Abstract:

Crystallization is a particle formation and purification techniques in all chemical industries, having particularly important role in the pharmaceutical industry. After the crystallization step, numerous downstream processes are performed such as filtration, milling or granulation to obtain the desired product that meets the critical quality attributes (CQAs). The size and shape distribution of crystals are important CQAs, which may significantly affect the bioavailability, bulk density and manufacturability of the final drug product.[1] The prediction of the process behavior plays key role in the design of complex, nonlinear processes with tight CQA constraints. A predictive process model can be used for *in silico* design of experiments (DoE), optimizing design tasks, as well as in model-based control, which inherently decreases the time and effort put into laboratory work as well as reduces the active pharmaceutical ingredient (API) requirements to find the operating strategy required for the manufacturing of the desired product.

In this work, we present the development and validation of a two-dimensional morphological population balance model (2D-PBM) for the batch cooling crystallization of a commercial API (compound A), from Takeda Pharmaceuticals International Co., which forms needle-like crystals. It was observed that compound A shows significant growth rate dependency in the function of actual thermodynamic conditions in both width and length directions, leading to considerable dynamic changes in the aspect ratio distribution of the crystals, which cannot be accurately captured with a single characteristic dimension used in 1D-PBM. Therefore a morphological population balance model that described the dynamic variations of both the length and width distribution of particles was developed using a two-dimensional (2D-PBM) that involves secondary nucleation, growth and dissolution of particles. The model-equations are solved with a graphical processing unit (GPU) accelerated high resolution finite volume method.[2] The kinetic parameters are estimated based on batch crystallization experiments, by defining and solving a process optimization problem that minimizes the deviations between the measured concentration, as well as distributions of particle length, width and sphere-equivalent mean diameter. The calibrated process model was able to reproduce the crystallization behavior of compound A, both in terms of the dynamics of crystal size and shape distributions as well as the variation of concentration. For the identification of the model parameters, several experiments were performed with temperature cycling to promote the growth of the crystals for kinetic studies by using process analytical technology tools (PAT) for data acquisition and monitoring the system, including ParticleTrack G400 (Mettler-Toledo) for in-line chord length distribution (CLD), for detecting the crystallization events like nucleation and dissolution, as well as for the crystal count measurement, the Zeiss MCS621 ATR-UV/Vis spectrophotometer for concentration measurement, and the ParicleView v19 (Mettler Toledo) for in-situ microscope images. First, a model was developed using size independent growth kinetics, but experimental evidence and parameter estimation studies showed that both length and width exhibit apparent size dependent growth mechanisms. A new semi-empirical size dependent growth formulation was developed, which covers both the exponential growth behavior of the small crystals (< 10 μm) and the slower growing trend of the larger crystals. The new size-dependent growth (SDG) expression was compared to two common size-dependent growth expression from literature, by performing parameter identification and model validation for each of the SDG models. Sum of square error results from the validation and training experiments showed that the new SDG rate model proposed provides a better description of the growth behavior for this system and provides a more accurate predictive model for this system without over fitting the parameters. In addition, a novel parameter estimation formulation was proposed that incorporates the measured crystal count term in the objective function by correlating the crystal density to the crystal count from focused bean reflectance measurement (FBRM), as well as including additional mechanistic constraints to restrict growth rates within physical meaningful values during the parameter estimation. It was shown that the novel parameter estimation formulation enhances the convergence of the optimization process and provides kinetic parameters with smaller uncertainty bounds, hence a more reliable predictive model. The developed model with the highly efficient numerical PBM solution approach developed, can be used for in silico design of experiments, crystallization process design and optimization, and implementation of nonlinear model predictive control.[4]

References:

[1] Bag, P. P., Chen, M., Sun, C. C., & Reddy, C. M. (2012). Direct correlation among crystal structure, mechanical behaviour and tabletability in a trimorphic molecular compound. CrystEngComm, 14(11), 3865–3867.

[2] Szilágyi, B., & Nagy, Z. K. (2018). Aspect Ratio Distribution and Chord Length Distribution Driven Modeling of Crystallization of Two-Dimensional Crystals for Real-Time Model-Based Applications. Crystal Growth and Design, 18(9), 5311–5321.

[3] Mydlarz, J., & Jones, A. G. (1993). On the estimation of size-dependent crystal growth rate functions in MSMPR crystallizers. The Chemical Engineering Journal, 53, 125–135.

[4] Szilagyi, B., & Nagy, Z. K. (2019). Real-time feasible model-based crystal size and shape control of crystallization processes. Computer Aided Chemical Engineering (Vol. 46). Elsevier Masson SAS.

596673 Determining Recipes for Seeded Batch Crystallization for Many Chemical Systems with Optimal Control Theory and a Dimensionless Framework

Abstract:

An adapted version of a dimensionless framework for batch crystallization processes (Ward et al., 2011) is developed to solve optimization problems for 32 chemical systems (Garside and Shah, 1980; Tseng and Ward, 2014) with different crystal growth and nucleation kinetic parameters. The optimization variable is the supersaturation trajectory during the batch while nucleated volume and nucleated number are selected as objective functions. Constraints are placed on the seed-grown volume, maximum admissible growth rate, and batch time. Optimization problems formulated with differential equations are solved nearly-analytically by application of a coordinate transformation and optimal control theory (Vollmer and Raisch, 2006; Hofmann and Raisch, 2010; Tseng and Ward, 2019).

The results suggest that seed properties have a greater effect on the nucleated mass and number of nuclei than the supersaturation trajectory. However, since product mean size depends strongly on seed mean size, there is a significant trade-off between objectives concerning nucleation and product mean size. The trade-off between nucleated number and nucleated volume is strong especially when nucleation rate is much more sensitive to supersaturation than growth rate as shown by plotting Pareto-optimal fronts. Objective values under constant growth trajectory are also evaluated since this control strategy can be implemented without prior knowledge of the kinetic parameters (Nagy, 2017). Results show that the difference in objective values is small when the nucleation rate is proportional to magma density or sensitivity of nucletion to crystal growth is small.

If the seed mean size is 50 microns and batch time is 1 hour, 22 of the 32 systems require 10% or lower seed loading to effectively inhibit nucleation (so that the nucleated mass is 1% of the seed-grown mass) after application of a constant growth trajectory. However, when the seed mean size is 200 microns, nucleation be suppressed with a seed loading less than 10 % in only four cases. Furthermore, increasing the batch time does not provide a significant benefit for most systems.

Though the kinetic models considered are relatively simple (size- and temperature-dependence of kinetic parameters as well as breakage and aggregation are neglected), this work provides understanding for optimal seeding and growth rate control policy for many chemical systems which can serve as a basis for understanding general trends and outliers in the synthesis of batch crystallization recipes.

References

Garside, J.; Shah, M. B., Crystallization kinetics from MSMPR crystallizers. *Industrial & Engineering Chemistry Process Design and Development* **1980,** *19* (4), 509-514.

Ward, J. D.; Yu, C. C.; Doherty, M. F., A new framework and a simpler method for the development of batch crystallization recipes. *AIChE journal* **2011,** *57* (3), 606-617.

Randolph, A. D., *Theory of particulate processes: analysis and techniques of continuous crystallization*. Academic Press: New York, 1971.

Vollmer, U.; Raisch, J., Control of batch crystallization—A system inversion approach. *Chemical Engineering and Processing: Process Intensification* **2006,** *45* (10), 874-885.

Hofmann, S.; Raisch, J. In *Application of optimal control theory to a batch crystallizer using orbital flatness*, 16th Nordic Process Control Workshop, Lund, Sweden, 2010; pp 25-27.

Tseng, Y.-T.; Pan, H.-J.; Ward, J. D., Pareto-optimal fronts for simple crystallization systems using Pontryagin's minimum principle. *Industrial & Engineering Chemistry Research* **2019,** *58* (31), 14239-14251.

Nagy, Z. K., Crystallization Control Approaches and Models. In *Engineering Crystallography: From Molecule to Crystal to Functional Form*, Springer: 2017; pp 289-300.

Tseng, Y. T.; Ward, J. D., Critical seed loading from nucleation kinetics. *AIChE Journal* **2014,** *60* (5), 1645-1653.

605997 Improvement of Drug Substance Bulk Powder Properties through Process Crystallization

Abstract:

The existing recrystallization process for a drug substance produced crystalline needle-shaped aggregates with low bulk and tap density and very poor flow properties. The particle properties led to poor performance of the drug substance during formulation. The goal of this work was to develop a process to yield drug substance with improved bulk powder properties for future formulation development and manufacturing work.

The approach taken for this work involved the use of computational and experimental solvent screens to rapidly narrow down the choice of solvent systems and working conditions. The computational work also had the advantage of significantly shortening development timelines.

Ultimately a crystallization process was developed which produced drug substance with enhanced particle morphology and significantly increased bulk density.

611726 Singly-Twinned Growth of Si Crystals upon Chemical Modification

Abstract:

During crystallization a continuum of patterns could emerge due to the interplay of growth kinetics, material or solution chemistry, and crystallographic defects. The coherent twin boundary is widely known to catalyze growth in pristine crystals including polycrystalline Si. Much remains unknown about the impact of changing the chemical environment of the crystallization process through the deliberate addition of trace metallic species – termed chemical modification. Pristine Si has been reported to grow through the classical model of two parallel twin planes acting in concert to enable steady-state propagation of the solid-liquid interfaces. Here, we achieve a new vision on the growth process *via in situ* synchrotron X-ray microtomography and further corroborated by *ex situ* crystallographic investigation. We find that steady-state growth is impossible in chemically-modified alloys that consist of trace (0.1 wt.%) Sr. This is because the Sr modifier poisons the concave re-entrant grooves, thereby de-activating the advantage of the twin-plane re-entrant edge mechanism and leading to a singly-twinned interface. This study may serve as a proxy to chemically-modified crystallization pathways of eutectic Si in Al-Si alloys and, more broadly, as a framework for the crystallization-mediated synthesis of materials.

597305 Elucidating the Influence of Hydroxycitrate on Brushite Crystallization Via Bulk and Interfacial Studies

Abstract:

Kidney stone diseases (nephrolithiasis) affect 12% of the world population and can occur on all ages, sexes and races. Of all the cases, calcium stones are the most prevalent and 20% of these are calcium phosphates (CaP). Brushites, or dicalcium phosphate dihydrate, are one of the types of CaP stones and known to induce calcium oxalate mineralization, and to be converted to less soluble form hydroxyapatite. Kidney stone incidence has increased about three times for the past decades. Efforts in identifying alternative inhibitors of stone growth remain one primary strategy in preventing the increasing occurrence of kidney stones. Once identified and obtained clinical clearance, this may provide patients more choices for therapeutics against pathological calcification. Fundamentally, carboxylates are known to inhibit crystal growth of these minerals and one example is hydroxycitrate (HCA) with have effectively curbed calcium oxalate mineralization. To tackle this goal, we employed a combination of bulk and atomic force microscopy studies to elucidate the impact of HCA on the crystal size, morphology, number density of CaP. Analysis shows a 50 % reduction in the number of crystals formed and an increase in the aspect ratio at a very small inhibitor concentration (Ca/inhibitor=25000). Amplitude and Frequency Modulated AFM studies show that surface roughness increased through time in the presence of the HCA. Changes in the periodic features at near atomic resolution was also observed in the presence of HCA implying possible modifier-crystal surface interactions. Results obtained in this study may serve as a step in understanding the inhibitors' effect on CaP crystals and in designing preventative drugs for kidney stone disease.

602826 Islatravir Penultimate Step Crystallization: Purification of an API in the Presence of High Protein Loading from an Enzymatic Cascade

Abstract:

Islatravir is a novel nucleoside active pharmaceutical ingredient (API) currently in clinical trials for both HIV treatment and prevention. As reported recently in *Science (*Huffman et. al. 2019*)*, the API synthesis under development involves a 9-enzyme 3-step in-vitro cascade without intermediate isolations until the final API molecule is assembled. This process represents an incredibly efficient synthesis of the API molecule, but introduces significant challenges to the downstream purification of the product. Enzymes, host-cell proteins, bacterial endotoxins, residual nucleic acids, and organic impurities/related substances all must be removed to generate a drug substance of adequate purity for human dosing.

The primary challenge in the purification development has been on enzyme/protein removal. The presence of high levels of protein tends to make filtrations very slow, so extractive work-ups are often preferable, but the relatively low solubility of the API in most organic solvents was a formidable obstacle to this approach. Several extractive work-ups were developed but each posed stability concerns.

Additionally, enzyme immobilization was pursued for the enzymes with the highest loadings in the reaction, which improved downstream processing but added a significant cost and supply-chain complexity.

In the final reaction, the API crystallizes from the reaction mixture, and even wth several enzymes removed via immobilization, filtration was still very challenging. Filtration was optimized by adding filter aid (diatomaceous earth), reducing shear in the reaction, adding seed crystals to the reaction, and agitating the batch intermittently during filtration. Optimizing the filtration step to isolate the crude API made it possible to produce quality API without the need to develop an extractive work-up or use immobilized enzymes.

601168 Development of a Crystallization Process for the Final Step of the Biocatalytic Synthesis of Islatravir

Abstract:

Merck recently reported a biocatalytic cascade for the manufacture of Islatravir, a potent HIV reverse transcriptase translocation inhibitor [1]. The API step of the synthesis requires removing protein and other impurities accumulated throughout the synthesis. Also, the low dose of the drug requires API particle size control in the final step to mitigate content uniformity concerns in the tablet. This industrial case study will provide an overview of the development of a pure step that meets purity and particle size targets. The presentation will highlight the development of a novel crystallization process to meet the desired particle size distribution while maintaining a rapid filtration rate of the product crystals. The scale-up crystallization performance and impurity rejection of the process will also be compared to a more 'traditional' seeded crystallization using jet milled seed. The development of the process was motivated by the desire to eliminate the cost and industrial hygiene liabilities associated with dry milling API as seed.

[1] Huffman et al. "Design of an in vitro biocatalytic cascade for the manufacture of islatravir." Science, 2019, 366 (6470), 1255-1259.

598747 Impact of the Process Conditions of an Evaporation to Dryness Operation Performed in a Conical Screw Vacuum Dryer on the Isolated Product Psd: Lab- and Pilot-Scale Studies

Abstract:

Conical Screw Vacuum Driers (CSVD) are frequently used in the pharmaceutical industry for drying heat-sensitive powders to a very low volatile matter content in a low shear mixing environment. This type of dryer can also be used to perform evaporation to dryness operations. Evaporation to dryness operations, routinely executed in laboratory preparations but less frequently at industrial scale, aim solely to recover solids from solutions by complete evaporation of the solvent. A manufacturing process industrialized by UCB-Pharma includes an evaporation to dryness operation executed in a CSVD. It follows a multi-column chromatographic separation, yielding the purity level required and it comes before a re-slurry of the powder in a suitable solvent. After solid-liquid separation and drying, the powder isolated from the re-slurry operation is milled to meet the API granulometry specifications of a mono-modal distribution, characterized by a d(v, 0.5) and a d(v, 0.9) within defined target ranges.

Following a process optimization exercise to increase productivity of the chromatographic unit, the solvents used were changed. This also resulted in a significant change in the typical Particle Size Distribution (PSD) observed after the evaporation to dryness operation. Risks associated to not knowing to what extent the final dry milling operation was able to mitigate those unpredicted changes in the control of PSD, triggered this comprehensive investigations on the CSVD process at different laboratory and manufactory scales. Given the change of performance driven by the solvent changes, an additional aim of this work was also to identify process conditions for the CSVD that might have been able to yield directly material within specifications.

A Design of Experiment (DoE) approach was used to determine the impact of the process conditions and the scale-up of the evaporation to dryness operation investigated here on the PSD of the isolated powder. A first DoE plan was carried out using a lab-scale CSVD (300 g scale) under well controlled process conditions. A second DoE was also reproduced at pilot plant-scale (30 kg scale) using appropriate scale-up criteria for those process parameters identified as critical, like the evaporation rate. The heat transfer characterisation of the equipment at laboratory and pilot plant scale allowed for establishing suitable procedures to control critical scale-up process parameters like jacket temperature and operating pressure.

The complexity of the mass and heat transfer phenomena occurring during the evaporation to dryness were elucidated and rationalised in a multiphase process steps while executing the experimental plan (Figure 1). Each process step was clearly defined by specific rheological behaviours. Evidence of patterns and reproducible mechanistic behaviours were found to be linked to specific operating conditions independently from the scale used for the operation. However, macroscale phenomena linked to the batch size of the operation together with exacerbation of the mixing conditions at scale led to mixing zone segregation. These differences ultimately significantly impacted the reproducibility of the performance of the operation under the same operating conditions at scale.

The complexity of the phenomena and phases involved was rationalised in a model behaviour and optimal process operating conditions and procedures to maximise process robustness. The operation was successfully validated at final production scale (300kg) with material produced according specifications over a number of validation batches.

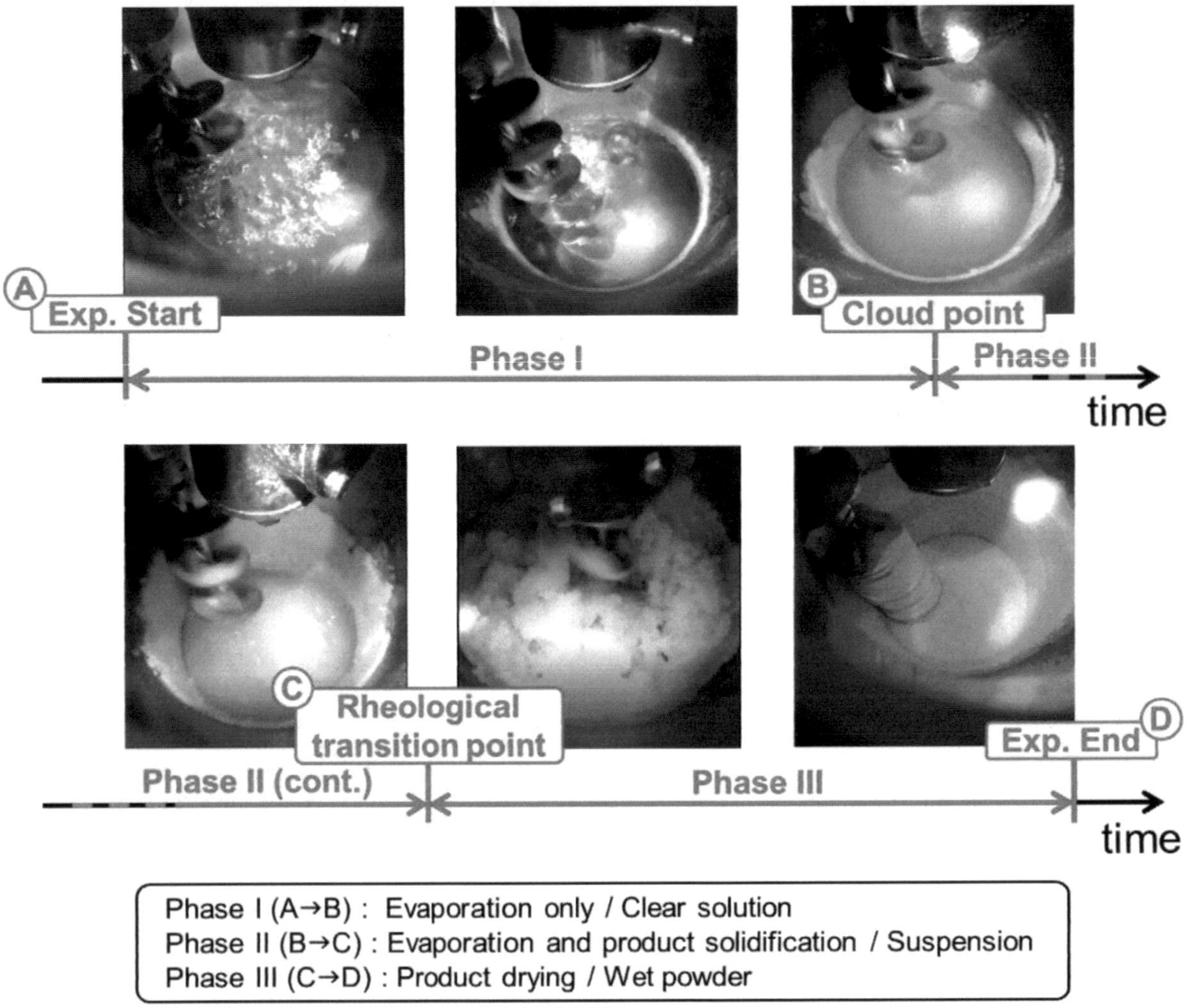

Figure 1 : Observed process phases during the evaporation to dryness experiments

610946 A Generalized Approach for Rapid Aqueous MOF Synthesis By Controlling Solution pH

Abstract:

Metal organic frameworks (MOFs) are a promising class of crystalline materials with potential to significantly advance the fields of catalysis, separations, biotechnology, and sensing, amongst others. Comprised of metal ion nodes and organic linkers, lab-based syntheses have excelled at producing novel MOFs with tuned pore size and chemical activity. However, developing a crystallization technique that meets high throughput scaling requirements and is environmentally benign remains a challenge. Currently, applying classical crystallization techniques to MOF formation are limited due to the multi-component nature. In this work, we approach understanding MOF formation through a reaction-based perspective to better understand the non-classical crystallization pathway. We demonstrate that tuning the reactant speciation, namely labile metal ion nodes and deprotonated organic acid linkers, via pH has a significant effect on MOF yield. Using this approach, we develop a generalizable technique to rapidly produce a range of MOFs, including UiO-66-NH_2, UiO-66, HKUST-1, and ZIF-L, under ambient, aqueous conditions. We further this understanding by tuning the labile metal node and deprotonated linker concentrations to produce space-time yields (STY) exceeding 2250 kg m^{-3} day^{-1}, which is among the highest reported STYs for HKUST-1 and ZIF-L and an order of magnitude higher for aqueous zirconium-based MOF synthesis. Finally, we demonstrate UiO-66-NH_2 completes crystallization in 5 minutes at room temperature with 70% of crystallization of UiO-66-NH_2 occurs within 30 seconds, which is among the fastest reported reaction times in literature. With this work, we demonstrate a generalizable pathway for scaling MOF crystallization to high STY under mild and benign synthesis conditions and lay ground work for understanding the complex multi-step crystallization kinetics of MOFs.

618106 Sponsored Technology Workshop - Hydrogenation in Pressure Reactors and Automation in Kilo Labs - Buchiglas

Abstract:

Hydrogenation in pressure reactors - Turnkey solution for reliable gas dosing & Automation in kilo labs - Solutions to master your challenges

614731 On Processes Delivering a Pure Enantiomer from a Racemic Feed

Abstract:

There are several processes delivering the target enantiomer of a chiral compound from a racemic feed; they have different prerequisites: (i) the availability of a Chiral Stationary Phase for chiral chromatography based processes; (ii) the conglomerate-type phase behavior in the case of a class of interesting crystallization based techniques.

Simulated Moving Bed chromatography can achieve 100% recovery only by recycling the undesired enantiomer through racemization, and we provide a comparative assessment of the two cases, i.e. SMB with and without racemization.

Preferential crystallization can be operated continuosly in an MSMPR (Mixed Suspension Mixed Product Removal) crystallizer, and we provide design criteria for this.

For chiral compounds prone to nucleation of the undesired enantiomer batchwise deracemization via temperature cycles is a simple and effective alternative to continuous crystallization.

Racemization is key to achieve 100% recovery in all these cases, and the rate of racemization is crucial in determining the process performance.

Comparative assessment across technologies is very challenging and case-specific, but the approach that we present here is a valuable methodology for this purpose.

614727 Analytical and Preparative Gradient Chromatography

Abstract:

Besides the widespread application of chromatography as a powerful and flexible analytical technique, there is a large interest in the pharmaceutical industry and in biotechnology to isolate and purify value added products using preparative liquid chromatography. This led in the last years to remarkable progress in the areas of developing and commercializing highly selective and efficient stationary phases, understanding and quantifying the consequences of overloading chromatographic columns and exploiting more efficiently optimized driving forces for the separation using advanced operating regimes [1-3].

Focus of this lecture will be the application of forced modulations of certain process parameters during the process. In contrast to classical isocratic elution such gradient operation offers the potential to improve certain performance criteria. However, the design and optimization of gradient chromatography requires knowledge regarding numerous problem specific parameters.

In the main part of the lecture we will report recent progress regarding the derivation of instructive and simple to apply expressions of the moments of elution profiles under gradient conditions [4]. These expressions provide a useful tool for parameter estimation.

[1] Guiochon G., Felinger A., Shirazi D.G., Katti A.M.,

Fundamentals of Preparative and Nonlinear Chromatography, 2nd Edition,

Academic Press, New York, 2006

[2] Carta G., Jungbauer A.;

Protein Chromatography: Process Development and Scale-Up, 2nd Edition,

Wiley-VCH, 2020

[3] Schmidt-Traub H., Schulte M., Seidel-Morgenstern A. (Eds.),

Preparative Chromatography, 3rd Edition,

Wiley-VCH, Weinheim, 2020

[4] Qamar, S., Rehman, N., Carta, G., Seidel-Morgenstern, A.,

Analysis of gradient elution chromatography using the transport model,

Chemical Engineering Science, 2020, 225,115809

614730 Determination of Binary Adsorption Isotherms of Gases Using Concentration Pulse Chromatography

Abstract:

Mixture adsorption isotherms of gases need to be determined for their use in designing separation systems. Some techniques used to experimentally determine these isotherms include the gravimetric, the volumetric, the breakthrough and the concentration pulse chromatographic techniques. Although it has not been used very frequently in the literature, the latter is an experimental technique that is easy to set up, to use and analyze.

In this presentation, the concentration pulse chromatographic technique will be reviewed for determining binary adsorption isotherms of gases. Some examples from the literature will be given and its versatility to be used for ideal, as well as non-ideal gas mixtures will be demonstrated. Binary adsorption isotherms determined by this method will be compared to the ones determined by other methods and the corresponding observed agreements will be shown.

607856 On the Use of the Absolute, Excess and Net Amounts Adsorbed in Dynamic Mass Balances

Abstract:

The definition of the amount adsorbed dates back over a century with the vision of W. J. Gibbs and his dividing surface (i.e., partition or interface) between the gas and adsorbed phases. Since that time, numerous articles have devoted significant effort into the interpretation and definitions of the absolute, excess and net amounts adsorbed to describe the equilibrium relationship, and more importantly the partitioning or demarcation, between the gas and adsorbed phases. The excess amount adsorbed has prevailed in the literature as the only true experimental quantity because it is the only one that can be measured unambiguously and universally within all adsorption laboratories. What is interesting, however, is that most dynamic mass balances utilize an absolute amount adsorbed formulation because the volume within the adsorbent accessible to helium is generally not accounted for, even though the isotherms are based on this helium volume, i.e., the excess amount adsorbed. This presentation will sort through some of these issues by addressing one or more of the following: 1) clearly defining these three definitions of the amount adsorbed and discussing their subtle differences (as done by others), 2) showing how to use each definition in a dynamic mass balance, 3) suggesting a universal solution to the definition of the adsorbed phase volume as recently initiated by Brandani et al. [*Adsorption*, 2016], and 4) treating some of the general issues, potential pitfalls and revelations that arise in modeling gas phase adsorption processes under dynamic conditions like breakthrough curves when using the three definitions of the amount adsorbed.

614728 Protein Preferred Binding Regions and Ligand Clustering in Multimodal Chromatographic Systems

Abstract:

Recent investigations into protein-ligand and protein-surface interactions carried out in our lab have helped us to develop a deeper molecular level understanding of the binding of biological products to Multimodal (MM) chromatographic systems using Nuclear Magnetic Resonance (NMR) spectroscopy and molecular simulations. In the current work, we extend this combined biophysical and simulations approach to a broader range of proteins and surfaces. A combination of covalent labeling and LC/MS analysis is developed to identify preferred regions on proteins that are involved in binding to chromatographic surfaces. Diethyl pyrocarbonate (DEPC) is employed as a versatile covalent label that can readily react with several nucleophilic amino acid residues on the protein surfaces. A comparison of the protein labeled in solution and bound to the resin is then used to identify preferred binding regions. This approach, which requires small amounts of sample and does not require NMR assignments, has the potential for dramatically increasing the sets of proteins that can be evaluated for preferred binding regions to chromatographic surfaces. Recent advances in our understanding of ligand clustering on these multimodal surfaces using molecular simulations are also presented, as well as the implications of these patch distributions for determining free energies of interactions and for impacting selectivity in these multimodal systems.

602141 The Essential Role of Chromatographic Science in Column Modeling of Protein Separations

Abstract:

The mechanisms underlying chromatography of proteins span a multi-scale set of phenomena that have been modeled individually and collectively for the better part of a century. Modeling of molecular-level behavior is more recent but quantitative prediction of adsorption behavior is still not readily accomplished, due to factors such as the size and complexity of protein molecules and uncertainties regarding resin structure at high resolution. Even at a more coarse-grained level, column modeling has not yet matured into a tool that can be used routinely for detailed process design in biopharmaceutical manufacturing. Nevertheless, the availability of efficient and versatile modeling packages has facilitated more concerted efforts to refine chromatographic modeling of protein separations to a point where predictive modeling is becoming achievable. This has been enabled to a large extent by careful chromatography science studies, many of them from the Carta group, to elucidate the underlying particle-level behavior. This presentation will examine the interplay between such chromatography science findings and column-level modeling, with applications to downstream processing of monoclonal antibodies (mAbs). Examples include cation-exchange chromatography of single- and multi-component mAb systems, Protein A capture, and the exploitation of sample displacement to enhance purification.

599515 The Impact of Alginate Physiochemical Properties on the Adsorption Performance and Reusability of Calcium Alginate Beads.

Abstract:

The establishment of a rigorous formulation methodology considering physiochemical properties of alginate systems has been widely overlooked within current biosorption research utilizing calcium alginate beads. Here, we present for the first time, the effects of molecular weight and viscosity on bead sphericity and its impacts on adsorption and regeneration efficiency using different calcium alginate sorbents. Findings show an increase in molecular weight (MW) results in increased viscosifying properties, which correlates closely to sphericity of the alginate bead. The relationship between sphericity and concentration followed a 2nd order polynomial, where maximum sphericity was maintained over an ideal concentration range of 2-6% (w/v) for respective brands of sodium alginate (MW 300-670 kDa). This increase in sphericity resulted in adsorption enhancement; 1-15%, 81-93% and 88-95% for Pb^{2+}, Cu^{2+} and Cd^{2+} respectively over non-spherical counterparts. Furthermore, compared to a commercial resin (Amberlite), the beads showed similar adsorption performance for Pb^{2+} with reductions of up to 40% and 70% observed for Cu^{2+} and Cd^{2+} respectively. This decline was attributed to the opposing affinity of the resin (Cd>Cu>Pb), whereas the alginate beads had an affinity following Pb>Cu>Cd. These results aligned well with the findings for sorbent regeneration, with subsequent adsorption efficiency largely decreasing (>20% reduction), ascribed by the alginate mass loss due to chemical and mechanical instability. Thus, future work is needed to further optimize the formulation methodology to promote greater adsorption and regeneration considering key performance indicators established in this study.

594580 Dynamic Operation of Batch Reverse Osmosis and Batch Pressure Retarded Osmosis

Abstract:

Optimal control theory is applied to dynamic operation of batch reverse osmosis (RO) for seawater desalination and batch pressure retarded osmosis (PRO) for power generation. It is proved that the water flux should be constant during the entire operation in order to minimize specific energy consumption (SEC) in RO or to maximize specific energy production (SEP) in PRO. A dimensionless parameter $\gamma = A_m L_p \pi_0 t_f / V_0$, similar to the one used in continuous osmotic membrane processes, is proposed to characterize the energy performance. While the SEC in RO is a monotonic function of recovery and γ, the SEP in PRO is solely a monotonic function of γ. It is shown that batch RO/PRO excels continuous RO/PRO because it mitigates the spatial variation of flux in a long pressure vessel observed in the latter. The batch operation provides an elegant way to implement continuous multi-stage RO with interstage pumps and an energy recovery device (ERD) as well as continuous multi-stage PRO with interstage turbines. Some tradeoff between energy efficiency and flux is deemed necessary. A preliminary cyclic diagram to operate the batch PRO is proposed.

References:

[1] Li, M. Dynamic Operation of Batch Reverse Osmosis and Batch Pressure Retarded Osmosis, *Industrial & Engineering Chemistry Research*, **59**, 3097-3108, 2020.

[2] Li, M. Systematic Analysis and Optimization of Power Generation in Pressure Retarded Osmosis: Effect of Multi-Stage Design, *AIChE Journal*, **64**, 144-152, 2018.

[3] Li, M. Reducing Specific Energy Consumption in Reverse Osmosis (RO) Water Desalination: An Analysis from First Principles, *Desalination*, **276**, 128-135, 2011.

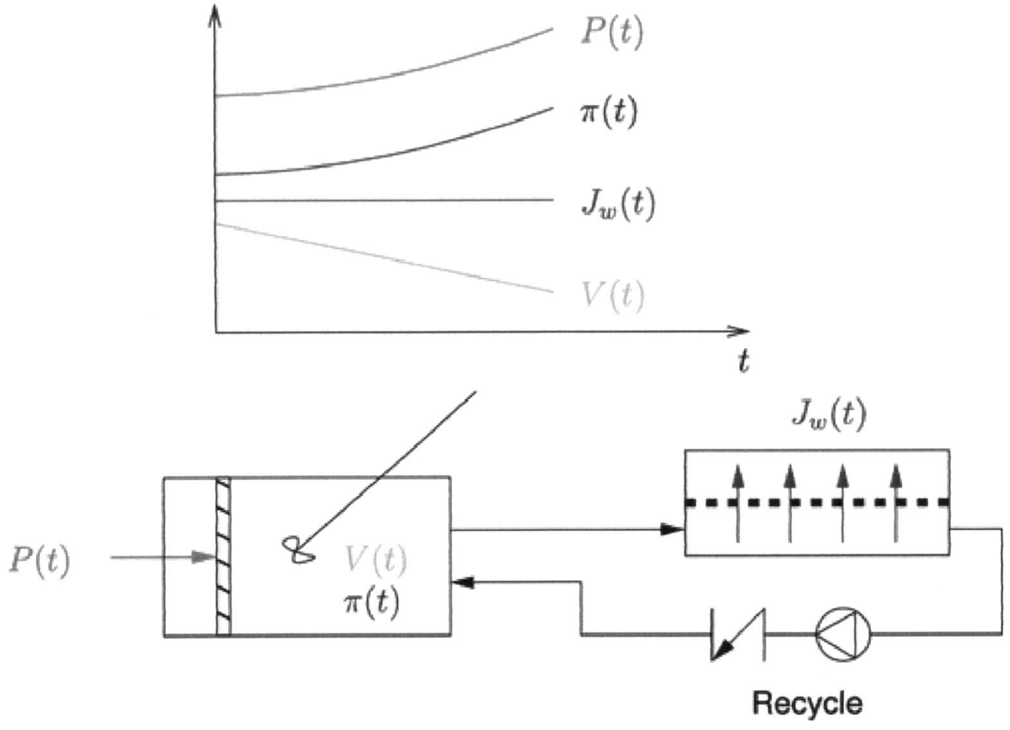

598954 Highly Permeable Supported PDMS Ultrathin Films As a Gutter Layer for Gas Separation Composite Membranes

Abstract:

Engineering useful membrane materials into ultrathin selective layers is the key to maximize membrane gas permeance for energy-efficient separations. The fabrication of a defect-free ultrathin selective layer usually requires a porous substrate to provide mechanical strength and a gutter layer to offer smooth coating surface and prevent pore penetration, forming a multilayer thin film composite (TFC) membrane. Polydimethylsiloxane (PDMS) is the most commonly used gutter layer material due to its high permeability and non-aging property. To minimize gas transport resistance introduced, the supported gutter layer should be 10-fold or more as permeable as the thin selective layer. However, current PDMS gutter layers can hardly provide more than 10,000 GPU CO_2 permeance, which are insufficient to support high-permeance TFC membranes, especially those for CO_2 capture from flue gas. The low permeance is most likely caused by the PDMS material intrusion into the porous substrate with limited porosity. To address this issue, we report a supported 100 nm-thick PDMS gutter layer with a record CO_2 permeance of 12,600 GPU and CO_2/N_2 selectivity of 11.5 at 25 °C. This extremely permeable yet defect-free gutter layer was achieved on a high-porosity (~20%), nanoporous (pore diameter of ~40 nm), and chemical-resistant polymer substrate that was developed through a traditional phase-inversion method. This presentation will discuss the formation of the high-performance porous substrate as well as the subsequent fabrication of the ultrathin PDMS gutter layer with minimal material intrusion. The sought-after performance and good scalability make this supported PDMS thin film a promising gutter layer for high-performance and large-scale TFC membranes for gas separation.

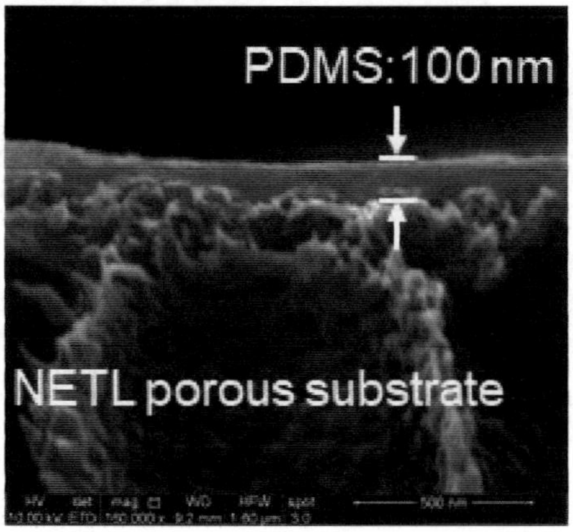

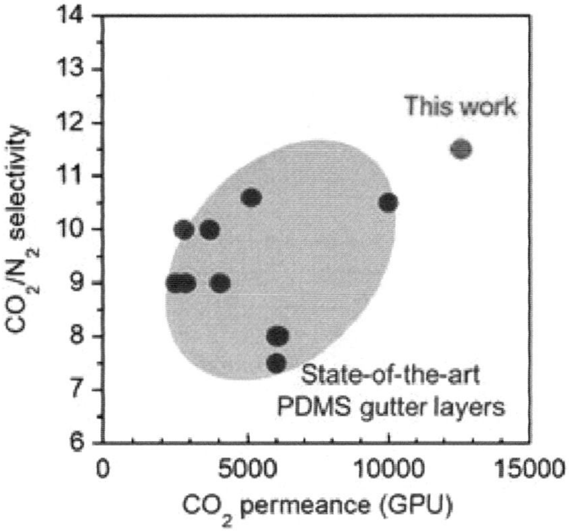

605924 Polymeric Membranes for Moisture-Driven Reversible CO_2 Capture from Ambient Air

Abstract:

To address the greenhouse gas problem, novel efficient and inexpensive carbon capture technologies are needed to be utilized for direct air capture (DAC). Moisture-swing CO_2 sorption is a novel process in which the capture and release of CO_2 is triggered by the changes in the atmospheric humidity, and the CO_2 adsorption/desorption cycle is driven by the evaporation energy consumption of water. Therefore, due to its low energy requirements compared to thermally driven sorption technologies, moisture-swing CO_2 sorption can be a new approach for economical and large-scale DAC. The sorbents that are usually used for this purpose have a fixed positive side of quaternary ammonium groups that hold the mobile anion counterions on their molecular chain. CO_2 reacts with the anionic counterions (hydroxide or carbonate) and produces bicarbonates. Subsequently, carbon capture is realized through the transformation of carbonate and bicarbonate ions and is eliminated from air. Accordingly, the aim of the current research is to investigate the performance of a new polymeric sorbent in the moisture-swing CO_2 sorption process with quaternary ammonium side groups on the polymeric backbone, and anion counterions. For this purpose, we synthesized a series of ammonium-functionalized poly(arylene ether sulfone) sorbents with hydroxide and carbonate counter-ions. Next, we fabricated dense membranes using these sorbents, and evaluated their performance in consecutive humidity-driven CO_2 adsorption/desorption cycles. The obtained results confirm the capability and effectiveness of the prepared novel sorbents in capturing CO_2 from air via moisture swing cycles and can be regarded as a new promising choice for economical large-scale DAC processes.

603658 Effect of Composition on the Transport Properties of Sulfonated Polymeric Ionic Liquid Block Copolymers Membranes

Abstract:

Polymeric ionic liquids (PIL's) block-copolymers are compounds that have at least one PIL block in their structure. PIL's are a type of polyelectrolytes which have ionic moieties as the repeating unit, and exhibit exceptional properties such as high ionic conductivity, a wide range of glass transition temperatures, and moderate chemical and thermal stability. These copolymers are commonly synthesized using controlled radical polymerization techniques or by post-polymerization modification of uncharged polymers. They are potentially suitable in separation processes and fuel cell applications to name a few. Fuel cells are ubiquitous in the rapidly growing market of portable devices and automobile industry. One of the key components of fuel cells is the proton exchange membrane (PEM), whose ionic conductivity is integral to the performance of the fuel.

In this study, a series of sulfonated PILs block-copolymersbased on poly(styrene–isobutylene–styrene) (SIBS) were synthesized for direct methanol fuel cell applications (DMFC). The synthesis consisted of the chloromethylation of SIBS via a Friedel–Crafts reaction, followed by the incorporation of n-alkylimidazole (*i.e.*, butylimidazole, hexylimidadole) through chemical grafting where the PIL block copolymer was obtained. Ananion exchange reaction was carried out, after which sulfonation of the resulting polymers was completed.Properties of the membranes were assessed as a function of PIL block composition due to the presence of different N-alkylimidazoles and anions. A series of materials characterization techniques (*e.g.,*FT-IR, HNMR AFM, SAXS) was performed to provide a fundamental understanding of the synergism between different ionic domains at different concentrations and their effect on the morphology and transport properties of the polymeric membranes.

Keywords: Polymeric ionic liquid block-copolymer, Sulfonation, Polyelectrolyte, Chemical grafting, Transport properties, Direct methanol fuel cell.

605346 Improved CH_4/N_2 Separation Performance of Zirconium-Based Metal-Organic Framework By Introducing Various Functional Groups

Abstract:

Shale gas, low-grade natural gas, landfill gas, and industrial by-product gas contain considerable amounts of CH_4. However, in order to utilize CH_4 as fuel, various impurities such as CO_2, SOx, H_2O, and N_2 should be removed. Most impurities can be easily removed by existing techniques, but N_2 is very hard to be removed due to its very similar physical and chemical properties to CH_4. Cryogenic distillation has been used for the separation of CH_4/N_2 mixtures, but it is costly and energy intensive. Therefore, adsorptive CH_4/N_2 separation has been studied as a cost- and energy-effective alternative.

Metal-organic frameworks (MOFs) have been considered as potential adsorbents for various applications due to their large surface areas, high porosities, and well-defined pore sizes. In this study, we developed MOF-based adsorbents with excellent CH_4/N_2 separation performance through the introduction of various functional groups into a zirconium-based MOF with high hydrothermal stability. In addition, CH_4 adsorption mechanism was identified through grand canonical Monte Carlo (GCMC) and density functional theory (DFT) calculations. Moreover, for the best candidate material, CH_4/N_2 separation under dynamic mixture flow conditions and cyclic CH_4 adsorption-desorption for 15 cycles were performed to check its potential for pressure swing adsorption (PSA) process.

Acknowledgments

We would like to acknowledge the "Next Generation Carbon Upcycling Project" (Project No.2017M1A2A2043449) through the National Research Foundation (NRF) funded by the Ministry of Science and ICT, Republic of Korea.

603768 Modeling Cation Dependence of Gas Adsorption in Zeolites

Abstract:

Natural gas mixtures comprised of CH_4, N_2, CO_2, and hydrocarbons (among other components) are of great industrial interest, most notably for CH_4 which can be burned for fuel. It is, however, highly inefficient to burn the raw mixture, resulting in the need to separate CH_4 from the rest of the components. Gas separation is a field with many avenues that strive for affordable, efficient, and accessible separation of target gases. To this end, this study examines the adsorption of various gases in zeolites using Monte Carlo methods. Zeolites are cheap aluminosilicate materials with varying cation content that are capable of thermodynamic and kinetic separation of gases. The thermodynamic adsorption of CH_4, N_2, and CO_2 in CLI, RHO, and LTL-type zeolites with varied cation mixtures of Na^+, K^+, Cs^+, Mg^{2+}, Ca^{2+}, Sr^{2+}, and Ba^{2+} were calculated from 0.1-60 bar at 298.15 K using the Continuous Fractional Component Monte Carlo technique. Upon examination of adsorption in single cation and selected binary cation zeolite systems, a linear trend in adsorption was observed. Using a randomized mixture of cations, adsorption was both simulated and calculated with a linear model using the aforementioned adsorbates and zeolites, yielding a high level of agreement between the two methods. Future directions for this study include seeking experimental confirmation of this model and studying its application to kinetic properties, such as diffusion.

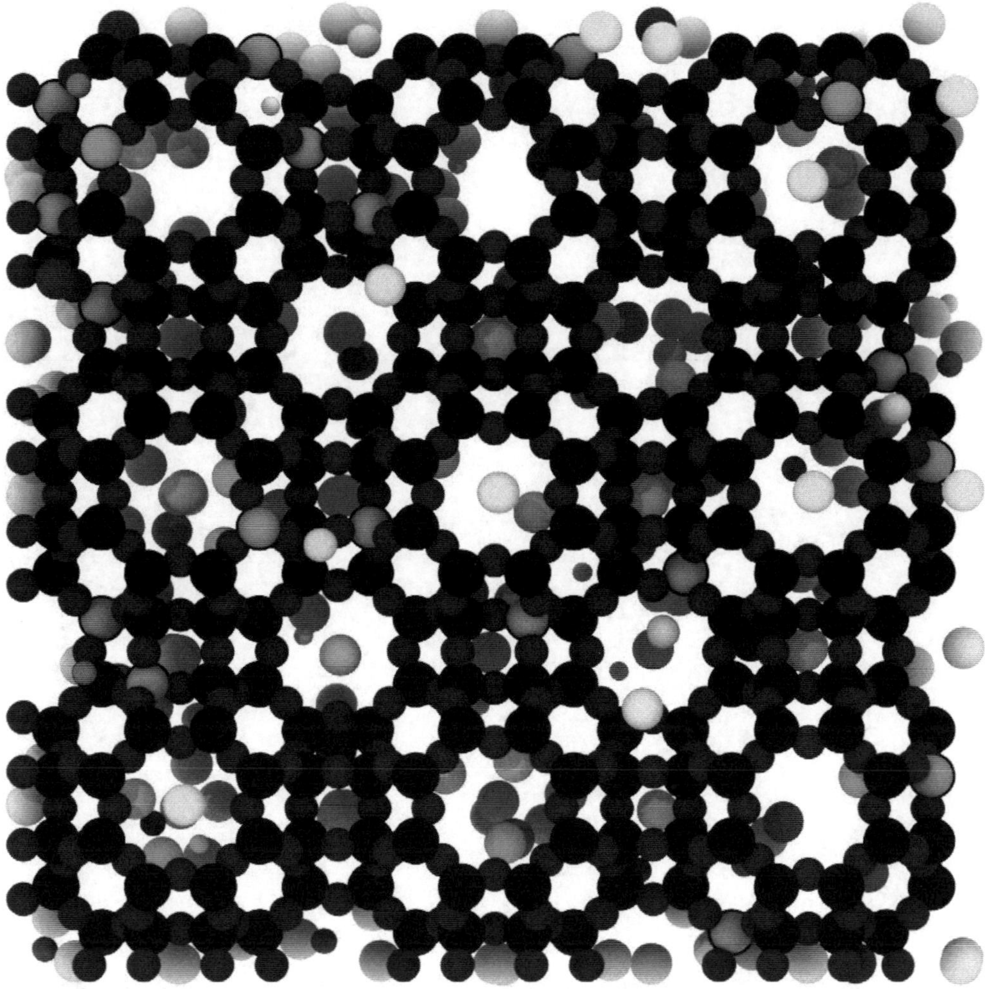

600631 Exploiting Dynamic Opening of Apertures in Partially Fluorinated MOF for Enhancing H2 Desorption Temperature and Isotope Separation

Abstract:

Deuterium has been recognized as an irreplaceable element in industrial and scientific research. However, hydrogen isotope separation still remains a huge challenge due to the identical physicochemical properties of the isotopes. In this paper, a partially fluorinated metal-organic framework (MOF) with copper, so-called FMOFCu, was investigated to determine the separation efficiency and capacity of the framework for deuterium extraction from hydrogen isotope mixture. The unique structure of this porous material consists of a tri-modal pore system with large tubular cavities connected through a smaller cavity with bottleneck apertures with the size of 3.6 Å plus a third hidden cavity connected by an even smaller aperture of 2.5 Å. Depending on temperature these two apertures show a gate opening effect and the cavities get successively accessible for hydrogen with increasing temperature. The thermal desorption spectroscopy (TDS) measurements indicate that the locally flexible MOF can separate D2 from isotope mixture efficiently, with a selectivity of 14 at 25 K and 4 at 77 K.

604798 Efficient Separation of Olefin and Paraffin Using Anion-Pillared Metal-Organic Frameworks

Abstract:

Olefin/paraffin separation is a momentous and challenging process for the production of polymer-grade C_2H_4 and polymer-grade C_3H_6 with high energy penalty. In 2016, about 0.3% global energy was used to purify C_2H_4 and C_3H_6. However, similar molecular size and physical properties of olefin/paraffin lead to great challenge for the separation of C_2H_4/C_2H_6 and C_3H_6/C_3H_8, which usually realized by energy-extensive cryogenic distillation. Metal organic frameworks (MOFs) and mixed matrix membranes (MMMs) are promising approaches for gas purification, while it is still very rare for energy-efficient materials without side effects to overcome the trade-off phenomenon (between selectivity and uptake capacity or between selectivity and permeability).[1,2,3] In the presentation, we will discuss strategies trying to address the problem mentioned above and recent advances of our group in the context of C_2H_4 and C_3H_6 purification. Firstly, we report the separation of C_3H_6 and C_3H_8 with two isostructural anion-pillared metal-organic frameworks (GeFSIX-2-Cu-i (also termed as ZU-32) and SIFSIX-2-Cu-i). Interestingly, GeFSIX-2-Cu-i shows a high uptake ratio of C_3H_6/C_3H_8, outperforming most of the existing materials. Furthermore, the breakthrough experiments for C_3H_6/C_3H_8 (50/50, v/v) binary mixtures were performed on GeFSIX-2-Cu-i and SIFSIX-2-Cu-i, respectively, which exhibits good dynamic separation and recyclability. The binding sites of GeFSIX-2-Cu-i for C_3H_6 and C_3H_8 were investigated by the first-principles density functional theory (DFT) calculations to give insights into the interaction mechanisms. After that, we incorporated SIFSIX-2-Cu-i in 6FDA-DAM polyimide to create MMMs for the separation of C_2H_4/C_2H_6 and C_3H_6/C_3H_8. Gas permeation experiments showed that the permeabilities of C_2H_4 and C_3H_6 as well as the selectivities of C_2H_4/C_2H_6 and C_3H_6/C_3H_8 have increased significantly compared with pure 6FDA-DAM membrane. Our work indicates that the anion-pillared metal organic frameworks can be considered as excellent candidates for the separation of olefin/paraffin purification.

Acknowledgement:

This work was supported by the National Natural Science Foundation of China (Grant No. 21725603, 21938011, and U1862110), the National Program for Support of Top-notch Young Professionals (H.X.), and the Research Computing Center in College of Chemical and Biological Engineering at Zhejiang University.

References:

1. Cui XL, Chen KJ, Xing HB, et al. Pore chemistry and size control in hybrid porous materials for acetylene capture from ethylene. *Science*. 2016; 353(6295): 141-144.
2. Zhang ZQ, Yang QW, Cui XL, et al.Sorting of C4 Olefins with Interpenetrated Hybrid Ultramicroporous Materials by Combining Molecular Recognition and Size-Sieving. *Angewandte Chemie-International Edition.* 2017; 56 (51): 16282-16287.
3. Wang XB, Zhang PX, Zhang ZQ, et al.Efficient Separation of Propene and Propane Using Anion-Pillared Metal-Organic Frameworks. *Industrial & Engineering Chemistry Research.* 2020; 59 (8): 3531-3537.

A supramolecular metal-organic framework with permanent porosity for Xe/Kr separation

Ying Liu†, Zhiguo Zhang†, Qiwei Yang†, Yiwen Yang†, Qilong Ren†, Zongbi Bao*†

Key Laboratory of Biomass Chemical Engineering of Ministry of Education, College of Chemical and Biological Engineering, Zhejiang University, Hangzhou 310027, P. R. China

Abstract

Xenon and krypton have been widely used in various fields including medical imaging, anesthesia, lighting, laser, and ion propulsion. Typically, a gas mixture of Xe/Kr (20/80, v/v) can be obtained as a by-product of air separation, and the pure Xe and Kr can be collected after distillation processes. Another promising Xe source is the reprocessing off-gases of UNF (used nuclear fuel), but it remains a challenge to separate radioactive noble gas isotopes due to their chemical inertness. The cryogenic distillation based on the difference in boiling points to separate Xe and Kr from air or UNF reprocessing off-gases is energy and capital-intensive, which finally leads to the high prices of pure products. The adsorption methods using porous materials have been regarded as a promising method to achieve noble gases separation. Numerous porous materials such as activated charcoal, zeolite, metal-organic frameworks (MOFs) and porous molecular solids have been reported to demonstrate excellent performance for Xe/Kr separation. Hydrogen-bonded organic frameworks (HOFs) with advantageous features have presented potential in various aspects. However, there is limited number of reports on HOFs performing well in noble gas separation.

Herein, we report a supramolecular metal-organic framework, SMOF-SIFSIX-2, self-assembled from cation metal complexes and inorganic anions through hydrogen-bonding interactions combined the advantages of MOFs and HOFs. SMOF-SIFSIX-2 not only possesses permanent porosity, but also realizes highly Xe/Kr separation. (Scheme 1). Single-component adsorption isotherms of Xe and Kr on SOMF-SIFSIX-2 revealed that the uptake of Xe can reach 2.73 mmol/g at 298 K and 1.0 bar, while the Kr uptake was 1.64 mmol/g (Figure 1). The IAST selectivity for Xe/Kr (20/80, v/v) at 298 K and 100 kPa were calculated to be 16.8, and Henry's constant were calculated to be 18.4. The column breakthrough experiments using a binary gas mixture of Xe/Kr (20/80, v/v) and a quinary gas mixture (Xe 400 ppm and Kr 40 ppm balanced with dry air without CO_2) were conducted to simulate the practical industrial separation at 298 K and 1.0 bar on SMOF-SIFSIX-2. For the binary gas mixture, Kr broke through immediately after feeding while Xe was not eluted until 35 min (Figure 2). For the gas mixture containing a low concentration of noble gases, the Xe eluted at around 48 min after the feed gas in with a total Xe capacity of 25.8 mmol/g (Figure 3). This work demonstrated that SMOF-SIFSIX-2 was a promising adsorbent for Xe/Kr separation.

References:

1. P. Häussinger, R. Glatthaar, W. Rhode, H. Kick, C. Benkmann, J. Weber, H. J. Wunschel, V. Stenke, E. Leicht, H. Stenger, Noble gases, **2000**.

2. Frank G. Kerry, Industrial gas handbook: gas separation and purification, **2006**.
3. D. Banerjee, A. J. Cairns, J. Liu, R. K. Motkuri, S. K. Nune, C. A. Fernandez, R. Krishna, D. M. Strachan, P. K. Thallapally, Acc. Chem. Res. **2015**, 48, 211-219.
4. R. T. Yang, Gas separation by adsorption processes, Butterworth-Heinemann, **2013**.
5. R. E. Bazan, M. Bastos-Neto, A. Moeller, F. Dreisbach, R. Staudt, Adsorption **2011**, 17, 371-383.
6. P. K. Thallapally, J. W. Grate, R. K. Motkuri, Chem. Commun. **2012**, 48, 347-349.
7. U. Mueller, M. Schubert, F. Teich, H. Puetter, K. Schierle-Arndt, J. Pastré, J. Mater. Chem. **2006**, 16, 626-636.
8. L. Chen, P. S. Reiss, S. Y. Chong, D. Holden, K. E. Jelfs, T. Hasell, M. A. Little, A. Kewley, M. E. Briggs, A. Stephenson, K. M. Thomas, J. A. Armstrong, J. Bell, J. Busto, R. Noel, J. Liu, D. M. Strachan, P. K. Thallapally, A. I. Cooper, Nat. Mater. **2014**, 13, 954.
9. D. Banerjee, C. M. Simon, A. M. Plonka, R. K. Motkuri, J. Liu, X. Chen, B. Smit, J. B. Parise, M. Haranczyk, P. K. Thallapally, Nat. Commun. **2016**, 7, ncomms11831.
10. J. Liu, P. K. Thallapally, D. Strachan, Langmuir **2012**, 28, 11584-11589.

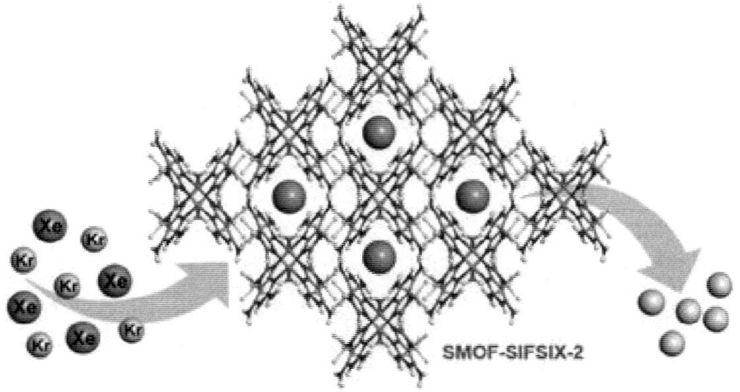

Scheme 1. Schematic illustration of the adsorption separation of Xe over Kr on SMOF-SIFSIX-2.

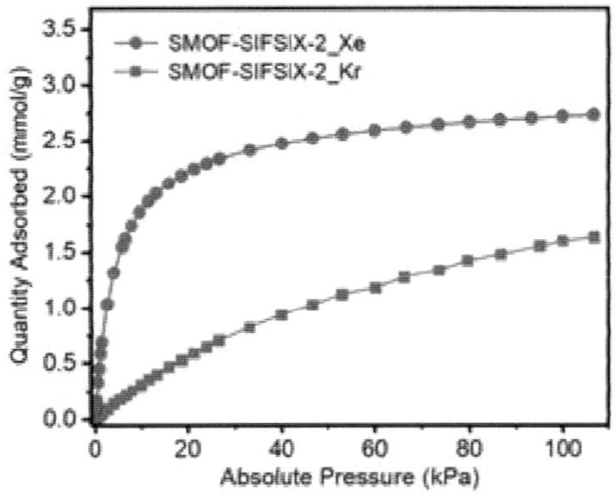

Figure 1. Single-component adsorption isotherms of Xe and Kr on SMOF-SIFSIX-2 at 298 K.

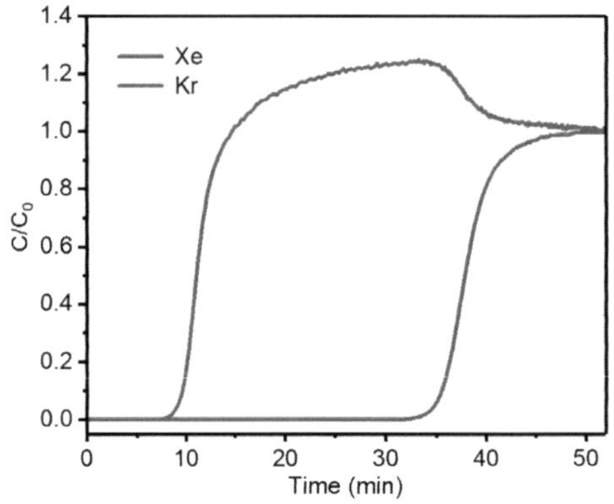

Figure 2. The binary mixture breakthrough curves on SMOF-SIFSIX-2 for Xe/Kr (20/80, v/v) at 298 K.

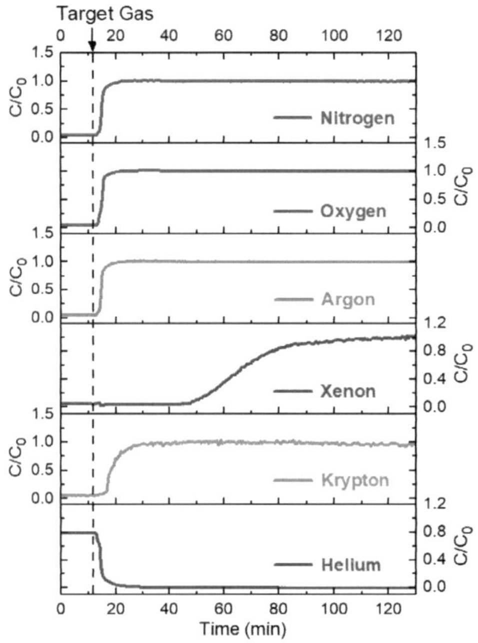

Figure 3. The breakthrough curves on SMOF-SIFSIX-2 for Xe 400 ppm and Kr 40 ppm balanced with dry air without CO_2 at 298 K.

601576 An Assessment of IAST for Predicting Selectivity in Binary Mixed Gas Adsorption" to "An Assessment of IAST for Predicting Mixed Gas Adsorption in MOFs

Abstract:

The ability to understand and design separation systems for complex mixtures under various realistic conditions will be a turning point for the separations community and is a key to transforming separation science. Adsorptive separation is an established technology and currently an active research area within separation science. However, to date its study has relied heavily on the Ideal Adsorbed Solution Theory (IAST) to predict mixture adsorption from pure component adsorption experiments while neglecting gas mixture adsorption experiments, especially for mixtures of more than two components. It has been noted that IAST is particularly limited in its ability to predict gas mixture adsorption in metal-organic frameworks (MOFs) for mixtures containing adsorbates of disparate sizes or sphericities, and especially for adsorbates of different polarities. Improving the ability to predict when mixture adsorption will deviate from IAST predictions is key to designing separation systems for complex mixtures. This knowledge gap in the separations field has persisted in large part due to the technically challenging nature of mixture adsorption measurement. Developing a better understanding of IAST requires measurement of mixture adsorption. This work is a study of how binary and ternary mixture adsorption on the MOFs UiO-66 and HKUST-1 deviates from IAST predictions for mixtures of CO_2, CH_4, and C_4H_{10}. In this work, a home-built, automated volumetric instrument is used to measure multicomponent gas adsorption.

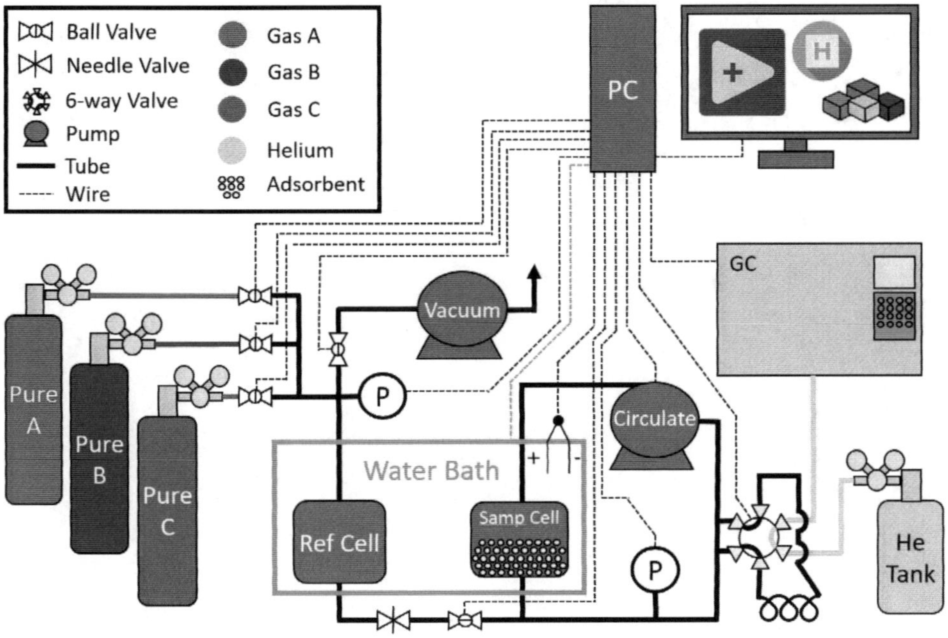

Figure 1. Simple schematic of the Multicomponent Gas Adsorption System (MC GAS) instrument used in this work to conduct binary and ternary adsorption measurements.

607572 Zero Length Column Measurement of Adsorption Equilibria in MOFs for Biogas Separation

Abstract:

Biogas is a promising alternative to natural gas as the global energy sources shift from fossil fuels to renewables. The main components in biogas are CH_4 (35-70%), CO_2 (15-60%), and N_2 (15-50%), with trace impurities such as NH_3, H_2O, and H_2S. Release of biogas into the atmosphere causes rising levels of greenhouse gases, and flaring biogas produces SO_X due to the H_2S present in biogas. It's critical to remove H_2S and CO_2 in biogas as H_2S causes equipment corrosion and poisons catalysts for downstream processes, and CO_2 reduces the heating value. Purifying biogas for electricity generation or biomethane production has gained interest among researchers. Many are exploring less energy-intensive separation processes compared to traditional methods involving liquid scrubbing. Adsorption using microporous materials such as zeolites, metal-organic frameworks (MOFs), and porous organic cages (POCs) is a promising technology as it provides tuned selectivity and requires less regeneration energy. However, the majority of research is on single-component adsorption, which does not well predict the competitive interaction in gas mixtures. This work focuses on the design of a Zero Length Column (ZLC) for rapid binary adsorption equilibria measurement. The binary adsorption equilibria are studied in both pristine and acid-gas exposed MOFs and zeolites such as MIL-125-NH_2, ZIF-8, mmen-Mg_2(dobpdc), and zeolite 13X. With a better understanding of competitive adsorption behavior, sorption materials for complex gas separation can be screened more rapidly, and processes such as pressure swing adsorption (PSA) can be modeled with better accuracy.

607829 Measurement of Competitive CO_2, H_2o and N_2 Adsorption on Zeolite 13X for Post-Combustion CO_2 Capture

Abstract:

Single component CO_2 and N_2 equilibrium loadings were measured on Zeochem zeolite 13X from 0 to 150°C and 1 mbar to 5 bar using volumetry and gravimetry [1, 2]. H_2O equilibrium data was collected from 22 to 100°C and 1 to 21 mbar using thermogravimetric analysis. A series of single and multicomponent CO_2, H_2O and N_2 dynamic column breakthrough (DCB) experiments were measured on zeolite 13X at 22°C and 0.98 bar. The adsorption breakthrough experiments were able to provide accurate equilibrium data for the heavy component (CO_2 in CO_2/N_2 mixtures and H_2O in $H_2O/CO_2/N_2$ mixtures), while failing to provide reliable light component (N_2 in $H_2O/CO_2/N_2$ mixtures and CO_2 in H_2O/CO_2 mixtures) data. It was shown that desorption experiments from a bed fully saturated with the desired composition provides the best estimate of the competitive light component loading. H_2O and CO_2 equilibrium data was fit to a dual-site Langmuir (DSL) isotherm. The equilibrium data for N_2 was fit using four DSL isotherm schemes: the perfect positive, the perfect negative, the equal energy site and the unequal energy site pairings. A detailed mathematical model that used input from the batch equilibrium experiments was able to predict the composition and thermal breakthrough curves well. The dual-site Langmuir isotherm with unequal energy sites was shown to predict the competitive loading and breakthrough curves well for CO_2 and N_2. The impact of the chosen adsorption isotherm model on process performance was evaluated by optimizing a 4-step vacuum swing adsorption process to concentrate CO_2 from dry post-combustion flue gas. The results show that the energy and the productivity are significantly affected by the choice of the adsorption isotherm proving the need to measure competitive adsorption data for reliable process design and optimization.

607843 Multi-Normal Energy Distribution Model for Correlating Single Gas Adsorption Isotherms for Carbon Dioxide, Methane, Ethane, Ethylene, Propane and Propylene on 5A Zeolite

Abstract:

A new multi-normal energy distribution (MNED) model was developed for correlating single gas equilibrium adsorption isotherms with any number of NED functions. This new model utilizes the single process Langmuir model in the integrand of the adsorption integral equation, along with the MNED function. The MNED model was correlated with experimental single gas isotherm data for CO_2, CH_4, C_2H_6, C_2H_4, C_3H_8 and C_3H_6 on 5A zeolite at 25, 50 and 75 ^oC and up to 110 kPa using one, two and three ND functions. For comparison, the same data were also correlated with the single, dual and three process Langmuir (SPL, DPL and TPL) models. All these models incorporated a temperature dependence on the saturation loading capacity that provided a better correlation. The goodness of the fit in each case was based on a specially defined weighed sum of log-error function. The single gas correlations of the isotherms, along with the corresponding isosteric heats of adsorption, from all these models will be presented and discussed.

593521 Removal of Metal Ions from Wastewater By Surface Activated Carbon Microsphere

Abstract:

Carbon Micro-spheres (CMS) were synthesized via hydrothermal reactions and then surface activated by different agents, respectively. The CMS samples were characterized using various instruments and tested for the adsorption of Ag^{+}/Cu^{2+} in aqueous solutions. It was found that CMS sample activated with NaOH presented the better adsorption capacity towards metal ions, while CH_3COOH-activated CMS sample had a comparable capacity to the native CMS. Experiments also revealed that longer time of reaction (19 hours) resulted in CNS with a more amorphous structure but a lowered adsorption capacity, the batch and fixed bed operation modes were tested and it was found the CMS can be used to remove the metal ions at low concentration with high removal efficiency.

595625 Adsorption Characteristics of Hexavalent Chromium from Aqueous Solution Onto Biochars

Abstract:

Chromium contamination of water is a growing concern due to its toxic nature. Chromium can exist in two major oxidation states: trivalent and hexavalent. Hexavalent Chromium (Cr(VI)) is more toxic than trivalent Chromium (Cr(III)). The allowable limit of Cr(VI) in surface water is 0.1 mg/L and in potable water it is 0.05 mg/L. High concentration of Cr(VI) can cause severe human health consequences. It is therefore important to remove Cr(VI) from wastewater before disposal. Various methods have been used for Cr(VI) removal including precipitation, ion exchange, membrane separation and adsorption. However, most of these treatment methods have limitations such as poor removal efficiencies, high operational and maintenance cost, or toxic sludge production. Recently, biochar has received considerable attention as an adsorbent because of its high pollutant removal efficiency, low cost, and availability.

This study focuses on the synthesis of biochars from wood shavings, sugarcane bagasse, and goat manure and their application to remove Cr(VI) from aqueous solution. The biomasses were pyrolyzed in a Non-stirred Parr Pressure Vessel at 450°C to synthesize biochars. All biochars used consisted of micrometer sized particles. Biochars were characterized using Scanning Electron Microscopy (for morphology), Brunauer-Emmet-Teller nitrogen physisorption method (for specific surface area) and Total Organic Carbon Analyzer (for carbon content). SEM micrographs reveal that biochars are amorphous and porous. Surface areas of wood shavings, sugarcane bagasse, and goat manure were determined to be 169.3 m^2/g, 66.7 m^2/g, and 68.3 m^2/g, respectively whereas carbon contents were found to be 78.3%, 73.8% and 53.9%, respectively. Batch tests were conducted to evaluate the adsorption characteristics of Cr(VI) onto biochars by varying pH and contact time. Tests were performed at pH values of 2, 5 and 9 for a biochar dosage of 10 g/L on all tests. The tests were continued for 5 days and samples were collected at each day to determine the residual Cr(VI) concentration using the Graphite Furnace Atomic Absorption Spectrophotometer. Fourier Transform Infrared (FTIR) spectroscopy was also used to explore the changes of surface functional groups on biochars.

The results indicate that both sugarcane baggase and goat manure biochars are highly effective (90%) in removing Cr(VI) although at different pH values. For sugarcane bagasse it is pH-9 whereas for goat manure it is pH-2. Woodchip biochar is not effective at all in Cr(VI) removal (30%). The general characteristics for Cr(VI) removal as a function of time are observed to be similar in all cases. The adsorption increases rapidly in the first 2 to 3 days, then increases asymptotically to the maximum value over the next few days. FTIR results indicate insignificant changes in prominent functional groups before and after its use in removing Cr(VI) confirming that physical adsorption has taken place. The FTIR spectra reveal the presence of C-O, C=O, and O-H functional groups on the surface of biochars derived from woodchips, sugarcane bagasse and goat manure. In addition, C=C stretch was found on the surface of the biochar derived from woodchips and sugarcane bagasse whereas C-C triple bond stretch was observed on the surface of the biochar derived from goat manure. Hence, biochars derived from sugarcane bagasse, and goat manure can be promising adsorbents for Cr(VI) removal from water contaminated with Cr(VI).

611830 Ammonia Removal/Recovery Using Electrokinetic Separations

Abstract:

The water scarcity and resource shortage due to accelerated population growth and industrial development are the major challenges for the society. Managing available resources such as nutrients and new water in the municipal area become emerging practice. The municipal and industrial wastewater contains large amount of nutrients (i.e., nitrogen and phosphorous), which are deemed to the reusable resources in the future. The nutrient-rich wastewater discharging into the nature environment can result in the eutrophication in the water bodies and deterioration in ecosystem. From 30% to 55% severe eutrophic lakes and reservoirs were found around the world and it spends more than 2.2 billion UDS of potential economic loss with freshwater eutrophication in the U.S. However, nitrogen (especially for ammonia) resource is the most important indigents for human life. The ammonia production for fertilizer ironically accounted for 1.5~2.5% global energy consumption and 1.6% carbon dioxide (CO2) emissions annually. By 2018, the global ammonia production reached 140 million tons per year, in which China accounting for 31.4%, followed by Russia (10%), US (8.9%), and India (7.8%). The excess discharged nutrients (e.g., ammonia) in the wastewater has resulted in the environmental problem, thus, to recover ammonia from wastewater can remediate the contaminated environment and mitigate the greenhouse gas emissions by reducing the demand of traditional ammonia manufactures.
The US Environmental Protection Agency upgraded national criteria for ammonia nitrogen discharge to improve the water quality. The removal of ammonia nitrogen from wastewater streams is important in preventing eutrophication of water bodies. The biological nitrification and denitrification are the conventional wastewater treatment processes on removal of ammonia, followed by a combined processes, Anammox, which can anaerobically convert ammonia and nitrite to N2. However, the biological processes do not recover nitrogen resource but release it into the atmosphere; thus, recovering the valuable nitrogen such as ammonia from waste streams would be an high energy efficient approach instead of fertilized product from atmospheric nitrogen.

This presentation discusses the developing electrokinetic and stripping technologies for ammonia removal/recovery from wastewater. A brief summary of various state-of-art technologies such as air stripping, gas-liquid contactor using modified packed bed and membrane contactor for ammonia removal compared to the electrokinetic processes including electrodialysis (ED), electrodeionization (EDI), capacitive deionization (CDI), electrochemical cell (EC) and bio-electrochemical system (BESs) will be presented. We will discuss the experimental assessment of using an in-situ pH adjustable electrokinetic separation technolog to reduce the energy consumption and processing cost to remove/recovery ammonia from waste water.

611516 Microfluidic Ionic Separations Using Selective Electrodes

Abstract:

Separation of highly soluble ionic constituents is a useful tool in the treatment of industrial wastewaters. In this study, a microfluidic electrochemical deionization cell has been employed for the treatment of ionic mixtures representative of certain wastewaters. Symmetric silver/silver chloride electrodes, in combination with microscale flow, separate Cl^- and SO_4^{2-} ions present in an input stream, enriching them in alternate streams. This electrode system exhibits a few useful characteristics including excellent reversibility and stability which allow rapid cycling.

Concentration gradients created in the cell during operation are maintained against diffusion by design of the advective flow field. During separation, the silver anode is oxidized and absorbs Cl^- ions from solution, while the silver cathode releases Cl^- ions. Electrode polarity reverses, and electrode function switches, when the oxidized electrode becomes saturated.

Ion chromatography is used for characterization of separation. The system shows increases, from input to enriched streams, of Cl^- and SO_4^{2-} ions of 42.8% and 21.8%, respectively. We implement an electrochemical transport model including the effect of ionic properties and flow and electrical parameters to elucidate factors influencing system performance characteristics including separation and charge efficiency. Modeling suggests potential changes to cell design and operational regimes for enhanced performance. The method considered offers useful alternative separation approaches for ionic mixtures containing Cl^-.

597108 Functionalization of Electrospun Membranes with Polyelectrolytes for Separation of Oil-in-Water Emulsions

Abstract:

Membrane fouling, which is partially determined by the electrostatic interaction between the membrane and emulsions, is one of the major challenges for the widespread of membrane technology in filtration field. To mitigate the effect of fouling, the layer-by-layer (LbL) technique was applied to alter the electrostatic properties of electrospun fiber membranes. This technique consists of plasma pretreatment to charge the pristine membrane and depositions of polycations and polyanions from solutions to build different charged polyelectrolyte layers on top of the pristine membrane. This technique is able to alter the surface charge of the electrospun membranes, while maintaining their interconnected pore structure and high porosity, both of which are preferred to achieve good robustness against fouling and high permeability. The LbL composite membranes were challenged by two types of emulsions, stabilized by anionic and cationic surfactants respectively in dead-end and cross-flow configurations, and the results were compared to those from the untreated membranes. When separating emulsions stabilized by anionic surfactants, the permeate flux of the plasma treated and polyanion-coated membranes increased by 3.2 fold and 2.5 fold respectively, after 4 hours of cross-flow filtration. When separating emulsions stabilized by cationic surfactants, the permeate flux of the polycation-coated membrane was increased by 3.3 fold. This work suggests that electrostatic repulsion can be used to mitigate the effect of fouling, and modification of membrane surface using LbL assembly technique is effective to improve membrane robustness against fouling.

603558 Continuous Water Purification Using Diffusiophoresis in Membraneless Tubular Separators

Abstract:

Water purification is a critical global need in developing nations, isolated communities, and specialty applications. The technological standard in potable water treatment is activated charcoal filtration, with some applications using micro-, ultra-, nano- or reverse osmosis filtration technologies. Such methods are energy intensive, bulky, and require expensive components that are prone to damage or fouling. Diffusiophoresis is an observable phenomena in which particles respond to an ion concentration gradient, inducing physical separation. This phenomenon is exploitable to drive the separation of charged contaminants in water. Using a tube-in-tube-in-tube configuration, a carbon dioxide gradient is developed radially with contaminated water passing through axially. The dissolved carbon dioxide draws the charged contaminant particles towards the outer wall of the reactor, leaving the water near the inner walls purified. This purified water is collected using a syringe and analyzed using a UV/Vis spectrometer. This study demonstrates the effectiveness of a membraneless water purification technique to remove charged polystyrene particles using only a carbon dioxide gradient as a means of contaminant separation.

601050 Linking Microstructure of Membranes and Performance

Abstract:

Even though synthetic membranes have been around since the early 20th century, predicting their separation performance for liquids remains a substantial challenge. This work addresses the important link between the microstructure of a membrane and its filtration performance.

2D computational fluid and particle drag mechanics are combined with particle and membrane force measurements in aqueous solutions containing inorganic ions to study particle intrusion and capture in microporous commercial polymer and computer-generated teardrop membranes. Fits of the DLVO theory to force-distance profiles obtained membrane surface potentials needed for the computations. In silico predictions of particle intrusion for a commercial membrane qualitatively agree with experimental filtration measurements using scanning electron microscopy with particle tracking via energy dispersive X-ray spectroscopy. Highlighting the poor flow field, several dominant inhomogeneous 2D flow conduits with large unused regions of the internal pore structure are discovered.

To guide improved design, new computer-generated microporous teardrop structures that can equalize the flow field, adjust the tortuosity of the flow path and vary the reactivity of the surface were tested in silico. The main assumptions of the computational model were that 2D flows are a valid description of 3D flows, all forces were applied at the particle center of mass, forces were calculated based on the physical diameter of the spherical particles. Relatively large pores (~5 μm) and large particles (~1 μm) were selected for easy detection and analysis. However, the computational fluid and particle flow analysis and the inter-surface forces scale independently with size and should apply at all classical dimensions (i.e. for nano-, ultra- and micro-filtration). Assumptions for the intermolecular force measurements were that electrostatic and van der Waal's forces dominated and hence that the DLVO theory was valid and that the zeta potential values were close to those at the wall (i.e. surface potential). In particular the DLVO was applied to ideal geometries: a sphere (i.e. AFM probe) near to a flat surface (i.e. either a silica wafer or a hot-pressed PES membrane).

To our knowledge, this is the first attempt combining particle drag mechanics with intermolecular force measurements to help explain particle dynamics in synthetic membranes. This computational fluid mechanics-based tool can be used to characterize membranes for separation performance and guide improved design, synthesis and testing of new microporous membranes.

600949 Polymeric Vs High Performance Inorganic Membrane Materials for of Biogas Upgrading Membrane Systems

Abstract:

Introduction

Membrane separation is a key technology for biogas purification. Multistaged processes based on either cellulose acetate (CA) or polyimide (PI) materials are classically used for this application [1]. In this study, a systematic process synthesis optimization is performed in order to identify the most cost effective solution for three different membrane materials (CA, PI, and zeolite) and three different outlet pressure levels (5, 10, and 15 Bar). More specifically, the objective is to investigate the possibility of advanced inorganic membranes such as zeolite to compete with polymers; a highly permeable (CO_2 permeance 3500 GPU) and selective (CO_2/CH_4 selectivity = 160) zeolite membrane is compared to polymeric materials (CO_2 permeance 60 PGU, CO_2/CH_4 selectivity 60), with a large cost difference (2000 €/m^2 for zeolite vs 60 €/m^2 for polyimide). The key question is to evaluate to what extent high performance materials generate different process configurations and whether they can provide cost effective solutions.

Methods

A general and systematic optimization model for membrane process proposed in [2] is used, which takes into account all the mentioned possibilities as a degrees of freedom in a membrane system (vacuum pumping, self recycling loops) (Figure 1). Moreover, three different outlet pressure levels are taken as constraints, for three different membrane materials. The overall target is to identify to what extent improved commercially available membrane materials, extended connection possibilities and supplementary equipment options (vacuum pump, expander) impact the biogas purification cost.

Results

It is shown that a costly (i.e. 2000 EUR per square meter vs 50 for CA and PI) but high performance membrane material such a zeolite offers the best cost effective solution compared to commercially available polymeric membranes (Figure 2.a). Increasing the outlet pressure increases the purification cost. Two stage processes with recycling loops offer the best balance between purity, recovery, complexity and cost, whatever the outlet pressure level (Figure 2.b). The use of vacuum pumping is shown to improve the process economy, while expander and extra feed compression do not show an interest.

[1] Angelidaki, L. Treu, P. Tsapekos, G. Luo, S. Campanaro, H. Wenzel, P. G. Kougias, Biogas upgrading and utilization: Current status and perspectives, Biotechnology Advances 36 (2018) 452–466.

[2] A. A. Ramirez-Santos, M. Bozorg, B. Addis, V. Piccialli, C. Castel, E. Favre, Optimization of

multistage membrane gas separation processes.Example of application to CO2 capture from blast furnace gas, Journal of Membrane Science 566 (2018) 346–366.

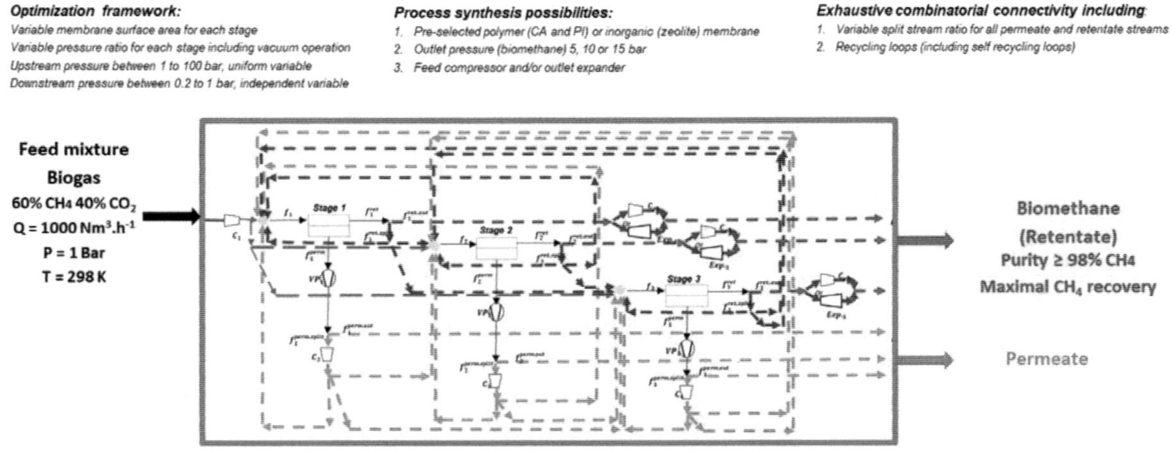

Membrane biogas upgrading process (1 to 3 stages)

601135 Onboard Membrane Carbon Capture and Liquefaction Design of LNG-Fueled Vessels

Abstract:

Shipping has the lowest carbon emissions compared to other modes of commercial transportation. However, to satisfy the Paris Climate Agreement in 2015, International shipping is making efforts to reduce carbon emissions through regulations on Energy Efficiency Design Index (EEDI) for ships established in the 62nd sessions of the International Maritime Organization's (IMO) Marine Environment Protection Committee (MEPC). To meet the IMO's greenhouse gas targets, alternative fuels such as liquefied natural gas(LNG), ammonia, and hydrogen are being explored by the shipbuilding and shipping industries. However, except for LNG, these fuels have several challenges to be adopted on vessels. For this reason, this study proposes the onboard membrane carbon capture and liquefaction system as a feasible alternative. Most previous studies concerning carbon capture mainly addressed the land-based power plants. Thus, there is a lack of research and awareness about onboard carbon capture and liquefaction. However, it consumes less energy and occupies less space than the land-based system because it only needs to satisfy the EEDI regulations. Therefore, this study suggests the onboard membrane carbon capture and liquefaction system as a feasible alternative technology by comparing the energy consumption.

606626 Fine-Tuning of Ionene Backbones with High-Free-Volume Elements for High-Performance CO_2 Separation Membranes

Abstract:

The development and commercialization of CO_2 sequestration or capture methods have become an important part of many industries as well as laboratory applications such as foam blowing, the production of urea, fertilizers, dry ice, and food/beverages, and in the supercritical state as a supercritical solvent and enhanced oil recovery (EOR) as a miscible gas to reduce the interfacial tension between oil and water. We investigate the use of CO_2 solubilizing ionic liquids in polymeric membrane systems for high degrees of membrane tunability and effect/control over polymer nanostructure with enhanced CO_2 separation properties. We have focused on the design and synthesis of various imidazole functionalized monomers having kinked or bridged bicyclic linking groups with a site of contortion that can rigidify polymer chain and generate imidazolium-ionenes with microporosity in the polymer matrix, which anticipated to share similar characteristics of polymers of intrinsic microporosity (PIMs). Spirobisindane, Tröger's base, and benzoxazole monomers having diimidazole functionalities were newly developed to synthesize various Ionene-PIMs. We also successfully developed V-shaped sulfonimide monomers for the synthesis of corresponding anionic forms of the ionenes which are unprecedented in the literature and will open-up broad new possibilities in polymer design with applications far beyond just gas separation membranes. Overall, the membranes prepared from these newly developed high free volume ionene polymers displayed extraordinary CO_2 separation properties, together with excellent mechanical and thermal stabilities. This presentation will detail our "bottom-up" approach of new ionene polymer architectures which contain the high free volume elements (e.g., Spirobisindane, Tröger's Base, Benzoxazole, Sulfonimides, etc.) covalently bonded to ionic groups and associated with extraordinary CO_2 separation over other light gases such as H_2, N_2, and CH_4.

609977 Mixed-Gas Sorption of Ternary Mixtures of CO_2, CH_4 and C_2H_6 in PIM-1: Experimental Measurement and Thermodynamic Modelling

Abstract:

Gas separation with dense polymeric membranes follows the solution-diffusion mechanism. Often only pure-gas solubility is measured and the very few measurements of solubility of gas mixtures in polymers reported are for the case of binary mixtures. In real applications, however, the presence of third components and contaminants can significantly alter the performance of the membrane, due to plasticization and competitive exclusion.

To characterize these effects, we measured the sorption isotherms of C_2H_6, CO_2, and CH_4 in PIM-1, in binary and ternary conditions, using a pressure decay apparatus, working at constant composition and variable pressure of the gas phase. A new measurement protocol was designed to measure sorption of ternary gas mixtures for the first time.

The results show strong competitive sorption effects in all binary and ternary cases analyzed. In the binary cases, both CO_2 and C_2H_6 have a strong exclusion power over CH_4, whereas the effect of CH_4 on the two gases is limited. Surprisingly, C_2H_6 has a strong exclusion power over CO_2 as well, even though these gases have very similar condensability and pure-gas sorption in PIM-1. In the ternary case, the superposition of the effects highlighted in binary tests is observed: both CO_2 and CH_4 are excluded due to the presence of C_2H_6, while C_2H_6 solubility is barely altered compared to pure-gas conditions. Overall, the effect of C_2H_6 on CO_2/CH_4 solubility-selectivity is modest, compared to the corresponding binary case.

The Non-Equilibrium Lattice Fluid (NELF)[1] model predictions for binary and ternary sorption, performed using only pure-gas parameters as input, are in very good agreement with the data, indicating that the model is a reliable tool to study mixed-gas sorption, even in complex scenarios.

References

[1] F. Doghieri, G.C. Sarti, Macromolecules, 24 (1996) 7885–7896

[2] O. Vopička, M.G. De Angelis, N. Du, N. Li, M.D. Guiver, G.C. Sarti, J. Membr. Sci., 459 (2014) 264–276

Figure. Third component (C_2H_6) effects on (a) CO_2 and (b) CH_4 solubility in PIM-1 at 35 °C. Empty symbols: pure gases. Striped symbols 19-81 mol% CO_2/CH_4 binary mixture.[2] Filled symbols 15-15-70 mol% $CO_2/C_2H_6/CH_4$ ternary mixture. Solid lines represent NELF model predictions of pure- and mixed gas sorption isotherms.

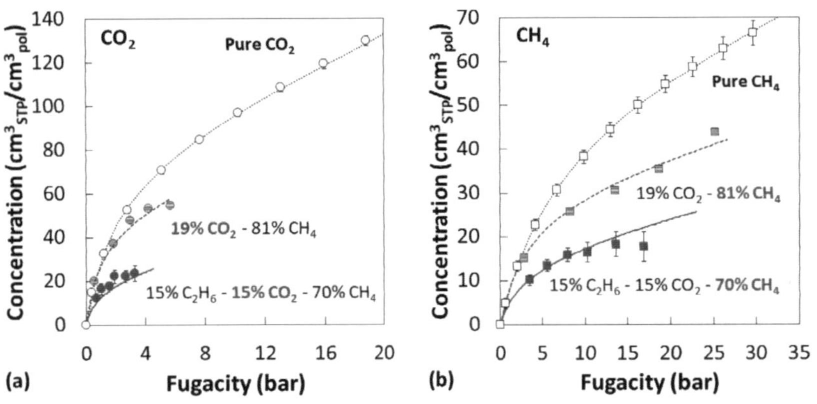

592196 Coupling of 1-Ethyl-3-Methylimidazolium Cation Based Room Temperature Ionic Liquids with Membrane Contactor; Post Combustion CO_2 Capture

Abstract:

Post-combustion CO_2-capture is a convenient way to control the anthropogenic increase of CO_2. The coupling of a hollow fiber membrane contactor (MC) with ionic liquids (ILs) for CO_2 extraction from gaseous fluxes is an innovative intensified process coupling the advantages of membrane technology with unique properties of ILs such as high CO_2 solubility and negligible vapor pressure.

Room temperature ILs namely 1-ethyl-3-methylimidazolium methyl sulfate ([emim][$MeSO_4$]), 1-ethyl-3-methylimidazolium dicyanamide ([emim][DCA]), 1-ethyl-3-methylimidazolium ethyl sulfate ([emim][$EtSO_4$]) and 1-ethyl-3-methylimidazolium acetate ([emim][AC]) were coupled with mesoporous hydrophobic polypropylene MC (Liqui-Cel®, USA), for CO_2 absorption and desorption considering a closed loop recirculation of ILs on the shell side. IL was recirculated during the absorption operation until reaching pseudo-steady state. ILs and membranes were characterized before to confirm no wetting conditions following Laplace-Young equation. The setup was tested against various operating conditions. Absorption performance of the coupled IL-MC setup, variation of absorption kinetics with CO_2 concentration, experimental mass transfer coefficients (K_{exp}) and absorption capacities of the ILs were thoroughly investigated in this work. Finally, a pseudo-steady state dynamic mathematical model was developed to compare (with experiments) and extend the study.

The Initial/pseudo-steady state K_{exp} recorded for ILs [emim][$MeSO_4$], [emim][DCA], [emim][$EtSO_4$] and [emim][AC] were $9.28*10^{-6}/0.22*10^{-6}$ $m.s^{-1}$, $9.27*10^{-6}/0.32*10^{-6}$ $m.s^{-1}$, $23.31*10^{-6}/0.59*10^{-6}$ $m.s^{-1}$ and $46.40*10^{-6}/0.44*10^{-6}$ $m.s^{-1}$, respectively. Gradual increase in the CO_2 concentration of the IL has strongly influenced the absorption process which is very fast in the beginning and becomes very slow at pseudo-steady state. Based on the overall CO_2 capture performance of the coupled system the ILs can be ranked as [emim][AC]>[emim][$EtSO_4$]>[emim][DCA]>[emim][$MeSO_4$]. The developed dynamic model showed very close agreement with the experiments (Figure 1 below).

As conclusions it can be asserted that findings of experimental and modelling results suggests that in spite of the difference in the absorption performance between ILs presented here, all are potential absorbents for CO_2 in MCs.

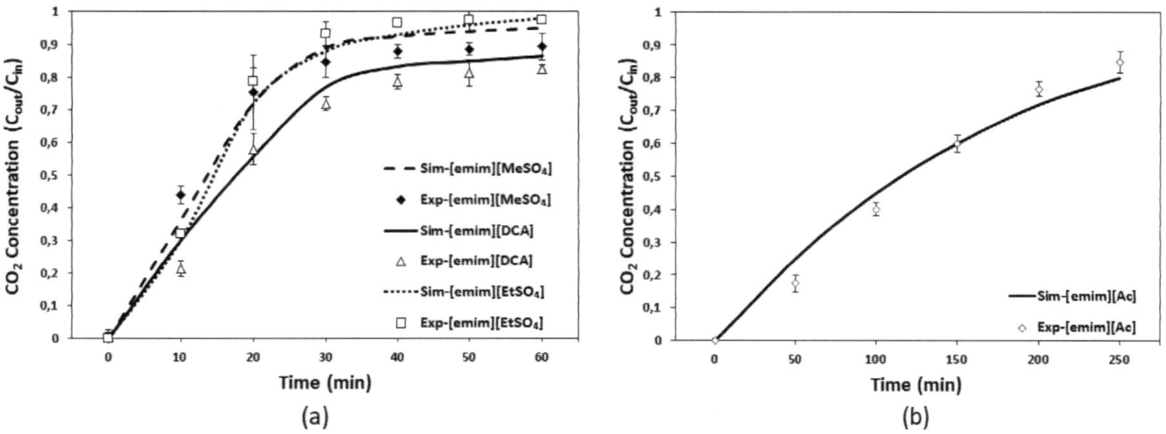

Figure 1 Agreement of experiments and simulations (a) [emim] [$MeSO_4$], [emim] [DCA] and [emim] [$EtSO_4$] (b) [emim] [Ac] at T= 288 K

611854 Molecular Mechanism of Gas Separation through Polymer-MOF Mmms

Abstract:

In this talk we will discuss our recent results of separation studies of complex gas mixtures such as CH_4/H_2 through polymer-MOF MMMs performed using microsecond long all-atom nonequilibrium molecular dynamics (NEMD) simulations. We have used MMMs of several rigid and flexible polymers with IRMOF-1 and HKUST fillers. The simulation trajectories were analyzed to unravel the effect of polymer rigidity, polymer chain length, and MOF geometry on their separation performance.

606074 Membrane Assisted Antisolvent Crystallization: Process Simulation and Control

Abstract:

Membrane Assisted Antisolvent Crystallization: Process Simulation and Control

Lei Sheng, Jin Li, Gaohong He, ***Xiaobin Jiang****

State Key Laboratory of Fine Chemicals, School of Chemical Engineering, Engineering Laboratory for Petrochemical Energy-efficient Separation Technology of Liaoning Province, Dalian University of Technology, Dalian, 116024, China

E-mail: xbjiang@dlut.edu.cn

ABSTRACT: Antisolvent crystallization is an important purification technology for the pharmaceutical and fine chemical industry. Herein, we investigated the mass transfer mechanism of membrane-assisted antisolvent crystallization (MAAC) and developed the multistage operation to reinforce the manufacture feature of antisolvent crystallization. Computational fluid dynamics simulation results via developed mathematics model and the experimental images via in-situ detecting technology jointly illustrated the advantages of MAAC on accurate mass transfer control. Multistage MAAC that intensifying the product capacity under a stable control is a promising direction for the new generation antisolvent crystallization.

Figure 1. (a) Schematic diagram of MAAC process with key parameters; (b) observed images of the antisolvent liquid layer, the membrane outer surface at different crystallization with increasing suspension crystal density (from left to right).

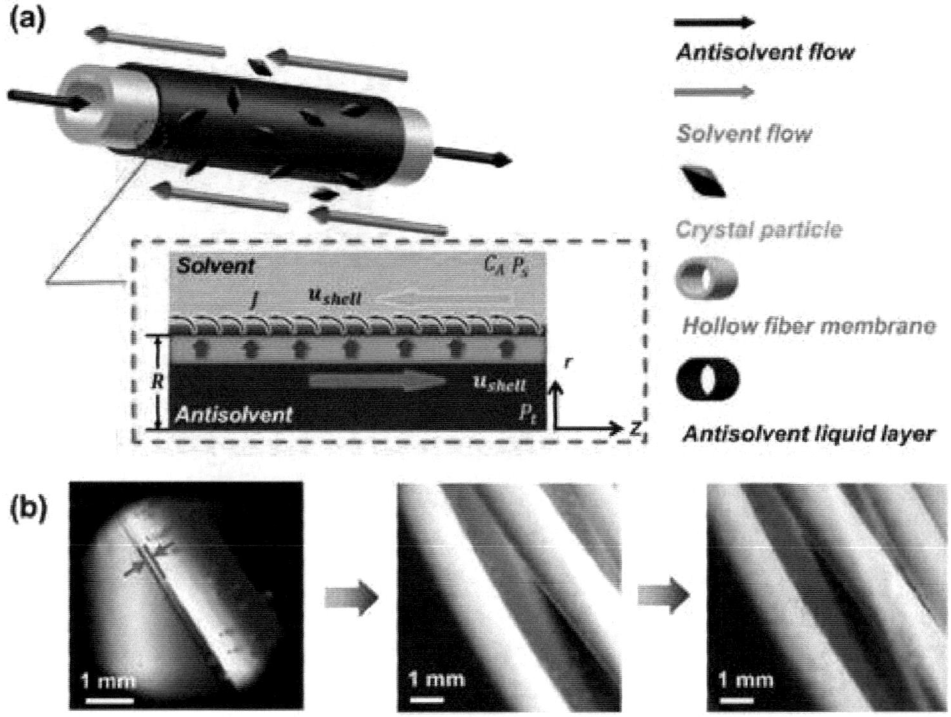

596667 Diffuse Membrane Via Pyrolysis of a Coating of Fish Scale Colloids for Organic Solvent Nanofiltration

Abstract:

This study utilizes nano grains of Barramundi fish scales (BFS) to construct a filtration membrane, which manifests organic solvent nanofiltration efficacy through the self-sustained percolation mechanism, i.e. adsorption of solute promotes retention of solute. A coating of BFS colloids is first developed on cellulose filtration paper. The coating sheet is then subjected to partial pyrolysis under slight compression to attain the filtration membrane. The membrane demonstrates > 90% retention for the Sudan IV and > 95% retention for rose bengal in isopropanol solution (30 ppm) with the average permeance (LMBH) of 1.9 and 6.1, respectively. It is also noticeable that when using tetracycline to examine the membrane, the initial negative rejection (-50%) followed by rejection (>70%) of this probe solute with an average permeance of 4.3. This phenomenon suggests that the concentration polarization triggers rejection via restricting surface diffusion and adsorption known as the dynamic separation way.

612650 Application of Supported-Ionic-Liquid Membranes to a Binary Separation of Nucleobases

Abstract:

Supported-ionic-liquid membranes (SILMs) are an exciting technology due to their use of tunable ionic liquids and their ability to conduct specific separations. In this work, the separation of nucleobases was studied to gain insight on whether SILMs could be used to perform separations via differing hydrogen bonding interactions with solutes. SILMs were fabricated using pressure-driven filtration to fill the pores of a microfiltration membrane with 1-hexyl-3-vinylimidazolium bistriflimide. These membranes were used in a diffusion-cell setup to separate an aqueous mixture of thymine and cytosine (nucleobases); membrane stability in water was also assessed. For a separation of a feed initially containing 21.8 mM of each solute, mass-transfer coefficients of $14.3 \cdot 10^{-9}$ m/s and $2.7 \cdot 10^{-9}$ m/s were calculated for thymine and cytosine (separation factor of 4.95). It is likely that differing ionic-liquid/solute hydrogen-bonding interactions accounted for this separation as both solutes and the ionic liquid can hydrogen bond. The mass-transfer coefficients and other experimental parameters were used to model a separation of a different mixture of thymine and cytosine. A stability test indicated significant leaching of ionic liquid from the membrane for the first 115 hours of the test, but the calculated separation factor remained in the range of 4.95-5.82 over the course of 48-hour and 96-hour separation tests. The demonstrated separation suggests the potential of supported ionic liquid membranes to be used for specific separations of nucleobases and similar compounds that can hydrogen bond.

603657 Effect of Amine Groups Incorporation on Morphology and Transport Properties of Sulfonated Poly(styrene-isobutylene-styrene) Membranes for Fuel Cell Applications

Abstract:

The growing demand in the search for energy sources independent of fossil fuels, has generated a considerable interest in the use of fuel cells as an alternative energy source with low environmental impact and high electrical efficiency. Even with their advantages, fuel cells exhibit some severe limitations like high permeability to fuels and high cost. These limitations are associated with the composition of the proton exchange membrane (PEM) which is considered a key component on these devices. In order to overcome these downsides, PEMs based on aromatic polymers have been synthesized and used because of their availability and easy modification.

In this study, blend membranes based on sulfonated poly (styrene-isobutylene-styrene) (SIBS) with isopentylamine (IPA) were synthetized as a potential candidate for direct methanol fuel cell (DMFC) applications. Properties of the resulting membranes were determined as a function of critical parameters such as sulfonation level (57-93 mol%) and percentage of IPA incorporation (1, 3, and 5 wt.%). FTIR analysis proved that isopentylamine was successfully incorporated into the sulfonated polymer matrix and confirmed the interaction between the sulfonic and amine groups. That interaction is attributed to the hydrogen bonds formed with the presence of both groups inside the membranes which generatessignificant morphological changes in the nanostructure of the membranes that are evident through results of small angle x-ray scattering (SAXS) and atomic force microscopy (AFM).On the other hand, proton conductivity and methanol permeability were measured to analyze the transport properties of the membranes. The results suggest different transport mechanisms for proton and methanol in the membrane, since IPA incorporation showed different patterns in both properties for each sulfonation level. Methanol permeability seems to be related to the water content in the membrane while proton conductivity seems to be more sensitive to the morphological changes produced after the IPA incorporation and the interconnection between the ionics domains. The selectivity (proton conductivity over methanol permeability) of membranes revealed an optimum IPA incorporation between 3-5 wt.%, which depends onthe sulfonation level. The results also indicated that blended membranes show a better chemical stability compared to the sulfonated membranes.

Keywords: Amine-sulfonic groups interaction, Proton exchange membrane, Morphology, Transport properties, Direct methanol fuel cell.

605609 Upgrading Polymer Hollow Fiber Membrane Modules to Mixed-Matrix Hollow Fiber Membrane Modules for Propylene/Propane Separation

Abstract:

Mixed-matrix membranes (MMMs) have shown promising performances for gas separation, in particular, propylene/propane separation. Their commercial applications require to fabricate MMMs in scalable forms, i.e., asymmetric mixed-matrix hollow fiber membranes (MMHFMs). Up until now, however, there have been few reports on MMHFMs due to the fundamental engineering challenges associated with current fiber spinning processes using filler-suspended dope solutions. Recently, we proposed a scalable MMM fabrication strategy, namely polymer-modification-enabled *in-situ* metal-organic framework formation (PMMOF). The PMMOF has potential to overcome some of the engineering challenges since it decouples hollow fiber spinning step and MMM formation step. Herein, we report asymmetric 6FDA-DAM/ZIF-8 MMHFM modules consisting of up to seven individual fiber strands by transforming 6FDA-DAM coated hollow fiber membrane (HFM) modules using the PMMOF. The resulting asymmetric MMHFM modules showed a promising propylene/propane separation performance (i.e., propylene permeance of ~ 2.15 GPU and separation factor of ~ 23.4) without additional defect healing steps. The membrane modules showed stable separation performance over a period of up to 25 days and at the total feed pressures of up to 6 bar. To the best of our knowledge, this is the first multi-stranded MMHFM modules showing promising propylene/propane separation performances, thereby opening up the possibility of commercial applications of MMMs.

606094 Comonomer Effects on Co-Permeation of Methanol and Acetate in Cation Exchange Membranes

Abstract:

Co-permeation in a hydrated dense membrane is crucial in many applications from energy conversion (i.e. photoelectrochemical CO_2 reduction cells) to liquid separation (i.e. pervaporation cells). One of the major challenges of CO_2 reduction cells is to design an ion exchange membrane that minimize the permeation of CO_2 reduction products (i.e. methanol and acetate). Previously, the transport behavior of Nafion® 117 and crosslinked PEGDA-AMPS cation exchange membranes to methanol and sodium acetate was investigated and distinct changes in permeabilities of membranes to sodium acetate was observed in co-permeation with methanol. To further investigate this co-permeation behavior, we modified PEGDA-AMPS structure by varying negatively-charged AMPS content with three different charge-neutral comonomers, acrylic acid (AA, n = 0), 2-hydroxylethyl methacrylate (HEMA, n = 1), and poly(ethylene glycol) methacrylate (PEGMA, n = 5). While the permeability to sodium acetate in co-permeation with methanol was increased for membranes with shorter comonomers, PEGDA-AA-AMPS and PEGDA-HEMA-AMPS, it was consistent for PEGDA-PEGMA-AMPS membranes. From these experiments, we deduce that charge-neutral molecules, both moving (e.g. methanol) and membrane-bound (e.g. ethylene glycol), can interfere with the electrostatic repulsion during ionic transport in ion exchange membranes and this shielding of electrostatic repulsion leads to variance in transport behavior.

607118 Ionic Liquid-Polymer Gel Electrolytes for Electrochemical Olefin Separations in Gas-Diffusion Electrodes

Abstract:

The separation of olefin/paraffin mixtures is one of the most energy intensive processes in the chemical industry, accounting for more than 0.3% of total global energy consumption. Currently, light olefins (e.g., ethylene and propylene) are purified via energy-intensive cryogenic distillations. Alternative non-thermal separation methods that rely on the difference in size (i.e., size-sieving membrane separations) are not effective due to the small size difference between small olefins and paraffins. Instead, significant energy savings could be realized by targeting differences in chemical reactivity between alkenes and alkanes. Separation of alkenes can be achieved electrochemically by reversible, selective absorption by a transition metal complex which is promoted at one oxidation state and reversed at another. Such separations have demonstrated high selectivity but low throughput with the potential for significant improvement with rigorous electrochemical engineering. Given the rapid separation mechanism (electrochemical reduction/oxidation), mass transport of gasses and ionic conductivity are the key limiting parameters. In this study, we develop gas-diffusion electrodes (GDEs) for high-throughput electrochemically modulated olefin/paraffin separation. Use of hybrid ionic liquid-polymer gel (ion gel) thin films containing transition metal complexes in the GDE will promote large gas contact area, short gas diffusion pathways, and high ionic conductivity in the electrode. In this work, I will discuss early results in the synthesis of Nickel dimercaptomaleonitrile complexes (i.e., $[Ni(S_2C_2(CN)_2)_2]$) as reversible olefin complexing agents, and their electrochemical characterization in the presence of olefins. Furthermore, I will discuss ongoing and future work on the fabrication, transport properties, and redox modulated olefin binding properties of ion gels composed of poly(vinylidene fluoride-co-hexafluoropropylene) films swollen with imidazolium-based ionic liquids and $[Ni(S_2C_2(CN)_2)_2]$ complexes, to be implemented in membrane-electrode-assembly swing absorbers for electrochemical separation of olefin/paraffin mixtures.

607345 Synthesis and Characterization of Poly(arylene ether) Sulfone Copolymers for Fuel Cells Applications

Abstract:

Proton Exchange Membrane Fuel Cells (PEMFC) have been an alternative to fossil fuel-based energy production, due to its higher efficiency and environmentally benign nature. Nafion® has been the standard proton exchange membrane (PEM) of choice for decades. Despite Nafion's high proton conductivity, issues related to water management, cost, and processability, limit its performance and further commercial expansion. This investigation pursues PEM's capable of enhancing the proton conductivity, while having the ability to operate at high temperatures. The objective of this study is to evaluate an alternative copolymer for PEMFC based on poly(arylene ether sulfones). Synthesis, materials characterization, and proton conductivity measurements for these multiblock copolymer membranes were studied. The copolymers were synthetized by a coupling reaction between decafluorobiphenyl (DFBP) end capped non sulfonated poly(arylene ether sulfone) (BPS0) and sulfonated poly(arylene ether sulfone) (BPS100). The reactive behavior of DFBP facilitates the low temperature coupling reaction (below 105 °C). Additional multiblocks are being synthesized to explore the ionic interconnection and the morphological variations of these PEMs.

607798 Synthesis and Characterization of Fully Aromatic and Semi-Aromatic Polyamide Reverse Osmosis Membranes

Abstract:

Reverse osmosis (RO) membranes are the gold standard for water desalination and have been in use for over 30 years. In last three decades, an ample amount of work has been done to enhance the performance of RO polyamide membranes by changing the membrane surface chemistry, adding nanomaterials in and to, and altering the morphology of the polyamide separation layers. The morphology and structure of the polyamide layer is governed by the interfacial polymerization reaction between the amine monomer and the acid chloride monomer. The solubility and diffusivity of the monomers during the interfacial polymerization reaction affects the morphology and network structure. Thus, gaining fundamental knowledge on how the monomers affect the structure and morphology of the separation layer on molecular scale is very important because we do not fully understand the separation mechanism on a molecular level of these RO membranes. Additionally, current RO membranes do not reject neutral, small uncharged organic molecules to a level to produce potable water, especially at near neutral pH, which requires the synthesis of RO membranes with novel chemistries. In this work, we synthesized aromatic and the semi-aromatic polyamide RO membrane separation layers via the interfacial polymerization reaction between an amine monomer (m-phenylenediamine and 1,3-diaminopropane were used for aromatic and semi-aromatic polyamide layers, respectively) and an acid chloride monomer (trimesoyl chloride). The synthesized membranes were characterized using SEM, ATR-FTIR, contact angle, and positron annihilation lifetime spectroscopy (or PALS). Membranes were performance tested for water permeance, salt rejection, and neutral, small uncharged organic molecule (urea and boric acid) rejection. Polymerization conditions such as monomer concentration and polymerization time were varied to determine how each affects the polyamide separation layer. By understanding the fundamental reason(s) why different amine monomers and polymerization conditions affect polyamide separation layer morphology and structure, and therefore the desalination performance on the molecular level, RO membranes can be engineered to make membrane-based separation processes more energy-efficient and cost-effective.

609339 Assessment of Low Cost Adsorbents for Removal of Siloxanes from Landfill Gas

Abstract:

Landfill gas (LFG) is increasingly used and researched as a feedstock for a variety of traditional and proposed Waste-to-Energy (WTE) technologies, which includes electricity generation, compressed natural gas, or liquid hydrocarbon fuels. In these various scenarios, contaminants in the LFG can have substantial economic and environmental consequences in the WTE processes [1-2]. Siloxanes are an increasing contaminant of LFG as many consumer products being land-filled contain this compound [3]. Siloxanes in biogas cause damage to machines if not removed because it thermally decomposes to silica [3-5]. This leads to high maintenance cost of WTE technologies thereby increasing the cost of the process. Current purification techniques available for siloxanes removal are too expensive [5]; it costs less to repair damaged engine parts than to adopt current siloxane purification techniques. In order to accelerate adoption of WTE processes, we need more economical methods for removing siloxanes from LFG.

The goal of this research project is to develop low cost strategies for siloxane removal from LFG. This research will adopt two approaches towards its goal of developing low cost strategies for siloxane removal from LFG. First, natural zeolite that can be purchased for as low as $100/ton, as compared to activated carbon that is factors of 3+ more expensive, as an adsorbent and other low cost adsorbents such as diatomaceous earth, crushed glass and biochar will be evaluated on the performance, in terms of capacity, selectivity, regeneration ability, and stability, and compared to activated carbon which is the current state of technology. Secondly, taking advantage of the siloxanes inclination to undergo thermal degradation, low cost/waste inorganic materials will be evaluated as catalytic adsorbents to capture only the silicon/silica of the siloxanes. An advantage of this approach is that the siloxanes permanently change phases into one that will remain trapped. The performances of these materials will be evaluated in terms of capacity and longevity. In both cases, the experimental results will be used to conduct a techno-economic feasibility study that can be compared to previous results found in literature using activated carbon as the adsorbent.

This presentation will focus on the first strategy which involves testing selected low-cost adsorbents for the adsorption capacity of selected siloxanes. Four low cost and/or waste materials were selected for examination as adsorbents for this study: clinoptilolite, diatomaceous earth (DE), biochar, and crushed glass. Selected siloxanes for this project are hexamethyldisiloxane (L2) and octamethylcyclotetrasiloxane (D4). Characterization techniques such as N_2 physisorption and XRD analysis were adopted to determine the physical characteristics (such as surface area) and chemical composition of the samples. Water vapor sorption experiments and CO_2 chemisorption experiments were used to examine the competition of adsorption sites by these compounds. Results of the experiments stated above for clinoptilolite are displayed in figure 1.

From the adsorption isotherms derived for clinoptilolite as displayed in figure 1, clinoptilolite was observed to possess a water vapor adsorption capacity of 102.9 mg H_2O/g clinoptilolite at room temperature and a CO_2 adsorption capacity of 96.3 mg CO_2/g clinoptilolite at room temperature. The N_2 adsorption isotherm was used to determine the BET surface area of clinoptilolite which resulted in 17 m^2/g. N_2 physisorption, CO_2 chemisorption and water vapor sorption experiments will be carried out on all other selected low cost adsorbents. We will also present results of vapor sorption experiments to generate breakthrough curves and adsorption isotherms to determine the adsorption capacity and regeneration ability of selected low cost adsorbents for selected siloxanes in inert and surrogate LFG. Finally, we will present results of a techno-economic analysis to compare the cost in comparison with conventional adsorbents.

References

[1] Ajhar, M., Travesset, M., Yüce, S., & Melin, T. (2010). Siloxane removal from landfill and digester gas–a technology overview. *Bioresource technology*, *101*(9), 2913-2923.

[2] Kuhn, J. N., Elwell, A. C., Elsayed, N. H., & Joseph, B. (2017). Requirements, techniques, and costs for contaminant removal from landfill gas. *Waste Management*, *63*, 246-256.

[3] Finocchio, E., Garuti, G., Baldi, M., & Busca, G. (2008). Decomposition of hexamethylcyclotrisiloxane over solid oxides. *Chemosphere*, *72*(11), 1659-1663.

[4] Sonoc, A. C., Thurgood, C., Peppley, B., & Kelly, D. G. (2017). Kinetic study of the thermal decomposition of octamethylcyclotetrasiloxane on activated gamma alumina. *Journal of environmental chemical engineering*, *5*(5), 4858-4865.

[5] Urban, W., Lohmann, H., & Gómez, J. S. (2009). Catalytically upgraded landfill gas as a cost-effective alternative for fuel cells. *Journal of Power Sources*, *193*(1), 359-366.

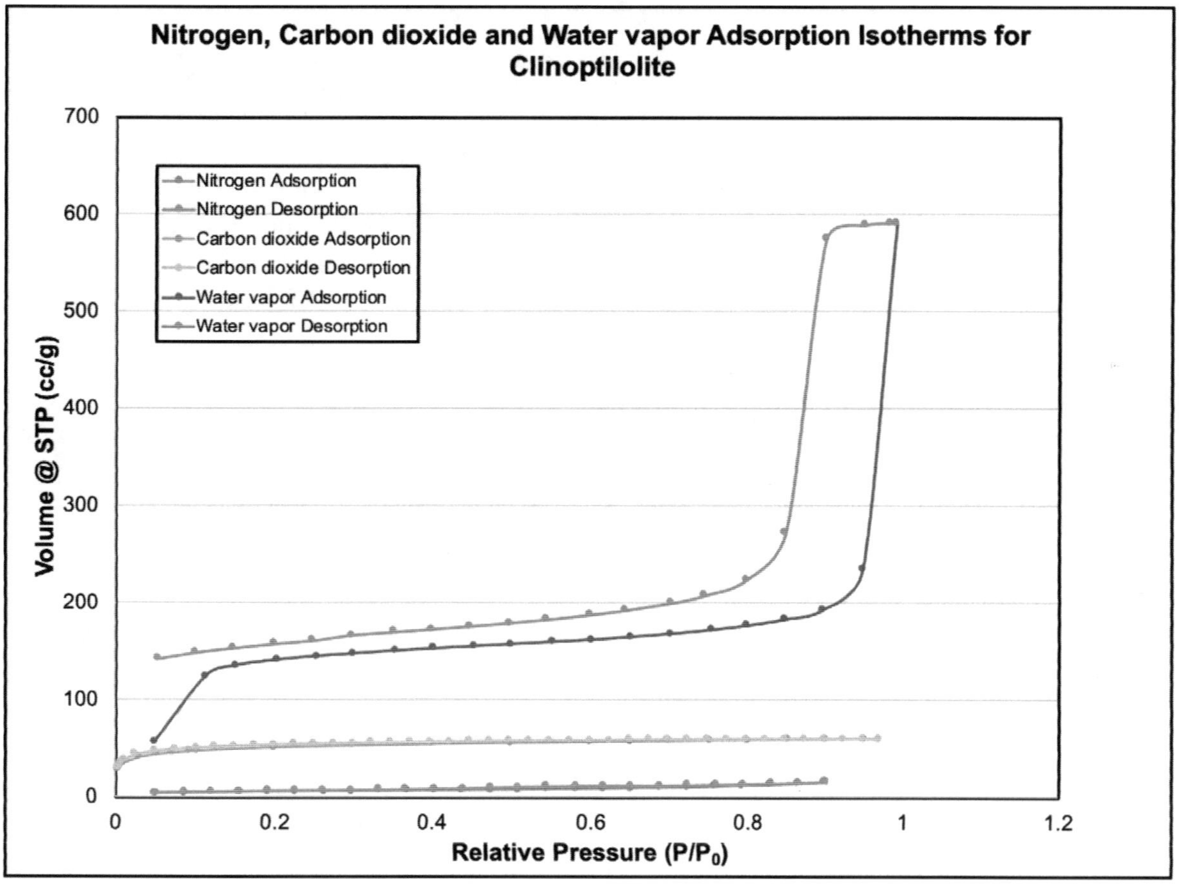

Figure 1: Nitrogen, carbon dioxide, and water vapor adsorption isotherms for clinoptilolite

609966 Exploiting Highly Anti-Fouling and Biofouling Activity through in-Situ Growth of Silver-Based MOFs on Zwitterion Incorporated TFC Polyamide Membranes

Abstract:

In this work, the surface of the polyamide (PA) forward osmosis (FO) membrane was modified with zwitterions followed by the in-situ growth of silver-based metal-organic frameworks (Ag-MOFs). First, the zwitterion 3-bromopropionic acid is grafted to the membrane surface using an initiator solution; then, the in-situ growth of Ag-MOFs is obtained by immersing the membranes in a metal solution followed by a ligand solution treatment. The successful surface functionalization is verified through surface characterization techniques. The contact angle measurements indicate an enhanced membrane wettability. The transport performance of the membranes is evaluated in a forward osmosis setup, in terms of water and salt transport parameters. The results show roughly 50% higher water flux than that of pristine membranes, with no significant change in salt selectivity. In order to investigate the antibacterial activity of the membrane, confocal, colony-forming unit (CFU), and scanning electron microscopy (SEM) tests were employed. The results show an over 77% increase in the bacterial inhibition rate of the modified membrane over the pristine membrane. The SEM images of the modified membrane treated with E. coli confirms the deterioration of bacteria cell. The long-term fouling and biofouling tests of the membranes reveal that the highest water flux drop for sodium alginate and E. coli do not exceed 25% and 20%, respectively. This proves the promising antifouling and anti-biofouling properties of the membranes. Finally, the stability of the membrane was tested over the measurement of silver ion release in 30 days. All modified membranes suggest negligible depletion of silver, confirming the strong immobilization Ag-MOFs on the surface of modified membranes.

610169 Characterization γ-Alumina Porosity By Monte Carlo Slit-Pore Kernel

Abstract:

Despite its importance in the area of catalysis and adsorption, γ-alumina does not have a specific kernel for its characterization. We report adsorption isotherms of nitrogen at 77 K in γ-alumina calculated using Grand Canonical Monte Carlo method. Nitrogen molecules were represented by a single-site model and γ-alumina was modeled as a limited collection of individual slit-shaped pores. Validated forcefield parameters for alumina-nitrogen interaction were applied. Adsorption analysis of specific individual pores is also presented. Our collection was able to reproduce the experimental isotherm of a porous γ-alumina sample. When compared to models based on cylindric-shaped pores or on DFT calculations, our model performed better, especially at very low-pressure range ($P/P_0 < 10^{-2}$) where the adsorption in the micropores is more significant. We also found that the microporous volumes of γ-alumina were systematically underestimated by existing methods.

610404 Thin-Film Polymer Membranes for Gas Separation Prepared By Interfacial Polymerization of Pillar[5]Arenes

Abstract:

Thin-film composite (TFC) membranes are considered as having good membrane architecture for membrane-based separations. Thin selective layer, of which thickness is usually less than one micrometer, is good for membrane productivity, and porous support membranes provide enough mechanical strength against high pressure operating conditions. For last few decades, a facile chemistry of interfacial polymerization has provided a great avenue for TFC membranes in liquid separation applications such as desalination or nanofiltration. However, the use of TFC membranes in gas separation has been limited. This is mainly attributed to the lack of knowledge that bridge chemistry-to-microstructure relationship of interfacial polymerization and, thereby, to the difficulty of tuning the microstructure of the TFC selective layer at molecular scale. Here, we propose to use pillar[5]arene nanoarchitectures to control the microstructure of these membranes by molecular design. Pillar[5]arenes have well defined molecular pore with ~4.7 angstrom in diameter. Also, a lot of functional groups enable pillar[5]arenes to directly participate in traditional interfacial polymerization process for TFC membrane fabrication. As a result, pillar[5]arene integrated TFC membranes showed good molecular sieving performance for light gas molecules; pure-gas selectivity were 26 for H_2/CO_2, 120 for CO_2/CH_4, 10 for O_2/N_2, and 3200 for H_2/CH_4, respectively. Also, cross-linked chemistry of interfacial polymerization provided substantial resistance to CO_2-induced plasticization for these membranes.

610592 Measuring Intrinsic Membrane Permeability Using in Situ FTIR

Abstract:

Membranes enable low energy separations in a variety of applications ranging from water purification to alternative energy devices. Membranes achieve solute separation because solutes transport across the membrane at different rates. Thus, accurate determination of membrane permeability, the intrinsic property that quantifies solute transport through the membrane, is essential to predicting membrane performance. Organic solute transport through a polymer electrolyte membrane is important in applications including artificial photosynthesis and direct methanol fuel cells. Permeation studies of organic solutes through commercial polymer electrolyte membranes have been conducted in aqueous diffusion cells equipped with in situ ATR-FTIR. In situ ATR-FTIR provides chemical specificity within multicomponent systems without the labor and complications of isolating samples for analysis. In these experiments, time-resolved concentration data on the downstream side of the membrane are collected, and by fitting a solution to the flux equation to this experimental data, the membrane permeability can be determined. Herein, we develop the methodology and data analysis utilized in in situ ATR-FTIR aqueous permeation experiments to account for boundary layers and determine the intrinsic membrane permeability. We demonstrate that the experimental conditions used herein are such that the apparent permeability is equivalent to the intrinsic membrane permeability. The development of this methodology improves our knowledge of a fundamental membrane property and could enable the study of higher complexity membrane systems in the future.

610916 Adsorption of C_8 Aromatics Isomers in Porous Clay Heterostructures Using the Headspace Technique

Abstract:

The separation of C_8 aromatics isomers is an important issue in the petrochemical industry. The C_8 aromatics fraction consists of four isomers: *o*-xylene (OX), *m*-xylene (MX), *p*-xylene (PX) and ethylbenzene (EB). They are the raw material for plastics, dyes, synthetic fibers and other products. The faujasite structure, especially Y zeolite, has been intensively studied and used for the separation of xylenes, mainly for p-xylene separation. However, a group of porous materials that appear as promising in the liquid phase separation is the porous clay heterostructures (PCH). In this study, the selective separation of xylene isomers in different PCHs samples was investigated. Porous clay heterostructures with Zr, Ti and Si (PCH-Zr, PCH-Ti and PCH-Si) were evaluated. A headspace technique was applied to acquire equilibrium data for adsorption of xylenes using small samples. Meta-xylene selectivity in relation to all the other C_8 aromatics was observed for the PCH samples. For PCH-Si, meta-xylene/ethylbenzene selectivity of 4.3 was obtained.

610951 Regenerability of Activated Carbon Used for Adsorption of Polycyclic Aromatic Hydrocarbons

Abstract:

The polycyclic aromatic hydrocarbons (PAHs) are priority pollutants due to their toxic, carcinogenic, and mutagenic effects. The PAHs presence in several hydrocarbon streams is directly dependent on the origin and characteristics of the petroleum and on the processes to which the raw material was submitted. Activated carbons may deal with PAHs compounds present in the heavy mineral oil. Some advantages such as low-energy demand, possibility of adsorbent regeneration, broad availability of adsorbents and possibility of adsorption from the liquid phase on activated carbons make this an interesting field of research. In this study, successive column experiments were carried out using a granular activated carbon (Norit 830W) to evaluate the reduction in capacity along continuous cycles and optimal operational conditions. A column simulation model, using the particle parameters obtained from batch studies, was applied to predict breakthrough curves for adsorption and desorption runs. After several cycles, using solvent or inert gas, a decrease in PAH adsorption capacity was observed. The adsorption capacity of 185 mg/g, assessed in terms of PAHs, was obtained for the fresh activated carbon sample. A slight decrease (*ca.* 6%) in adsorbent capacity was observed after three adsorption/desorption cycles using n-hexane as solvent in the desorption step. For desorption with inert gas at 150°C, the capacity was kept essentially constant after the 5th cycle, an interesting characteristic for a continuous operation process. A conceptual continuous process was simulated using the proposed model and evaluated based on productivity in terms of volume of treated oil per mass of adsorbent per time, showing a maximum value of 0.53 m^3/(kg ads.).h at space velocity of *ca.* 4 h^{-1}.

597143 Synthesis and Characterization of Highly Permeable Polyethersulfone Substrates with Bicontinuous Structure for CO_2-Selective Composite Membranes

Abstract:

The substrate pore structure has a significant effect on the CO_2 transport performance of a thin-film-composite (TFC) membrane. In this study, a novel solvent system comprising 2-pyrrolidone (2PD) and 2-methoxyethanol (2-ME) was used to prepare polyethersulfone (PES) substrates with bicontinuous structure by vapor-induced phase separation, followed by immersion in a nonsolvent, water. Compared with the common solvent, e.g., *N*-methyl-2-pyrrolidone (NMP), 2PD is a solvent with a lower solubility. However, the solubility was improved by the addition of 2-ME into 2PD. Moreover, the 2PD/2-ME system is more hydrophilic and can decrease the thermodynamic stability of the casting solution significantly. As a result, the phase separation during the membrane formation process was induced via the spinodal decomposition mechanism. The process parameters, including PES concentration, 2PD/2-ME weight ratio, water vapor exposure time, and water bath temperature, were optimized by characterizing the membrane morphologies and the CO_2 transport performance. Compared with the baseline PES substrate optimized previously with NMP, the new substrate showed a high surface porosity of 19.9%, indicating a 48.5% increase. Also, the new substrate exhibited a very high CO_2 permeance of 133,226 GPU at 57°C, which was 5 times more permeable than the baseline substrate. This high permeance was contributed mainly by the absence of a top dense layer and the presence of highly interconnected pores in the substrate. By using this substrate, the prepared TFC membrane containing amines showed a CO_2 permeance of 908 GPU, which was 48 GPU higher than that with the baseline substrate at 57°C. This improvement can be explained well using the resistance-in-series model. The improved CO_2 permeance was attributed to both the reduced lateral diffusion and substrate transport resistances. In addition, the scalability of the PES substrate with the bicontinuous structure was demonstrated by the successful roll-to-roll fabrication of a 50-foot long scale-up PES substrate with a width of 21″.

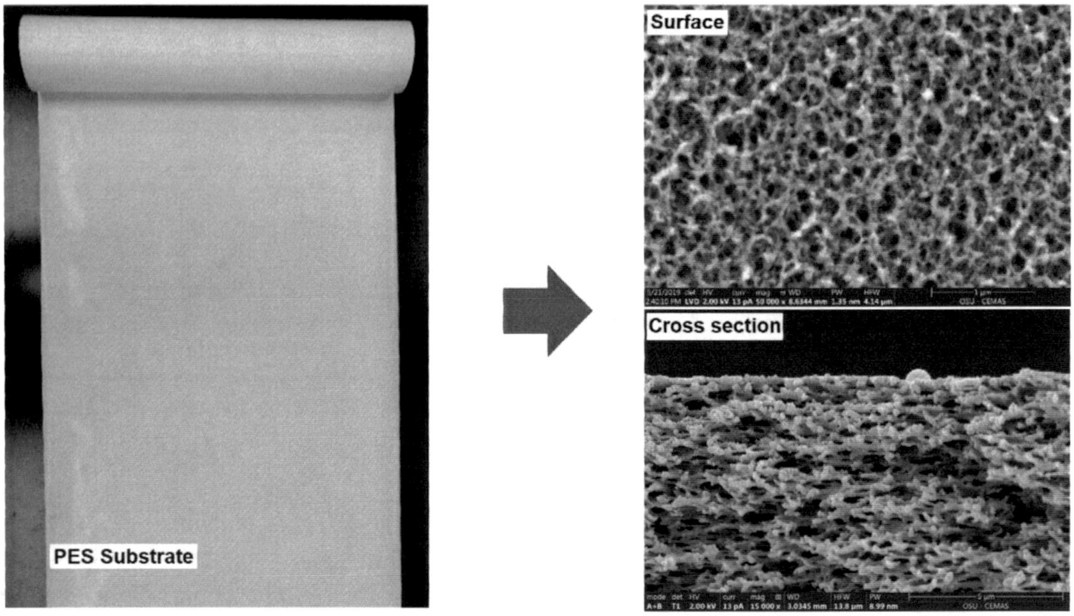

599052 Experimental Study of CO_2 Solubility in Aqueous Solution of Monoethanolamine and 2-Amino-2-Methyl-Propanol Blended Amine Using Electrolyte-NRTL Model

Abstract:

The technology of reducing CO_2 is a critical issue in various industries. Some kinds of alkanolamines have been used for CO_2 absorption: primary (monoethanolamine, MEA), secondary (diisopropylamine, DIPA), tertiary (methyldiethanolamine, MDEA) and steric hindrance (2-amino-2-methyl-1-propanol, AMP) amines. MEA has been used to process of CO_2 sequestration because of high reactivity and reasonable cost. AMP is applied as a stable amine for high CO_2 absorption capacity. In this study, equilibrium solubility of CO_2 in MEA and AMP mixtures were evaluated by using experimental data and thermodynamic models. The solubility of CO_2 was measured at various temperatures and amine mass fractions. The Electrolyte-NRTL model is used to estimate interactions between chemical species in liquid phase. To consider the non-ideality, activity coefficients, Electrolyte-NRTL model, were regressed from experimental data.

599053 Modeling of CO2 Solubility in Aqueous Solutions of Three Blended Amine Using Activity Coefficients

Abstract:

It is important issue that Reducing emissions of Carbon Dioxide (CO_2) without decreasing fossil fuel usage. some kinds of alkanolamines has been used for CO_2 absorption processes : primary, secondary, tertiary, steric hindrance amines and their binary mixtures. Among them, MEA (Monoethanolamine) is widely used as the solvent for carbon capture and sequestration (CCS). And AMP (2 – amino – 2 – methyl - propanol) is utilized as a stable amine for high CO_2 absorption capacity. Also, MDEA (N-methyl-diethanolamine) is uesd as a sweetening agent in chemical, oil refinery, syngas production and natural gas. In this study, equilibrium solubility characteristics of CO_2 in 3 Blended amines (MEA, AMP and MDEA) and their mixtures were evaluated by using experimental data and thermodynamic models. To consider the non-ideality, binary parameters of activity coefficients and equilibrium constants were regressed from experimental data. The DESHMUKGH – MATER models are used to estimate interactions between solute species in the liquid phase. Calculations of solubility and optimizations (parameter regression) were conducted by MATLAB® 2019a version. This frame work could be applied to processes of SOx, NOx.

599826 Improved Synthesis of Sterically Hindered Poly(N-methyl-N-vinylamine) Membrane for Post-Combustion Carbon Capture

Abstract:

Membrane separation technology requires highly CO_2-selective and permeable membranes for a cost-effective post-combustion CO_2 capture from flue gas. Amine-containing facilitated transport membranes are able to realize an increased permeability due to the reversible reaction between CO_2 and amino groups. Here, we have developed an improved method for the synthesis of a high-molecular-weight sterically hindered polyvinylamine (SH-PVAm) as the new fixed-site carrier in our ultra-thin CO_2 selective membranes.

Commercial polyvinylamine (PVAm) was mono-methylated into SH-PVAm using the stepwise reductive amination with a highly polar fluorinated alcohol as solvent to enhance the equilibrium shift to the imine formation. The method was found to prevent the over-alkylation, resulting in the increased yield of the target product. The synthesized SH-PVAm was characterized and confirmed by FTIR and ^{1}H NMR. The SH-PVAm solution retained a sufficiently high viscosity after incorporating the aminoacid salt, 2-(1-piperazinyl)ethylamine sarcosinate, as the mobile carrier for the membrane coating without any penetration issue. The SH-PVAm exhibited a significantly enhanced performance in the blended membrane of about 170 nm thickness over the unmodified PVAm for CO_2/N_2 separation at 57°C. The former membrane demonstrated a CO_2 permeance of 1071 GPU and a CO_2/N_2 selectivity of 183 whereas the latter showed 975 GPU CO_2 permeance and 163 selectivity. The improved membrane performance corresponded well with our computational results on the reaction chemistry of CO_2 between the sterically hindered vinylamine monomer (SH-VAm) and the pristine vinylamine monomer (VAm) by density functional theory (DFT). SH-VAm showed a preference for the more efficient bicarbonate formation pathway and contributed to a higher CO_2 loading capacity. Therefore, the steric hindrance effect of SH-PVAm enhanced the solubility of CO_2 in the polymer matrix and resulted in the higher CO_2 permeance.

602371 Facilitated Transport Membranes with Tunable Amine–CO2 Chemistry for Hydrogen Purification

Abstract:

CO_2-selective, amine-containing facilitated transport membranes (FTMs) are of great interest for syngas purification since high-pressure H_2 can be retained upon CO_2 removal. Various FTMs have shown decent chemical and thermal stability at aggressive conditions, but their CO_2/H_2 separation properties are largely limited by the severe carrier saturation at high syngas pressure. Herein, we report a new approach to enhance the CO_2 permeability by manipulating the steric hindrance of the amine carrier. A series of α-aminoacids with different alkyl or hydroxyethyl substituents were deprotonated by 2-(1-piperazinyl)ethylamine, resulting in nonvolatile amine carriers with different degrees of steric hindrance. For hosting the low MW amine carriers, a water-swellable polymer network was synthesized from poly(vinyl alcohol) crosslinked by a bidentate tertiary aminosilane. In the presence of moisture, a bulkier alkyl substituent increased the steric hindrance and hence destabilized the carbamate adduct to afford bicarbonate through hydrolysis. Thus, this drastically increased the chemisorption of CO_2. Further, density functional theory (DFT) calculations were conducted to study the function of the hydroxyethyl substituent, which indicated that the hydroxyl group stabilized the bicarbonate through strong hydrogen bonding, thus further improving the CO_2 sorption. The enhanced CO_2 solubility significantly mitigated the carrier saturation, and an unprecedented CO_2/H_2 selectivity greater than 130 was demonstrated at 107°C and 12.5 atm of CO_2 partial pressure. As the CO_2 partial pressure reduced to 0.4 atm, a less hindered amine yielded a higher reactive diffusivity of CO_2, resulting in a CO_2 permeance of 435 GPU with a selectivity greater than 500. The CO_2/H_2 separation performance of these reaction-mediated polymeric membranes is well above the theoretical upper bound, and they open up a new avenue for designing a highly selective membrane process for syngas purification.

602373 Amine-Containing CO2-Selective Membrane and Process for Carbon Capture from Flue Gas

Abstract:

Large-scale application of membrane in post-combustion carbon capture has been limited by the trade-off between CO_2 permeance and CO_2/N_2 selectivity of most polymeric membrane materials. In order to overcome this limitation, research efforts on facilitated transport membrane (FTM) have been devised with the objectives of (1) developing carriers with high CO_2 loading capacity and reactive diffusivity, and (2) designing membrane processes that can capitalize on the outstanding selectivity shown by FTM. In this presentation, a novel FTM was synthesized in a composite membrane configuration with a 170-nm selective layer coated on a nanoporous support. In the selective layer, polyvinylamine was used as fixed-site carrier and an amino acid salt, synthesized by deprotonating sarcosine with 2-(1-piperazinyl)ethylamine, was blended as mobile carrier. The membrane was used to fabricate 1.4-m^2 spiral-wound modules, which exhibited a CO_2 permeance of 1450 GPU and a CO_2/N_2 selectivity of 185 at 67°C with actual flue gas at the National Carbon Capture Center in Wilsonville, AL, USA.

In addition, for the range of CO_2 partial pressure relevant to carbon capture from coal-derived flue gas, significant increase in CO_2 permeance can be achieved upon the bulk removal of CO_2. For instance, at 67°C, the CO_2 permeance can increase from 1464 to 1918 GPU when the CO_2 partial pressure reduces from 74.1 to 3.9 kPa. Such a carrier saturation phenomenon has been modeled and incorporated into a two-stage membrane process featuring partial retentate recycle. It has been demonstrated that the bulk CO_2 removal reduces the CO_2 partial pressure gradually in the membrane module. This feature mitigates the carrier saturation and results in an uprising CO_2 permeance upon the CO_2 removal. For the membrane performance at 67°C, an attractive capture cost of $41.75/tonne can be achieved for 90% CO_2 capture in addition to offering the reduced system footprint.

603434 Self-Diffusion of Small Alcohols inside ZIF/Polymer Hybrid Membranes By Pulsed Field Gradient (PFG) NMR

Abstract:

Mixed matrix membranes (MMMs) formed by dispersing zeolitic imidazolate frameworks (ZIFs) crystals in a polymer have emerged as promising materials for gas and liquid separations. ZIFs represent a class of metal organic frameworks (MOFs) with zeolite-like structural properties. ZIF/polymer MMMs benefit from molecular sieving properties of ZIFs and advantages associated with a fabrication of pure polymeric membranes.

^{13}C pulsed field gradient (PFG) NMR was used to study self-diffusion of methanol and ethanol inside loosely packed beds of ZIF-71 crystals and ZIF-71 / Torlon MMM under conditions of sorption equilibrium with the saturation vapor pressure of each alcohol. Diffusion measurements were performed at 253 K by utilizing a high magnetic field (14 T) and high magnetic field gradients (up to 18 T/m) to obtain intra-ZIF self-diffusivities for the length scales of displacements smaller than and comparable to the average size of the ZIF-71 crystals. To confirm an absence of any measurement artifacts, additional measurements were performed at 17.6 T. It was observed that the intra-ZIF self-diffusivities of the same liquid in ZIF-71 crystal beds and ZIF-71 / Torlon MMM were the same, within the experimental error, for the smallest root mean square displacements (root MSDs) corresponding to no or little influence of the crystal external surface on the diffusion process. For larger root MSDs, the intra-ZIF self-diffusivities in ZIF-71 / Torlon MMM were observed to decrease with increasing root MSD values. This observation is tentatively explained by the presence of diffusion barriers on the interfaces separating individual crystals in ZIF-71 interconnected clusters, which are believed to form in the studied MMM. The diffusion data for liquid sorbates is discussed in the context of the results previously measured by PFG NMR for gas diffusion in a similar ZIF / Torlon MMM.

591570 Product Yield and Purity in Stripping Crystallization of *L*-Menthol from the Enantiomer Mixture

Abstract:

Stripping crystallization (SC) is a new separation technique which combines distillation and crystallization. Basically, SC is operated at a triple-point condition, in which the liquid is simultaneously vaporized and crystallized due to the three-phase equilibrium. In essence, the process is continued until the liquid phase is completely eliminated and only the pure solid crystals remain in the feed. Thus, crystal washing is not required since only pure solid crystals remain in the feed and no impurities are adhered on the crystal surfaces at the end of SC. In the previous research, SC has been successfully applied to separate p-xylene from the mixed xylenes by Shiau et al. (2008, 2013, 2018).

Stripping crystallization (SC) is introduced in this work for chiral purification of *L*-menthol from the enantiomer mixture. As SC is operated in a series of the three-phase transformation conditions, SC combines melt crystallization and vaporization to produce the *L*-menthol crystalline product and mixture vapor from the mixture melt. The operating pressure for the three-phase transformation during the SC cooling process is determined based on the thermodynamic calculations. Sweating is applied at the end of SC to expel the mixture melt remained with the final *L*-menthol product. A model based on the mass and energy balances is proposed to determine the product yield and enantiomeric purity of *L*-menthol for SC.

References:

1. 1. Shiau, L. D. “Purification of styrene from a styrene/ethylbenzene mixture by stripping crystallization”, *Eng. Chem. Res*, 2018, 57, p6759-6765.
2. Shiau, L. D.; Liu, K. F. ”Investigations into the effects of the cooling rate on stripping crystallization”, Ind. Eng. Chem. Res, 2013, 52, p1716-1722.
3. Shiau, L. D., Wen, C. C. and Lin, B. S. “Separation of p-xylene from the multicomponent xylene system by stripping crystallization”, AIChE Journal, 2008, 54(1), p337-342, 2008.

Keywords: crystallization; distillation; purification

596596 Separation of Hydrofluorocarbon Azeotropic Mixtures Using Polymeric Membranes

Abstract:

Project EARTH (Environmentally Applied Research Towards Hydrofluorocarbons) is focused on identifying sustainable processes for the selective separation of hydrofluorocarbon (HFC) refrigerant mixtures. HFCs and HFC mixtures were developed to replace chlorofluorocarbons (CFCs), which were linked to the depletion of the Earth's ozone layer. However, the current concern is some HFCs and HFC mixtures have high global warming potentials (GWPs) and recent legislation such as the Kigali agreement are calling for restrictions in certain applications with an eventual phase-out of HFCs. Currently, millions of kilograms of HFCs and HFC mixtures are in use with no efficient method for their disposal. In order to avoid incinerating or venting HFC refrigerants into the atmosphere—processes that are harmful to the environment and would waste millions of kilograms of refrigerants—Project EARTH aims to develop environmentally responsible methods for separating azeotropic refrigerant mixtures so that low-GWP components can be repurposed into next-generation refrigerants. The initial focus of Project EARTH is the separation of R-410a, a binary azeotropic mixture of 50 wt.% R-32 (CH_2F_2) and 50 wt.% R-125 (CHF_2CF_3), commonly used in air conditioning and heating pump systems. R-32 has a much lower GWP (675) compared with R-125 (3,500) and, if separated, could be reused in new refrigerant blends with hydrofluoroolefins (HFOs). This project focuses on the use of polymeric membranes for the separation of R-410a. Membranes are an attractive method for separating HFC refrigerant mixtures due to lower energy consumption and smaller capital requirements compared with alternative separation methods such as distillation. This presentation will cover current results on the solubility, diffusivity, permeability, and selectivity of R-32 and R-125 in a variety of membrane materials.

604125 Investigations of Mixed-Mode Cation Exchange Chromatography for Control of Aggregates in Bispecific Antibodies

Abstract:

Bispecific antibodies are a class of engineered antibodies with the ability to bind to two distinct antigens. Compared to monoclonal antibodies, single chain variable fragment-IgG bispecific antibodies are prone to high levels of aggregate (10-30%), which can impact product yields and must be removed during downstream polishing steps. Mixed mode chromatography (MMC) provides a potential method of aggregate control due to its multiple modes of protein-ligand interactions, resulting in enhanced selectivity compared to traditional ion exchange and hydrophobic interaction chromatography resins. Ceramic hydroxyapatite (CHT) is an MMC resin often used as a polishing step for many high aggregate forming species; however, its scalability issues raise the need for an alternative aggregate removal step. Initial studies showed that mixed-mode cation exchange (CEX) resins clear aggregates with high selectivity and minimal breakthrough while operating in flow-through mode. To further investigate potential alternatives to CHT, we evaluated prototype POROS™ mixed-mode CEX resins from Thermo Fisher Scientific to study the impact of ligand density, operating pH and arginine concentration on aggregate removal ability.

606031 Droplet Evaporative Crystallization on Regular Micro-Structure Platform

Abstract:

Droplet Evaporative Crystallization on Regular Micro-structure Platform

Xiaobin Jiang, Mingguang Han, Gaohong He*

School of Chemical Engineering, Dalian University of Technology, Dalian 116024, China

E-mail: hgaohong@dlut.edu.cn

Abstract

Droplet evaporative crystallization on microscale heterogeneous surface is a vivid topic in chemical engineering, bioengineering, nanomaterials, etc. Here, an interfacial platform with regular pillar convexity and tunnel structure is fabricated to reveal the mechanism of the interfacial micro droplet crystallization. Element-based rotation volume model is established to simulate the concentration and nucleation barrier distribution during the microscale process. Sodium urate monohydrate and NaCl crystallization on the pillar convex structure both confirm that confined capillary flow in the micro droplet and initial nucleation condition dominate the nucleation, growth control and particle distribution. Droplet crystallization stretches over the tunnel structure reveal an interesting phenomenon that two regions possessing distinct-different nucleation barriers can isolated obtain the crystal particles from nanoscale to even millimetre scale. The fabricated platform and the capillary circulation transfer theory unfold a potential approach to develop novel micro-scale crystallization.

Figure 1 Nucleation energy barrier (J/mol) distribution (A-1 to A-4) and the corresponding crystal formation (B-1 to B-4) of SUM droplets with different initial saturated degree.

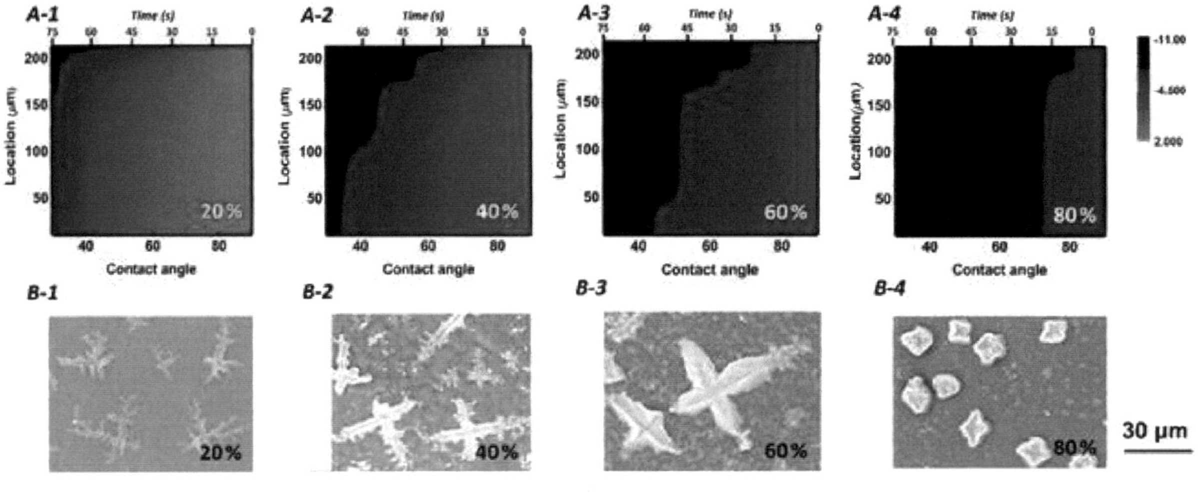

610410 Para-Crystallization Behaviors in Organic Solvents

Abstract:

Crystallization from solution underlies myriad processes in nature and industry. Solvents crucially determine the crystal habit and solvent-surface interactions impose distinct crystal morphologies. The association of solute molecules to crystal growth sites drives away solvent associated to solute and structured in the vicinity of growth sites. Thus, insight into the solvent structure and dynamics at the crystal-solution interface is an essential prerequisite for understanding of growth mechanisms. In recent years, the crystallization of organic materials from the organic solvents has been a focus of attention as a major step in the manufacture of pharmaceuticals and fine chemicals. Despite extensive efforts, a molecular perspective of the fundamental thermodynamics and kinetics aspects of the growth mechanisms remain elusive. We employ all-atom molecular dynamics simulations and advanced sampling techniques to investigate the structure, dynamics, and energetics that govern the molecular behaviors during organic crystal growth. We address the solvent structuring at the interface, solute oligomerization in the solution and the adsorption of the solute monomers and oligomers on the crystal surface, processes that run parallel to crystallization and largely determine its rates and outcomes. We found that the weak intermolecular bonds between organic solvents and the crystal surface did not preclude the build-up of layered solvent structure along the surface. The degree of order in the solvation layer is determined by the chemical and topological features exposed on the crystal surface. A solvation layer nearest to the surface may impeded or promote the solvent and solute dynamics and the layer characteristics drive disparate growth rates on the anisotropic faces of β-hematin crystals, a central part of the survival of malaria parasites. Furthermore, despite common beliefs, a majority species of solute in the solution may or may not be the incorporating species for the crystal growth. The (010) faces of etioporphyrin crystals, an organic semiconductor, grow by direct incorporation of the dimers with dimers present in the solution as the majority component. On the other hand, the (002) face of olanzapine, an antipsychotic drug for schizophrenia, grows by the surface diffusion with incorporation of the dimers even though dimers represent a minority component in the solution. We show that growth by dimers is preferred owing to their preferential adsorption on the crystal surface, complemented by additional dimerization on the surface. These findings are crucial for molecular-level understanding of the fundamental thermodynamic and kinetic parameters of growth that guide the optimization of crystallization from the organic solvents.

611939 Errors in the Prediction of Irrigated Pressure Drop, Holdup, and Flooding Velocity By Stichlmair, Bravo and FAIR (GAS SEP AND PURIF. 3,19-28(1989)

Abstract:

The reinterpretation of the equations for friction factor for dry pressure drop in terms of a simplified power law model proposed by Stichlmair et al (1989) has been shown to be incorrect and unnecssary as it does not simplify the model equations developed for the prediction of wet pressure drop, liquid holdup, and flooding velocity substantially nor the computations involved get simplified. Straight forward prediction of the above parameters following the basic model equations of Stichlmair et al but avoiding the use of power law model have been developed and their experimental data have been reinterpreted. New experimental results for dry pressure drop, wet pressure drop, holdup, loading and flooding velocities obtained by using IMTP and Mellapak 350Y packings in a 30 cm dia packed column in the present work have also been evaluated by using the corrected model of Stichlmair et al (1989)

598530 Separating Mixture Glycols in DWC with Flow Rate – Composition Cascade Control Structure

Abstract:

Dividing wall column (DWC) is an enhanced distillation technology which can separate multicomponent mixtures in one column with only a single condenser and reboiler. It can save up to 30% energy and reduce 30% capital costs over conventional distillation column sequences. However, there are only a few DWCs applied in industry since its first application in 1985 by BASF. The design of control structure is one of the keys to limit its industrial application.

In this work, glycols (1,2-propanediol+ 1,3-butanediol+ 1,4-butanediol) were taken as the feed mixture, an optimized 50-stages DWC was built in Aspen Plus, and four control structures were designed in Aspen Dynamics. In order to investigate the dynamic responses for the control structures in DWC, ± 10% feed disturbances would be applied to the system after 1h operation.

The first control structure proposed for DWC was temperature control structure (CS1). Reflux ratio (RR), side stream flow rate (S) and reboiler heat duty (Q_R) were selected as manipulated variables. Sensitivity analysis and singular value decomposition (SVD) criteria were used to obtain the appropriate reference stage. Temperature controllers were tuned through the relay-feedback test and Tyreus-Luyben tuning method. The dead time was set as 1min and the control loops were tuned sequentially. The results showed that CS1 had difficulty in maintaining the product purities under ± 10% feed disturbances, which included feed flow rate disturbances and feed composition disturbances.

Temperature difference control structure (CS2) was also investigated in this work. Different from CS1, the system could be stabilized in 3hrs for CS2 and the maximum deviations for the purities of 1,2-propanediol and 1,3-butanediol were 0.04% and 1.03%, respectively, while the purity of the bottom product 1,4-butanediol couldn't become steady after 10hrs' operation yet.

For Composition control structure (CS3), composition controllers were tuned through the relay-feedback test and Tyreus-Luyben tuning method. The dead time was set as 3mins. The results showed that the maximum deviations for the product purities of 1,2-propanediol, 1,3-butanediol, and 1,4-butanediol were 0.35%, 2.78% and 0.41%, respectively. The feed composition disturbances could be handled well in 4hrs while it took about 7hrs to become steady for the feed flow rate disturbances.

A flow rate-composition cascade control structure (CS4) was presented in this paper finally. A cascade loop was used to control the purity of side stream product by manipulating S. A temperature controller was used to control the temperature of reference stage in the rectifying section by manipulating RR. And a composition controller was used to control the purity of bottom product by manipulating Q_R. The maximum deviations for the product purities of 1,2-propanediol, 1,3-butanediol, and 1,4-butanediol were 0.07%, 1.44% and 0.46%, respectively. The DWC could recover to steady state in 4hrs. It showed that CS4 had a better performance on dealing with the disturbances than CS3.

In conclusion, the feed disturbances couldn't be handled well for CS1; the settling time was shortest while the bottom product purity couldn't become steady after 10hrs' operation yet for CS2; the system could be stable with a long settling time for CS3; and the feed disturbances could be dealt with small deviations and short settling time for CS4. Both CS3 and CS4 could handle the disturbances, while CS4 had the best performance among the four control structures proposed in this work.

596492 Exploration of the Gas-Liquid Vibration Mechanism for Industrial Float Valve Tray

Abstract:

Gas-liquid vibration is a unique operating phenomenon caused by the flowing of gas and liquid phases when the distillation tray is operating at an extremely low gas load. The mechanical fatigue is easily caused when the distillation column is operating under the condition of gas-liquid vibration for quite a long time. It is easy to induce the occurrence of equipment resonance when the frequency of gas-liquid vibration is close to the natural frequency of the equipment. At the same time, the high energy of vibration can easily cause mechanical damage, even cause major safety accidents such as explosions and fires. Therefore, it is essential to carry out research work on the gas-liquid vibration phenomenon of distillation columns.

In this paper, the mechanism and hydraulic characteristics of gas-liquid vibration have been investigated in detail based on an industrial-scale two-pass rectangular simulator (6400×800 mm^2) with the air/water system combining with multiple-point pressure measurement. The main results are as follows:

Based on the energy analysis of gas-liquid phases in the VCFC model of the float valve tray and combined with the principle of minimum energy, and it is concluded that the continuous bubbling and weeping region cannot be able to exist stably at the stage of gas-liquid vibration but change with the time being. At the same time, it is proved that zero hydraulic gradient on the tray is one of the necessary conditions for the occurrence of gas-liquid vibration.

It is confirmed that the formation of the large size of bubbles with low-frequency is the second necessary condition for the occurrence of gas-liquid vibration through the experimental research on the Glitsch V1 float valve and Super V1 float valve.

The specific reason that the gas and liquid phase fluctuated and the formation process of gas-liquid vibration are explored by analyzing the fluctuation pressure signal and the process of energy conversion of gas-liquid phases.

By setting the vibration experiments on the counter-flow column, it is proved that the essence of gas-liquid vibration is the interaction between the liquid layer pressure on the tray and the gas chamber pressure below the tray, at the same time, the liquid that jumps through the downcomer intermittently owning to vibration has little effect on vibration characteristics.

The experimental results show that the liquid height on the tray and weeping fraction in each section are distributed homogeneously due to gas-liquid vibration. However, the uniformity of distribution of liquid height and weeping fraction on industrial float valve tray becomes worse under the conditions of high weir height and high liquid flow intensity compared with the experimental results of the small-scale float valve tray, which is the scale-up effect also exists in the region of gas-liquid vibration. Furthermore, the heterogeneity of hydraulic characteristics becomes more evident with the increase in the number of valves and liquid flow intensity.

594574 Understanding of Fluid Flow Characteristics in Macrocellular Foam Materials for Enhanced Distillation Process

Abstract:

Due to macrocellular foam materials have great application prospects in medium-low pressure drop with low liquid load operation, it has been a hot spot in this field to explore the liquid phase flow inside the pores of macrocellular foam materials. The previous research found that under certain operating conditions, when liquid phase flows through the macrocellular foam porous channels, liquid phase special flow pattern will form in the SiC open-cell foam porous channels. The liquid-phase special flow pattern increases the residence time of the liquid in the packing and enhances the reaction efficiency between the liquid phases. Therefore, applications using macrocellular foam material as distillation packing or trays have demonstrated higher performance than classical mass transfer units, except for the operation range. To overcome these disadvantages of classical units, a series of many novel mass transfer units based on the macrocellular foam materials have been developed, such as Foam Monolithic Tray, Foam Ring Random Packing (FRRP), Structured Corrugation Foam Packing (SCFP) and so on. In this paper, the hydrodynamic performances and mass transfer efficiency of a series novel mass transfer units are examined in lab-scale equipment, with special emphasis on the effect of the foam cell size and sheet thickness. In general, based on the comparison with the traditional novel mass transfer units, the experimental results indicate that the hydrodynamics and mass transfer performances of novel mass transfer units meet the requirements for the mass transfer elements with high theoretical stages and low pressure drop in the distillation column.

597189 Ultra-Permeable Wafer-Scale SWCNT Membranes for Efficient Dye/Salt Fractionation

Abstract:

Enhanced fluid transport in single-walled carbon nanotubes (SWCNT) promises to enable major advancements in many membrane applications, from efficient water purification to advanced protective garments. So far, practical realization of these promises has been hampered by the difficulty of fabricating large-area, defect-free membranes with a high density of open, small diameter SWCNT pores. To simultaneously increase permeability, selectivity, and membrane area, we fabricated membranes from wafer-scale (100 mm) SWCNT forests optimized for high CNT number density (2×10^{12} tubes/cm^2) and narrow diameters (average < 2 nm) and ensured complete opening of all conducting nanotubes in the composite. Opening all the nanotubes enabled us to quantify pressure-driven transport enhancement for gases and liquids more accurately than previous experimental reports. Measured gas permeances exceed Knudsen diffusion by more than 100-fold, whereas water transport rates surpass the no-slip Hagen-Poiseuille prediction by 6200 times. In addition, we demonstrate the ability to scale up membrane area 60 times without loss of performance. Fabricated CNT membranes at both small- and large-scale display water permeances in excess of 200 L/m^2hbar, significantly larger than those of commercial loose nanofiltration/tight ultrafiltration membranes of similar pore size and reject nanometer-sized molecules while simultaneously permitting the complete passage of concentrated salt solutions (NaCl and Na_2SO_4). Taken together with their excellent chemical resistance, these membranes offer opportunities for energy-efficient nanofiltration/ultrafiltration processes in chemically demanding environments.

610075 Electrochemical Water Desalination Using an Iron-Based Redox Couple

Abstract:

Water desalination is a promising technological solution to water scarcity. While several technologies are currently utilized to produce freshwater from saline water sources, there are remaining challenges, including intensive energy usage (e.g., thermal distillation), fouling (e.g., reverse osmosis), and high material costs (e.g., electrodialysis). The use of capacitive or battery electrodes allows for electrochemically driven capture and release of ions, efficient water desalination using inexpensive electrode material, and low voltage operation without producing byproducts. However, the separation can only be achieved using materials available on the electrode surface, which leads to undesirable mixing between cycles and limited magnitude of desalination. To overcome this challenge, we demonstrated a new strategy for water desalination using an iron-based redox couple. Water desalination was achieved in a 4-channel electrochemical flow cell separated by an alternative array of anion and cation exchange membranes. The redox couple was recirculated between two channels at each end of the cell. Electrochemical reactions occurring at carbon electrodes created a charge imbalance and enabled the transport of Na^+ and Cl^- through cation and anion exchange membranes to maintain electroneutrality. Using 20 mM NaCl as a model brackish water, we continuously produced a desalinated water stream (15 mM NaCl) with energy consumption that was less than one third (< 0.02 kWh m^{-3}) of that required in electrode-based systems (> 0.06 kWh m^{-3}) at a productivity of ~ 40 L m^{-2} h^{-1}. Continuous production of freshwater at a drinking water level (0.5 g/L) was achieved using 50 mM NaCl as a feed solution. Other aspects of this system, including electrochemical properties, environmental compatibility, and water recovery ratio, will be discussed as well. Collectively, we show that the use of a redox couple can represent an energy-efficient method for electrochemical water desalination.

611494 CVD Graphene Based Membrane for Water Desalination

Abstract:

Although almost 70% of the world is covered by water, less than 3% of it is fresh. The remainder—the vast majority of Earth's water—is saline. To meet humanity's increasing and unavoidable need for water, we must convert saline water into fresh water through desalination. Many desalination technologies are available, and membrane-based technologies, such as reverse osmosis (RO), are widely used. Though RO is especially common because of slightly fast transport of water molecules through a membrane while obstructing all ions, but desalination through RO faces several challenges. For example, high operation cost due to high pressure during the process or its degradable polymeric active layer in contact with Chlorine-containing compounds. Polymeric membranes are also susceptible to various foulants, such as biofouling, which become more problematic in the absence of treatments like chlorine. In terms of general functionality, water transport through membranes could still be improved by increasing membrane permeability, but progress in this area has been impeded by a lack of data about how polymer membrane chemistry and structure affect fundamental transport properties. Consequently, improvements in membrane permeability have been relatively slow. Current composite membranes have permeability less than two times higher than those produced twenty years ago.

One of the most famous and recently studied candidates for use in RO or NF membranes is Graphene which has excellent chemical, and mechanical stability and it is the most prominent thinnest possible membrane with its one atom thickness acting as the membrane. In a further benefit, graphene manifests greater resistance to chlorine than current polyamide membranes. This research focused on developing graphene membranes with higher flux and better ion/molecule selectivity compared to the RO/NF membranes currently available on the market.

In this study, the process of transferring Chemical Vapor Deposition (CVD) graphene on to a hydrogel will be discussed. The process of transferring the membrane to the substrate is challenging: CVD graphene is fragile and can easily be torn if it is directly transferred to the support structure. To allow transference without damage, a hydrogel substrate—polyvinyl alcohol (PVA)—has been synthesized and cross-linked. The degree of crosslinking, the thickness of the casting, can affect the permeability of a PVA membrane. The graphene is transferred onto the PVA support by a simple but unique approach to decreasing the chance that defects will form.

Making tunable nanometer pores by plasma: As reported by O'Hern et al. , using plasma through the graphene can create sub-nanometer pores that make the graphene more selective. Consequently, it will be discussed the results of the plasma cleaner with the different condition that was used to apply nanometer pores on the graphene surface.

Experimental work combined with membrane characterization methods (FESEM, AFM, and, LEXT) and membrane performance studies by using a RO system to examine the graphene as RO/NF membrane will be presented. These results will be compared with the other synthesized membrane and different types of supports.

This research will provide insights into developing CVD graphene-based membranes with high water permeability and excellent size selectivity, membranes that will be highly attractive for water purification and desalination because of their high energy-efficiency.

609511 Anti-Bacterial Chitosan-Copper Supramolecular Assembly for Coating Thin-Film Composite Membranes

Abstract:

Membrane-based desalination has become a standard practice for providing water for the growing demand. Membranes used in water desalination are prone to Biofouling and approach to mitigate membrane biofouling is at high demand. To impart antibiofouling properties to membranes various approaches has been used. Among these approach metallic biocides, such a silver and copper nanoparticles, has shown promising results. Nonetheless, these particles are prone to dissolution in water. Here we demonstrate the immobilization of copper on Thin Film Composite Reverse Osmosis membrane surface by including copper ions in chitosan complex deposited on the surface. The chitosan-copper chelate was formed by dissolving different concentrations of Copper sulfate in 1000 mg/L Chitosan solution at pH 2.85. The Commercial Thin Film Composite Reverse Osmosis was subjected to oxidizing agent, creating active sites on polyamide for binding with the chitosan-copper complex. After immobilizing the supramolecular complex on membranes, they were cured at 80°C for 30 minutes. Subsequently, the excess Copper was removed by putting the membrane in 0.1M Sodium hydroxide overnight. The film uniformity, metal ion concentration, metal ion leaching rate, antibacterial performance, and transport properties of the membrane were evaluated. The modified membrane demonstrated anti-bacterial properties, proving that the current scheme is successful in introducing biocides into TFC membranes. Our data suggests that copper ions leach out from the surface, results in the loss of antibacterial activity. Thus, we developed a process for rejuvenation of the surface through post-treatment.

Acknowledgement: Authors acknowledge Indo-US Science and Technology Forum - Water Advanced Research and Innovation Fellowship under which the research work was undertaken and University of Nebraska Lincoln, the host university.

607845 Selective Desolvation of Functional Groups Determines the Crystal Structure during the Process of Crystallization

Abstract:

Crystallization involves self-assembly of solute molecules which results in a crystal with specific structure and morphology. The molecular events in such self-assembly are driven by supersaturation. Understanding of which molecular events lead to the certain outcome of crystallization can significantly improve the efficiency of the crystallization process and reduce the need of post-processing steps. Recently, the dynamics of solvation shell during the process of crystallization were uncovered which were then used to calculate the growth rates. To effectively answer which crystal structure will arise during nucleation, it is necessary to understand the functional group specific desolvation. In this study, molecular dynamics simulations of glutamic acid molecule in water at different supersaturations were performed and the interaction energy phase space arising due to different interactions of different functional groups was explored. Findings of this study suggests that glutamic acid molecules attain a dense phase of energy minima before the formation of crystallization process, similar to the non-classical nucleation mechanism observed for many organic and protein molecules. Further, the groups specific analysis allows mapping the molecular events to the polymorphs of glutamic acid. At high supersaturation, the energy minima in the phase space matches with the metastable polymorph of the crystal structure suggesting the asymmetric distribution of water around different functional groups determines the crystal structure.

600017 Ultrasound-Assisted Intensified Crystallization of L-Glutamic Acid: Crystal Nucleation and Polymorph Transformation

Abstract:

Solution crystallization is an important and energy-efficient separation process for crystalline material production. In the past few years, several external fields were introduced to optimize the crystal product, in which the ultrasonic irradiation was proved to be one of the most effective technologies to tune crystal properties and intensify the crystallization process. In this work, the crystallization of L-glutamic acid with ultrasound applied was explored in detail, including the process of nucleation, polymorphic control and polymorphic transformation (Fig. 1). The induction time and metastable zone widths (MSZWs) were measured with and without ultrasound during the nucleation process. The induction time and MSZWs were decreased by ultrasound and the induction time was further decreased by higher ultrasonic power. The calculated nucleation parameters (such as interfacial energy, critical nucleus size and critical Gibbs energy) showed an obvious decrease in the presence of ultrasound, indicating that the nucleation was enhanced with ultrasound applied. By adjusting the ultrasonic power in the quench cooling process, the difference in nucleation temperatures would determine the distribution of polymorphs. In further, the polymorphic transformation was investigated quantitively, and to the best of our knowledge, it was the first time to study the transformation kinetics with ultrasound using Avrami-Erofeev model. In the transformation process, the crystallization mechanism of the stable form was modified by ultrasound. The ultrasound eliminated the nucleation element in the rate-limiting step and facilitated the crystal growth of stable form. Thus, the ultrasound has a profound influence on L-glutamic acid crystallization.

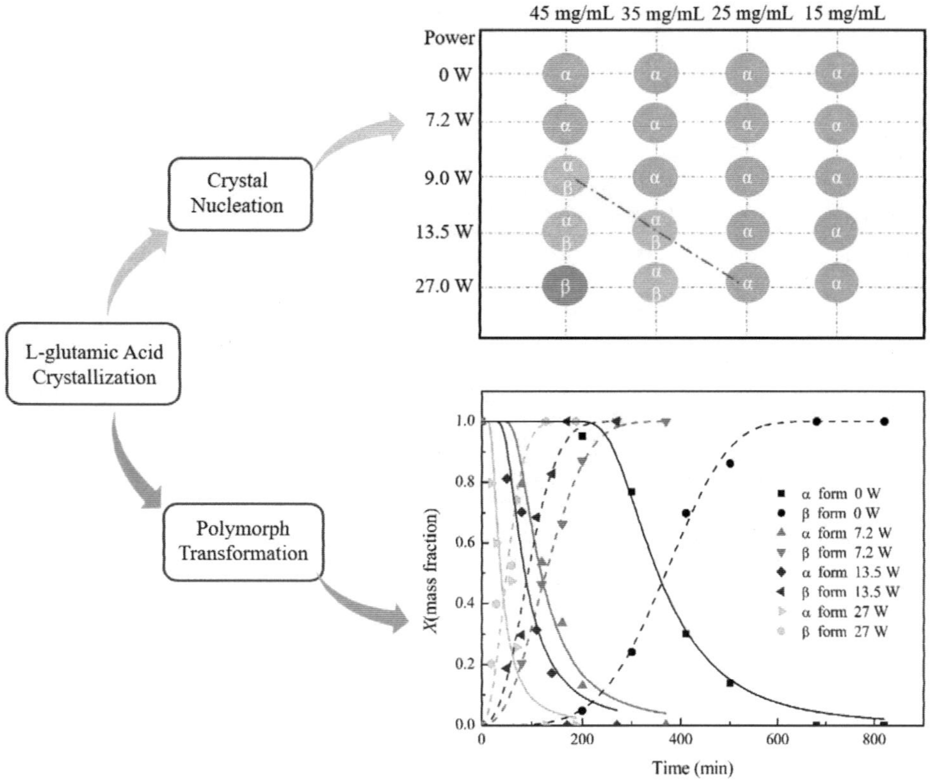

Fig. 1. The logic diagram of L-glutamic acid crystallization with the ultrasound assisted.

610065 The Effect of Extensional Flow on the Evaporative Assembly of Donor-Acceptor Semiconducting Polymers

Abstract:

Because they are lightweight, mechanically compliant, and easily tunable through chemical synthesis, polymer semiconductors (PSCs) are excellent candidates for use in electronics with functionalities currently underserved by conventional inorganic semiconductors. The ability of PSCs to be easily deposited over large area at near ambient conditions makes them particularly promising for low-cost large area electronic applications. Given the sensitivity of charge carrier mobility in PSCs to the nature of their deposition from solution, a mechanistic understanding of deposition processes is essential for enhancing and controlling a PSC's electrical performance. Polymer chain alignment has been largely posited as a strategy for improving electrical performance of PSCs. In this work the ability of a micropillar shearing blade to align polymer chains of a DPP-based polymer, poly(2,5-bis(2-octyldodecyl)-3,6-di(thiophen-2-yl)diketopyrrolo[3,4-c]pyrrole1,4-dione-alt-thieno[3,2-b]thiophen) (DPP-TT), during solution shearing by increasing extensional flow is investigated. A sufficiently strong extensional flow field is found to be capable of aligning polymer chains and inducing crystallization. Field effect transistors (FETs) are fabricated and tested to explore the relationship between induced alignment/crystallization and charge transport. The effect on charge carrier mobility is hypothesized to be a function of deposition parameters. The semiconducting characteristics of a PSC are uniquely dependent on chain length. The molecular weight of a PSC plays a critical role in determining both the final thin film morphology as well as the dominant mechanism of charge transport (i.e. inter- vs. intra- chain transport). Moreover, conformation/orientation changes resulting from flow are anticipated to be sensitive to chain length. We extend our investigation to polymers of variable molecular weight and observe that the extent of alignment, as well as the effect of such an alignment on subsequent charge transport, is highly influenced.

597299 Pressure-Dependent Ion Rejection in Nanopores

Abstract:

It is generally considered that ion rejection of a desalination membrane is independent of the operation pressure drops (ΔPs), which is typically not higher than 10 MPa. However, this may not be true for pressures as high as hundreds of megapascals usually used in simulations. Therefore, simulation results at high ΔPs cannot be directly used to predict real-world ion rejections, which is often overlooked. Herein, we investigate the ion rejection of carbon nanotube membranes in a large scale of ΔPs via molecular dynamics simulations. With effective pressure drops (ΔP_e's) increased from 2.85 to 996 MPa, the ion rejection drops from 100% to nearly zero. Rather than directly investigating the rejection, the relationships of ion and water fluxes with ΔPs are separately investigated. With rising ΔP_e's, the water flux increases linearly, while the ion flux undergoes a two-stage increase: firstly, an exponential increase at $\Delta P_e \leq 53.4$ MPa and then a linear increase. An equation of describing the ΔP_e-dependent ion rejection is then developed based on these observations. Moreover, the rejection mechanism is also discovered that the enhanced input energy makes ions easier to overcome the energy barrier rather than the molecular-configurational reasons. These findings are expected to fill the big gaps between simulations and experiments, and may also be helpful for the rational design of the next-generation desalination membranes.

597322 Transport Mechanism of Water Molecules Passing through Polyamide/Cof Mixed Matrix Membranes

Abstract:

The mixed matrix membranes (MMMs) have gained a great deal of attention due to their higher water permeability without cost of salt rejection. However, the mechanism of water transporting through MMMs is still controversial. Herein, a kind of two dimensional covalent organic framework (COF), TpPa-1, was selected as blending additives mixed with polyamide (PA) due to its good compatibility with polymeric matrices. We performed nonequilibrium molecular dynamics simulations to investigate the transport properties of water molecules passing through MMMs. The permeance of MMMs is obviously enhanced compared to the pure PA membrane. However, such permeance enhancement does not originate from the intrinsic pore of COFs. The PA atoms confined in pores have extremely low diffusivities, the water molecules are thus trapped inside COF pores. Outside COFs pores, there is an interfacial region between COFs and the PA, in which the water molecules aggregates to form a high-density region. In this region, the average flow velocities of water molecules are much higher than those in pure PA or PA outside the mixed region of MMMs. The water transport mechanism discovered in this work offers an alternative and more likely explanation for the high permeance results observed in PA/COF MMMs, and these understandings can be helpful to design more MMMs.

602269 Carbon Molecular Sieve Hollow Fiber Membranes for High Purity Oxygen Production

Abstract:

Hydrocarbon fuel combustion in oxygen enriched air (OEA) is studied as a pathway to improve energy efficiency of carbon capture process. Membrane technology is a good alternative to industry standard cryogenic methods for OEA production owing to their practical merits including continuous process operations, good scalability, and cost-effectiveness with small footprint. In particular, carbon molecular sieve (CMS) membranes have gained attention for gas separation including air separations. The gas separation performance of the CMS membranes can be controlled for specific gas pairs (e.g. O_2/N_2, CO_2/CH_4, H_2/CO_2) to meet application specific needs with proper polymer precursors selections and programmable pyrolysis protocols. Here, we explore CMS hollow fiber membranes (HFMs) prepared from rigid and rod-like aromatic polybenzimidazole (PBI) materials for OEA productions. PBI mambrane material highly-packed interchain configurations endowed the resulting CMS HFMs with more oxygen selective characteristics than previously reported CMS membranes derived from other polymers. Comprehensive investigations of pyrolysis protocols for highly oxygen selective PBI-based CMS HFM fabrications and their performance relationships will be discussed. In addition, practical feasiblity of energy efficient OEA productions using CMS HFMs is assessed via process modelling simulations with the selected performance data.

607878 Fabricating Freestanding Thin-Film Membranes with Carbon Nanodots Via Layer-By-Layer Assembly

Abstract:

The application of membranes in liquid and gas separation is attractive because of their energy efficiency. Synthesis of novel membranes with well-defined microstructure is necessary to achieve highly permeable and selective membranes for separation processes. Recently, carbon-based materials such as carbon nanodots, graphene, and carbon nanotubes have emerged as an exciting class of nanomaterials due to their unique topologies and tailorable functionalities. Incorporation of these nanomaterials in the membranes has shown to improve the performance. Carbon nanodots (CNDs) have shown great promise in many applications because of their low toxicity, biocompatibility, and photoluminescent behavior. Here, we report the fabrication of highly crosslinked membranes by using aminated CNDs as a building block for organic nanoseparation applications. These CNDs are synthesized from m-phenylenediamine by using a solvothermal method. We use a layer-by-layer framework to obtain nanoscale membranes by covalently crosslinking trimesoyl chloride and CNDs. The synthesized membranes are stable in various organic solvents and have manifested high selectivity (up to 90%) and permeability for dye molecules such as brilliant blue and disperse red in methanol. As the CNDs synthesized here are fluorescent under UV light, the resultant film is also fluorescent. This property can be harnessed for diagnostic purposes, such as tracking mechanical failure and fouling of the membranes.

612116 Hydrothermal Liquefaction of Unhydrolyzed Solids for Fuels and Value-Added Products

Abstract:

Integrated biorefineries utilizing lignocellulosic biomass feedstock for fuel production will require additional revenue generation from the derived waste streams to demonstrate technoeconomic feasibility. One of the ways of revenue generation will be the efficient valorization of waste streams for the production of value added products. To understand technoeconomic feasibility of integrated bioprocessing, we investigated hydrothermal liquefaction (HTL) of unhydrolyzed solids (UHS) derived from some preprocessing of a lignocellulosic biomass and subsequent valorization of biocrude waste streams at a laboratory scale. Both solid and liquid side products were selectively enriched and their yields were taken into account to estimate revenue generation. A scale-up of the envisioned pilot scale integrated bioprocessing plant, and preliminary process and cost analysis were performed. Biomass processing steps, energy and mass balances, energy recovery, waste stream valorization, and initial assessment of technoeconomic feasibility will be presented.

608027 Understanding of Metabolic Pathways in *Pseudomonas Putida* for Biosynthesis of PHA Using Sugar, Glycerol, Benzoate, and Their Combinations

Abstract:

Efficient utilization of all available carbons from lignocellulosic biomass is critical for economic efficiency of a bioconversion process to produce renewable bioproducts. However, the metabolic phenotypes enabling bacteria such as *Pseudomonas* to utilize mixed carbon sources to generate energy and stored bioproducts remain poorly understood. In this study, *Pseudomonas putida* KT2440 was employed to investigate the polyhydroxyalkanoates (PHA) production from mixed carbon source of glucose/benzoate or glycerol/benzoate. A multi-omics approach involving label-free proteomics and NMR metabolomics was taken to compare the metabolic profiles of different mixed carbon sources. Results showed different patterns of reductant generation and fatty-acid metabolism pathways with different mixed carbon sources, which resulted in enhancing or inhibiting biomass and PHA productions. Overall, our study provides new insights into the metabolic architecture of *P. putida* with different mixture carbon substrates and reveals the potential regulatory metabolic nodes that may guide further system biology design to improve the PHA production.

594776 Lignin-Based Drop-in Jet Fuel Production

Abstract:

Sustainable aviation fuels remain the only near-, mid-, and likely long-term solution for lowering the carbon emissions of commercial and military aviation. Utilizing lignin feedstock for production of high energy density jet fuel along with cellulosic ethanol offers an unparalleled opportunity to enhance the overall operation efficiency, carbon conversion efficiency, economic viability, and sustainability of biofuels and chemicals production. In addition, determination of sustainable aviation fuel's chemical compositions and prediction of their properties is the critical first step for further research and development that eventually lead to the final certification process. Several catalytic processes with bifunctional catalyst Ru/H^+-Y, Ru/Super Lewis acid, and Ru-M/H^+-Y (M= Fe, Ni, Cu, Zn) were recently developed by our lab. Biomass-derived lignin was successfully converted into the lignin-based jet fuel, which consists of mainly paraffinic hydrocarbon species. They can be further classified into several classes, including n-paraffins, iso-paraffins, mono-, di-, and tri-cycloparaffins of which the majority contains carbon numbers in the range of 7 to 20. However, the "impurities" in the lignin wastes such as sugars, furans, and their derivatives were found to suppress metal catalysts and inhibit lignin depolymerization and hydrodeoxygenation reactions. Furthermore, the possibility of meeting drop-in specifications such as the low aromatic concentrations, which illuminates favorable energy content, energy density, possible low emissions, and high-performance characteristics, will be discussed in the presentation.

613064 Toward Feedstock Design for Biomaterial: Novel Biomass Structure Characteristics Determining Properties of Lignin-Based Carbon Fiber

Abstract:

Lignin has been explored extensively as a renewable feedstock for biomaterials, considering its abundance as a major component of plant cell wall and its sustainability as a waste of lignocellulosic biorefinery and paper-making industries. Despite the extensive efforts on defining process-property relationship, it remains largely unknown how lignin biosynthesis and precursor chemistry in planta would impact on the properties of biomaterials. Current knowledge about reducing lignin content to improve saccharification efficiency does not necessarily inform the feedstock for enhanced properties of biomaterials. Such inadequate understanding fundamentally limited the feedstock design and selection to improve carbon material properties toward broader commercial application. Using lignin from a broad range of biomass feedstock for carbon fiber manufacturing, we have fundamentally explored the structure-function relationship between lignin chemistry and carbon fiber properties, for both mechanical and electro-conductive performance. Specifically, lignin extracted from hardwood (sugar maple), softwood (loblolly pine and red cedar), and herbaceous plant (corn stover and switchgrass) were used for carbon fiber manufacturing, considering the very different lignin and biomass structures from these feedstock. Linear regression models were established to define the relationship between carbon fiber mechanical properties (Elastic Modulus, Tensile Strength, and Elongation) with lignin structure characteristics (content, composition, linkage profile and others), along with carbon fiber microstructures (crystallite content and size). The results highlighted that traditional biomass characteristics such as lignin content and composition don't impact the properties of lignin-based carbon fiber. Instead, the content of β-O-4 linkage correlates significantly with tensile strength and elastic modulus of lignin carbon fiber, indicating that the more linear β-O-4 linkage would promote carbon fiber performance. Besides the mechanical property, electro-conductive property is essential for broader energy application of lignin-based carbon fiber, yet the mechanisms controlling the electrocoductive properties of lignin-based carbon fiber is largely unknown. We hereby demonstrated that a higher β-O-4 content also promotes electroconductive performance of carbon fiber. Moreover, we also found that higher β-O-4 linkage profile defines the crystallite content and size of the resultant carbon fiber, which could be due to the better miscibility with the guest polymer. Overall, the study established lignin linkage profile, in particular, β-O-4 content as the new structure determinants for biomaterial property, which could guide the future biomass feedstock development.

611655 3D-Printed Zeolite Monoliths with Metal Dopants for Methanol Conversion

Abstract:

Additive manufacturing by 3D printing has gained worldwide attention in formulation of porous materials into practical contactors as a result of flexibility in design alteration and preparation of robust and complex geometries which requires fewer steps and minimal resources in comparison to the previously established extrusion process.[1-4] In this study, 3D-printed ZSM-5 monoliths doped with Ga_2O_3, Cr_2O_3, CuO, ZnO, MoO_3, and Y_2O_3 were synthesized using the state-of-art 3D printing technique. The physicochemical properties of the catalysts were characterized by X-ray diffraction, N_2 physisorption, NH_3 and CO_2 temperature-programed desorption and H_2 temperature-programmed reduction. The promotional effect of doped metals on catalytic performance of 3D-printed ZSM-5 monoliths in methanol to hydrocarbon (MTH) reaction in the presence and absence of CO_2 was investigated. Results indicated that both metal dopants type and reaction atmosphere greatly influence catalyst stability and product distribution. The yield of light olefins was enhanced over all metal-doped 3D-printed ZSM-5 monoliths in N_2 atmosphere (absence of CO_2), however, CO_2 atmosphere did not favor the production of light olefins. Although selectivity toward ethylene slightly decreased, the propylene yield was almost constant after switching N_2 to CO_2 in MTH reaction at 400 °C. Furthermore, it was found that Y- and Zn-doped ZSM-5 monoliths exhibited higher yield of light olefins and BTX compounds in the in the absence and presence of CO_2, respectively.

References:

[1] F. Magzoub, X. Li, J. Al-Darwish, F. Rezaei, A.A. Rownaghi, 3D-printed ZSM-5 monoliths with metal dopants for methanol conversion in the presence and absence of carbon dioxide, Appl. Catal. B Environ. 245 (2019) 486–495.

[2] X. Li, A. Alwakwak, F. Rezaei, A.A. Rownaghi, Synthesis of Cr , Cu , Ni , and Y - Doped 3D-Printed ZSM-5 Monoliths and Their Catalytic Performance for n-Hexane Cracking, ACS Appl. Energy Mater. 1 (2018) 2740–2748.

[3] X. Li, F. Rezaei, A.A. Rownaghi, 3D-printed zeolite monoliths with hierarchical porosity for selective methanol to light olefin reaction, React. Chem. Eng. 3 (2018) 733–746.

[4] X. Li, F. Rezaei, A. Rownaghi, Methanol-to-olefin conversion on 3D-printed ZSM-5 monolith catalysts: Effects of metal doping, mesoporosity and acid strength, Microporous Mesoporous Mater. 276 (2019) 1–12.

596901 Fluid Flow Control Devices with 3D-Graded Permeability

Abstract:

Chemical engineering systems often rely on components that involve intimate contact between a fluid and a solid, such as in catalyzed reactive flows, fluid purifiers, and chromatographic separations. To accomplish this, it is often necessary to distribute the fluid from a relatively narrow tube to a broader cross section, to an array of tubes, or into a porous medium that can be modeled by Darcy's Law. (Darcy's Law states that the volumetric flow rate is proportional to the permeability, an intrinsic material property related to its porosity.) Ideally, the flow rates in the tube array or throughout the porous medium are uniform, and the size, weight, and cost of this portion of the system are a small fraction of those of the chemically reactive portion.

Additive manufacturing techniques raise the possibilities that porous media can be fabricated in which the permeability can be arbitrarily specified in three dimensions, and that a broader range of permeabilities can be achieved than by traditional methods used to manufacture porous media.[1] We are using optimization algorithms[2] to design devices that distribute a fluid from a narrow inlet to a broad outlet, where the outlet flow rate is spatially uniform, and device geometry and/or pressure drop are constrained, by spatially varying the permeability in the device. We have considered a Darcy's law model, as well as a modification of the Navier-Stokes equations with a term representing permeability.[3] Numerical models show that designs varying permeability in three or two dimensions can achieve greater uniformity than designs that vary permeability in only one dimension (such as with stacked flow elements).

Meanwhile, we are establishing and evaluating methods to fabricate stainless steel structures with spatially varying porosity on conventional additive manufacturing tools, as a path to build the designed devices for future experimental testing.

Our work is supported by the Laboratory-Directed Research and Development program at Sandia National Laboratories, a multimission laboratory managed and operated by National Technology & Engineering Solutions of Sandia, LLC, a wholly owned subsidiary of Honeywell International Inc., for the U.S. Department of Energy's National Nuclear Security Administration under contract DE-NA0003525.

SAND2020-3764 A

[1] V. P. Palumbo et al. "Porous Devices Made by Laser Additive Manufacturing." US Patent Application 2017/0239726 A1, Mott Corporation, 2017.

[2] M. A. Heroux, R. A. Bartlett et al. "An overview of the Trilinos project" ACM Trans. Math. Softw. 31(3), 397-423, 2005.

[3] T. Borrvall, J. Petersson. "Topology Optimization of Fluids in Stokes Flow." Int. J. Numer. Meth. Fluids 41, 77–107, 2003.

605743 3D Printed Gyroid-like Structured Packings

Abstract:

Structured packings are an integral part in distillation columns, absorption columns, and chemical reactors. An effective packing requires large surface area, low pressure drops, and even liquid distribution. Standard structured packings have been established for decades and typically consist of corrugated sheets stacked in alternating arrangements. Recently, triply periodic minimal surfaces (TPMS) have received attention for potential use in heat exchangers, mass exchangers, and chemical reactors. This class of structures, including the familiar 'Gryoid', uniquely requires additive manufacturing for production. Here, we demonstrate the use of several TPMS geometries as structured packings fabricated using additive manufacturing. These packings exhibit improved liquid distribution, reduced wall channeling, and increased fluid mixing in comparison to conventional structured packings.

This work was performed under the auspices of the U.S. Department of Energy by Lawrence Livermore National Laboratory under Contract DE-AC52-07NA27344. LLNL-ABS-809142.

610417 Coupled Numerical and Experimental Analysis of Transport Properties in 3D-Printed Periodic Cellular Media to Enable Catalytic Reactors Design

Abstract:

Process Intensification is a key topic in academic and industrial research aiming at the development of more compact and energy-efficient processes [1]. In this context, structured catalysts have been considered as a promising solution to overcome the limitations of conventional catalytic supports. Currently, cellular materials (foams, Periodic Open Cellular Structures – POCS) are envisioned as potential candidates for enhanced catalyst substrates. Thanks to the possibility of advanced manufacturing and design offered by 3D printing, POCS allow for extreme flexibility in the geometrical optimization (cell shape, strut size, solid fraction) to tailor the design to the process needs [2]. In this view, the fundamental understanding of the coupling between the geometrical properties and the transport phenomena is essential for the optimization of the supports and their industrial application.

In this work, we propose a combined experimental and numerical investigation of the transport properties of POCS aimed at the derivation of engineering correlations to enable the development and optimization of enhanced catalytic reactors. Computational Fluid Dynamics (CFD) allows for the analysis of the transport properties as a function of the geometrical parameters, enabling the understanding of the effect of single morphological parameters (e.g. cell shape, porosity, cell diameter) on the overall performances and providing deep insight in the transport mechanisms.

The significant computational burden of CFD simulations hinders their direct application to the industrial reactor scale. Hence, a hierarchical approach following the procedure proposed by Rebughini et al. [3,4] is employed to investigate representative structure volumes by means of CFD.

While numerical simulations allow to evaluate the transport properties of structures regardless of the limits imposed by their manufacturability, they are unable to provide a broad range of data capable of completely define the performances of a specific structure. Contextually, experimental analysis performed on 3D-printed replicas, while requiring specialized manufacturing infrastructures, can provide enough data for the complete characterization of the most promising geometries and cross-validation with CFD simulations.

In this view, high precision 3D printing technology enables the fabrication of highly accurate structures. The precision and accuracy of currently available SLA 3D printers, combined with the wide range of available materials, allows for their effective deployment in the experimental analysis of fluid-solid interactions. The possibility to create faithful physical replicas of digital structures allows for the direct comparison between simulations and experiments hence providing a reliable tool for the cross-validation of experimental and computational results.

As a result, the numerical and experimental analyses can be combined and exploited to formulate engineering correlations able to predict the structures behaviour as a function of the geometrical and physical parameters [5].

As an example of this approach, an analysis of pressure drops and mass transfer over TKKD cellular structures is presented. Experimental investigation is performed both in cold flow (pressure drops) and reactive conditions (mass transfer) over 3D printed resin samples with different geometrical characteristics. Pressure drops are evaluated by flowing air and measuring the differential pressure across 3D printed resin samples. Transport properties are assessed by reactive runs performed on samples realized with a High Temperature 3D printed resin ($HDT_{@0.45MPa}$ = 289°C). The oxidation of H_2 over Pd/CeO_2 in rich conditions is used as the test reaction to allow for the onset of external mass transfer limited regime at temperatures compatible with the resin. The conversion of O_2 (the limiting reactant) in external mass transfer limited regime is employed to evaluate the mass transfer coefficients due to the similarity between its diffusivity and CO while the large excess of H_2 prevents back-diffusion problems.

In parallel, CFD simulations are performed on the same geometry. The working fluid is CO diluted in air at 298K, 1 atm, the test reaction is the total oxidation of CO. Hence, the mass transfer properties are assessed in steady state conditions by evaluating CO conversion (limiting reactant). Pressure drops and conversion are together estimated for a representative elementary volume (REV, [6]) made of a set of unitary cells.

Figure 1 (left) shows the result of the numerical analysis (lines) along with experimental results (dots) for the TKKD geometry, showing a very good agreement between CFD and experiments. The analysis shows the effects of the different geometrical parameters (cell size, porosity) on the pressure drops. In this view, the envisioned approach offers a reliable tool for assessing the pressure drop functional dependency on the structures geometrical properties. Figure 1 (right) shows the outcome of the analysis of the gas-to-solid transport properties of the same geometry. Experimental and numerical data (dots) are exploited to develop a gas/solid mass transfer correlation for the TKKD unit cell able to predict its behaviour as a function of the geometrical and the physical parameters. The proposed dimensionless correlation (line) shows a good agreement with respect to the available data.

In conclusion, we propose a methodology for the fundamental investigation of transport properties in periodic cellular substrates based on coupled numerical and experimental analysis. The combined use of CFD and experiments allows for the fast screening of a wide variety of different structures while enabling the in-depth characterization of the most promising geometries without the limitations of either numerical or empirical independent approaches The illustrated case study highlights the potential of such approach in the development of engineering correlations able to describe substrates transport properties regardless of the specific geometries. Thanks to the gathered knowledge and to commercial 3D printing solutions, it is then possible the manufacture catalytic substrates with tailored properties for each specific applications with materials such as metals (Steel FeCrAlloy, Copper, Titanium, and others) or ceramics, with details as small as 200μm. In this view, the proposed methodology allows for the proficient exploitation of the extreme design flexibility of advanced 3d printing, thus enabling effective process intensification.

References

[1] A.I. Stankiewicz at al. Chem. Eng. Prog. 96 (2000) 22–33.

[2] C. Busse at al. Chem. Eng. Process. Process Intensif. 124 (2018) 199–214.

[3] S. Rebughini, et al., React. Chem. Eng. 3 (2018) 25–33.

[4] S. Rebughini,et al., Chem. Eng. J. 289 (2016) 471–478.

[5] M. Bracconi, et al., Chem. Eng. J. 377 (2019.)

[6] M. Bracconi et al., Chem. Eng. J. 352, pp. 558-571 (2018).

Acknowledgements

This project has received funding from the European Research Council under Grant Agreement no. 694910 (INTENT).

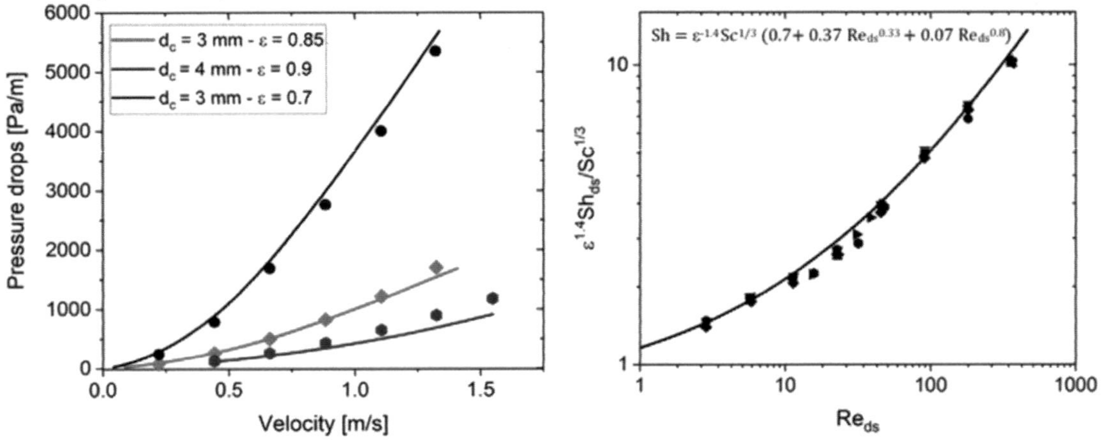

Figure 1: Left – Pressure drops of TKKD cells evaluated with experiments (points) and CFD (lines); right: correlation for Sh number of TKKD unit cells (line) derived from CFD simulations (points)

605946 Development of Polymer Monolith Pd Catalysts for Batch, Modular Continuous Flow and Novel 3-D Printed PEEK Reactor Platforms for Hydrogenation Reactions

Abstract:

Direct Hydrogenation is one of the most heavily utilised and essential reactions within the synthetic routes of many active pharmaceutical ingredients (APIs). Continuous hydrogenation as a result is becoming a topic of significant interest throughout the chemical industry with the development of novel reactor technologies[1] and catalysts[2], to optimising mass transfer, residence time distribution, and surface area limitations often associated with heterogeneous catalysis.

In this study, novel polymer monolith Pd catalysts have been developed for both batch and continuous flow hydrogenations. Modular reactors based on standard tubes and fittings utilised the commercially available H-Cube. Comparison with 5% Pd/C for the direct hydrogenation of nitro, alkene, and imine functional groups were investigated. Furthermore, novel 3-D printed polymer reactors fabricated in the chemically and mechanically resistant material PEEK (poly(etheretherketone)) have been designed and printed for heterogenous hydrogenation reactions[3]. The monolithic catalysts developed are ideally suited for incorporation into Fused Filament Fabrication printed reactors as they can be covalently attached to the reactor walls, grown within complex printed geometries and catalyst synthesis conditions are compatible with standard PEEK processing conditions.

Acknowledgements:

This publication is supported by Science Foundation Ireland (SFI) through SSPC, The SFI Research Centre for Pharmaceuticals (Grant number 12/RC/2275_P2) and I-Form, the SFI Research Centre for Advanced Manufacturing (Grant number 16/RC/3872)

References:

[1] M. O'Brien, N. Taylor, A. Polyzos, I. R. Baxendale, S. V. Ley, *Chem. Sci.* **2011**, *2*, 1250–1257.

[2] CH. Pélisson T. Nakanishi, Y. Zhu, K. Morisato, T. Kamei, A. Maeno, H. Kaji, S. Muroyama, M. Tafu, K. Kanamori, T. Shimada, K. Nakanishi, *ACS Appl. Mater. Interfaces*., **2017**, *1*, 406–412 .

[3] M. J. Harding, S. Brady, H. O'Connor, R. Lopez-Rodriguez, M. D. Edwards, S. Tracy, D. Dowling, G. Gibson, K. P. Girard, S. Ferguson, *React. Chem. Eng.*, **2020**, *5*, 728-735

Figure 1: (left) PEEK 3-D printed reactor with a straight channel (2 mm ID) housing polymer monolith. (right) X-Ray scan of polymer monolith grown throughout the inner channel of the reactor.

609526 Finite-Size Effects on the Chemical Potentials of Molecular Solids

Abstract:

Numerous important properties of solid oral therapeutics are associated with the chemical potential of the active pharmaceutical ingredient (API) in the solid form. These properties include aqueous solubility, dissolution rate, and physical stability in its crystalline form among others. Molecular modeling of the crystal structures of APIs to estimate the chemical potentials has become an increasingly popular approach in recent years for confirming the suitability of API solid forms for further development *in silico*. However, due to the computational expense of these approaches, the models are limited to time and length scales far smaller than macroscopic materials, and rely on simplifications such as microscopic cells with periodic boundary conditions, as well as cut-offs for long-range interactions. These approximations can have a significant effect on the chemical potential of organic small molecule crystals, particularly at small system sizes.

Here we systematically examine the effect of supercell size and long-range cutoff distance on the computed chemical potentials of pharmaceutically relevant molecular crystals using classical molecular dynamics simulations. The Gibbs free energies of the crystal polymorphs and liquid phases of carbamazepine and sulfamerazine were computed from 0 K to 600 K for various system sizes and long-range cutoff radii. We compare the estimated chemical potentials with small systems and short cutoffs to those of the corresponding larger systems and show how these differences change with temperature. Furthermore, we show the effects that these small perturbations to the chemical potentials have on the macroscopic observables of enantiotropic transition temperature and solid-liquid melting temperature. Finally, we make recommendations for future molecular simulations of API crystal structures to balance the trade-off of chemical accuracy and computational efficiency.

609123 Computer-Aided Design of Solvent Blends for the Crystallisation of Ibuprofen – Regulating Crystal Morphology By the Spiral Growth Model

Abstract:

Solid dose pharmaceuticals continue to be the most common means of drug delivery worldwide; because of this, pharmaceutical production is dependent on effective crystallisation systems. The majority of industrial crystallisers are solvent based – hence, decisions surrounding solvent choice can drastically affect the outcome of the crystallisation process. Although attaining a high crystal yield of the active pharmaceutical ingredient (API) is typically the dominant objective during the solvent selection process, solvent and crystal properties can also influence health, safety, environmental, and operational performance. Indeed, the shape of the crystalline product can greatly impact downstream processes[1]; filtration can be hampered by needle- or plate-like particles that form dense cakes, high aspect ratio crystalline material may not readily flow during transportation, and tabletting via direct compression can fail if the solid product fractures during compaction. Moreover, crystal morphology can affect the *in-vivo* efficacy of the final pharmaceutical product.

Despite the many material, time and cost constraints present during initial design stages, solvent selection for the crystallisation of novel API molecules is currently performed via time-consuming and expensive experiments, heavily reliant on past experience and rules-of-thumb[2]. Over the last decades, computer-aided molecular design (CAMD) methods have been developed to overcome such difficulties[3], with the aim of guiding lab-based experiments towards optimal candidate molecules for crystallisation[4,5]. Whilst it is generally agreed that crystal shape has a significant role when designing optimal solvents, the integration of shape considerations within CAMD methods has principally focused on empirical approaches[4] combining an understanding of the underlying chemistry with prior experimentation. Direct, quantitative methods, where mechanistic interpretations of crystal growth are considered[6], have not yet received significant attention in formulating the solvent design problem. This is likely due to the absence of applicable group contribution approaches to these models, as well as the additional complexity of formulating and resolving a mixed-integer optimisation problem to represent these design choices.

We extend the recent work of Watson et al.[7], in which the SAFT-γ Mie equation of state[8] was employed in a CAMD framework to identify optimal solvent blends and process conditions for integrated cooling and antisolvent crystallisation, to include the target morphology of the pharmaceutical product. By assuming a low supersaturation growth environment – a standard condition for industrial crystallisation – it is possible to calculate the relative growth rates of corresponding crystal faces[6], thus providing a quantitative measure of aspect ratio of the crystalline API. The design formulation is implemented in gPROMS and is applied to the design of solvent mixtures for the crystallisation of ibuprofen, whereby the impact of crystal shape on the optimal solvent molecules and process conditions is investigated.

1. Dandekar, P., Kuvadia, Z.B., Doherty, M.F., 2013. Annual Review of Materials Research 43, 359-386.
2. Brown, C.J., McGlone, T., Yerdelen, S., et al., 2018. Molecular Systems Design & Engineering 3(3), 518-549.

3. Gani, R., 2004. Computers & Chemical Engineering 28(12), 2441-2457.
4. Karunanithi, A.T., Achenie, L.E., Gani, R., 2006. Chemical Engineering Science 61(4), 1247-1260.
5. Jonuzaj, S., Gupta, A., Adjiman, C.S., 2018. Computers and Chemical Engineering 116, 401-421.
6. Li, J., Tilbury, C.J., Kim, S.H., Doherty, M.F., 2016. Progress in Materials Science 82, 1-38.
7. Watson, O.L., Galindo, A., Jackson, G., Adjiman, C.S., 2019. Computer Aided Chemical Engineering 46, 949-954.
8. Papaioannou, V., Lafitte, T., Avendano, C., et al., 2014. The Journal of chemical physics 140(5), 054107.

596083 Machine Learning for Thermodynamic Properties of Materials: Towards the Prediction of Solubility

- Download/View Extended Abstract

Abstract:

Introduction

Solubility is one of the most important properties in the development and design of drugs. It plays a big role in pharmacokinetics and pharmacodynamics, where it relates to dosing, toxicity and the effect of drugs on their targets. In the field of process design, knowledge on the solubility of drugs and other intermediate products are crucial in the design of separation techniques and for the transitioning from batch to flow processes. Computer-aided screening of the solubility of API's can dramatically shorten the drug discovery cycle and the costs related to the development and manufacturing of new drugs.

State of the art solubility models focus on the prediction of octanol-water partitioning coefficients (logP) and solid aqueous solubility (logS) because of their role in the evaluation of the drug-likeness of API's in an initial screening process. Although models for octanol-water partitioning coefficients have improved a lot the last couple of years, the prediction of aqueous solubility for a broad range of drug-like compounds remains challenging. In view of downstream manufacturing and process intensification, the interest in solubility prediction is not limited to octanol or water as a solvent. Computer-aided screening of potential solvents can contribute significantly to a reduction of waste streams and ease the upscaling towards continuous processing. However, models for solubility predictions in other organic solvents are limited by the scarce availability of experimental data.

This works aims at using machine learning techniques and physical relationships to predict the solubility of gases, liquids and solids in a variety of solvents. Theoretical quantum calculations are used to compensate for scarce and biased experimental databases. Furthermore, the physical relation between gas-liquid solvation energies, liquid-liquid partitioning coefficients and solid solubility is used to aid solubility predictions for drug-like molecules in a broad range of solvents.

Methodology

Graph convolutional neural networks are used to encode the molecular structure of both the solute and solvent. A molecular embedding is learned separately for the solute and solvent with the directed message passing neural network, as implemented in Chemprop. The solute and solvent embeddings are concatenated and fed through another dense layer for the property prediction.

Quantum chemical calculations are done to determine the gas-liquid solvation free energies for various solute and solvent combinations. These calculations are used to pretrain the message passing neural network, such that physical interactions between solvents and solutes can be learned. Contrary to the experimental gas-liquid solvation free energies, this dataset is not biased towards certain solvents (eg. water). Using theoretical data to pretrain the model also reduces the effect of aleatoric errors, caused by data noise, on the final model predictions. The dataset can be extended towards classes of solutes and solvents for which no or limited experimental data is available, improving the generalizability of the model.

For the prediction of liquid-liquid partitioning coefficients and solid solubilities, the physical relationship with the well-established gas-liquid solvation free energies is enforced. Liquid-liquid partitioning coefficients are directly related to the ratio of the gas-liquid partitioning coefficients of the same solute in both solvents. Some error is introduced by the mutual solubility of the solvents, which will be accounted for in the loss function. The solid solubility is related to the gas-liquid partitioning coefficients and the fusion free energy of the crystal structure. To account for the fusion free energy, an additional model is trained on an experimental dataset of melting temperatures and fusion enthalpies.

Results

A preliminary theoretical dataset is constructed with COSMO*therm*. It consists of >20k data points and includes 78 solvents and 244 solutes. Solvation free energies of the theoretical dataset are predicted by the model with a RMSE of 0.08 kcal/mol and a MAE of 0.04 kcal/mol. Analysis of the solute and solvent molecular embeddings demonstrate the learned physical interactions. A principal component analysis of the solvent molecular embeddings clusters solvents according to their physical properties relevant to solvation. Analysis of the atom messages for some solute molecules highlights atoms with an important hydrogen bonding/donating character.

The model trained on the theoretical dataset is further refined using experimental data for solvation free energies. The experimental dataset is gathered from different sources including the Minnesota solvation database, the Compsol database and different publications related to the regression of Abrahams solute and solvent parameters. The total experimental dataset has more than 8900 entries for solute-solvent combinations with over 300 different solvents. The machine learning model uses molecular embeddings for the solute and solvent molecules that are learned from the theoretical dataset. The dense layer for property prediction is initialized by the learned weights from the theoretical dataset and optimized with the experimental solvation free energies. Solvation free energies in a randomly selected test set can be predicted with a RMSE of 0.60 (±0.04) kcal/mol and MAE of 0.27 (±0.02) kcal/mol. These results are similar compared to those of a model without pretraining. However, if no pretraining is done on a theoretical dataset, analysis of the molecular embeddings do not directly demonstrate learning of any physical interactions. Moreover, the predictions on "unseen" solvents, i.e. those not included in the training set, improve significantly if pretraining is applied.

For the prediction of partitioning coefficients, a dataset available in OChem is used. This dataset includes 3600 entries for partitioning coefficients between water and various solvents including chloroform, cyclohexane, toluene and hexadecane. The model trained on theoretical gas-liquid solvation energies is used only for prediction of the experimental logP data with a RMSE 1.18 and MAE 0.66. Further efforts on improving predictions of partitioning coefficients will include (1) collection of a broader dataset (2) refine a pretrained model on experimental data enforcing the physical relation with gas-liquid solvation energies.

Solid solubility can be related to gas-liquid solvation energies of the solute in the respective solvent and the solute fusion free energy. The fusion free energy of a crystal structure can be approximated with a known melting temperature and fusion enthalpy. Experimental databases on latter two properties are gathered from different sources, including OChem and NIST. The database is used to train a machine learning model with a directed message passing neural network to construct the molecular embeddings, as implemented in Chemprop. Fusion enthalpies are predicted with a RMSE of 1.69 (±0.14) kcal/mol and melting temperatures with a RMSE of 33.7 (±1.6) K. These predictions will be combined with the model for gas-liquid solvation energies to create a model for solid solubilities in a broad range of solvents.

Conclusions

A machine learning model is developed for the prediction of gas-liquid solvation free energies. The molecular embeddings for solvent and solute are done with a directed message passing neural network. Experimental solvation free energies are predicted using the new model, pretrained on theoretical data, with a RMSE of 0.60 kcal/mol. The pretraining on a theoretical dataset ensures that the model captures the physical interactions between solvent and solute molecules and improves the performance of the model on unseen solvents.

The physical relationship between gas-liquid solvation energies, partitioning coefficients and solid solubility will be used to enable the prediction of those properties in a broad range of solvents. The solute fusion enthalpy and melting temperature are required for the solid solubility predictions. Those are predicted by another model with a RMSE of 1.69 kcal/mol and 33.7 K respectively.

Figure

Figure 1. Parity plots for the predictions of gas-liquid solvation free energies ΔG_{solv}, fusion enthalpies ΔH_{fus} and melting temperatures T_m.

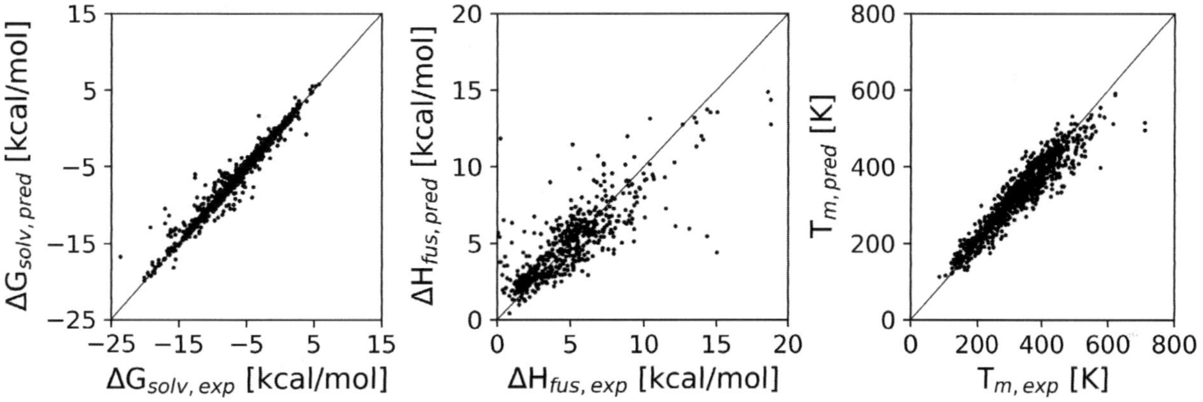

604581 Approaching Coupled Cluster Accuracy with a General-Purpose Neural Network Potential

Abstract:

In this talk, we will present a fully transferable deep learning potential that is applicable to complex and diverse molecular systems well beyond the training dataset. Recently we introduced ANAKIN-ME (Accurate NeurAl networK engINe for Molecular Energies) or ANI in short. ANI is a new *method and sampling procedure* for training neural network potentials that utilizes a special kind of symmetry functions to build single-atom atomic environment vectors (AEV) as a molecular representation.

The AI methods that focuses on the use of large and diverse data sets in training new potentials, has consistently proven to be universally applicable to systems containing the atomic species in the training set. Focusing on parametrization for organic molecules (with CHNOSFCl atoms so far), we have developed a universal neural network potential which is *highly accurate compared to reference QM calculations at speeds* 10^7 *faster*. The potential is shown to accurately represent the underlying physical chemistry of molecules through various test cases including: *chemical reactions (both thermodynamics and kinetics), thermochemistry, structural optimization, and molecular dynamics simulations*. The results presented in this talk will provide evidence of the universal applicability of deep learning to various chemistry problems involving organic molecules.

604074 Kinetic Modelling and Risk Assessment of an Exothermic Fed-Batch Reaction Using Calorimetry

Abstract:

The digital design of manufacturing processes using mechanistic models has been playing an increasingly important role in drug substance process development. It enables the study of unknown kinetics using experimental data, risk assessment at extreme process conditions where experimentation may be infeasible, and efficient exploration of decision space for critical process parameters.

Calorimetry studies are an integral part of process development of synthesis steps in the pharmaceutical industry. Experimental kits, such as the Mettler-Toledo RC1 are used for this purpose with the typical aim of understanding the heat of the reaction, heat evolution, adiabatic temperature rise and therefore, safety of the chemical reactions under scrutiny. They can also see usage in reactive systems to regress kinetics of reactions, which are otherwise difficult to detect through offline composition analysis. In order to assess the thermal risks associated with a chemical reaction, both the desired and undesired (i.e., decomposition) reactions must be studied. Synthesis steps in pharmaceutical processes can be accompanied by significant heat release and must be understood well to manage them successfully at pilot and plant scales. In this study, we use mathematical modeling to predict the behavior of a reaction under different heat transfer profiles and simulate what would happen in the event of process disturbances such as a cooling jacket failure, or loss of control of fed batch (e.g. reactant added too quickly).

The reaction was carried out in a 1L stirred tank reactor and the reactor temperature was maintained by a cooling jacket. When the reactor temperature is well regulated, only the primary reaction occurs, however in the event of a cooling failure, an undesired decomposition reaction occurs. Experimental data from RC1 heat flow measurements were obtained using experiments executed at different process temperatures. A fed-batch reactor model was developed and simulated in gPROMS FormulatedProducts. Reaction kinetics for the primary reaction were extracted using parameter estimation by fitting the RC1 data (Figure 1).

In order to study the effect of temperature control failure, a pre-defined decomposition reaction was added to the model. Global system analysis (GSA) is an effective tool to explore design space, determine safe operating regimes for exothermic reactions, quantify effects of uncertainty, manage risk and identify opportunities for optimization. This tool was used to perform a sensitivity analysis by varying the heat transfer coefficients and initial temperatures within a specific range. The results of GSA give a clear picture of: 1) what the maximum temperature in the reactor would be if temperature control fails and hence the operation conditions that can trigger synthesis of undesirable products and impurities; 2) when the cooling failure occurs during the process, i.e. during the reactant addition or following reaction. The output from this may help to inform the control strategy for this reaction process, such as the setting of maximum reactant addition flow rates to ensure safe operation.

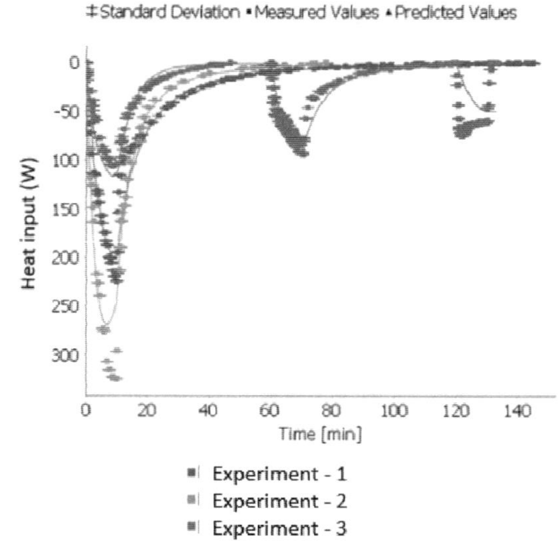

603492 Kinetic Modeling of a Non-Homogeneous Metal-Catalyzed Reaction System for Drug Synthesis

Abstract:

Mathematical modelling of chemical reaction kinetics has proven to be a very effective tool for pharmaceutical drug synthesis development, where complex organic reaction networks are often encountered. The chemical and molecular complexity of modern drug molecules has been rapidly increasing, and this requires design and optimization of extensive synthesis routes which presents a challenge to process scientists. First-principles modeling can play an important role in streamlining this process. It helps with analyzing, interpreting and discerning the relationship between various process parameters and reaction profiles with minimal experimentation requirements.

In this presentation, a modelling workflow to understand the Buchwald-Hartwig cross-coupling reactions is described. These reactions are ubiquitous in process chemistry. This is a non-homogeneous, metal-catalyzed reaction system encompassing multiple partially soluble reactants, a biphasic liquid mixture, and homogenous catalytic cycle in the organic phase. Challenges arise from subtle case-to-case variability, with different influential factors like mass transfer limitations, catalyst activation and water sensitivity.

Using this case, the presentation will illustrate the following steps in a typical mechanistic modeling lifecycle: 1) Design and perform kinetic experiments; 2) Recreate experiments *in silico*; 3) Estimate reaction kinetic parameters; 4) Evaluate model predictions; 5) Postulate new mechanisms, kinetic rate expressions and/or perform additional targeted experimentation until predictions are satisfactory

HPLC measurements were available for key reactants and products from experiments conducted for a range of operating conditions. A mathematical model was developed in gPROMS FormulatedProducts starting from preliminary fitting of primary reaction pathway(s) assuming simple reaction mechanisms and solubility correlations.

Using a systematic parameter estimation approach this was further expanded to a more rigorous model that includes effect of particle size distribution, mass transfer effects and relative concentration effects. The regression of kinetic parameters takes into account uncertainty in measurements and process parameters, which provides a robust model that can be used to make process predictions, and aid scale-up/technical transfer decisions.

The model can be further used to understand the behavior of the system over a wide range of input conditions. This becomes particularly important to make scale up decisions and identify critical operating regimes early in the development of a pharmaceutical compound. Global system analysis was used to explore the design space and identify key factors that influence reaction selectivity and impurity generation. This allows for better understanding of the process and thus provides useful insights to support control strategy, optimization goals, and risk assessment.

Keywords: reaction kinetics, parameter estimation, mechanistic modelling, digital design, catalyst

601227 The Flory-Huggins Parameters of Binary Ionic Liquid Mixtures with Ammonia

Abstract:

Phase equilibrium measurements and modeling are essential for the design of chemical processes such as separations, reactions, fluids flow, to name a few. One of the industrial processes that received increased attention in the last decade involves ionic liquids (ILs), which are low melting point salts composed entirely of ions. The field of ILs requires a more in-depth understanding of the interaction between ILs and other substances so that novel applications can be developed. In this work, our goal is to investigate the non-ideality of a binary mixture of ammonia and four imidazolium-based ionic liquids using the extended Flory-Huggins model, where both entropic (combinatorial) and enthalpic (residual) contributions to non-ideality are included. We previously reported the solubility of NH_3 in four ILs using the gravimetric method (1-butyl-3-methylimidazolium hexafluorophosphate, 1-butyl-3-methylimidazolium tetrafluoroborate, 1-ethyl-3-methylimidazolium bis(trifluoromethylsulfonyl)imide, and 1-ethyl-3-methylimidazolium 1,1,2,2-tetrafluoroethanesulfonate). Here, the Flory parameters (χ) of these four ILs and ammonia systems were regressed with the assumption that the χ parameter is dependent on temperature and independent of the concentration. The Flory-Huggins model with the combinatorial term only, and the extended Flory-Huggins model with both combinatorial and residual terms are compared. The results showed that the NH_3+IL systems have a strong negative deviation from the ideal solubility that cannot be solely described with entropic impacts, specifically at high ammonia concentrations in the ILs. This suggests an increase in the size of the IL molecule might improve the ammonia sorption capacity of an IL at low pressures. However, the ammonia sorption capacity could be significantly improved by considering the molecular interaction between ammonia and ILs (i.e., hydrogen bonding) in addition to the variation in size of the molecule.

601430 Predicting Hydrogen Bond Donating Ability between Nitroaniline Dyes and Type III DES Using Tzvp Based COSMO Models

Abstract:

Type III deep eutectic solvents (DES) are formed by hydrogen bond donors and acceptors (HBD and HBA respectively). These HBA's and HBD's form a liquid that has a lower melting point then either of the individual compounds. The main interaction that drives this depression in melting point is the hydrogen bonding between the HBD and the halogen cation of the HBA. In addition to this, the HBA itself may have hydrogen's able to participate in bonding; such as the primary alcohol in the common HBA choline chloride. These complex hydrogen bonding systems have been qualitatively described with methods such as pulse-field gradient nuclear magnetic resonance spectroscopy. This method does not necessarily describe the changes in hydrogen bonding potential difference between the individual HBA and HBD's and the resulting DES. Solvatochromic measurements are a quick experimental tool that can utilize dyes with UV-Vis spectra that change based on their environment. Dye pairs such as 4-nitroaniline and N,N-diethyl 4-nitroaniline can be used with Type III DES to measure hydrogen bonding acceptance capacity; so how saturated the DES is with hydrogen bonds.

In addition to experimental methods, computational modeling can attempt to predict the ability of a DES to form additional hydrogen bonds. COSMO (Conductor-like Screening Model) uses electron distribution calculations to determine thermophysical properties and phase properties such as liquid-liquid equilibrium of binary and ternary systems. One output of these electron density calculations for Type III DES is the surface potential of the molecules. This potential is directly related to the molecules ability to interact with potential bonding sources, such as hydrogens available for bonding. Knowing how different HBA's and HBD's interact will be useful for designing DES for target specific applications (such as liquid-liquid extraction or gas absorption) where hydrogen bonding may be desirable.

This study evaluates Type III DES's ability to form hydrogen bonds through both experimental measurement with solvatochromic dyes and computational prediction through COSMO calculations with a TZVP basis set. DES made from HBA (quaternary ammonium and polyphenol phosphonium salts) and HBD (glycol, glycerol, and carboxylic acids) were used with dyes of 4-nitroaniline, N,N-dimethyl 4-nitroaniline, and N,N-diethyl 4-nitroaniline to examine hydrogen bonding ability of the DES with dyes of different polarity and basicity. Electron distribution based modeling in COSMO was used to predict the relative effect of each component used to form DES in the solvent's ability to form hydrogen bonds with the selected dyes.

610647 Thermally Stable Ionic Liquid As Media for Separating Aliphatic and Aromatic Compounds

Abstract:

The process of separating aromatic and aliphatic hydrocarbons is energy intensive due to similar boiling points and formation of azeotropes. The use of ionic liquids (ILs) for these separations has the potential to lower these energy costs. We show that thermally-robust peraryl ILs especially, triphenyl-*p*-phenoxylphenylphosphonium bistriflimide [Ph_3P-*p*-$POPNTf_2$] have a high affinity towards aromatic compounds while remaining immiscible with aliphatic compounds. Furthermore, when mixed with excess aromatic the IL forms two distinct phases: an IL-phase saturated with aromatic and a nearly pure aromatic phase. The molecular-level interactions leading to this behavior, however, still remain unclear. Here, we use molecular dynamics (MD) simulations to provide a detailed molecular-level understanding of these systems. The fundamental interactions between ILs and model compounds such as *ortho*-, *meta*-, *para*- xylene, toluene, hexafluorobenzene, cyclohexane, naphthalene, ethylbenzene, pentylbenzene, bromobenzene, cyanobenzene and *n*-heptane are assessed. MD correctly predicts the formation of two phases and there is close quantitative agreement with the saturation limit of aromatics in the IL phase, while *n*-heptane, cyclohexane and pentylbenzene were completely immiscible with the ILs. The presence of aromatics in the ILs disrupt the IL network by forming cation-aromatic, anion-aromatic while weakening ion-ion repulsive interactions, meanwhile counter ions network remained intact. The change in IL-aromatic structure is further analyzed through radial/spatial distribution functions, nearest neighbors, pair-interaction energies between several atom pairs. The results showed formation of cage like structures (liquid clathrate) which are dynamic in nature. These efforts provide an improved insights into molecule-level understanding of the solvent environment provided by the cation and anion during separation and further help in tuning the ILs for efficient separation applications.

604062 Effects of Physical Structure of Alkylammonium Hydrogen Sulfate Ionic Liquids on Acidity

Abstract:

Alkylammonium [HSO_4] protic ionic liquids (ILs) are a promising class of ILs, shown to be effective in a range of applications such as biomass pretreatment, catalysis, chemical synthesis, and protein stabilisation [1]–[5]. Their main advantage is their extremely low cost ($0.78/kg [6]), due to a simple one-step synthesis of two cheap reagents. Previous studies on ILs have found that the chemical activity of the IL is dominated by the anion. Therefore, while a number of physical properties of hydrogen sulfate ILs have been investigated, they have tended to involve a small number of cations, with relatively little investigation into the effect of cation structure on chemical behaviour. Additionally, most applications have focused on the use of tertiary alkylammonium cations. Changing the physical structure of the cation could potentially further reduce costs, as well as widen the range of potential applications by modifying chemical behaviour. Four alkylammonium cations were screened in this study, with three different degrees of substitution (shown in Fig. 1). The acidity of the resulting IL solutions (containing 20% wt. water) was quantified using the Hammett acidity function.

Decreasing degree of alkylammonium substitution was found to significantly enhance solution acidity. The acidity of [MBA][HSO_4] was found to be equivalent to that of [DMBA][HSO_4] with 11% excess sulfuric acid, while that of [HA][HSO_4] and [BA][HSO_4] was found to be equivalent to that of [DMBA][HSO_4] with 31% and 32% excess acid respectively. Addition of up to 20% excess base to [BA][HSO_4] expectedly made the IL solution less acidic, although its Hammett acidity was still substantially lower than that of [DMBA][HSO_4], as well as lower than that of [MBA][HSO_4]. Such a strong trend in acidity was unexpected from aqueous acid-base theory, due to the narrow range of constituent amine pKa's and small differences in pH of heavily diluted IL solutions. This was attributed to increased solvation of primary and secondary alkylammonium cations, decreasing the solvation of the [HSO_4^-] anion and thus increasing its proton activity. The effect of solution density was found not to play a significant role. When adding excess sulfuric acid to [DMBA][HSO_4], the IL solution displayed a buffering effect at low excesses (1-5% wt.), showing a higher Hammett acidity than purely aqueous sulfuric acid.

Biomass pretreatment was identified as an application which would benefit from enhancement of alkylammonium [HSO_4] solution acidity. Pretreatments using primary and secondary alkylammonium cations were carried out and compared to those using [DMBA][HSO_4], a current gold-standard protic IL. Pretreatment with primary alkylammonium [HSO_4] ILs led to overly severe pretreatments with very accelerated kinetics, while pretreatment with [MBA] [HSO_4] ILs showed equally high performance to [DMBA][HSO_4] but with accelerated kinetics. Addition of up to 20% excess base to primary alkylammonium ILs had relatively little effect on pretreatment severity. Additional pretreatments were performed using [DMBA][HSO_4] with excess acid, matching the Hammett acidities of [HA][HSO_4] and [MBA][HSO_4]. Despite equal Hammett acidities, pretreatments using excess acid were significantly more severe than those using the primary and secondary alkylammonium HSO_4 ILs. The observed differences in pretreatment severity suggest different sources of acidity between the two types of solutions, which cannot be distinguished using the

Hammett acidity parameter. Pretreatment findings add credence to the theory that the acidity of the primary and secondary alkylammonium ILs is due to the increased activity of protons from lower solvation of the $[HSO_4^-]$ anion, rather than increased proton dissociation in solution. In addition, the development of $[MBA][HSO_4]$ as a promising pretreatment solvent highlights the possible improvements from screening of ion structure in ILs.

[1] F. J. V. Gschwend, C. L. Chambon, M. Biedka, A. Brandt-Talbot, P. S. Fennell, and J. P. Hallett, "Quantitative glucose release from softwood after pretreatment with low-cost ionic liquids," *Green Chem.*, vol. 21, no. 3, pp. 692–703, 2019.

[2] F. J. V. Gschwend, F. Malaret, S. Shinde, A. Brandt-Talbot, and J. P. Hallett, "Rapid pretreatment of Miscanthus using the low-cost ionic liquid triethylammonium hydrogen sulfate at elevated temperatures," *Green Chem.*, pp. 1–3, 2018.

[3] P. Attri and P. Venkatesu, "Exploring the thermal stability of α-chymotrypsin in protic ionic liquids," *Process Biochem.*, vol. 48, no. 3, pp. 462–470, Mar. 2013.

[4] E. T. Kermani, H. Khabazzadeh, and T. Jazinizadeh, "Friedländer synthesis of poly-substituted quinolines in the presence of triethylammonium hydrogen sulfate [Et3NH][HSO4] as a highly efficient, and cost effective acidic ionic liquid catalyst," *J. Heterocycl. Chem.*, vol. 48, no. 5, pp. 1192–1196, Sep. 2011.

[5] M. R. M. Shafiee, B. H. Najafabadi, and M. Ghashang, "Triethylammonium Hydrogen Sulfate as a Catalyst for the Preparation of 1,5-diaryl-3-(arylamino)-1 *H* -pyrrol-2(5 *H*)-one derivatives," *J. Chem. Res.*, vol. 35, no. 11, pp. 634–636, Nov. 2011.

[6] Baaqel et al, *Green Chemistry*, 2020 (in press)

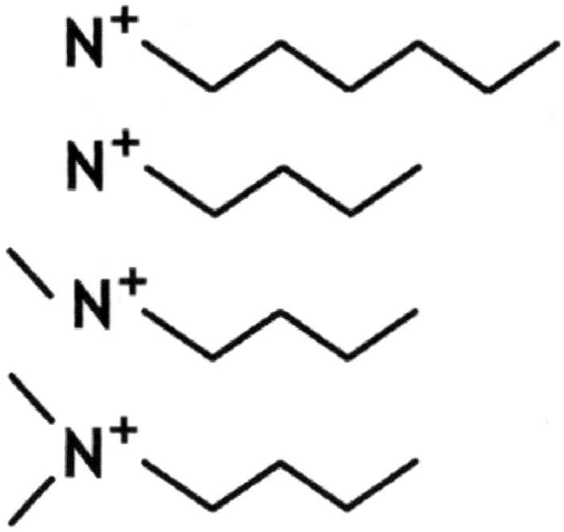

Hexylammonium [HA]

Butylammonium [BA]

MethylButylammonium [MBA]

DiMethylButylammonium [DMBA]

Figure 1: Cations screened, all using the $[HSO_4^-]$ anion.

608640 Highly Selective Electrochemical Reduction of CO_2 to CO Via a Novel Catalyst with Mn-N_3 Site

Abstract:

Electrochemical CO_2 reduction reaction (CO_2RR) is one of the most promising ways to deal with the excess CO_2 and resolve the energy crisis [1]. However, due to the stability of CO_2 molecule and reaction involves multiple electrons and protons, poor selectivity and low current density are the main challenges of this filed [2]. So, efficient catalysts are in great need.

As the third abundance of transition metals in Earth's crust, manganese (Mn) has shown great potentials on many electrocatalytic fields. Nevertheless, the performance of Mn based catalysts is largely limited for CO_2RR. Herein, we firstly prepared Mn single-atom catalyst (SAC) with Mn-N_3 site supported by g-C_3N_4. The prepared catalyst exhibited a maximum CO FE of 98.8% with 14.0 mA cm^{-2} CO current density (j_{CO}) at overpotential of 0.44 V in $KHCO_3$ electrolyte, outperforming all reported Mn SACs and being one of the best reported electrocatalysts for CO_2 conversion CO. Moreover, a higher j_{CO} value of 29.7 mA cm^{-2} was obtained at overpotential of 0.62 V, when ionic liquid was used as electrolyte. In-situ experiment and DFT results confirmed that Mn-N_3 site is the active site, on which the free energy barrier of key intermediate formation is greatly reduced. It can be anticipated that Mn SACs may also exhibit excellent performance in other electrochemical reactions through changing supporting materials to form exclusive active site.

References

[1] Feng JQ, Gao HS, Feng JP, Liu L, Zeng SJ, Dong HF, Bai YG, Liu LC, Zhang XP. Morphology modulation-engineered flowerlike In_2S_3 via ionothermal method for efficient CO_2 Electroreduction. *ChemCat Chem*, 2020, 12: 926-931.

[2] Feng JP, Zeng SJ, Liu HZ, Feng JQ, Gao HS, Bai L, Dong HF, Zhang SJ, Zhang XP. Insights into carbon dioxide electroreduction in ionic liquids: Carbon dioxide activation and selectivity tailored by ionic microhabitat. *ChemSusChem*, 2018, 11: 3191-3197.

608885 Thermodynamic Studies/Analysis of NH_3 Absorption in Pyridinium-Based Protic Ionic Liquids

Abstract:

Ammonia (NH_3) is undoubtedly another major threat to environment and human health since it is considered as a main precursor of PM2.5 (Particulate matter ≤ 2.5 μm)[1]. Meanwhile, NH_3 is an important chemical raw material and widely applied in some fields, such as fertilizers, nitric acid, freezing medium and so on. Therefore, the recovery of ammonia from tail gas is of great significance for pollution control and resource utilization. Although the water scrubbing has been widely used in industry, it possesses some inherent drawbacks, such as high energy consumption, huge water consumption. Hence, it is extremely essential to develop new absorbents with high efficiency, reversibility and low energy consumption for NH_3 recovery.

In our previous research, we developed the hydroxyl-functionalized ionic liquids[2] and protic ionic liquids[3] for absorption NH_3, respectively. They all displayed effective absorption performance for NH_3 through the hydrogen bonds. In this work, a novel type of pyridinium-based ionic liquids (ILs) with acidic protons and hydroxyl groups were designed and synthesized for NH_3 absorption. The NH_3 solubility in five pyridinium-based protic ILs at temperatures from 303.15 to 343.15 K and pressures up to 600 kPa were measured using gas-liquid equilibrium method, and the thermodynamic properties including reaction enthalpy, entropy, Gibbs free energy were further obtained by a Reaction Equilibrium Thermodynamic Model (RETM)[4] to better understand the NH_3 absorption process. The results indicated that the PIL [4-MeOHPy][NTf_2] exhibits the maximum NH_3 solubility of 3.43 mol NH_3/mol IL at 313.15 K and atmospheric pressure, surpassing any nonmetallic ILs previously reported. Furthermore, the characteristics of isotherms under low pressures behaved an obvious chemical reaction between ILs and NH_3. This work provides an important reference for the thermodynamic properties of NH_3 absorption process in practical application.

612216 Aqueous Ionic Amines for CO2 Capture in Air Revitalization

Abstract:

Air revitalization in closed environments is a mission critical task for NASA, with the chief technical challenge being the removal of CO_2 from the recirculating air supply. Specifically, efficient removal of CO_2 from closed space craft environments is a NASA research and development priority as stated in the NASA Technology Roadmap Technical Area (TA): Human Health, Life Support, and Habitation Systems document. The practical significance of this requirement was recently illustrated in a NASA study that showed that 7% of the genes changed their expression during spaceflight, which was attributed to metabolic stress on the body due to elevated CO_2 exposure in the spacecraft.[1-2] Additionally, NASA is concerned with cognitive and behavior effects on humans exposed to increased levels of CO_2.[3]

On the international space station, CO_2 removal is currently accomplished using a thermally regenerable solid zeolite, Grace 13X. While effective, the solid zeolite has a number of logistical problems including the production of dust from the absorbent bed and sensitivity to ambient humidity. A similar separations challenge exists on submarines, where closed environment CO_2 removal is a crucial part of the life support system. On submarines, CO_2 is removed via chemical capture with an aqueous ethanolamine solution. This is also an effective process, but with significant disadvantages; chiefly, the volatilization of the amine, which creates a foul odor, can damage electronics and presents long term heath effects. Furthermore, regeneration of the solution (desorption of the CO_2) is somewhat energy intensive as the large thermal mass of the solution (often 70 mass % water) must be heated to promote desorption. While the energy requirement is moot on a nuclear submarine, power consumption is a key concern for spacecraft.

A related area of research is the removal of CO_2 from post-combustion flue gas streams, such as those from coal and natural gas-powered electrical generation plants. Although the goal in this application is also the removal from of CO_2 from ambient/near ambient pressure gas streams, the system temperature is somewhat higher, and the partial pressure of CO_2 is much higher than for air revitalization. However, the same technology, non-volatile aqueous amines, may present a solution for both challenges. The authors have worked with industrial partners to develop concentrated aqueous solutions of non-volatile (ionic) amines which have shown promise as viable post-combustion CO_2 capture agents. From this work, three joint process patents focused on CO_2-capture efficiency and minimizing regeneration energy requirements have been issued[4-6] in addition to the composition of matter patents already owned by South Alabama.[7-8] While the previous work was focused on optimizing the solutions for the industrial setting at higher temperatures and higher CO_2 partial pressure, the work described below focuses on the development and evaluation of these systems for used in closed environment air revitalization. Furthermore, as these are aqueous systems, they are inherently compatible with humid air and as hygroscopic salts, the potential exists for them to be used as part of an overall humidity control system.

The work presented here focuses on the thermophysical and thermodynamic properties of aqueous solutions of ionic amines for use in CO_2 capture for air revitalization, thermodynamic of cycling and thermal stability of the compounds. These properties are probed using a combination of DSC and SDT techniques along with data collected from a lab-scale absorber.

1. Drake, N., No, Scott Kelly's Year in Space Didn't Mutate His DNA. National Geographic 2018.
2. Witze, A., Astronaut twin study hints at stress of space travel. Nature 2017.
3. Stankovic, A.; Alexander, D.; Oman, C. M.; Schneiderman, J. A review of cognitive and behavioral effects of increased carbon dioxide exposure in humans; 0499-9320; Massachusetts Institute of Technology: 2016; pp 219277/1-219277/22.
4. Chinn, D.; Cooper, R. E.; He, Z.; Davis, J. H., Jr.; West, K. N.; Timken, H. K.; Driver, M. S. Aqueous solutions of amine functionalized ionic compounds for carbon capture processes. US20120171094A1, 2012.
5. Chinn, D.; Cooper, R. E.; He, Z.; Timken, H. H.; Driver, M. S.; Davis, J. H., Jr.; West, K. N. Process for the separation of carbon dioxide from flue gas. WO2012092176A2, 2012.
6. Cooper, R. E.; Chinn, D.; He, Z.; Davis, J. H.; West, K. N.; Timken, H. H.; Driver, M. S. Method for improving total energy demand in a post-combustion carbon capture process with ionic absorbent. WO2012092204A2, 2012.
7. Davis, J. H., Jr. Functionalized ionic liquids for removal of acid components from sour natural gas. WO2003086605A2, 2003.
8. Davis, J. H., Jr. Carbon dioxide scrubbing using ionic materials. WO2008122030A2, 2008.

602555 Separation of Hydrofluorocarbon Mixtures Using Ionic Liquids

Abstract:

Hydrofluorocarbons (HFCs) are organic compounds that are commonly used as refrigerants. HFCs were introduced as a replacement to chlorofluorocarbons (CFCs) in response to the Montreal Protocol due to their low Ozone Depletion Potential (ODP). Although HFCs have low ODP, many have a high Global Warming Potential (GWP) and are now being phased out as a result of the Kigali agreement and replaced with hydrofluoroolefins (HFOs). The phasing out of HFCs and HFC mixtures alone is leading to thousands of tons of refrigerants being unusable with the only options for disposal being to vent or incinerate them. It is important that new techniques are developed to sustainably separate and recycle HFC mixtures so they can be disposed of properly or repurposed in new, more environmentally friendly refrigerants such as HFO mixtures.

Project EARTH (Environmentally Applied Research Toward Hydrofluorocarbons) is a collaborative project between multiple universities and industry companies that is focused on developing efficient and selective separation processes and recycling methods for HFC mixtures. The separation of HFC mixtures is difficult due to the fact that many of the mixtures are azeotropic or near azeotropic. For this reason, Project EARTH is looking into multiple approaches to this challenge, one of which being the use of ionic liquids (ILs). We believe that IL's properties (no measurable vapor pressure, dissolution of many organic and inorganic compounds, variable solubility of gases and liquids, and high thermal, chemical, and electrochemical stability) make them uniquely suited to creating selective separation processes for various HFC mixtures.

This presentation will cover results on the use of various ILs for the separation of refrigerant mixture R-410a which is a 50/50 mixture by weight of R-125 (CHF_2CH_3) and R-32 (CH_2F_2).

608807 Functionalized Ionic Liquids Based on Vegetable Oils for Rare Earth Elements Recovery

Abstract:

Rare earth elements (La-Lu, Y and Sc) and their compounds are critical materials in computers, audio systems, electric and hybrid vehicles, cell phones, wind turbines, magnetic resonance imaging machines, et al. To extract rare earth metals from aqueous solution is an extremely important topic in the world today and is anticipated to be of even greater future importance due to the increasing demand for these critical metals. Unfortunately, producing rare earths is a problem of economical extraction and always requires using large volumes of environmentally harmful chemicals. In this work, functionalized ionic liquids (FILs) based on vegetable oils have been directly synthesized and used to effectively extract rare earth elements. Carbon dioxide gas is introduced to successfully strip rare-earth-loaded FILs in the presence of water. The novel extraction process reveals great advantages of accessibility, biocompatibility and sustainability as well as cost efficiency.

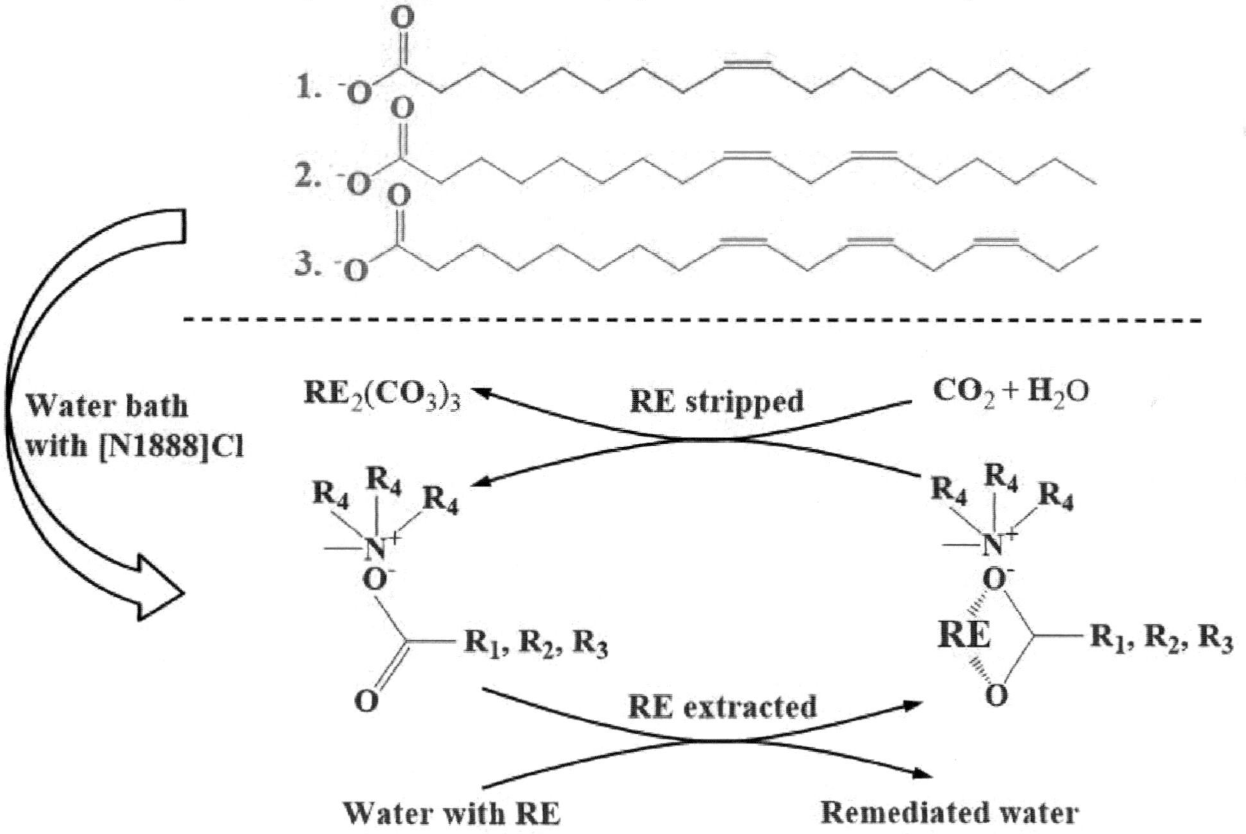

596060 Downstream Process with Countercurrent Chromatography to Recover Bioproducts from Sustainable Feedstocks

Abstract:

Separation of bioproducts produced from renewable feed streams is challenging in downstream process due to complex feed composition and low feed concentration. Current technologies such as ion exchange, adsorption, and liquid-liquid extraction are often faced with large solvent consumption, limited choice of resins, additional pretreatment and regeneration steps, and limited pH operation range. CounterCurrent Chromatography (CCC) is an emerging approach that uses at least two immiscible liquids for both mobile and stationary phase and is a potentially attractive separation method to recover bioproducts from complex process streams. Although there are multiple advantages of CCC for its liquid stationary phase in the context of natural products, there is little attention on employing CCC into downstream process for commodity bioproducts and biochemicals. This talk will discuss the feasibility of incorporating CCC into downstream process to produce high-value products from lignocellulosic feedstocks. Separation of bioproducts from lignocellulosic feedstocks such as alkaline pretreatment liquor and reductive catalytic fraction oils will be presented as demonstrations. The energy and solvent demands of the system correlated with the physicochemical properties of solvent systems, CCC operation mode, and feed composition will be discussed.

610487 Separation of Lignocellulosic Sugars from Pyrolytic Bio-Oil Using Simulated Moving Bed

Abstract:

Anhydrosugars are the major carbohydrate products of autothermal pyrolysis of lignocellulosic biomass that can be upgraded into biofuels and bio-based chemicals. Our research group has developed a fast pyrolysis process, namely autothermal pyrolysis, which can produce high yields of bio-oil by thermal depolymerization of pretreated biomass. A significant fraction of the heavy ends of bio-oil contain sugars, mainly levoglucosan, but also xylose as the major carbohydrate products. The anhydrosugars in bio-oil are acid-hydrolyzed to produce glucose, which can serve as a fermentation substrate for manufacturing common renewable fuels such as ethanol.

The bio-oil contains some phenolic monomers generated from the thermal depolymerization of lignin during the pyrolysis process. These phenolic monomers end up in the hydrolyzed, neutralized, and filtered water-soluble sugar fraction of the bio-oil due to a moderate solubility of phenol and other substituted phenolic monomers in water. The presence of phenolic species in the aqueous extraction of pyrolytic bio-oil poses significant toxicity challenges in employing fermenting microorganisms downstream and therefore must be removed from the aqueous solution of bio-oil sugars prior to fermentation. Our previous report shows effectiveness of hydrophobic polymeric resins in preferential adsorption of phenolic species compared to levoglucosan and glucose from an aqueous fraction of bio-oil.

In this work, we have developed a simulated moving bed technology (SMB) to employ continuous scale separation of sugars from phenolic species for an aqueous bio-oil fraction using the principle of adsorption chromatography. Sugars were separated from phenolics in the SMB processing a hydrolyzed, neutralized and filtered water-soluble fraction of bio-oil containing 225 g/L glucose, 10 g/L xylose and 21.1 g/L phenolics at the rate of 4 mL/min. Although it is a binary separation, the SMB is modeled to separate two classes of chemicals (sugars and phenolics) rather than two individual chemical species. The SMB uses a 3-1-1-1 configuration of six columns each having 53 cc bed volume and can operate at 0.5-8.0 bed volumes per hour (BV/h) flow rates. The columns are packed with hydrophobic resin Dowex L493 to adsorb the phenolic compounds preferentially on the resin surfaces as they are more non-polar compared to the sugars in aqueous solution. Ethanol is used to desorb the adsorbed phenolic species on the column surfaces while DI water is used to purge any adsorbed sugars on columns into a purified sugar solution. Water is also used to regenerate the columns after each cycle of feeding, desorption and purging before the next cycle of feed flow begins. The entire unit periodically switches the entry point locations of feed, purge, desorption, and regeneration steams to simulate a continuous flow of solid beds of adsorbent resin against the liquid solution of water-soluble sugars. The purified sugar product contained 154 g/L glucose and 0.8 g/L phenolics after 6 h of operation at 4 BV/h using 20 min switch interval time and 1:1 ethanol/water consumption. We have further explored the effects of switch interval time and flow rates of the streams on the separation efficiency of the SMB unit to optimize its design.

603910 Black Liquor Gasification Chemical Recovery and Combined Cycle Power at Pulp and Paper Mills Biorefinery

Abstract:

In the areas of chemicals and allied products and energy process engineering, it is of particular interest to design processes with low energy requirements (energy intensive) (preferably renewable materials, e.g., biomass) and the use of minimum raw materials or use of a byproduct as feedstock (materials intensive) and minimum number of equipment (i.e., lean manufacturing). This is the basis from which the area of process intensification (PI) arises. The main objective of PI is to propose production and manufacturing processes in plants as small as possible and with low total energy, materials, water, and other resources.

Pulp and paper are primarily made out of wood fibers originating from natural forests or pulpwood plantations. Recycled fiber and other fiber sources such as agricultural residue are also utilized and recycled fiber is becoming more commonly used in pulp and paper making. There are two significant pulping technologies available that differ greatly in terms of process, i.e., mechanical and chemical pulping. Approximately 30% of the total pulp production in European Union is from mechanical pulping while the rest is produced by means of chemical pulping. North America has major pulp and paper industry, about 21% of the total pulp produced is from mechanical pulping and rest is produced chemically.

A pulp mill that produces bleached kraft pulp generates 1.7-1.8 ton of black liquor (measured as dry content) per ton of pulp. Black liquor thus represents a potential energy source of 250-500 MW per mill. As modern kraft pulp mills have a surplus of energy, they could become key suppliers of renewable fuels in the future energy systems. Today, black liquor is the most important source of energy from biomass. It is thus of great interest to convert the primary energy in the black liquor to an energy carrier of high value. A key advantage of black liquor compared to biofuels and fossil fuels is that it is already at the mill; the handling infrastructure already exists and there are no collection and transport costs.

An extended case study is presented in which black liquor, a byproduct of wood-pulping process, is utilized to make chemicals and biofuels, in particular methanol, dimethyl ether and syngas. Different feedstock, use of byproduct and alternatives and principles of PI are commented (Das, 2020).

Reference: Das, T.K. (2020). Industrial Environmental Management: Engineering, Science and Policy, Wiley, Hoboken, NJ.

600048 From Biomass to Power Using Integrated Compact Pyrolysis, Combustor, and Stirling Engine

Abstract:

The feedstock logistic challenges, high feedstock procuring and shipping costs, high capital and operating costs, and a stringent requirement of making gasoline and diesel blends are the major barriers to commercialize biomass conversion pathways such as pyrolysis to make renewable fuels. These barriers can be overcome by developing portable technologies with a low weight, small footprint, and a high energy efficiency that can upgrade biomass to power. In this work, the viability of converting wood chips and plastic to power via integrated compact catalytic fast pyrolysis, combustor, and stirling engine technologies was determined through simulation and modeling analysis. The catalytic fast pyrolysis component consists of auger reactor followed by a catalytic upgrading reactor and the air-cooled condenser that upgrades wood chips and plastic to a high energy dense fuel. The ASPEN simulations followed by sizing analysis were performed to determine the size, weight, and net energy efficiency of producing 42.5-Watt equivalent of power generation from wood chips and plastic using the integrated portable technology. The impact of catalytic fast pyrolysis oil yields, oil quality, condenser type and its efficiency, type of fuel combustor, and combustor efficiency on the size, weight, and net energy efficiency of upgrading biomass to power was determined.

602795 Techno-Economic Analysis and Life-Cycle Assessment of Emerging Technologies for Bioprocessing Separations

Abstract:

Limited availability, rising costs, and environmental concerns about fossil fuels have generated considerable interest in finding alternative, renewable sources including biomass, which can be converted into a number of biofuels and bioproducts. In comparison to petroleum-based products, high processing costs, mainly associated with bioprocessing separations, limit widespread implementation of biofuels and bioproducts. Bioprocess-related separations are also complicated, regardless of the conversion pathway, due to the dilute nature of products and the chemically complex mixtures that result from biomass deconstruction. In many cases, bioprocessing separation approaches lack a technology baseline, or definition of the state of technology (SOT). This work focuses on evaluating the technology readiness of novel separations technologies for the conversion of biomass into biofuels and bioproducts and will address three key topics: 1) SOT description, 2) economic viability analysis of both SOT and innovative separations, and 3) environmental impact assessment of both SOT and novel processes. The current SOT for lignin valorization, dilute carbon recovery, and impurity removal is used to identify potential opportunities for improvement and provide a baseline for comparison with the emerging technologies. Detailed techno-economic analysis (TEA) and life cycle assessment (LCA) are applied to understand the key drivers and challenges related to the economic feasibility and environmental impacts, respectively.

609525 Professor Lin's Contributions to Molecular Sieve Membranes

Abstract:

Since the early 90's Professor Lin has made important contributions in many fronts of adsorption and membrane science and engineering. I will highlight his group's influential contributions on zeolite and MOF membranes and adsorbents including zeolite membrane microstructure evolution during thermal treatments (Dong et al., MICROPOROUS AND MESOPOROUS MATERIALS, 34, 241-253 (2000)), novel synthesis approaches for para-xylene selective zeolite membranes (JOURNAL OF THE AMERICAN CHEMICAL SOCIETY , 126, 4776-4777 (2004), and stability of ZIF-8 membranes and crystals in water (JOURNAL OF MEMBRANE SCIENCE, 485, 103-111 (2015). New directions that this work opened will be discussed.

600263 Novel Materials for Membrane Applications

Abstract:

Lytotropic Liquid Crystals (LLCs) are amphiphilic molecules containing a hydrophobic organic tail section and a hydrophilic headgroup, which self-assemble in the presence of water or other hydrophilic species into highly ordered, yet fluid phase-separated assemblies with specific nanometer-scale hydrophilic domains. Depending on the shape of the constituent LLCs and the interfacial curvature energy of the system, monodispersed hydrophilic domains ranging in structure from lamellae to cylindrical channels to three-dimensionally interconnected pores with dimensions in the 1–5 nm size range can be formed. By making polymerizable analogs of LLC mesogens, it has been possible to stabilize the various LLC assemblies with retention of phase microstructure via photo-initiated radical cross-linking. These membranes have charged pores and reject ions in solution analogous to reverse osmosis and small neutral solutes based on size similar to nanofiltration membranes. These membranes have demonstrated exceptional performance to treat fracking water, reject chemical warfare agents, and operate for electrochemical applications.

601565 Methanol Synthesis in a High-Pressure Membrane Contactor Reactor Employing an Ionic Liquid As the Sweep

Abstract:

Introduction

Methanol (MeOH) finds broad use as a transportation fuel and as an important raw material used in the production of other fuels and chemicals. There are several commercial methanol synthesis (MeS) processes[1,2], but they are all challenged by low per-pass conversion that necessitates the recycling of the unreacted syngas. This is especially problematic for MeOH production from renewable biomass, where the use of oxygen-blown gasifiers is not economic, and the available syngas has a large nitrogen content. One way to overcome the thermodynamic limitations of MeS, and thus increase the per-pass conversion is to utilize a membrane contactor reactor (MCR). In this study, a novel high-pressure MCR is used, in which a membrane with the desired characteristics serves as an interface contactor between the MeS environment in the shell-side and a sweep liquid flow in the permeate-side. Two different sweep liquids have been employed in the study: A high boiling point petroleum-derived solvent[3] tetraethylene glycol dimethyl ether (TGDE) and an ionic liquid (IL) 1-Ethyl-3-methylimidazolium tetrafluoroborate ([Emim][BF4]). The choice of TGDE and the IL as sweep liquids is because MeOH has high solubility in both of them, while the permanent gases like H_2 and CO have negligible solubility. Compared to TGDE, the IL has higher decomposition temperature (400 ^{o}C), which implies less demanding downstream processing for MeOH recovery; its lower vapor pressure means lower solvent loss, and its higher thermal capacity affords a broader range of MeS operating conditions and more efficient thermal management. TGDE, on the other hand, is at present more readily available and less costly In this study, MeS is studied in a MCR using IL as the sweep liquid and the results are compared with the MCR system employing TGDE. The data are used for validating a model which is then used for process scale-up and technoeconomic analysis (TEA).

The experimental set-up is described elsewhere[3]. It consists of the syngas delivery system, the MCR, the sweep liquid delivery system, and the analysis section using GC. A mesoporous alumina membranes, whose surface is rendered hydrophobic via an appropriate modifying agent, is installed in the MCR. During operation, the syngas is fed into the shell-side where it contacts under high pressure/temperature the MeS Cu-ZnO-Al_2O_3 catalyst to convert into MeOH. The sweep liquid is pumped through the membrane tube-side to remove the produced MeOH, and help increase the MeS conversion/yield. Experiments were conducted in a broad range of pressures and temperatures, sweep liquid flow rates and for two different sweep liquids, TGDE and IL. The lab-scale system operates also as a packed-bed reactor (PBR) when the inlet and outlet of the membrane tube-side are closed.

Selected Results

In this project, we studied the performance of the MCR under different pressures, sweep liquid flow rates and syngas feed compositions and compared it with the PBR under the same conditions. Figure 1 shows the effect of sweep liquid flow rate and reactor shell-side pressure on the carbon conversion at a temperature of 220 ^{o}C, and at a fixed value of W/F (catalyst weight/total molar flow rate) = 170 kg*s/mol with a syngas with SN = $(F_{H2} - F_{CO2})/(F_{CO} + F_{CO2})$)=1.96 and CF= $F_{CO}/(F_{CO} + F_{CO2})$)=0.625. As expected, the reactor conversion increases with increasing sweep liquid flow rate, with all MCR conversions exceeding the thermodynamic values. This is because higher sweep liquid flow rate helps remove a larger amount of MeOH from the reaction zone. The IL shows a higher conversion rate, likely related to the higher solubility of MeOH in IL compared to TGDE. The MCR exhibits conversions higher than the PBR. MCR and PBR conversions both increase with increasing pressure, since MeS is thermodynamically favored at higher pressures. An added reason for the increase in MCR conversion with pressure, is because a higher amount of MeOH transports through the membrane and is removed by the sweep liquid, thus shifting the equilibrium towards the MeOH generation side. For the whole range of pressures studied, the IL attains higher conversions than the TGDE, which relates to its higher MeOH solubility. Constrains with the size of the lab-scale system limit the gains in conversion afforded by the MCR. Modeling studies, however, indicate that conversions >90% are readily attainable. TEA indicates that such conversions would allow the MCR process to operate on biomass-derived syngas in an one-pass configuration without recycle of unreacted syngas needed.

Key Conclusions

In this study a high-pressure MCR was employed to carry-out the MeS reaction. MCR performance was investigated under a variety of experimental conditions with the results revealing improved carbon conversions compared to the conventional PBR.

Literature Cited

1. Aasberg-Petersen, K., Nielsen, C.S., Dybkjær, I., Perregaard, J. "Large scale methanol production from natural gas" Haldor Topsoe, **22**. (2008).
2. Behr, A. Methanol: The Basic Chemical and Energy Feedstock of the Future. Asinger's Vision Today. Edited by Martin Bertau, Heribert Offermanns, Ludolf Plass, Friedrich Schmidt and Hans-Jürgen Wernicke. Angewandte Chemie International Edition, **53(47)**, pp. 12674-12674 (2014).
3. Li, Z.; Tsotsis, T. T. "Methanol synthesis in a high-pressure membrane reactor with liquid sweep" *J. Membr. Sci*. **570**, pp. 103– 111 (2019).

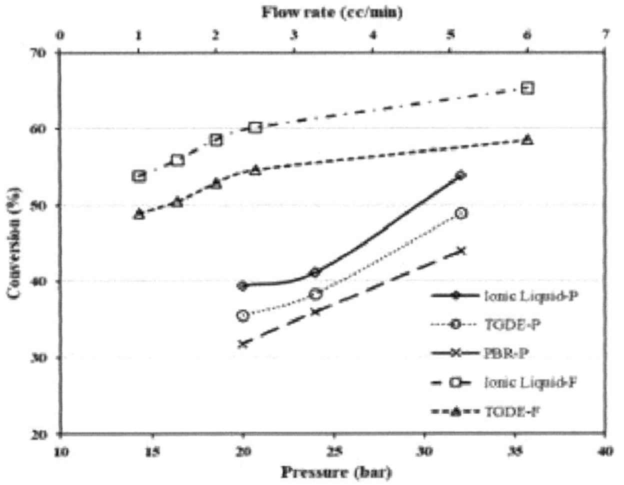

Figure 1. Effect of liquid sweep flow rate (F subscript) for pressure =32 bar, and effect of pressure (P subscript) on conversion

600662 Novel Membrane Performing Blood Separation and Sensing Synchronously

Abstract:

Membrane technology has been widely applied in industrial fields such as water treatment, product extraction, air purification, etc. However, in the medical and healthcare fields, blood separation which is vital for the diagnosis and rescue still relies on traditional centrifugation method and remains great challenge for current membrane technology. Recently, our group proposed a new type of separation-sensing membrane for blood simultaneous separation and sensing, achieving immediate serum analysis during the blood drawing. This presentation will review our latest progresses in the design strategy, materials preparation and nanostructure control for coupling membrane separation and biosensing abilities into one. We will demonstrate that this new separation-sensing membrane can be of universal ability in monitoring blood glucose, lactate, glutamate, transaminase, cancer markers, proteins and so on. It has successfully and accurately detected the key blood components in the patients suffering the various diseases, such as liver transplantation, portal vein thrombosis, endometrial cancer, femoral neck fracture and rectal place. Our results demonstrate that the proposed separation-sensing membrane design, showing great potential in the online monitor of vital components during surgery, emergency rescue and disaster relief, decreasing the mortality, is expected to provide an inspiration to accelerate the development of membrane in clinic science.

612255 Polymer Hydroxide Exchange Membranes: From Fuel Cells to Electrolyzers

Abstract:

One of the grand challenges facing humanity today is the development of an alternative energy system that is safe, clean, and sustainable. A distributed renewable electrochemical energy and mobility system (DREEMS) that is enabled by affordable wind and solar energy can meet this challenge. At the foundation of this new energy system, we have chosen to study fuel cells for zero-emission vehicles and electrolyzers for green hydrogen generation. For these devices, polymer electrolytes play a critical role in controlling their cost, performance, and durability, and consequently their economic viability. In this presentation, I will focus on our recent work on hydroxide exchange membrane fuel cells (HEMFCs). More specifically I will show the roadmap we have developed for this technology, the progress we have made in developing the most stable membranes. I will also report our latest results on hydroxide exchange membrane electrolyzers (HEMELs) with pure water feed.

600540 Microstructural Engineering and Architectural Design of Metal–Organic Framework Membranes

Abstract:

Membrane based separation is developing quickly due to low energy consumption, high efficiency, and easy operation. Metal–organic frameworks (MOFs) are a new class of molecular-sieve materials composed of metal ions or metal-oxide clusters coordinated by organic linkers to form highly regular porous networks. Owing to their adjustability in both pore apertures and functionality, MOFs in the form of membranes have offered unprecedented opportunities for efficient gas separations. A number of MOF membranes have therefore been fabricated and shown superior performance in H_2 purification, CO_2 capture, and C_3H_6/C_3H_8 separation. Moreover, in comparison with other candidates like zeolites, graphene oxides (GOs), and carbon molecular sieves, the presence of coordinatively unsaturated metal sites, guest-induced gate-opening phenomenon and framework flexibility in some MOF materials further endows them with infinite possibilities for efficient separation of a series of gas mixtures with similar kinetic diameters (like CO/N_2, C_2H_2/CO_2, and C_2H_4/C_2H_6 separation), which is otherwise quite challenging. Unceasing pursuit of materials with superior selectivity, permeability, and long-term operation stability remains the permanent themes in the field of membrane science and technology. In the case of MOF membranes, pore aperture (selectivity), functionality (selectivity), grain boundary defects (selectivity), thickness (permeability), and binding strength (operation stability) represent the most critical factors influencing their performance. Therefore, microstructural engineering, which typically includes orientation manipulation, interfacial synthesis, and construction of mixed-phase membranes at a mesoscopic level, and architectural design, which typically contains aperture size adjustment, cage modification, and post-decoration at a microscopic level, have become indispensable for performance improvement of MOF membranes.

599864 Connecting Physical and Mathematical Models of Carbon Molecular Sieve Membranes

Abstract:

Carbon molecular sieve (CMS) membranes combine outstanding separation performance for many different important applications. The special combination of ability to discriminate between similarly sized penetrants, coupled with scalable manufacturing possibilities makes CMS membranes attractive for next generation membranes. Properties of polyimide-derived CMS membranes, whose transformation from flexible polymer coils into rigid molecular sieves during pyrolysis will be discussed within a mathematical framework. During pyrolysis, entangled random coil polymer chains experience stress, causing them to aromatize and fragment into shorter rigid strands. Similar to isotropic-nematic entropically-driven phase transitions occurring in liquid crystals, the aromatic strands are believed to favor arrangement into plate-like structures. Ultramicroporous gaps between the individual aromatic strands provide ideal molecular sieving sites. Imperfect packing of such plates creates larger microporous volumes for sorption, characterized by a Langmuir sorption model. This presentation will focus upon the connection between the formation process and functional separation properties. Experimental results suggest that a small but important disordered continuous phase characterized by an effective Henry's Law model applies, resulting in dual modes of sorption in local equilibrium. To complement this dual mode sorption model, a dual mode transport mode is also discussed in the context of the CMS formation process.

605148 Microporous Membranes with Gated Pore Structure

Abstract:

Inorganic membranes for molecular separations include non-porous and microporous membranes. For non-porous, dense membranes the perm-selectivity is determined by the interaction of the membrane material with specific gases and the permeance is controlled by both the thickness of the dense membrane layer and surface reaction kientics. Non-porous membranes are only perm-selective to oxygen, hydrogen or carbon-dioxide. Microporous membranes are versatile and can be designed for the separation of many gas or liquid mixtures. For microporous membranes the perm-selectivity depends on the pore size and solubility (adsorption equilibrium) and the permeance is determined by the pore size, porosity and thickness of the microporous layer. Pore size reduction has been studied for the improvement of perm-selectivity of microporous membranes but such effort often leads to a reduction in porosity and hence the permeability for the permeating molecule. The more effective way to improve the perm-selectivity with minimum reduction in permeability is to reduce the pore size without much change in porosity. Experimental results of chemical vapor deposition modification of mesoporous alumina membranes will be presented to demonstrate these two structures. Modification of microporous membranes to improve perm-selectivity with a minimum reduction in permeance for the perm-selective molecule can be achieved with a design of gated-pore-structure, i.e., inserting a gate in the micropores to reduce the pore size in a narrow region of the membrane. This strategy will be demonstrated using two examples. One is the catalytic cracking deposition of silica on microporous MFI zeolite membranes to form a gate that effectively narrows the MFI zeolitic pores. The modified MFI zeolite membranes show a 5 to 30 fold increase in hydrogen/carbon-dioxide selectivity with only about 30% reduction in hydrogen permeance. The other example is the membrane surface ligand exchange modification of ZIF-8 membranes (exchanging partially the original ligand linkers in ZIF-8 with a bulkier one). The ligand exchange is designed to narrow the aperture of the crystallites on the surface of the ZIF-8 membrane. Such post-synthesis modification can improve propylene/propane selectivity of ZIF-8 membranes from about 100 to 200, with virtually no reduction in propylene permeance. However, challenges remain including controlled mofidication and characterization of the structure of modified membranes. All these results show promise of post-synthesis modification based on the design of the gated-pore structure to obtain membranes with high perm-selectivity and permeance (or permeability).

605395 CO_2 Recovery from Ultra-Dilute Streams: Bridging a Novel Temperature Vacuum Swing Adsorption Model with Materials Screening

Abstract:

Separation is the single most important Chemical Engineering unit operation that is responsible for 15% of our total energy consumption, and with the increase of the global population this number is expected to rise to as much as 45% (Sholl and Lively, 2016). Increasing the efficiency of separation processes will therefore have an enormous impact on society (National Academies of Sciences, Engineering, and Medicine, 2019). As an example, separation of a dilute component from a gas mixture (e.g., the removal of carbon dioxide from a confined space or mercury vapour from air) implies processing of extremely large volumes of gas to capture adequate amounts of the dilute component. This fact has significant implications on the energy consumption of the process and the plant size (Santori et al., 2018).

Studies attested that adsorption processes are best suited for the efficient removal of trace gas impurities aiming to achieve extremely high purities (Sircar et al., 1996). The two main reasons for this are: (a) the availability of a large spectrum of microporous adsorbents with varying pore structures and surface properties, and (b) the possibility of designing many different process schemes by tailoring generic adsorption separation methods. Materials like metal-organic frameworks (MOFs), whose chemical composition and pore shape can be optimally tuned for particular applications are good candidates, but we are currently lacking a meaningful systematic way, from a process viewpoint, to rank them for particular applications.

In our previous work, we developed a molecular simulation tool which allows us to screen thousands of materials for adsorption using their physicochemical and adsorptive properties (Boyd et al., 2019). In this work, we extend that study by coupling the molecular simulation tool with process modelling to rank materials for dilute gas separations and for a given set of process metrics, including the overall energy efficiency and process productivity. A novel 5-step temperature-vacuum swing adsorption (TVSA) process was developed, and its performance was evaluated by using an equilibrium-based shortcut approach similar to that described by Joss et al., 2015. Our TVSA process model is then coupled with molecular simulations, which allows the evaluation of both, physicochemical and adsorptive properties of thousands of microporous structures, including ~300,000 MOFs (Boyd et al., 2019).

We applied the model to a particular ultra-dilute binary mixture of N_2 and CO_2 (CO_2 concentration of 2000 ppm) for applications such as CO_2 capture from confined spaces like spacecraft, aircraft, and submarines, and by using the commercial adsorbent zeolite 13X. The CO_2 purity achieved after each step of the process, together with the total amount of CO_2 and N_2 species, is presented in Fig. 1. This novel process yields higher CO_2 purities than the ones obtained by standard previously used 5-step TVSA processes.

The performance of this process is expected to be further improved by coupling molecular modelling with process optimization. By doing so, we aim to screen the large pool of available MOF materials to find the best performing ones for this particular application. This analysis provides new insights into which relevant metrics need to be included when ranking materials for ultra-dilute gas separations, and identifies the top performing ones given the relevant process metrics.

Acknowledgement

The authors are thankful for the support from the PrISMa Project (No 299659), which is funded through the ACT programme (Accelerating CCS Technologies, Horizon2020 Project No 294766). Financial contributions made from: BEIS together with extra funding from NERC and EPSRC, UK; RCN, Norway; SFOE, Switzerland and US-DOE, USA, are gratefully acknowledged.

References

Boyd, P.G., Chidambaram, A., García-Díez, E., Ireland, C.P., Daff, T.D., Bounds, R., Gładysiak, A., Schouwink, P., Moosavi, S.M., Maroto-Valer, M.M., Reimer, J.A., Navarro, J.A.R., Woo, T.K., Garcia, S., Stylianou, K.C., Smit, B., 2019. *Data-driven design of metal–organic frameworks for wet flue gas CO_2 capture*. Nature 576, 253–256.

Joss, L., Gazzani, M., Hefti, M., Marx, D., Mazzotti, M., 2015. *Temperature swing adsorption for the recovery of the heavy component: An equilibrium-based shortcut model.* Ind. Eng. Chem. Res. 54, 3027–3038.

National Academies of Sciences, Engineering, and Medicine, 2019. *A Research Agenda for Transforming Separation Science*. National Academies Press, Washington, DC.

Santori, G., Charalambous, C., Ferrari, M.C., Brandani, S., 2018. *Adsorption artificial tree for atmospheric carbon dioxide capture, purification and compression*. Energy 162, 1158–1168.

Sholl, D.S., Lively, R.P., 2016. *Seven chemical separations to change the world.* Nature 532, 435–437.

Sircar, S., Golden, T.C.C., Rao, M.B.B., 1996. *Activated carbon for gas separation and storage.* Carbon N. Y. 34, 1–12.

Figure 1. CO_2 purity and total amount of each species at the end of each step of the novel TVSA processes, for removing CO_2 from confined spaces (CO_2 concentration of 2000 ppm) by using the commercial zeolite 13X. The system can reach 74% purity and 83% recovery by consuming 11.5 $MJ_{thermal}/kgCO_2$. *Main process parameters: 10 °C adsorption temperature, 100 °C desorption temperature, 1 kPa vacuum pressure level.*

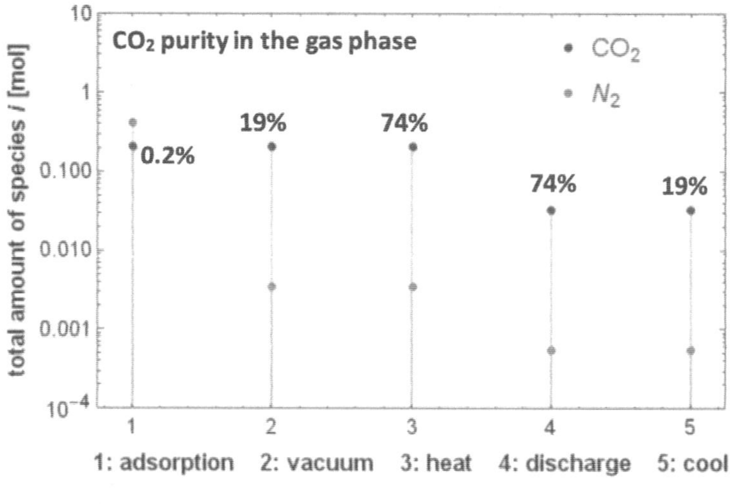

608589 Experimental Humid Post-Combustion CO_2 Capture Using a Metal-Organic Framework, Calf-20 By Vacuum Pressure Swing Adsorption

Abstract:

The presence of water in the flue gas from coal power plants makes CO_2 capture processes more difficult. Often, many studies assumed water is removed before entering the CO_2 capture unit. Characterizing the impact of H_2O on the CO_2 separation is essential and challenged due to a high affinity of H_2O onto most solid porous adsorbents. Particularly MOFs are considered to be sensitive to moisture. In this work, CALF-20, a water resistant and water stable metal organic framework, was studied and characterized for humid post-combustion CO_2 capture process. A mathematical model coupled with a process optimizer helped in identifying operating conditions that can guarantee high CO_2 purity and recovery. Two process cycles were experimentally tested in a lab-scale Vacuum swing adsorption rig under both dry and wet condition using a synthetic flue gas with 15% CO_2 and 85% N_2. Under the dry case, it was possible to obtain a CO_2 purity of 95% and a recovery of 90%. To study the impact of H_2O, the humidifier was introduced to create a stable humid stream. Various relative humidity (RH) values were examined: 15%, 25%, 45% and 70%. All wet PVSA experiments were run up to 400 cycles. The MOF demonstrated extraordinary stability and produced exceptional purity and recovery of CO_2.

595428 Modeling Hysteretic Isotherms and Parameter Estimation for Pressure Swing Adsorption Process Using Gate-Opening Metal-Organic Frameworks

Abstract:

Improvement in the performance of adsorbent may reduce the cost of CO_2 capture substantially. Metal-Organic Frameworks (MOFs) have been gaining attention in recent years, which are hybrid crystalline porous materials that consist of a regular array of positively charged metal ions surrounded by organic molecules. Among various MOF structures reported in literature, we investigate ELM-11, which has a sigmoidal isotherm shape. This material can captures and releases CO_2 with a small change of pressure[1], which would be a substantial advantage for applications in Pressure Swing Adsorption (PSA).

PSA process using ELM-11 has not been investigated in the past partly because of the difficulty of modeling the complex hysteresis of the isotherms due to the "gate adsorption" mechanism; the structure of ELM-11 changes as CO_2 molecules are captured in the framework. Figure 1 shows isotherms of ELM-11. By increasing the partial pressure of CO_2 from the state without CO_2 molecules in the framework, the space between layers of ELM-11 opens (gate opening) and the CO_2 molecules are captured in the space. This mechanism is reversed in desorption; when the partial pressure of CO_2 around ELM-11 is decreased, the captured CO_2 molecules are released, which leads to the closure of the space between the layers (gate closing). Note that there is a gap between the two trajectories of the adsorption and desorption isotherms.

In this research, we model the isotherms and estimate the parameters in the model from experimental data. We employ the Markov chain Monte Carlo (MCMC) method, a sampling algorithm for solving Bayesian estimation. This approach provides not only the estimate of parameter values but also the uncertainty in the estimates as probability distributions. The isotherm with the obtained parameters, which is a hybrid model, is incorporated into a PSA model to estimate the process performance.

Reference

[1] Tanaka, H. et al, 2015. J. Phys. Chem. C 119, 11533–11543.

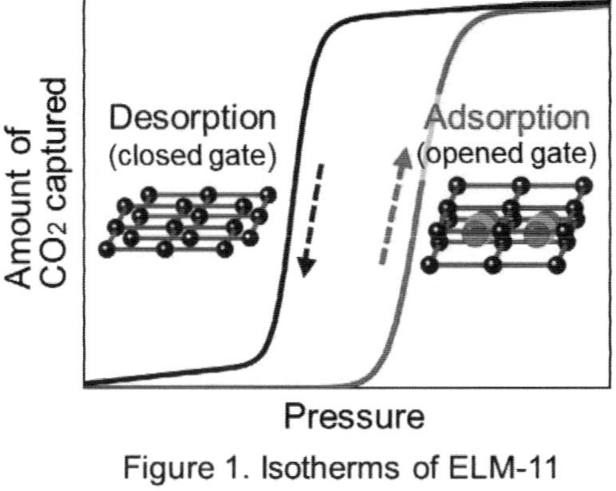

Figure 1. Isotherms of ELM-11

605065 3D-Printed Surface Topography to Achieve on-Demand Wettability and Separation

Abstract:

The biomimetic superhydrophobic and superoleophobic surfaces require the microscale and nanoscale hierarchical surface textures, combined with coatings with low surface energy. While it has always been difficult and expensive to fabricate the superwetting surfaces with well-controlled textures via conventional manufacturing techniques, 3D-printing serves as a time-efficient and cost-efficient alternative. In the current study, a novel microscale re-entrant surface topography is successfully fabricated by two-photon polymerization (2PP) 3D-printing. On-demand wettability of various liquids has been achieved on the single re-entrant surface. For a given liquid, the re-entrant angle can be precisely manipulated to be either smaller than the liquid contact angle (CA) on the flat surface to achieve dewetting, or larger than the liquid CA on the flat surface to achieve wetting. The 3D-printable microscopic repeating re-entrant surface, which is difficult to make by conventional techniques, has been demonstrated to further enhance the desired on-demand surface wettability. Moreover, efficient on-demand separations of liquid mixtures with high flux are enabled by the 3D-printed separation membranes with the repeating re-entrant structures, which is critical for the wastewater treatment in the chemical industry.

599289 Kidney-Inspired Membranes with Enhanced Anti-Fouling Properties, through Grafting of Stimuli-Responsive Polymer Brushes

Abstract:

Fouling represents one of the biggest bottlenecks in membrane technology. To tackle this problem, we seek inspiration in the kidney, a remarkable organ capable of producing over four million litres of effectively protein-free urine over a lifetime with no significant fouling (1). The glomerulus is a bundle of specialized blood vessels, which carries out the first stage of kidney filtration, the stage most comparable to membrane separations. Within the lumen of these blood vessels, a brush-like structure is present, composed of proteoglycans and glycoproteins, which are responsible for the overall negative charge and hydrophilicity of this layer. The combination of the brush structure, hydrophilicity and charge of the system are believed to be key causes behind the superior anti-fouling properties of the kidney.

To translate this source of inspiration into the context of artificial membranes, the design employed here involves the grafting of polyelectrolyte polymer brushes as an anti-fouling layer onto track-etched polyester (PET) membranes. This grafting is achieved through Surface-Initiated Controlled Radical Polymerization (SI-CRP) using Activators Regenerated by Electron Transfer Atomic Transfer Radical Polymerization (ARGET ATRP). The stimuli-responsive polymer brushes reduce fouling through the formation of a hydration layer, steric repulsion and electrostatic interactions, due to the negative charge of the layer at pH above its pKa. This has been demonstrated in the filtration of bovine serum albumin. The flux decline for the modified membranes showed a significant improvement over unmodified membranes with maximum flux declines of 24% and 60% over 0.5 h, respectively. This is related to the enhanced hydrophilicity of the membranes seen in Figure 1.

1. Hausmann R et al., The glomerular filtration barrier function: new concepts, Curr. Opin. Nephrol. , 2012, 21(4), 441–9.

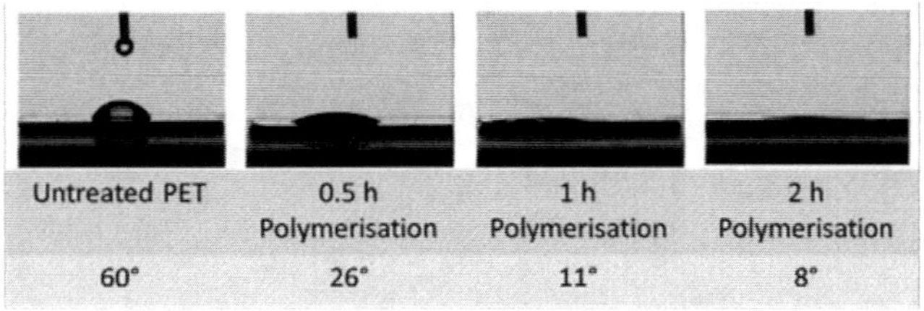

Figure 1: Water contact angle comparison of unmodified PET and subsequent grafting times of polyacrylic acid brushes.

597359 Nature-Inspired Graphene Oxide Membrane with Vertical Pores for Fast Water Treatment

Abstract:

Due to its green, energy-saving and continuous operation features, membrane technology could improve production of clean, fresh water. Graphene oxide (GO), bearing one carbon atom thickness and various oxidative groups, has become a popular membrane material, owing to its unique features, such as excellent mechanical strength, high surface area and flexibility, low friction for water and CO_2 transport, and the possibility of large-scale production. After its first report in 2007, GO-based membranes have drawn extraordinary attention for applications spanning from desalination to oil/water separation. This is because GO-based membranes possess a tuneable interlayer distance, which allows the selective passage of molecular or ionic species. However, every molecule that travels through GO-based membranes should bypass the GO nanosheets, which increases the mass transfer pathway and decreases the permeate flux. Herein, GO membranes with vertical pores are fabricated by using anodic aluminum oxide (AAO) membrane as template and etching with O_2 plasma. Such membranes possess short water transfer channels for water transport and, meanwhile, the hydroxyl groups formed by O_2 plasma could bond water molecules and generate a hydration layer on the membrane surface. The latter improves their antifouling properties, which is inspired by the structure of sugars on cell membranes, taking advantage of the hydration layer effect and steric hindrance effect to prevent foulants from adhering on the membrane surface. By controlling the etching time of the O_2 plasma, the pore size of the GO membrane is changed from 0.8 nm (the interlayer spacing) to 200 nm, which can be used in different fields for water treatment. This study may offer a simple and versatile approach to regulate the pore size of GO-based membranes with improved antifouling properties over a wide range of operating conditions.

607695 Membrane Protein Channel-Inserted Biomimetic Membranes

Abstract:

Biomimetic and bioinspired membranes containing membrane protein (MP) channels and MP-mimic functionalized materials have emerged as significant platforms to develop filtration membranes with specific solute selectivity. These membranes typically consist of highly permeable biological or bioinspired pore structures or channels with well-defined pore geometries embedded in a relatively impermeable synthetic membrane matrix. Nevertheless, membranes integrating MP-based membranes into current manufactured technology for separations has remained challenges. These challenges are possibly due to the amount of time needed to create these membranes, the use of vesicular morphologies and the low density of MPs in resulting membranes, which have resulted in much lower than anticipated improvements. In this presentation, we will describe a comprehensive approach to fabricate high-performance biomimetic filtration membranes that are based on densely packed MP channel in block copolymers (BCPs) to create robust and flat nanosheets. These highly packed crystalline structures and well-ordered nanosheets were constructed by a two-hour organic solvent evaporation method and further deposited on a porous substrate as high-throughput selective layers to form composite membranes. Three different types of β-barrel MP channels with pore sizes of 0.8 nm, 1.3 nm, and 1.5 nm were utilized in this work. The resulting three different MP-BCP nanosheet membranes created demonstrated high water permeability of ~300-1,000 ($l\ m^{-2}\ h^{-1}\ bar^{-1}$) and still maintained a sharp and tunable selectivity with three different pore sizes of MP channels. The approaches described could be adapted with the insertion of MP channels of different pore sizes or chemistries into polymer matrices to conduct specifically designed separations.

602396 Simulation Based Design of Polymeric Membranes with Biomimetic Water Channel

Abstract:

Inspired by nature's design of pore proteins embedded in cell walls, synthetic pore molecules embedded in self-assembled polymer surfactant membranes are the subject of intensive current research, as a possible route to more efficient reverse osmosis (RO) membranes. RO membranes are the key element in producing drinkable water from brackish or sea water; improved materials would help make this expensive process more widely applicable, increasing fresh water supplies worldwide.

In this work, we simulated polybutadiene-polyethylene oxide (PB-PEO) bilayers containing Peptide-appended pillar[5]arene (PAP5) channels. PAP5 channels are a biomimetic alternative to aquaporin embedded lipid membranes. Synthetic pore molecules and polymer bilayers are more physiochemically stable than pore proteins and biological membranes, promising greater ease of synthesis and scalability.

With a rigid pore dimension of ~5 Å and terminal carboxylate group, PAP[5] channels have high water permeability combined with excellent selectivity. In our simulations, we systematically varied the PB-PEO surfactant structure to maximize water mobility, while also noting the effect of surfactant design on the membrane.

We measured water diffusivity in our designs and compared our values to those inferred from experimental membrane permeability. In our best design, we obtained a water diffusivity of $30.38 \pm 0.19 \times 10^8$ cm^2 s^{-1} in the pore. To further analyze and describe our simulation results for water mobility in the pore, we devised a novel analysis method, describing the entrances and exits of water molecules from the pore channel to a kinetic model.

596464 Advancing NF Membranes: Role of Surface Charge and Support

Abstract:

Nanofiltration (NF) membranes have been reported to selectively separate charged small molecules and ions providing a variety of applications, as for example rejection of negatively charged ions/molecules or recovery of valuable metals (multivalent cations), simultaneously reducing the osmotic pressure of the system by letting pass non-desired ions. Therefore, tuning and understanding the role of surface charge will allow better NF design as well as suitable applications.

Moreover, the ultrafiltration (UF) membrane as support for NF membranes has been reported to impact both the formation and performance of the NF layer. In this study, we use 3 different UF supports (two polyethersulfone (PES) and one polysulfone (PSf – commercial PS35), and we synthesize the NF layer by rapid interfacial polymerization using two (commercially available) amine groups: traditional negatively charged NF membrane with piperazine (PIP); and another with modified chemistry by addition of polyallylamine hydrochloride (PAH) – more positively charged.

It was observed an impact on the performance of the synthesized NF membranes from both differences in the substrate as well as by the addition of PAH, making them part of the variables (direct or indirect) of the final NF performance. An increase in the zeta potential (around 2 points in the isoelectronic point) of the membranes is obtained from the addition of PAH for all substrates used. Using latest theoretical approach of zeta potential calculations for composite with porous materials (as UF membranes), a more real estimation of the difference in surface zeta potential between NF membranes with and without PAH addition is calculated through an algebraic approach.

In terms of performance, on one hand, water permeabilities were affected: up to 7% decrease by addition of PAH, and a more significant decrease of 2 and 3 folds by varying support. On the other hand, rejection of salt increased: up to 20 – 30% points of PAH addition, and 20 – 40% from different support. Deep characterization of substrates, as well as the synthesized NF membranes (including modeled pore size), is performed for correlation with the mentioned variables.

610164 Novel Assembly of Charge Mosaic Membranes for Salt Separation from Aqueous Solutions

Abstract:

Charge mosaic membranes are of interest for desalination as salt, the minor component, is preferentially transported instead of water. This contrasts to the selective transport of water, the major component, in reverse osmosis. The literature related to charge mosaic membranes is limited. Here, a novel method for assembly and use of charge mosaic structures is presented.

The method utilizes layer by layer deposition of an anionic and cationic polymers inside the pores of an ultrafiltration support. The sequential deposition of polyelectrolytes enables control over the thickness of the charge mosaic structure. Membranes were prepared using either poly(allylamine hydrochloride)/poly(styrene sodium sulfonate) or ε-poly-l-lysine hydrochloride/sodium alginate as counterion polyelectrolytes. EDS was used to confirm membrane formation.

To enhance salt transport in filtration, one of the polymers used to form the membrane was added to the feed solution to increase salt solubility in the membrane through Gibbs-Donnan equilibrium. Membranes were characterized for broad ranges of membrane formation conditions, feed salt compositions, and concentration of polymer added to the feed solution.

Salt concentrations in the filtrate could be increased by up to a factor of 9 over that of the feed. The enhancement (negative rejection) depended strongly on feed salt and added polymer concentration. The membranes proved stable over extended periods of operation.

The design of a process that removes salt using the membranes is discussed. After salt is preferentially removed in a first stage, a second ultrafiltration stage is required to yield the desalinated water and allow recycle of the added polymer to the first stage.

602796 Tuning La(III)-Binding Strength in Membrane Adsorbers Using Heterogeneous Polymer Brushes

Abstract:

Radiochromatography plays a key role in the selective extraction of lanthanides and actinides to separate fission products, medical isotopes and to concentrate radioactive waste. These separations are challenging as they often involve separating chemically similar neighbors on the periodic table. Radiochemists still rely on resin-based chromatography columns which are known to suffer from diffusion limitations—leading to long purification times and high elution volumes. An alternative to resins that may reduce separation time is membrane adsorbers (MAs) which are macroporous membranes coated with polymeric ion-binding ligands. Organophosphorus ligands, especially phosphonates and phosphonic acids, are widely used in radiochemical separations due to their high polarizability and affinity for the *f*-block elements. The selectivity and affinity of the organophosphorus ligand can be tuned by controlling the hardness of the phosphoryl oxygen. In typical adsorption media, ligand hardness (and ion-affinity) can only be changed by synthesizing a new brand new ligand. In this work, we aim to tune ion-affinity by changing the local chemical environment around the ligand by altering the composition of the polymer brush. Tuning the binding strength of ligands through copolymerization would be a simple, scalable tool for designing new chromatography media.

This work describes a simple MA system that has been designed to study the effect of co-monomers on the hardness of an amphoteric phosphonate (O-PO(OH)$_2$) ligand, ethylene glycol methacrylate phosphate (EGMP). Membrane adsorbers with varying film compositions were synthesized by UV-initiated free radical polymerization. Monomers in this study include: EGMP (La^{3+} binding), 2-hydroxyethyl methacrylate (HEMA: does not bind La^{3+}, H-bond donating), and butyl methacrylate (BuMa: does not bind La^{3+}, does not H-bond). Membrane surface chemistry was characterized using Fourier-transform infrared spectroscopy and contact angle measurements, before and after functionalization. The degree of grafting was calculated gravimetrically and the composition of the copolymer brushes was determined by digesting the membranes and performing total phosphorous measurements. The effect of brush composition on membrane capacity, affinity and binding kinetics were studied by performing batch adsorption experiments. The degree of grafting (mg polymer grafted/g membrane) for p(EGMP), p(EGMP-co-HEMA) and p(EGMP-co-BuMa) was 0.3 mg/g, 0.12 mg/g and 0.09 mg/g, respectively. Equilibrium binding capacities for La^{3+} were modeled using the Langmuir isotherm and normalized to the mass of EGMP per membrane, for comparison. MA capacity for La^{3+} (mg La^{3+}/g of EGMP) increases with decreasing H-bonding of the co-monomer following the trend: p(EGMP) = 91.2 mg/g < p(EGMP-*co*-HEMA) = 200.7 mg/g < p(EGMP-*co*-BuMa) = 653.8 mg/g. Even though p(EGMP) has the highest number of binding sites, as measured by degree of grafting, it has the lowest capacity. The data imply that the high degrees of inter- and intra-molecular interactions limit the MA capacity. Understanding the effect of polymer composition and how to control ligand hardness will guide the future design of chromatography materials.

602089 Relationship between Structural Properties and Self-Diffusion of Molecular Mixtures in Nafion By High Field Diffusion NMR in Combination with SANS and SAXS

Abstract:

Nafion is a commercially available perfluorosulfonic acid (PSA) polymer which exhibits properties attractive for applications, which can benefit from its use as an ion exchange membrane, a solid super acid catalyst, and a chemical sensor. It is therefore important to understand fundamentals of molecular transport within these membranes to better utilize their potential in these various applications. The Nafion structure in the presence of water consists of a hydrophobic crystalline matrix made of a polytetrafluoroethylene backbone and channels available for water diffusion. These channels are formed by the ionic side chains containing sulfonic groups. Several recent studies have focused on water transport mechanisms within these water channels. Similar studies of small organic molecules such as methanol have also been performed. However, there is still a noticeable gap in a detailed understanding of the role of the structural heterogeneity of Nafion on the intramembrane self-diffusion of water and organic molecules at the micrometer and sub-micrometer length scales. Here, we address this gap in knowledge by applying pulsed field gradient (PFG) NMR at high field up to 17.6 T and large magnetic field gradients up to 25 T/m to study self-diffusion on a broad range of microscopic length scales in Nafion 117 loaded with acetone or vanillic acid in the presence of water. Complementary studies were performed by small-angle neutron scattering (SANS) and small-angle x-ray scattering (SAXS) to study structural properties of the same Nafion samples.

H-1 and C-13 PFG NMR studies revealed that at high water concentrations (i.e. > 0.5 mmol water/g Nafion), the measured self-diffusivities of all studied guest molecules in Nafion did not vary with increasing root mean square displacements (root MSDs). The maximum range of the probed root MSDs was between 0.5 and 28 micrometers. However, at lower water concentrations and in the presence of acetone, water experiences a significant stepwise decrease in the self-diffusivity with increasing root MSD at around a 5 micrometer length scale. This is in agreement with Percolation Theory that, at some sufficiently small water concentration, the hydrophilic water channels of the membrane form finite clusters and are no longer interconnected into a spanning (infinite) percolation cluster. Our PFG MMR measurements indicate that the average size of such finite clusters under our experimental conditions was around 5 micrometers. In contrast, the self-diffusivity of acetone shows no dependence on root MSD in the same samples and under the same conditions when such dependence was observed for water. To better understand why water encounters different paths available for diffusion than acetone, SANS and SAXS experiments were performed. The SANS and SAXS data suggest acetone is located in the interfacial perfluoroether regions, i.e. between the sulfonic groups in the water channels and the crystalline matrix. As acetone can diffuse through these interfacial regions, its self-diffusion coefficient does not need to be sensitive to the finite size of domains of interconnected water channels, which explains our PFG NMR data for acetone.

607896 Diffusion of CH_4 and N_2 in a Reduced Pore-Zorite (Ba-RPZ)

Abstract:

Ba-RPZ is a titanosilcate adsorbent that shows excellent kinetic selectivity for CH_4/N_2 separations [1]. Single component N_2 and CH_4 equilibrium loadings were measured on crystalline Ba-RPZ between 1 mbar to 1.2 bar and 30, 50 and 70°C using low pressure volumetry. The isosteric heat of adsorption was calculated using the collected equilibrium data. The isosteric heat of adsorption shows that both CH_4 and N_2 sees the Ba-RPZ surface as energetically homogenous. The equilibrium data for all components were fit to a single or dual site Langmuir isotherm. Mass transfer parameters for Ba-RPZ and zeolite 4A were measured for N_2 (at -20, -10 and 0°C) and CH_4 (at 30, 40 and 50°C) using a commercial volumetric apparatus. The measured uptake profiles show that diffusion of N_2 into Ba-RPZ is very fast. The CH_4 uptake profiles were fitted to a range of micropore models that describe resistance at the pore mouth, resistance within the micropores of the solid and a dual resistance model that assumes both the pore mouth and micropores contribute to the mass transfer resistance. It was found that mass transfer resistance mechanism of CH_4 in Ba-RPZ is best described by the dual resistance model. Collected diffusion parameters for both surface resistance and micropore resistance show a strong concentration dependence.

604932 Modeling Chromate Removal for Drinking Water Application Using the Method of Moments to Identify Key Parameters

Abstract:

Introduction:

Chromium-based compounds, such as Chromates and dichromates, are widely used in the plating industry to protect metals against corrosion. Unfortunately, chromium is a highly oxidizing metal that can be carcinogenic at low parts per billion concentrations. The European Union has restricted its use since 2002 while the EPA limited total chromium in water to 100 parts per billion (Directive 2002/95/EC of the European Parliament and the council) [1]. According to the USGS report, the EPA proposed in 2010 to "classify Cr (VI) as likely to cause cancer in humans when ingested over a lifetime" [2]. As the chromate ion is a negatively charged oxyanion that needs to be below 10 ppb in drinking water, a quaternary amine anion exchanger resin is the standard purification step used. In this work, the gel type strong anion exchanger, Purolite A600E, was evaluated for chromates removal from water containing an abundant amount of typical anions such as chlorides, sulfates, and bicarbonates to reflect real-life situations based on the hazardous nature of chromate and the necessity in removing it at concentrations below 20 ppb.

Objective:

As several fixed-bed sigmoidal models exist to predict the ion exchange process such as the Yoon-Nelson, the Clark, and the Thomas models [3], it is necessary to understand how process/operating conditions such as flow rates and inlet chromate concentrations affect resin capacity and therefore the performance of the ion exchange system. The objective of this work is to identify a working dynamic model where the system and resin characteristics are reflected in the model to improve the predictability of the chromate removal system. This is accomplished by the following objectives: (i) understanding the kinetics of chromate removal by identifying a suitable model [4], (ii) integrating the kinetics to the process operations via a detailed moment-based modeling approach, and (iii) optimizing the process operation using the models identified in objectives (i) and (ii).

Methodology:

Most existing models follow the general sigmoidal function, seen in (eq. 1) representing the ion exchange system is usually in the form seen in (eq. 1).

Normalized concentration = $\psi = C/C_0 = 1/[1+\exp(a-bt)]$ (1)

Where C – outlet concentration, C_0 – inlet concentration, ψ – normalized concentration, and t - time

Model discrimination was necessary to reflect process characteristics such as flow rates and inlet concentrations. Based on that necessity, the Thomas model was chosen for reflecting such characteristics as seen in (eq. 2).

Thomas model = $\psi = C/C_0 = 1/[1+\exp(K_T\, q_m \tau - K_T\, C_0\, t)]$ (2)

Where K_T - Thomas constant parameter, q_m – resin maximum capacity parameter, τ – contact time (V/v_0), V – volume of resin in column, and v_0 – flow rate in column

The parameters in the model will be calculated by comparing the predicted calculated results to the experimental measurements by minimizing the residual sum of the squares error function (eq. 3).

Residual Sum of Square Error = $\sum[(\psi_{cal} - \psi_{exp})/\psi_{cal}]^2$ (3)

While the method of moments is limited in many publications to pulse data in chromatographic separations, the method was reformulated and used on the column and the resin systems. The moments were evaluated with respect to the amount of chromate in the effluent and on the resin respectively [4]. The moment equation used is seen in (eq. 3).

$\mu_i = \int_{t=1}^{\infty} t^i C \, dt$ (4)

$m_i = \int_{t=1}^{\infty} t^i q \, dt = \int_{t=1}^{\infty} t^i (C_0 - C) v_0 \, dt$ (5)

where μ_i – ith moment on the column system, m_i – ith moment on the resin system, and q – concentration on the resin system.

These moments will be used to calculate the parameters in the Thomas model and compared to the values obtained from minimizing the residual sum of square error.

Using the method of moments, different flow rates and chromate concentrations are used to confirm the accuracy of the moments in predicting the Thomas model representation of the chromate removal process.

Summary:

Due to the hazardous nature of chromates at low concentrations, it is important to use the appropriate modeling techniques to maximize the performance of the ion exchange purification step. The fixed-bed Thomas model was picked to predict the ion exchange process performance since it considers some important process characteristics such as inlet resin concentration and flow rates in its equation. To improve the control aspects of this process, the method of moments was used to display characteristics such as the maximum resin capacity (zeroth moment of the resin system) and longer processing time (first moment of the column system). The model was checked successfully for different inlet chromate concentrations and contact times.

Since the anion exchanger is typically used once due to the toxicity of chromium being removed, it is necessary to maximize the anion exchanger capacity. This is achieved by applying optimal control[6] on the system by maximizing the objective function associated with the resin loading capacity at a defined breakthrough point with the control variable being the flow rate into the resin process.

Keywords: Thomas Model simulation, model optimization, Chromate extraction, method of moments.

References:

1. "Directive 2002/95/ec of the European parliament and of the council" *(PDF). Eur-lex.europa.eu. Retrieved 3 July 2015.*
2. Gibbons, Catherine. "IRIS Toxicological Review of Hexavalent Chromium (2010 External

Review Draft), IRIS US EPA" 2010, 4.
3. Nur, T., W. G. Shim, P. Loganathan, S. Vigneswaran, and J. Kandasamy. "Nitrate Removal Using Purolite A520E Ion Exchange Resin: Batch and Fixed-Bed Column Adsorption Modelling." International Journal of Environmental Science and Technology 12, no. 4 (April 2015): 1311–20.
4. Li, Xue. "Meeting the New California MCL for Hexavalent Chromium with Strong Base Anion Exchange Resin", ProQuest 10124393, University of California – Davis, 2016.
5. Goltz, Mark N., and Roberts, Paul V. "Using the Method of Moments to Analyze Three-Dimensional Diffusion-Limited Solute Transport from Temporal and Spatial Perspectives". Water Resources Research, Vol 23, No 8, Pages 157
6. K. M. Yenkie and U. Diwekar, "Stochastic Optimal Control of Seeded Batch Crystallizer Applying the Ito Process," *Ind. Eng. Chem. Res.*, p. 120604103933002, Jun. 2012, doi: 10.1021/ie300491v.

611598 Adsorption Equilibrium and Kinetics of Cadmium (II) from Aqueous Solution Onto Activated Carbon Fiber. Hindered Diffusion

Abstract:

The contamination of water by heavy metals represents a hazardous problem for human health and the environment. The ingestion of water contaminated with metals can cause enzymatic alterations and affect the renal, respiratory and digestive systems, among others. One of the most efficient processes for the removal of these pollutants from aqueous solution is adsorption, which is a surface phenomenon that involves the accumulation of substances on a surface or interphase. The primary objective of this work is to study the adsorption equilibrium and kinetics of Cd(II) from aqueous solution on activated carbon fiber (ACF), as well as to model the adsorption rate data using a diffusional model.

The experimental adsorption equilibrium data were obtained in a batch adsorber consisting of 50 mL plastic vial, and the effect of pH, temperature and ionic strength of the solution on the adsorption capacity was studied thoroughly. The adsorption rate data was procured in a differential column batch adsorber composed of a polyethylene column packed with ACF, a feed vessel and a peristaltic pump.

The textural properties of ACF were the BET surface area of 1039 m^2/g, the average pore diameter of 1.1 nm, the total pore volume of 0.57 cm^3/g and the volume of the micropores of 0.50 cm^3/g, the latter was calculated by the Dubinin-Radushkevich's method. Thus, the ACF is a very microporous material since the micropore volume represents 88 % of the total pore volume.

The experimental adsorption equilibrium data were satisfactorily interpreted with the Prausnitz-Radke isotherm. It was found that the adsorption capacity of FCA was enhanced by raising the solution pH from 5 to 7. This behavior was attributed to the electrostatic interactions between the Cd^{2+} and the negatively charged surface of the ACF. The adsorption capacity of the ACF decreased markedly with increasing ionic strength of the solution, corroborating the importance of electrostatic interactions in the adsorption of Cd(II) on ACF. Adsorption equilibrium experiments at different temperatures revealed that the adsorption capacity of FCA toward Cd(II) does not vary by increasing the temperature when the concentration of Cd(II) is high. The electrostatic interactions and ion exchange play an essential role in the adsorption mechanism.

A diffusional model was proposed to interpret the adsorption rate of Cd(II) on ACF from a water solution. The system was comprised of two phases: a liquid solution of Cd(II) and a solid adsorbent. In deriving this model, mass balances of Cd(II) were performed in each phase, and the following suppositions were assumed: (i) the fibrils of ACF were cylindrical, ii) the external mass transfer was represented by a mass transfer coefficient, iii) the intraparticle mass transfer was only due to pore volume diffusion, and (iv) the adsorption rate at an active site is instantaneous.

The kinetic study revealed that the adsorption equilibrium of Cd(II) on FCA was reached in 45 min. The experimental data of the concentration decay curves were interpreted reasonably well by a numerical solution of the diffusional model, and the overall rate of adsorption was shown to be controlled by intraparticular diffusion. Additionally, it was found that the intraparticular diffusion of Cd(II) in the FCA is drastically affected by restricted diffusion due to exclusion, friction in the walls of the micropores and obstruction caused by the Cd (II) cations adsorbed on the surface of the pores.

598792 High-Throughput Screening of Alkane and Organophosphate Diffusion in Metal-Organic Frameworks

Abstract:

Metal-Organic Frameworks (MOFs) that contain zirconium (Zr_6) nodes catalyze the decomposition of organophosphate chemical warfare agents (CWAs), yet the ambient conditions required for practical CWA decomposition may introduce mass transfer limitations for bulky, polar adsorbates. These limitations may prevent reactants from reaching active sites or products from leaving the crystal. For a catalytic Zr_6 node with varying connectivity and an organic linker of interest, there are numerous unique topologies that can be formed, each with distinct pore structure. We developed a database of Zr_6 MOFs with 20 different topologies to systematically evaluate the role of MOF topology on diffusion. Molecular dynamics simulations were used to simulate diffusion of alkanes and organophosphates to identify MOFs with favorable diffusion properties. We used high-throughput and automated spatial analysis tools to map pore regions with similar framework distances or adsorption energies to quantitatively describe diffusion at finite loadings.

Simulated diffusion coefficients of alkanes and organophosphates vary by orders of magnitude depending on MOF topology, even for the similar linker sizes used here. This is due in part to the Zr_6 node connectivity, which can be 6-, 8-, or 12-connected to linkers. For topologies that are 12-connected, average diffusion coefficients of alkanes are lower at all adsorbate loadings relative to less-connected topologies, due to smaller average pore apertures. Average diffusion coefficients for both 6- and 8- connected topologies, however, do not differ from each other. The connectivity of the linkers does not significantly impact diffusivity. At high adsorbate fractional loading, diffusion coefficients correlate with pore size, but at low loadings, pore connectivity and pore shape control diffusion. For polar organophosphates, stronger adsorption to Zr_6 nodes reduces diffusion at low adsorbate loadings, which are most relevant for CWA decomposition. We found that the choice of topology and linker were crucial to control organophosphate diffusion at low loadings, as diffusion coefficients vary by 3 orders of magnitude among topologies for structurally similar nodes and linkers. Coulombic interactions between nodes and adsorbates were shown to be dominant by removing node partial charges and comparing the distribution of adsorbates between frameworks with and without partial charges. We then extended these results to develop design rules for the diffusion of bulky polar adsorbates in other crystalline microporous materials.

605841 GO-Based Membrane As Barrier Against Toxic Vapors/Gases

Abstract:

Traditional protective garments loaded with activated carbons to remove toxic gases are bulky. Novel graphene oxide (GO) flake-based composite lamellar membrane was developed for a garment for protection against chemical warfare agent (CWA) simulants: dimethyl methyl phosphonate (DMMP) (a sarin-simulant); 2-chloroethyl ethyl sulfide (CEES) (a simulant for sulfur mustard); yet it has a high moisture transmission rate. GO flakes (300-800 nm, 0.7-1.2 nm thick) dispersed in water were formed into a membrane by vacuum filtration on porous polyethersulfone (PES) or poly (ether-ether-ketone) (PEEK) support membrane for noncovalent π-π interactions with GO flakes. After physical compression, Upright Cup tests indicated: it can block toluene for 3-4 days and DMMP for 5 days while exhibiting excellent water vapor permeation. Further, they have low permeances for small molecule gases/vapors. The GO flakes underwent crosslinking later with ethylenediamine (EDA) during vacuum filtration followed by physical compression and heating. With a further spray-coating of polyurethane (PU), these membranes could be bent without losing barrier properties vis-à-vis DMMP for 5 days; a membrane not subjected to bending blocked DMMP for 15 days. For the PEEK-EDA-GO-PU compressed membranes after bending, the separation factors of H_2O over others for low gas flow rates in the dynamic moisture permeation cell (DMPC) were: $\alpha_{H2O\text{-}He}$, 42.3; $\alpha_{H2O\text{-}N2}$, 110; $\alpha_{H2O\text{-}ethane}$, 1800. At higher gas flow rates in the DMPC, moisture transport rate goes up considerably due to reduced boundary layer resistances and exceeds the breathable fabric threshold water vapor flux, 2000 g/m^2-day. The membrane displayed considerable resistance to permeation by CEES as well. This behavior has been compared with that of a nanopacked bed of membrane supported nanocrystals of UiO-66-NH_2 MOF. The PES-EDA-GO-PU compressed membrane shows good mechanical property under tensile strength tests.

599973 New Facilitated Transport Membranes for Hydrogen Purification from Coal-Derived Syngas

Abstract:

CO_2-selective, amine-containing facilitated transport membranes (FTMs) are of great interest for syngas purification since high-pressure H_2 can be retained upon CO_2 removal. Various FTMs have shown decent chemical and thermal stability at aggressive conditions, but their CO_2/H_2 separation properties are largely limited by the severe carrier saturation at high syngas pressure. Herein, we report a new approach to enhance the CO_2 permeability by manipulating the steric hindrance of the amine carrier. A series of α-aminoacids with different alkyl or hydroxyethyl substituents were deprotonated by 2-(1-piperazinyl)ethylamine, resulting in nonvolatile amine carriers with different degrees of steric hindrance. For hosting the low MW amine carriers, a water-swellable polymer network was synthesized from poly(vinyl alcohol) crosslinked by a bidentate tertiary aminosilane. In the presence of moisture, a bulkier alkyl substituent increased the steric hindrance and hence destabilized the carbamate adduct to afford bicarbonate through hydrolysis. Thus, this drastically increased the chemisorption of CO_2. Further, density functional theory (DFT) calculations were conducted to study the function of the hydroxyethyl substituent, which indicated that the hydroxyl group stabilized the bicarbonate through strong hydrogen bonding, thus further improving the CO_2 sorption. The enhanced CO_2 solubility significantly mitigated the carrier saturation, and an unprecedented CO_2/H_2 selectivity greater than 130 was demonstrated at 107°C and 12.5 atm of CO_2 partial pressure. As the CO_2 partial pressure reduced to 0.4 atm, a less hindered amine yielded a higher reactive diffusivity of CO_2, resulting in a CO_2 permeance of 435 GPU with a selectivity greater than 500. The CO_2/H_2 separation performance of these reaction-mediated polymeric membranes is well above the theoretical upper bound, and they open up a new avenue for designing a highly selective membrane process for syngas purification.

604852 Cross-Linking of Matrimid/Porous Organic Cage Molecularly Mixed Composite Membranes Using Cycloaddition Click Chemistry

Abstract:

Individual molecules with intrinsic porosity, like amorphous scrambled porous organic cages (ASPOCs), have the potential to "change the game" when it comes to creating composite and hybrid materials. A particularly exciting potential application is to incorporate these materials into polymeric membranes to form molecularly mixed composite membranes (MMCMs). Since ASPOCs are individual molecules instead of continuous 3-dimensional structures like metal-organic frameworks or zeolites, they should be able to form a "solid solution" with the ASPOC uniformly dispersed throughout the membrane at a molecular level. This molecular-level mixing overcomes many of the performance/processing issues encountered with other composite materials. Due to small window sizes, ASPOCs should be able to distinguish between aggressive, small molecule organic solvents effectively. Cross-linked polyimides like Matrimid® are also thought to be appropriate for this application due to high chemical stability. Unfortunately, the normal diamine crosslinking process for polyimides would cause the degradation of the ASPOC molecules since they are bound by relatively weak imine bonds. Here we propose an alternative method of cross-linking Matrimid® in MMCMs using "click" chemistry, specifically the Huisgen azide-alkyne cycloaddition reaction. Importantly, the reaction does not involve species that will degrade the ASPOC. To test MMCM separation performance, we fabricate thin film composite membranes and test them in a crossflow permeation system with standard styrene oligomers dissolved in a variety of organic solvents. We find that the presence of ASPOC increases both permeance and styrene dimer separation factor up to a loading of 20 wt%. Even in THF, known to dissolve uncross-linked Matrimid, we measure permeances of 0.96 LMHBar and a dimer separation factor of 27.5 at 10 wt% ASPOC loading. The crosslinked MMCMs are also able to distinguish between toluene and triisopropylbenzene, suggesting their suitability for solvent-solvent class separations.

610130 Membranes with Cross-Linked Zwitterionic Nanochannels Achieve Sub-Nanometer Selectivity

Abstract:

Self-assembly is the process by which ordered structures are spontaneously formed from simpler building blocks. Self-assembling copolymers have attracted great interest for membrane applications, due to facile processing and the exceptional selectivity provided by their ordered transport pathways. Achieving effective pore sizes less than 1.0 nm has been a significant challenge, however, and the sub-nanometer selectivity needed for the broad application space of nanofiltration has not yet been demonstrated. Random zwitterionic amphiphilic copolymers (r-ZACs) are strong candidates to meet this challenge. These copolymers can self-assemble to form zwitterionic nanochannels through which water and small solutes can permeate. Membranes derived from r-ZACs demonstrate exceptional fouling resistance, low ion retention, and a size-based cutoff of ~ 1.5 nm. Slight reductions in the effective pore size would enable r-ZAC membranes to filter divalent ions and sub-nanometer neutral solutes.

In the present work, we employed a novel cross-linking technique to reduce the size of the zwitterionic nanochannels. The effective pore size was less than 1.0 nm, as indicated by the rejection of sugars (sucrose rejection 99.6%; Stokes diameter = 0.96 nm). The membranes also demonstrated outstanding divalent-monovalent selectivity when challenged with artificial seawater (sulfate rejection = 99.2%, sulfate-chloride separation factor = 75). The membranes were also extremely fouling resistant, which was due to the strong hydration barrier surrounding the zwitterions. These results illustrate a promising route to achieving sub-nanometer separations with self-assembling copolymer membranes.

610920 Highly Selective Canal Ladder Polymers for Gas Separations: Tuning Backbone Conformation and Alkyl Functionalities

Abstract:

The development of materials with improved transport properties is necessary to further enable membrane technologies for emerging industrial applications. In this work, a new family of polymers based on Catalytic Arene-Norbornene Annulation (CANAL)[1] polymerization is discussed for its application as gas separation membranes. CANALs are rigid ladder polymers similar in structure to PIMs. However, from a design standpoint PIMs contain Tröger's base-[2] or dioxane-based[3] connectivity, while CANALs feature norbornyl benzocyclobutene backbones, thereby introducing a new high-performing ladder architecture to the gas separation field.

We have investigated how subtle changes in short alkyl substituents have an impact on transport properties.[4] The unique nonpolar, rigid backbone of CANALs that feature no heteroatoms, make them an exceptional model system for fundamental investigations of the structure-property relationships of ladder polymers. An analysis of the energetic contributions to sorption and diffusion interestingly showed that CANALs present features of both dense and microporous polymers.

Recent developments allowed the formation of a new high molecular weight CANAL polymer featuring five-membered rings in the polymer backbone. By making this modification on the backbone conformation, we have been able to achieve remarkable combinations of high selectivity and permeability. In particular, H_2/CH_4, H_2/N_2, and O_2/N_2 separations showed performance beyond the 2015 upper bounds[5], as well as 2008 H_2/CO_2 and CO_2/CH_4 upper bounds. Mixed-gas data were also performed for a broad range of compositions and pressures to assess the potential of this new material in more realistic conditions, and compared with pure- and mixed-gas upper bounds[6], showing very promising results. Sorption and diffusion experiments, as well as X-ray scattering, were analyzed to provide a fundamental understanding of the exceptional performance reported.

The structural diversity explored in this work, and the remarkable results obtained, can provide a rational to design new microporous polymers with enhanced transport properties. Thereby, CANALs represent a new platform of materials with potential to address fundamental limitations in current design strategies for polymer membranes.

References

[1] S. Liu, Z. Jin, Y. C. Teo, Y. Xia, *J. Am. Chem. Soc.* **2014**, *136*, 17434.

[2] M. Carta, R. Malpass-Evans, M. Croad, Y. Rogan, J. C. Jansen, P. Bernardo, F. Bazzarelli, N. B. McKeown, *Science (80-.).* **2013**, *339*, 303.

[3] P. M. Budd, E. S. Elabas, B. S. Ghanem, S. Makhseed, N. B. McKeown, K. J. Msayib, C. E. Tattershall, D. Wang, *Adv. Mater.* **2004**, *16*, 456.

[4] H. W. H. Lai, F. M. Benedetti, Z. Jin, Y. C. Teo, A. X. Wu, M. G. De Angelis, Z. P. Smith, Y. Xia, *Macromolecules* **2019**, *52*, 6294.

[5] R. Swaidan, B. Ghanem, I. Pinnau, *ACS Macro Lett.* **2015**, *4*, 947.

[6] Y. Wang, X. Ma, B. S. Ghanem, F. Alghunaimi, I. Pinnau, Y. Han, *Mater. Today Nano* **2018**, *3*, 69.

601977 Facile Hydrophobic Modification Strategy on Hydrophilic Metal Organic Frameworks to Improve Volatile Organic Compounds Removal from Air and Aqueous Environments

Abstract:

Clean air quality is essential to the wellbeing of all living species on the planet. Chemical emissions including volatile organic compounds (VOCs) and toxic aromatic hydrocarbons contribute strongly to the urban atmospheric air pollution as they pose great harms to the environment and human health. Adsorption technology in the form of commercial air filters has commonly been used for indoor VOCs abatement. These air filtration units are either equipped with activated carbon (AC), zeolites or hybrid materials among many others [1,2] but, these conventional technologies suffer from several limitations such as low adsorption capacity, structural amorphism and high regeneration costs [3]. As the pursuit for effective sorptive technology continues, the potent roles of metal-organic frameworks (MOFs) with their high surface area, chemical and thermal stability, tailorable functionalities for enhanced adsorption and reusability can help resolve the problems [4]. Of our interest is the ability to post-synthetically modify the structures to enhance adsorption or generate novel properties.

To enable efficient VOCs removal from air, the omnipresence of moisture must first be addressed by means of a systematically developed hydrophobic surface modification. Attempted strategies to improve hydrophobic character of MOFs include functionalising the MOF with fluoro-based linkers [5], co-synthesizing them with hydrophobic materials [6] or via an interfacial assembling pathway [7]. However, most reported methods are not suitable for industrial scalability. Therefore, by using a facile modification method, we will demonstrate that rationally designed MOFs have the upper hand for atmospheric VOCs capture over carbon-based adsorbents. Preliminary experiments indicate almost complete porosity retainment after the MOF has been modified.

To assert a realistic assessment, investigations will cover the sorptive behaviour of an environmentally benign MOF alongside its hydrophobically modified version towards a mixture of harmful representative airborne pollutants at ambient conditions (single species and competitive adsorption with co-existing pollutants), relative performance under varying humidity, desorption, breakthrough and dynamic reusability studies. Besides, the resulting water-repellent feature renders the modified MOF suitable for VOCs adsorption from aqueous phase systems too as supported from our ongoing experimental data. The outcomes from this in-depth materials characterization and development will provide excellent MOF protection from moisture and harsh operating conditions while offering a practical solution to many plaguing environmental issues.

Reference:

1. Xie L-H, Liu X-M, He T, Li J-R. Metal-organic frameworks for the capture of trace aromatic volatile organic compounds. Chem, 2018, 4(8): 1911–1927
2. Kim K-H, Szulejko J E, Raza N, Kumar V, Vikrant K, Tsang D C W, Bolan N S, Ok Y S, Khan A. Identifying the best materials for the removal of airborne toluene based on performance metrics-A critical review. Journal of Cleaner Production, 2019, 241: 118408

3. Sühnholz S, Kopinke F-D, Weiner B. Hydrothermal treatment for regeneration of activated carbon loaded with organic micropollutants. Science of The Total Environment, 2018, 644: 854–861
4. Vellingiri K, Kumar P, Deep A, Kim K-H. Metal-organic frameworks for the adsorption of gaseous toluene under ambient temperature and pressure. Chemical Engineering Journal, 2017, 307: 1116–1126
5. Yang C, Kaipa U, Mather Q Z, Wang X, Nesterov V, Venero A F, Omary M A. Fluorous metal-organic frameworks with superior adsorption and hydrophobic properties toward oil spill cleanup and hydrocarbon storage. Journal of the American Chemical Society, 2011, 133(45): 18094–18097
6. Karikkethu Prabhakaran P, Deschamps J. Doping activated carbon incorporated composite MIL-101 using lithium: Impact on hydrogen uptake. Journal of Materials Chemistry A, 2015, 3(13): 7014–7021
7. Zhang F, Sang X, Tan X, Liu C, Zhang J, Luo T, Liu L, Han B, Yang G, Binks B P. Converting Metal-Organic Framework Particles from Hydrophilic to Hydrophobic by an Interfacial Assembling Route. Langmuir, 2017, 33(43): 12427–12433

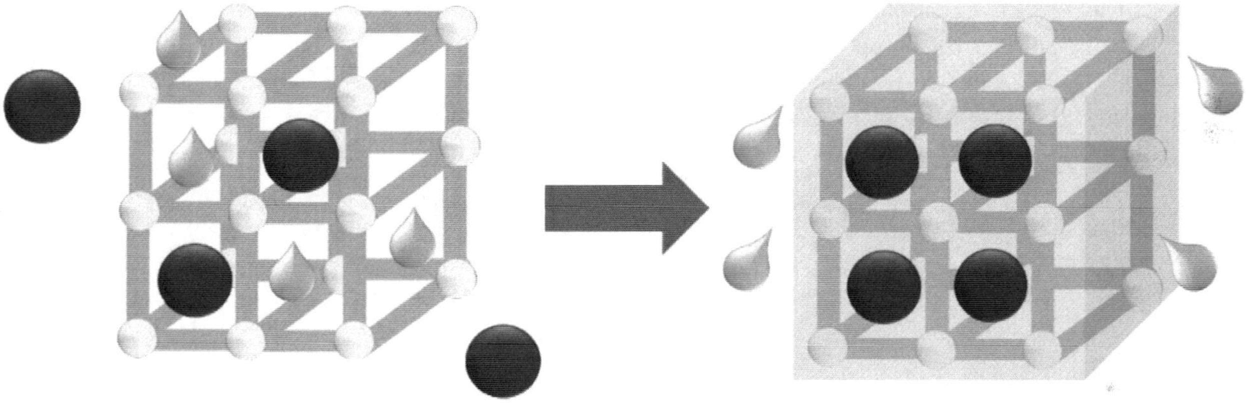

609539 Hierarchical Porous MOF@LDH Core-Shell Structures As Efficient Adsorbent for Phosphate

Abstract:

Phosphate is a major water pollutant from agricultural fertilizers, industrial wastewater and household products. For practical removal of phosphate, adsorption at sources should be the one of the most important routes, which however needs high adsorptive efficiency. From the viewpoint of material engineering, the adsorbents require large surface area, high adsorption capacity, and strong bonding towards phosphate. Layered double hydroxides (LDHs) are a class of synthetic anionic clays and have been studied for various applications including as phosphate adsorbents, however the flake structures are easy to overlap, thus reducing the contact area of materials with pollutants. In this work, an aqueous stable metal organic framework (MOF) is used as supporter to vertically grow LDH on it, resulting in Hierarchical porous MOF@LDH Core-Shell structures. Based on the different characterization techniques, a growth mechanism of LDH on MOF is proposed. Various factors, such as pH, competitive anions, temperature, and zeta potential, were discussed. The adsorption capacity of as-synthesized core-shell composites enhances and the adsorption to reach 95% phosphate reduction decreases from 10-16min to 4-6min for high concentration (500ppm) of phosphate. The study provides novel insights into pollutant adsorptive removal and could be beneficial for inspiring functional composite material synthesis for environmental applications.

604814 A Porous Copper Metal Organic Framework with Highly Nucleophilic Amine Links for CO_2 Gas Separation from Biogas

Abstract:

The rising interest in using biogas as an energy fuel has lower the consumption of fossil fuel and inclines the environment toward sustainability. Despite having low CO_2 emissions during the combustion of biogas, it still involves a significant challenge to overcome. Biogas consist of ~ 40% of CO_2 depending upon the substrate and conditions of the digester used during production [1]. This CO_2 is the major undesirable component for the energy recovery of the biogas as it lowers the calorific value of the fuel. Adsorption driven materials have set a benchmark intensively in CO_2 separation in the biogas upgrading process [2,3]. Wide variety of adsorbents have been studied for their separation performances and energy consumption to procure efficient, energy-saving and environmentally benign separation processes [4]. MOFs (Metal-organic framework) are an emerging class of porous materials, which has gained enormous attention in the past two decades and has a widespread potential application in gas separation and purification technology. MOFs outstand the other porous material due to their unique properties such as high surface area, tunable pore size and easy tailoring of chemical functionalities.

A new copper MOF was synthesized using facile solvothermal synthesis and resulted in a cuboid-shaped, 300-500 nm size particles. The resultant MOF showed the carbon dioxide uptake capacity of 1.8 mmol/g at 293 K and 1 bar. Furthermore, when the temperature was decreased up to 273 K and at 1 bar, 16 % increase in volumetric uptake was observed. The MOF displayed type I isotherm with permanent porosity and no hysteresis. The N_2 adsorption isotherm was recorded at 77 K and revealed BET surface area of 756 m^2/g with a pore volume of 0.27 cm^3/g calculated using density functional theory. The gravimetric high-pressure adsorption was carried out for both CO_2 and CH_4 gases at 10 bar. The Cu-MOF showed 215 % higher uptake of CO_2 than CH_4 at 10 bar and 293 K. The CO_2 binding energy of the adsorbent was evaluated by calculating isosteric heat of adsorption (Q_{st}) by interpolation approach from the temperature-dependent isotherms using Van't Hoff's equation and resulted in a significant low value,18 kJ/mole, which describes that the low energy is required to regenerate the MOF in desorption cycles and hence reduces the energy consumption. To investigate the CO_2 separation performance, the Cu-MOF powders were structured into hierarchical strong pellets and further used in pressure swing adsorption (PSA) equipment for breakthrough adsorption and desorption cycles. Kinetic study of the pellets was carried by further simulating the breakthrough data to obtain mass transfer coefficient and diffusivities value.

References

[1] I Angelidaki, L Treu, P Tsapekos, G Luo, S Campanaro, H Wenzel, et al., Biogas upgrading and utilization: Current status and perspectives, Biotechnology Advances. 36 (2018) 452-466.
[2] S Chaemchuen, NA Kabir, K Zhou, F Verpoort. Metal-organic frameworks for upgrading biogas via CO2 adsorption to biogas green energy, Chem.Soc.Rev. 42 (2013) 9304-9332.
[3] B Yuan, X Wu, Y Chen, J Huang, H Luo, S Deng. Adsorption of CO2, CH4, and N2 on Ordered mesoporous carbon: Approach for greenhouse gases capture and biogas upgrading, Environ.Sci.Technol. 47 (2013) 5474-5480.
[4] K Zhou, S Chaemchuen, F Verpoort. Alternative materials in technologies for Biogas upgrading via CO 2 capture, Renewable Sustainable Energy Rev. 79 (2017) 1414-1441.

604973 Hydrocarbon Recovery Using Ultra-Microporous Fluorinated MOF Platform Structure Properties Relationships for C_2H_2/C_2H_4 and C_4 Olefins Separation Separations

Abstract:

In this work, the reticular chemistry approach was implemented on a very stable ultra-microporous fluorinated **MFFIVE**-1-Ni MOFs to unveil the effect of subtle changes of the structure-employment of different fluorinated inorganic block on the adsorption of C_2H_2, C_2H_4 and C_4 olefins. A series of variable temperature single C_2H_2, $C_2H_4$and C_4 adsorption isotherms and mixed gas adsorption column breakthrough experiments for different $C_2H_2/C_2H_4$and C_4 isomers gas pair systems were carried out on the two isoreticular **NbOFFIVE**-1-Ni and **AlFFIVE**-1-Ni, containing respectively $[NbOF_5]^{2-}$ and $[AlF_5]^{2-}$ inorganic building blocks. The introduction of potential open metal site in a very confined pores led to favoring interaction of C_2H_2 and C_2H_4 but lowering the interaction with CO_2, which resulted in enhancement of C_2H_2/C_2H_4 selectivity at low C_2H_2 concentration but a decrease in CO_2/C_{2+}selectivity. The comparison of the C_2H_2/C_2H_4 and C_4 olefins separation performances with the structures of the **MFFIVE**-1-Ni MOFs provides useful information to shed light on the relationship between the structural features of this MOF platform and C_2H_2/C_2H_4 and C_4 olefins separation properties.

608451 Characterization of Calf-20: A Water Stable Metal-Organic Framework for Humid Post-Combustion CO_2 Capture

Abstract:

Calgary Framework-20 (CALF-20) is a water stable MOF with a unique potential for post-combustion CO_2 capture. In this study, the competitive adsorption of (CO_2 + N_2) and (CO_2 + H_2O) on CALF-20 was examined by dynamic column breakthrough (DCB) experiments. The single component isotherms of CO_2 were measured from 22°C to 100°C and 0 bar to 5 bar using volumetric and gravimetric methods. Dual-site Langmuir (DSL) isotherm model was employed to fit CO_2 equilibrium data. CO_2+N_2 competitive measurements were performed using DCB and were well-characterized by the DSL isotherm. The H_2O adsorption isotherms were measured up to 80%RH in the temperature range of 22°C to 100°C. At low temperatures, the H_2O isotherm showed an inflection point, indicating very low water adsorption at low RH values, a characteristic that can be very advantageous for CO_2 capture. The Quadratic-Langmuir isotherm model was applied to describe the H_2O equilibrium data. H_2O dynamic column breakthrough (DCB) experiments were performed at a variety of RH values and the breakthrough curves demonstrated a variety of shapes reflecting the complexity of the isotherm. To measure the CO_2 and H_2O competitive loadings, two components (CO_2/H_2O) DCB experiments were studied at various RH (12% to 90%). The results indicated that under the presence of CO_2, H_2O broke through 20 times faster than the pure H_2O at 45% of RH. All single component CO_2 and H_2O breakthrough profiles and binary CO_2/H_2O breakthrough profiles were simulated using a built in-house MATLAB simulator. A good agreement between the experiments and simulations were obtained. The potential of using this material for vacuum swing adsorption (VSA) separation is also explored using numerical simulations

600615 First Observation of Microscopic H_2 Isotope Dynamics in Flexible MOFs for Hydrogen Isotope Separation

Abstract:

Development of cost efficient and environmental friendly methods of hydrogen isotope separation is of a very high importance today due to the rapidly increasing demand in medical healthcare such as imaging and cancer therapy. Unfortunately, the current separation methods suffer from low separation efficiency due to their nearly same chemical properties and thus high purity isotopes are not easily achieved. Recently nanoporous materials have been proposed as alternatives, which are supported by the newly proposed separation mechanism, quantum sieving (D2, which has a shorter de Broglie wavelength than H2, can diffuse faster through the confined pore at a cryogenic temperature). A few research groups have been investigated for this hydrogen isotope separation using rigid nanostructure up to now.

Structural flexibility is a unique property of some metal−organic frameworks (MOFs) that clearly distinguishes them from other inorganic porous materials. A MIL-53(Al), one of the most well-known flexible MOF, leads to dynamic changes as closed pore transitions to open pore, also called 'breathing' phenomenon. During the flexible and reversible transition, the pore apertures are continuously adjusted, thus providing the tremendous opportunity to separate mixtures of hydrogen isotope that require precise pore tuning.

Here, we report quasielastic neutron scattering studies of H2-D2 diffusion in MIL-53(Al), demonstrating remarkable quantum effects, with significant diffusivity difference even at high temperature. Such flexible porous materials exhibit the its dynamic behavior of adsorbed hydrogen isotopes during 'breathing' phenomenon which will enable novel processes for kinetic molecular sieving of hydrogen isotopes.

597737 Investigating Indirect Ultrasound on Particle Size, Shape and Nucleation of APIs through Process Intensification

Abstract:

Background

Particle engineering of active pharmaceutical ingredients (API) is an increasingly studied area of research & development. During drug substance manufacturing, particle formation typically originates from crystallisation which is a critical multiphase unit operation for the purification and separation of APIs. It is imperative that product properties such as particle size, shape and polymorphic form are well defined and well controlled as they affect a drug's bioavailability, shelf-life and to effectively formulate the API into a finished drug product. [1]

However, achieving control over these properties is not necessarily met at the point of API crystallisation and isolation. Process intensified strategies through incorporation of ultrasound devices has shown wide interest in the last decade for manipulating particle properties during or immediately after the crystallisation step. For instance, the integration of ultrasound with crystallization allows for multipurpose functionality such as particle size & shape modification, [2] de-agglomeration, [3] scale-up, [4] accelerating the rate of crystallization kinetics and presents an overall more flexible approach to particle engineering. [5]

Ultrasound technology is commonly applied across several industries through insertion of a probe horn in direct contact with a liquid medium. For pharmaceutical crystallisation, this can be advantageous in process development especially when small seed particles <15 μm are preferred for controlling the crystal growth and thus, final product properties. [6] Whilst still a popular method, direct contact of the probe horn with particles has well reported limitations such as surface erosion, [7] substantial temperature deviations, [8] excessive noise in the vicinity of health and safety, rapid dissipation of energy transferred per unit volume and continuous operability. Therefore, it is worthwhile exploring alternative particle engineering strategies which can avoid the aforementioned limitations whilst ensuring the manufacturability and quality of the drug substance.

Case-Study

In this work, indirect ultrasound is explored as a viable and robust means for particle engineering of APIs. A semi-continuous platform which applies indirect ultrasound for inducing nucleation, particle size reduction and shape modification is experimentally investigated via process intensification. The process consists of an inline flow-cell which generates intense ultrasonic vibrations via electrical stimulation (Figure 1). These mechanical vibrations are then transferred to a glass tube which in turn sonicate the incoming process stream without direct contact.

Two APIs are selected for this study both of which have been found hard to nucleate with standard operating conditions under current industrial manufacturing processes. To establish key relationships between process parameters and product attributes, a series of isothermal desupersaturation experiments were conducted which monitored the impact of power amplitude (0, 20%, 40% and 80%) on inline count profiles, chord length distribution (CLD), particle size distributions (PSD) and solution concentrations. Nucleation rates affected from ultrasound impact were then extracted through population balance modelling as a proof-of-concept demonstration. To our knowledge, this is the first study to test and implement this indirect ultrasound flow-cell strategy on APIs for nucleation acceleration and control of particle attributes. The overall platform integration and ease of operation were additionally assessed for potential application in continuous manufacturing of pharmaceuticals.

References:

1. Variankaval, N., A.S. Cote, and M.F.J.A.J. Doherty, *From form to function: Crystallization of active pharmaceutical ingredients.* 2008. **54**(7): p. 1682-1688.
2. Zeiger, B.W. and K.S.J.J.o.t.A.C.S. Suslick, *Sonofragmentation of molecular crystals.* 2011. **133**(37): p. 14530-14533.
3. Cote, A. and E.J.A.P.R. Sirota, *CRYSTALLIZATION: the pursuit of a robust approach for growing crystals directly to target size.* 2010. **13**(7): p. 46.
4. Kim, S., et al., *Crystallization process development of an active pharmaceutical ingredient and particle engineering via the use of ultrasonics and temperature cycling.* 2003. **7**(6): p. 997-1001.
5. Ruecroft, G., et al., *Sonocrystallization: the use of ultrasound for improved industrial crystallization.* 2005. **9**(6): p. 923-932.
6. Jiang, M., et al., *Indirect ultrasonication in continuous slug-flow crystallization.* 2015. **15**(5): p. 2486-2492.
7. Price, C.J., *Application of Ultrasound in Crystallization (Sonocrystallization)*, in *Engineering Crystallography: From Molecule to Crystal to Functional Form*. 2017, Springer. p. 301-313.
8. Zhang, Z., et al., *Enhancement of crystallization processes by power ultrasound: current state-of-the-art and research advances.* 2015. **14**(4): p. 303-316.

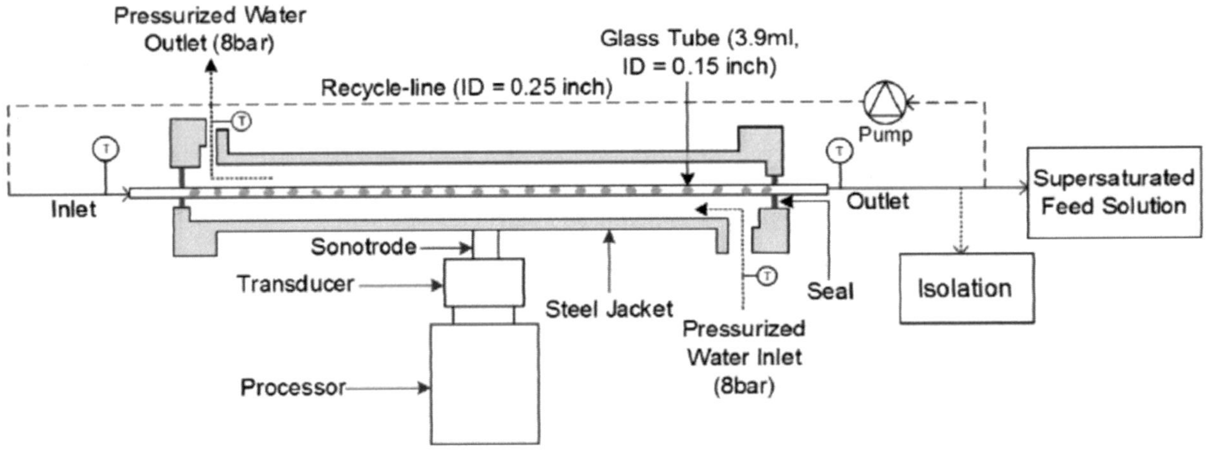

Figure 1: Sketch of the implemented indirect ultrasound flow-cell used in this study

607549 Experimental Investigation of Wet-Mill Integrated Crystallization with Temperature Cycles for Size and Aspect Ratio Control of a Commercial Pharmaceutical Ingredient

Abstract:

The control of the crystal properties has been in the focus of pharmaceutical industries and it represents a challenging aspect of the active pharmaceutical ingredient (API) production processes. To ensure that the desired critical quality attributes (CQAs) of the product are met, such as size, shape or polymorphic form, as well as to ease the downstream operations such as filtration or tableting, a careful design of the crystallization processes is critical. In this work we demonstrate the systematic design of a crystallization system for a commercial API, (compound A) from Takeda Pharmaceuticals International. The challenges related to the crystallization of the API is that the process is nucleation dominated and crystals are difficult to grow to the desired particle size (150-250 μm) by a simple cooling process. Moreover, the compound A tends to form high aspect ratio (AR) crystals, which can yield manufacturability problems. The aim of the crystallization design is to produce low AR and sufficiently large crystals with narrow distribution. Two methods were applied to reach these goals: (1) implementing temperature cycles to internally remove the fines, and (2) application of immersion milling to further control the shape and size of crystals.[1,2,3] These approaches were suggested in the literature based on the results of comprehensive model-based optimization studies.[4]

The work shows the implementation of a Quality-by-Control (QbC) guided wet-mill integrated crystallization design to produce large crystals with low AR. In this work, nine batch cooling crystallization experiments were performed. The first experiment was a seeded crystallization with very slow linear cooling rate. The remaining eight experiments were divided into two sets: (i) temperature cycling using simultaneous internal wet-mill; and (ii) crystallizations using temperature cycling but without milling. All experiments were performed with similar batch times but increasing the temperature cycle number from zero to four. Then, by using the same temperature profiles and operating procedure, simultaneous milling was applied for 75 % of the duration of the first temperature cycle as suggested in the literature, by using an immersion mill.[3]

A batch seeded cooling crystallization with very slow cooling (-0.02 °C/min) was performed, which showed that even this very slow cooling was insufficient to obtain the desired large crystals (> 150 μm) because of strong secondary nucleation in the process. Instead, applying internal fines removal by temperature cycles led to considerable increase of the product crystal size. Increasing the number of temperature cycles had a significant effect on producing large crystals (with mean size of 150-250 μm) by eliminating the fines produced by nucleation. These conclusions were also evidenced by using FBRM and in-line particle images. The results suggested that at least 2 temperature cycles were required to reach the desired product size for compound A. In the second part of the study, experiments with the same temperature profiles with wet-milling applied during the first cycle were performed. In these experiments, large crystals with low AR (>150 μm with mean AR around 2.5) could be successfully produced. The reason for using the wet-mill only at the beginning of the process is that it helps create large amount of small, low AR particles, which are partly removed by the first heat-up stage. In addition, it was shown with the 4 cycle-experiment that the milling rate does not have a significant effect on the CSD within the range applied (4000 and 8000 RPM) due to the achievable smallest size limit that characterizes the milling equipment. Therefore, the milling intensity was set to 4,000 RPM for all experiments. The experiments demonstrated that wet-mill contributed significantly to reaching the desired product crystal size and shape. This was quantified by 2D size distribution measurements based on optical microscopy images. Additionally, unseeded experiments were also performed, in which in situ seed generation was achieved by (1) cooling or (2) application of the immersion-mill as a high share nucleator. Both of these experiments demonstrated that applying wet-milling during the first temperature cycle not only significantly improves the CSD and AR of the product, but also considerably decreases the uncertainty caused by variation in the seed quality, and in combination with internal seeding via cooling or using the immersion mill as a nucleator can increase the overall efficiency and robustness of the process.

References:

[1] D. Ramkrishna and M. R. Singh, "Population Balance Modeling: Current Status and Future Prospects," Annu. Rev. Chem. Biomol. Eng., vol. 5, no. 1, pp. 123–146, 2014.

[2] D. Acevedo, V. K. Kamaraju, B. Glennon, and Z. K. Nagy, "Modeling and Characterization of an in Situ Wet Mill Operation," Org. Process Res. Dev., vol. 21, no. 7, pp. 1069–1079, 2017.

[3] Salvatori, F., Binel, P., & Mazzotti, M. (2019). Efficient assessment of combined crystallization, milling, and dissolution cycles for crystal size and shape manipulation. Chemical Engineering Science: X, 1, 100004.

[4] B. Szilagyi and Z. K. Nagy, "Model-based analysis and quality-by-design framework for high aspect ratio crystals in crystallizer-wet mill systems using GPU acceleration enabled optimization," Comput. Chem. Eng., vol. 126, pp. 421–433, 2019.

608902 Beyond Structure Determination: Crystallisation As a Purification Unit Process for Proteins and Peptides

Abstract:

The purification of high molecular weight proteins and peptides, for their use in biopharmaceutical drug therapeutics, can still account for a significant proportion of the drug development cost, which averages at between $1-6 billion. Crystallisation of these materials has mainly been used only in the field of structural determination, using small scale methods such as hanging drop vapour diffusion. However, advances in the optimisation of crystallisation conditions has seen significant increase in the scalability of protein crystallisation, including the development of continuous crystallisation approaches. This has seen crystallisation emerge as a more economic and greener alternative to the existing chromatographic methods typically utilised in peptide and protein purification.

The Made Smarter review of 2017 identified that manufacturing industries should embrace transformative digital technologies to streamline their unit processes. Mechanistic process modelling, based on population balance equations to represent the underpinning physical phenomena, has proved a computationally efficient method of predicting optimal crystallisation conditions of the purification of small molecule active pharmaceutical ingredients. However, this has yet to be expanded into the crystallisation of biopharmaceutical proteins.

Here, we utilise a combined process modelling and experimental approach to scale up the crystallisation of lysozyme and insulin from the μl to the tens of ml scale. We successfully optimise conditions, such as stirring rate and operation, pH, concentration and templating additives to not only scale up, but also increase the achieved crystallisation yield.

Insulin, being more susceptible to denaturing, utilised shaking to induce mixing of the protein rich buffer and salt precipitant. Small scale hanging drop screening experiments found that the addition of amino acids could reproducibly encourage the crystallisation of insulin, offering the promise of control over the nucleation. Scale up experiments to 15ml allowed the investigation of the impact of shaker mixing from 50-150 rpm. The insulin case study also highlights how to circumvent common issues with protein crystallisation, such as working with extremely low solubility materials and identifying the nucleation and growth points during the crystallisation process.

The lysozyme protein is more resistant to denature than insulin, whereby it can be subjected to mechanical stirring. Here we show successful batch and continuous crystallisations, utilising both shaking and mechanical stirring, up to the 30ml scale. Induction times were measured and classical nucleation kinetics parameters were derived from this data, with the pre-exponential factor and the surface energies being consistent with previous studies. Further optimisation of stirring methods and initial protein concentration was investigated to optimise the size and shape of the produced lysozyme crystals, whereby their purity was compared to the as purchased lipolyzed powder.

Finally, process modelling techniques were applied to predict the crystallisation of lysozyme and insulin in the batch reactor conditions. Such models are well parameterised to describe the crystallisation of small molecule active pharmaceutical materials. However, the classical nucleation and growth equations, as well as the models to describe the liquid state, have been less commonly applied to biological materials. Kinetic parameters to describe the processes that were extracted from the literature were used as a starting point for the model, where careful optimisation of these parameters were used to identify problems and where discrepancies between modelled and experimental particle size distributions and concentration vs time data originated. The kinetic parameters that were optimised against reliable experimental data were confronted with further experimental data to examine the wider applicability of protein crystallisation process model.

It was found that the classical Mullin nucleation and power law growth kinetic equations could be successfully regressed against experimental concentration vs time data, even for long nucleation times (Figure 1).

Figure 1: Concentration vs time for a 30ml batch crystallisation of lysozyme, using a sodium acetate buffer and sodium chloride precipitant. Blue dots are experimental data with approximate 10% error from nanodrop UV vis measurements and the red line in the regressed fit using Mullin classical nucleation kinetics and power law growth kinetics

Figure 1 shows that a good fit of the process model to the experimental concentrations vs time data was achieved. It is interesting to observe that the classical two step growth kinetics failed to find a fit, where we speculate that the diffusion term associated with the large lysozyme molecule limits the growth of this material in the model. Further investigation of the exact parameters in the equations will be presented.

This study demonstrates a novel experimental and process modelling workflow for optimising the crystallisation of large biological molecules, paving a path for this to be a viable unit process in biopharmaceutical drug development. This study gives guidance and insights into the challenges of protein crystallisation and modelling of this process, along with what are the next steps to make both processes more robust for further scale-up.

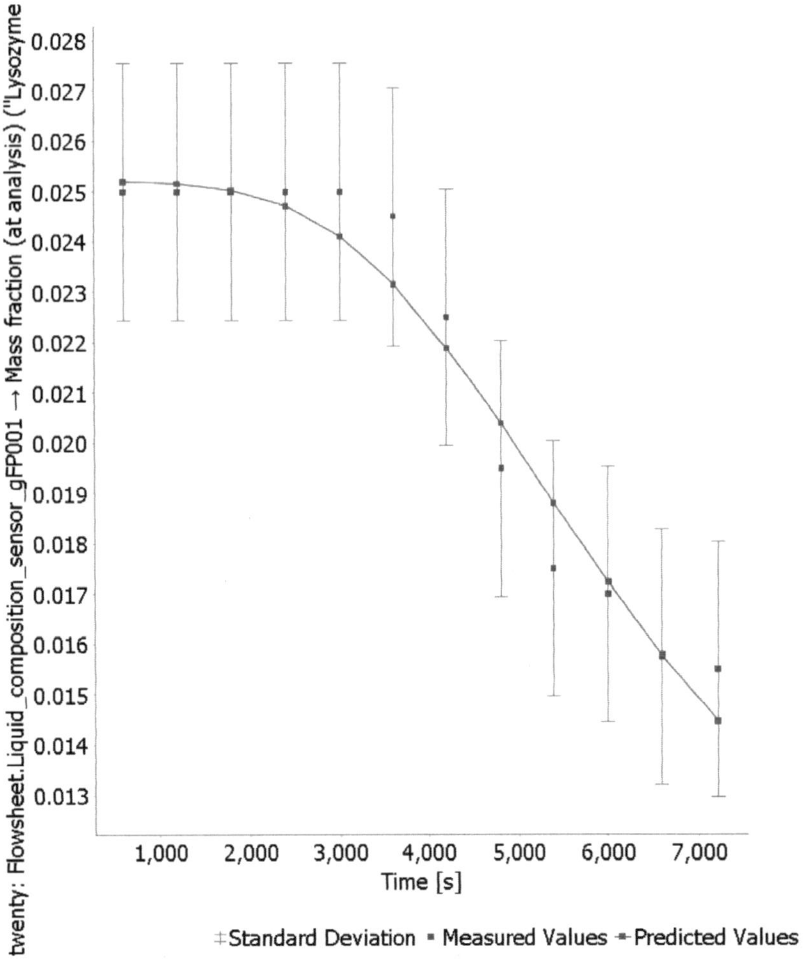

605706 Maneuvering Around Liquid-Liquid Phase Separation (LLPS) to Enable Development of Robust Crystallization Processes.

Abstract:

Anti-solvent and/or cooling crystallizations are commonly employed in purification of pharmaceutically relevant molecules. In addition, the crystallizations are also typically tasked with isolating solids with controlled physical properties. During development of multi-solvent crystallizations, it is not unusual to observe Liquid-Liquid Phase Separation (LLPS), commonly referred to as "oiling out". LLPS is generally undesirable due to its detrimental impact on purity and physical properties as well as due to the challenges it brings in during scale-up. LLPS is impacted by multiple parameters such as solute concentration, solvent composition, and temperature. Understanding of such a dynamic phase diagram, in addition to the kinetic parameters, is paramount to achieve a predictable process scale-up.

In this presentation, we detail a methodology to develop a hybrid anti-solvent / cooling crystallization process maneuvering around the LLPS boundaries. Use of automated Lab reactors (ALR) equipped with Process Analytical Technologies (PAT) is highlighted for performing data-rich experiments quickly. The generated data is further modelled with thermodynamic equations for enhanced process understanding. Finally, we will present a case study to showcase how the above-referenced methodology was utilized for enhanced process understanding and for development of a robust and scalable crystallization process.

All authors are employees of AbbVie and may own AbbVie stock. AbbVie sponsored and funded the study; contributed to the design; participated in the collection, analysis, and interpretation of data, and in writing, reviewing, and approval of the final publication.

603539 Utilizing Mechanistic Modelling to Support Crystallization Seeding Design

Abstract:

Introduction

Digital design of manufacturing processes using mechanistic models is fast becoming an essential tool during Active Pharmaceutical Ingredient (API) early process development activities. It enables rapid and effective exploration of the decision space for Critical Process Parameters (CPP), helping to reduce risk and product time-to-market, and aiding in the effective and safe production of high quality pharmaceutical products. In this work, we demonstrate how mechanistic modelling can be utilised to support the early stage design of seeding strategy for a batch cooling crystallization process. We outline a step-wise workflow consisting of model validation from which primary nucleation and growth kinetics are estimated, as a minimum list of crystallization mechanisms. The data used during the bespoken parameter estimation was experimentally measured in an unseeded small-scale crystallization process.

As part of this work, the model predictions for the seeded crystallization process was subsequently verified experimentally. This model-based approach along with rich experimental data yielded a more efficient workflow for seeding design, with reduced experimental effort compared to a more traditional purely empirical approach. A range of process data was utilized for the purposes of qualitative mechanism discrimination and quantitative model validation as follows:

- Qualitative mechanism & model discrimination
  - FBRM data to inform active crystallization mechanisms to consider
  - SEM and/or optical images of product crystals to inform crystallization mechanisms to consider within the mechanistic model
- Quantitative model validation
  - FBRM data to provide indication of onset of nucleation and access to the rate of nucleation in the unseeded crystallization runs.
  - Solute concentration data over time both online (IR/Raman) &/or offline (HPLC)
  - Particle size measurements of the product crystals from laser diffraction

The validated mechanistic model was subsequently used for model-based design of seeding policy, to assist in the seeding design (seed mass, seed addition point, and/or seed particle size distribution) required to achieve the target PSD specifications (defined by the volume-based quantiles d10, d50 and d90 of the product PSD). Uncertainty and sensitivity analysis were performed to understand the attainable region for the product PSD in terms of quantiles for the crystallization process and to identify the relative importance of process parameters, including seed mass and seed PSD.

Conclusions

The main conclusions of the work include the following:

- It has been shown that a model-based approach can be utilised to support the early stage development of crystallization processes to achieve the desired quality attributes of the product particles.
- A model-based approach has been shown to help to reduce the degree of experimental effort

required compared to a purely empirical approach.

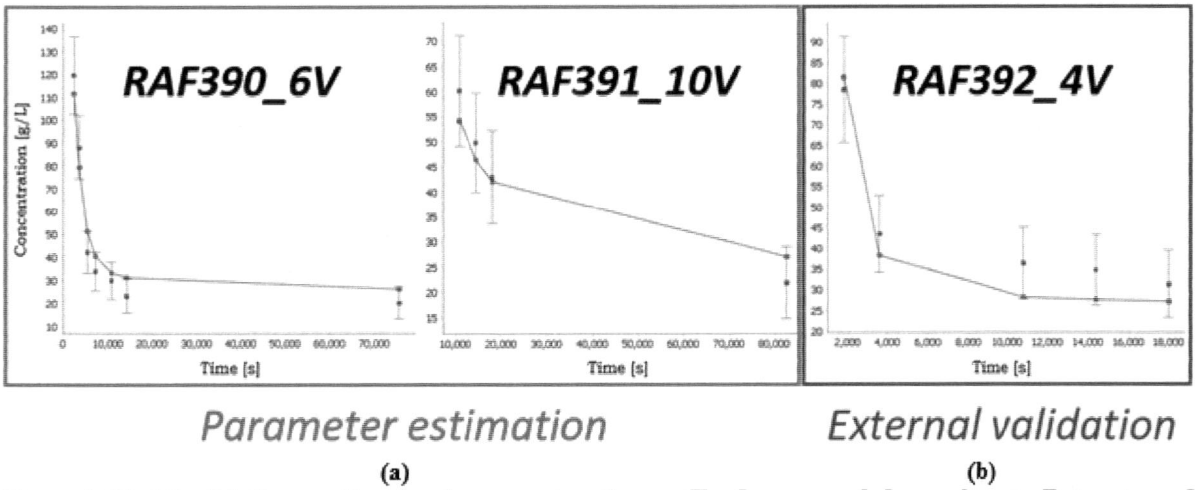

Figure 1: Model validation results using de-supersaturation profiles from unseeded experiments. Data sets used for (a) parameter estimation and (b) external model validation.

Figure 2: Sensitivity analysis results showcasing the impact of mass of seed added and seed mean particle size on the D50 quantile of the crystallized material.

611464 Process Optimization for Controlling the Morphology of Cof-5 Crystals

Abstract:

Covalent Organic Frameworks (COFs) have gained significant attention in the field of material science over last two decades. 2D boronate-ester linked frameworks like COF-5, are porous, crystalline, organic polymers with low density, that can be used for numerous applications such as gas storage and as membranes for various industrial purification applications. COF-5 exhibits high thermal stability, high surface area and permanent porosity widening the scope of its applicability. To maximize its applicability, it is essential to optimize and control the morphology of the crystallites by tweaking appropriate processing conditions. COF-5 can be synthesized by evaporative crystallization using 1,4-diboronic acid (PBBA) and Hexahydroxy Triphenylene (HHTP), in an equimolar solvent mixture of Dioxane and Mesitylene. This process is heterogenous, uncatalyzed and slow, needing three days to reach a considerable and acceptable yield. Researchers have previously made efforts to reduce the reaction barrier and reaction time to 20 hrs. However, a systematic understanding of the effects of alterations in processing conditions on crystallite yield and morphology remain unanswered. Effective and accelerated crystallization of COF-5 while controlling the structure of these crystals is the prime focus of our research. This study has addressed the optimization of processing conditions, catalyst content, and solvent environment to synthesize and control the morphology of COF-5 at a higher rate. Additionally, these conditions were utilized to grow COF-5 films on substrates, while ensuring that their crystallite properties remained intact. Experimentally derived kinetic models using the optimized conditions, have been coupled with population balance equations to get a deeper insight on the crystal growth mechanisms of COF-5.

608089 Crystallization Model Discrimination Using in Situ Chord Length Monitoring and a Size Conversion Model (CLD2PSD model).

Abstract:

A new framework has been recently introduced to correlate in situ chord length distribution (CLD) measurements and offline particle size distribution (PSD) data (Irizarry et al, 2017) and applied to different particle morphologies (Schoell et al, 2019). Subsequently the model was expanded to consider bimodal distributions and inferences of modality and morphology changes (Irizarry et al, 2020). This work applies the proposed approach to estimate the kinetics parameters for seeded batch crystallization processes of different APIs in combination with population balance modelling. First, a data-driven CLD2PSD model was built for the studied pharmaceutical compounds since the differences in optical properties and particle morphology require a new model for every single substance. Then, time-resolved PSDs have been extracted from in situ CLD data of seeded batch processes and the CLD2PSD model. Finally, the online PSDs have been combined with online solute concentration data to allow for the estimation of crystal growth, secondary nucleation and agglomeration kinetics, depending on the supersaturation level encountered during the process. The validity of this approach has been confirmed by comparison to kinetic data determined using a traditional approach relying on offline data. The main advantage of the proposed approach is that all data is generated online and in situ, thus avoiding sampling errors.

References

- Irizarry, R. et al., 2017. Data-driven model and model paradigm to predict 1D and 2D particle size distribution from measured chord-length distribution. Chemical Engineering Science, 164, pp. 202-218.
- Schoell, J. et al., 2019. Determining particle-size distributions from chord length measurements for different particle morphologies. AIChE Journal, 65, e16560

606939 Continuous Solid-State Deracemization Via Temperature Cycles – a Model-Based Process Development

Abstract:

Introduction

Although preferential crystallization is a standard industrial process to obtain enantiopure crystals of conglomerate forming compounds, its maximum theoretical yield is 50%, unless one adds to it recycle and racemization steps [1]. On the contrary, one can produce an enantiopure powder from an initially racemic one with 100% yield through solid-state deracemization by exploiting several techniques [1,2]. Among such techniques, one based on temperature cycling is a promising candidate for industrial applications due its simplicity of operation [1,3,4]. However, the process itself is complex and hence it warrants the need for a mathematical model to understand the effect of initial conditions and of operating parameters on the process outcome, i.e. the production of the desired enantiomer, the process time, etc.

Continuous processes have been shown to outperform batch operation. In addition, they can recover from process disturbances, because they run at stable steady state [5]. Continuous crystallization processes have been adapted for the separation of conglomerate forming enantiomers, in order to exploit the advantage of continuous crystallization techniques. In case racemization of enantiomers in the solution proves difficult, continuous preferential crystallization without a racemization reaction is a widely studied choice [6]. If the racemization reaction is available, continuous preferential crystallization with racemization reaction or continuous Viedma ripening can be performed. Alternatively, deracemization via temperature cycles in mixed suspension mixed product removal crystallizers (MSMPRC) can also be performed, but this is an unexplored method of choice.

Mathematical model

A population balance equation (PBE) -based mathematical model has been developed to provide a clear explanation of the phenomena involved in the deracemization process. The model describes the interplay of several phenomena involved in a temperature cycling process (size-dependent solubility, crystal growth and dissolution, agglomeration, attrition/secondary nucleation and racemization) accounting for the dependence of their thermodynamic and kinetic parameters on the crystal size and on the temperature [5].

Results and discussion- batch

Using the PBE-model that we have introduced, first, we have simulated temperature cycles in batch crystallizers. A mathematical model of deracemization is a useful tool to perform process analysis: it provides information on all the relevant physical quantities during the whole process and enables decoupling of the individual phenomena, and investigating the effect of variations in e.g., the breakage rate intensity, the agglomeration intensity, etc. After identifying the necessary phenomena to deracemize using temperature cycles, process characterization was performed using the developed model, in order to identify an optimal process. Parametric studies varying operating parameters like the temperature and cooling rate were performed and based on the outcome it is shown that the simulation outcome qualitatively describes experimental results [1,4].

Moreover, by investigating the effect of slight variation in the initial conditions, an explanation to the large variations in deracemization time and in process outcome — that are typically observed in experiments — is provided using the mathematical model.

Results and discussion- continuous

The inherent disadvantage of the batch crystallization process is that it suffers from batch-to-batch product variability and requires control techniques to obtain the product with the specific quality of interest. As tuning the many degrees of freedom blindly while performing the actual experiments is ill-advised [7], this contribution aims at providing a mathematical tool that can be used to design robust continuous processes.

Therefore, we have extended our batch non-isothermal population balance-based model of solid-state deracemization via temperature cycles to simulate the proposed continuous configuration. To avoid undesired nucleation of the counter enantiomer, different process variants are considered, e.g., utilization of recycle loop, wet mill, fines dissolution loop. We have investigated the steady state operation of temperature cycles carried out in MSMPRC. Periodic operation of the continuous crystallizer via temperature cycling affects the transient behaviour of the process and thereby the attainable steady state. We have run simulations to study the interplay of the cycle time and the residence time with respect to the physical properties of the system, i.e., how the process characteristic times compare to the characteristic times of growth, dissolution and racemization.

The aim of this analysis was to verify whether one of the continuous configurations envisaged performs better than the batch alternative. The effect of the initial conditions (variations in the initial enantiomeric excess, in the suspension density, etc.) and of the operating parameters on the periodically perturbed process and its steady state in terms of productivity and attainable enantiomeric excess are identified. Together with this assessment, we aim to provide strategies for the optimal implementation of the process.

Acknowledgments

This research has received funding as part of the CORE project (October 2016 – September 2020) from the European Union's Horizon 2020 research and innovation programme under the Marie Sklodowska-Curie grant agreement No 722456 CORE ITN.

References

[1] Breveglieri, G. M. Maggioni, M. Mazzotti, Cryst. Growth Des. 18 (2018) 1873-1881.
[2] L. Noorduin, W. J. P van Enckevort, H. Meekes, B. Kaptein, R. M. Kellogg, J. C. Tully, J. M. McBride, E. Vlieg, Angew. Chem., Int. Ed. 49 (2010) 8435-8438.
[3] Iggland and M. Mazzotti, Cryst. Growth Des. 18 (2011) 4611-4622.
[4] Suwannasang, A. E. Flood, C. Rougeot, G. Coquerel, Org. Process Res. Dev. 21 (2017) 623-630.
[5] Köllges and T. Vetter, Cryst. Growth Des. 17 (2017) 233-247.
[6] Vetter, C. L. Burcham, M. F. Doherty, AICHE J. 61 (2015) 2810-2823.
[7] Bodák, G.M. Maggioni, M. Mazzotti, Cryst. Growth Des. 18 (2018) 7122-7131.

$$\frac{\partial f_i}{\partial t} + \frac{\partial(G_i(S_{\infty,i},T,L)\,f_i)}{\partial L} = B_i(L,c_i,T) - D_i(L,c_i,T) + \frac{\dot{m}_{in}f_{in,i} - \dot{m}_{out}f_i}{m}$$

$$\frac{dc_i}{dt} = -k_v\rho\frac{d\phi_{3,i}}{dt} + R_i(c_i,c_j,T) + \frac{\dot{m}_{in}c_{in,i} - \dot{m}_{out}c_i + \dot{m}_{rec}c_i}{m} \qquad (i = D, L)$$

- $\phi_{n,i} = \int_0^{+\infty} L^n f_i(L,t)dL$
- $ee = \frac{\phi_{3,D}-\phi_{3,L}}{\phi_{3,D}+\phi_{3,L}}$

Initial and boundary conditions:

$$f_D(L,0) = f_D^0(L)$$
$$f_L(L,0) = f_L^0(L)$$
$$f_i(0,t) = 0$$
$$c_D(0) = c_L(0) = c_\infty(0)$$

Figure 1. Summary of the model equations that are the basis for the model of our continuous configuration. L: size of the crystal, T: temperature, S_∞: supersaturation, f_i: particle size distributions, G_i : growth rates, B_i and D_i: birth and death terms, $\dot{m}_{in}$, $\dot{m}_{out}$, $\dot{m}_{rec}$: mass flow rates, c_i: concentrations, R_i: racemization term

603007 Hydroxy Metal-Organic Framework-Based Mixed Matrix Membranes Enabling Enhanced CO_2 Capture and Hydrogen Purification

Abstract:

Membrane separation technology offers enormous energy- and cost-savings for molecular separations in chemical industries, complementing conventional thermally-driven approaches, such as distillation. However, current membranes face numerous challenges, primarily lying in Robeson upper bounds, i.e., tradeoffs between the separation productivity and selectivity. In this work, we demonstrate a new class of hydroxy metal-organic frameworks (MOFs)-based mixed matrix membranes, enabling high performance for applications including CO_2 capture and hydrogen purification, compatible with current membrane manufacturing processes. The mixed matrix membranes perform beyond present Robeson upper bounds with H_2 and CO_2 permeabilities of 907 and 650 Barrers, respectively, with H_2/CH_4 and CO_2/CH_4 selectivities of 45 and 32, respectively. The outstanding performance is attributed to the intimate interfacial interaction of hydroxy MOFs and carboxylic polyimide via intensive hydrogen bonds. This design principle provides a new approach to fabricating membranes for energy-efficient gas separations with enhanced performance and compatibility with state-of-the-art membrane manufacturing.

601054 Development of New Siloxane Terpolymers for Enhanced C_{3+} Hydrocarbon Recovery from Natural Gas

Abstract:

Natural gas plays a significant role in the global energy mix. Although raw natural gas consists primarily of CH_4, it can contain significant amounts of other gas components, including acid gases (CO_2 & H_2S), N_2, He, and C_{3+} hydrocarbons. These C_{3+} hydrocarbons are valuable chemical feedstocks and can be used as a liquid fuel for power generation and seawater desalination. Commercially available rubbery siloxane membranes are utilized in industry for the rejection of N_2 and for the concentration of C_{3+} hydrocarbons in the membrane permeate stream. However, current state-of-the-art PDMS membranes with $(Me_2SiO)_n$ backbones show lower C_{3+}/CH_4 selectivities under industrially relevant feed streams and testing conditions. In this work, modified crosslinked siloxane terpolymer $(VinylMeSiO)_p(Me_2SiO)_m(OctMeSiO)_n$ membranes (Ter-PDMS) were fabricated and evaluated for separations of C_{3+} hydrocarbons from natural gas. Results show that modified crosslinked siloxane Ter-PDMS membranes showed enhanced C_{3+} separation performance compared to conventional PDMS membrane under simulated typical field gas testing conditions (i.e. feed pressures up to 850 psi for multicomponent gas mixture consisting C_1-C_5 hydrocarbons, CO_2, N_2, and BTEX). At 800 psi, Ter-PDMS membranes exhibited C_4/CH_4 mixed gas selectivity and C_4H_{10} permeability of 17.2 and 15100 Barrers, respectively. These results emphasize the observations that membrane separation performance is strongly related to the testing conditions and feed compositions, highlighting the importance of exploring and designing better modified rubbery siloxane membrane materials for the actual industrial process.

612762 Ultrafiltration polyvinylidene fluoride (PVDF) membranes prepared by combined crystallisation & diffusion (CCD) technique

Abstract:

Polyvinylidene fluoride (PVDF) is one of the most commonly used materials for making ultrafiltration and microfiltration membranes. It is desirable because of its superior mechanical properties and stability in various chemical and thermal environments. It is one of the best suited membrane materials for waste water treatment applications. Unfortunately, existing commercial PVDF membranes in the ultrafiltration range suffer from low permeation fluxes (less than 150 LMH.bar^{-1}) [1]. These commercial PVDF membranes are prepared via traditional routes such as non-solvent induced phase separation (NIPS) or thermally induced phase separation (TIPS) or a combination of both [2-7]. To compensate for its poor permeation performance, large areas of membranes are required for a separation application, which drive the capital and operating costs upwards making it less desirable as a separation technology option.

To address this challenge, a new method of fabrication called combined crystallisation and diffusion (CCD), inspired by ice templating, was developed by our research group [1, 8]. This method involves unidirectional cooling of casted membrane using a solid media maintained at -30 °C. The objective is to freeze the solvent of the casted polymer solution into crystals and leach them out in an ice-cold water bath as it can then serve as pore templates in the final membrane. The permeation rates of these CCD membranes were found to be an order of magnitude higher than the commercial membranes having similar pore sizes.

The first part of this study involved performance comparison of PVDF flat sheet membranes prepared via the most common membrane fabrication method, NIPS and the innovative unidirectional freezing technique of CCD. The polymer dope concentration was varied from 10-20wt.% to produce PVDF membranes of varying pore sizes. The results indicated similar pore sizes for CCD and NIPS membranes prepared by specific dope solution concentrations, but significantly higher permeation rates for CCD membranes (10 times higher) than its NIPS counterpart. This was attributed to differences in the morphology of the two membrane types as seen from their SEM images presented in Fig 1. The CCD membranes also had superior mechanical and anti-fouling properties. The flux recovery of fouled membranes after cleaning was significantly higher for CCD membranes.

The second part of this study focused on devising an efficient way to influence the pore sizes of CCD membranes. So far, varying either the cooling conditions or the total polymer dope concentration would influence the pore size. At 20wt.% total polymer dope concentration and existing cooling conditions, the smallest pore size achieved for CCD membranes was about 30 nm. The purpose of this study was to decrease the pore size further, without increasing the total polymer concentration or changing the cooling conditions, as both present operational challenges. To achieve this blending modification was applied using poly(methyl methacrylate) (PMMA) as an additive.

The pore size of CCD PVDF membranes was reduced by 50% to 15nm by carefully adjusting the blend proportion of PMMA in the PVDF-DMSO dope solutions. Systemic study shows the pore size reduces when increasing PMMA concentration in the membranes; but after an initial drop compared to the pure PVDF membrane, the pure water flux doesn’t decline with PMMA concentration. The effect of different molecular weights of PMMA (120kDa and 35kDa) was also investigated, where the results showed a smaller pore size for membranes prepared using the higher molecular weight PMMA. Scanning electron microscopy (SEM), cloud-point temperature measurement and Fourier transform infrared spectroscopy (FTIR) were used to characterise the membranes and help to understand how PMMA affects membrane formation. From the evidence gathered, we attribute the effects of PMMA to the change in the number of nucleation centre, diffusional rate, as well as the gelation point based on a previously proposed CCD mechanism.

References

[1] B. Wang, J. Ji, and K. Li, "Crystal nuclei templated nanostructured membranes prepared by solvent crystallization and polymer migration," *Nat Commun,* vol. 7, p. 12804, Sep 19 2016.

[2] T.-T. Jin, Z.-P. Zhao, and K.-C. Chen, "Preparation of a poly(vinyl chloride) ultrafiltration membrane through the combination of thermally induced phase separation and non-solvent-induced phase separation," *Journal of Applied Polymer Science,* vol. 133, no. 5, pp. n/a-n/a, 2016.

[3] J. T. Jung, J. F. Kim, H. H. Wang, E. di Nicolo, E. Drioli, and Y. M. Lee, "Understanding the non-solvent induced phase separation (NIPS) effect during the fabrication of microporous PVDF membranes via thermally induced phase separation (TIPS)," *Journal of Membrane Science,* vol. 514, pp. 250-263, 2016.

[4] J. Lee, B. Park, J. Kim, and S. B. Park, "Effect of PVP, lithium chloride, and glycerol additives on PVDF dual-layer hollow fiber membranes fabricated using simultaneous spinning of TIPS and NIPS," *Macromolecular Research,* vol. 23, no. 3, pp. 291-299, 2015.

[5] J. Liu, X. Lu, J. Li, and C. Wu, "Preparation and properties of poly (vinylidene fluoride) membranes via the low temperature thermally induced phase separation method," *Journal of Polymer Research,* vol. 21, no. 10, 2014.

[6] T. Xiao, P. Wang, X. Yang, X. Cai, and J. Lu, "Fabrication and characterization of novel asymmetric polyvinylidene fluoride (PVDF) membranes by the nonsolvent thermally induced phase separation (NTIPS) method for membrane distillation applications," *Journal of Membrane Science,* vol. 489, pp. 160-174, 2015.

[7] H. Matsuyama, S. Berghmans, and D. R. Lloyd, "Formation of anisotropic membranes via thermally induced phase separation," *Polymer,* vol. 40, no. 9, pp. 2289-2301, 1999/04/01/ 1999.

[8] B. Wang, J. Ji, and K. Li, "Separation Membranes," United Kingdom, 2017.

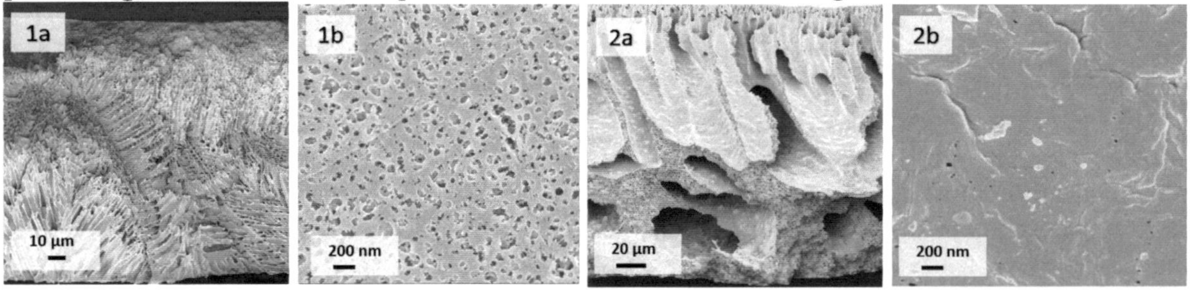

Fig 1. SEM images of (a)cross section and (b)separation side surface of (1)CCD and (2)NIPS membranes prepared with 20wt% concentration

606106 Transport Behavior of Crosslinked PEGDA Membranes to Methanol and Sodium Acetate

Abstract:

Understanding molecular transport behavior in hydrated membranes is important for various applications from desalination to energy conversion devices, such as photoelectrochemical CO_2 reduction cell. In such device, CO_2 is simultaneously being reduced into valuable chemicals, such as methanol and acetate, and a hydrated membrane is placed in between cathode and anode cells to minimize the permeation of CO_2 reduction products from cathode to anode. Due to the complex nature of capturing the transport of multiple solutes in liquid phase, multi-solute transport in hydrated membranes have not been studied thoroughly. Previously, our group proposed an in-situ ATR-FTIR method to measure the concentration of multiple solutes simultaneously. Here we use this method to study the transport behavior of a series of UV crosslinked poly(ethylene glycol) diacrylate (PEGDA, n = 13) membranes with different pre-polymerization water contents to two CO_2 reduction products, methanol and acetate, in co-permeation. Changes in membrane permeability to these solutes is observed with changing prepolymerization water content, and in the presence of the co-solute. We evaluate these changes in terms of polymer structures, water-polymer interactions, solute-polymer interactions, and solute-solute-polymer interaction.

602303 Design of Robust, High-Performance Ionenes and Ionic Composites for Membrane-Based Gas Separations

Abstract:

We have developed a collection of high-performance (HP) ionenes, or polymers which contain ionic groups along the backbone rather than as pendants. These materials combine structural elements typically associated with state-of-the-art gas-separation membranes with functionality utilized in ultra-high-performance engineering polymers (i.e. imide, amide, urea linkages). This work probes the effects of spacing and sequencing functional features and ionic moieties within these ionenes, which contain imidazolium cations (paired with fluorinated anions) integrated within the main chain. Structural segments known to impart free volume, including spirobisindane and Tröger's base units, have also been incorporated within these ionenes. These materials are robust, exhibiting good thermal and mechanical stability, and are extremely tailorable due to the design and modular synthetic method. Thus, the thermophysical properties and transport behaviors of these materials are highly tunable based on methodical combination of diverse monomers and linkages. These polymeric materials exhibit additional self-assembly and nanostructuring when impregnated with "free" imidazolium-based ionic liquids (IL) or more complex ionic fillers, which imparts another degree of tunability and coordinates the ionene chains. These HP-ionenes are thoroughly characterized in order to develop structure-property relationships which correlate functional and charged features with the resultant polymer properties. Composites are similarly analyzed for comparison with the corresponding neat polymers, to gain a better understanding of the coordination between polymer chains and dispersed additives. This presentation highlights the design and synthesis of these HP-ionenes, focusing on the performance of these materials as gas separation membranes.

605602 Novel Fouling Resistant Electrospun Membranes for Treating Hydraulic Fracturing Flow Back and Produced Water

Abstract:

Membrane distillation is an emerging technology for treating oily wastewaters especially from oil and gas manufacturing operations. Here we focus on the development of fouling resistant membranes for treating hydraulic fracturing flow back and produced water. These wastewaters are highly fouling as they contain not only dissolved salts but also dissolved organic compounds and surfactants. We have developed a bilayer electrospun membrane that consists of an omniphobic bottom layer to prevent low-surface tension solution wicking and an under-water oleophobic antifouling surface to suppress foulant deposition. Silica nanoparticles were grown on the bottom layer through electrochemical interaction followed by coating with a fluorine monomer on the fiber surface. The zwitterionic co-polymer poly (glycidyl methacrylate-sulfobetaine methacrylate) (GS) was grafted using self-assembly between hydroxy group on the surface generated by alkaline treatment and anchor segment epoxy group on GS. The novel electrospun membrane have been challenged with industrial real hydraulic fracturing produce water. Preliminary results shows that fouling in our bilayered membrane is significantly reduced. This bilayer electrospun membrane may enable the commercialization of membrane distillation for treating highly impaired hydraulic fracture and flow back wastewaters.

611070 Fully Functional Large-Area Nanoporous Single Layer Graphene Membranes for Desalination and Water Purification

Abstract:

An ideal desalination membrane should exhibit minimum thickness to maximize water permeance and narrow pore size distribution for efficient ionic /molecular separations. Monolayer graphene with uniform distribution of high-density of sub-nanometer pores has been considered as the perfect material offering ultra-fast water permeance and high solute rejection. However, scalable production of graphene membranes with control over sub-nanometer pores remains challenging. Here, we report the fabrication of large-area nanoporous graphene membrane for high-efficiency water desalination. The density of defects that manifest as small nanopores in graphene is significantly increased via facile and scalable oxidative etching of the graphene lattice. Novel interfacial polymerization is used to seal tears and unwanted defects in graphene. The resulting centimeter-scale monolayer graphene membranes with high-density of sub-nanometer nanopores not only allow high water permeance (upto 23 times higher than conventional membranes), but also show an excellent rejection of salt ions and organic molecules (~100% rejection). Our work provides a facile and scalable approach for atomically thin desalination and nanofiltration membranes.

References:

Cheng et al. (submitted)

Kidambi et al. Adv. Mat. 2018

Kidambi et al. Adv. Mat. 2018

Kidambi et al. ACS App. Mat. Int. 2018

Kidambi et al. Adv. Mat. 2017

Kidambi et al. Adv. Mat. 2017

Kidambi et al. Nanoscale 2017

611400 Finely Tuned Submicroporous Thin-Film Nanocomposite (TFN) Molecular Sieve Membranes for Water Purification

Abstract:

Water separation membranes formed from state-of-the-art polyamide thin-film composites (TFCs) show relatively low rejections to small neutral contaminants such as boron. The rapid increase in global demand for high purity water for municipal, agricultural, and industrial applications has stimulated substantial research in developing highly selective membranes for water purification. Aiming to improve the selectivity of the current TFC desalination membranes, metal-organic frameworks (MOFs), such as MIL-101(Cr), which have good water stability, high surface area, small particle size, and amine functional groups were successfully synthesized and incorporated into the polyamide selective layer of a TFC membrane to fabricate thin-film nanocomposite (TFN) membranes. Compared to the unmodified benchmark TFC membrane, the incorporation of amine-functionalized MOF nanoparticles during interfacial polymerization not only changed the properties of the polyamide matrix but also reduced the effective pore size of MOF fillers through a crosslinking reaction between the amine functional groups of the MIL-101(Cr) fillers and the trimesoyl chloride (TMC) monomer, thus leading to a significant improvement in membrane selectivity. The best performing TFN membrane showed outstanding rejections to NaCl and small neutral contaminants such as boric acid, which are much higher than those of the benchmark TFC membrane and even better than the state-of-the-art seawater desalination membranes. This study has demonstrated the promising potential of MOF–polymer composite materials for applications related to water purification.

601119 Separation of Organic Solvents with Uio-66/Matrimid® Dual-Layer Hollow Fiber Membranes

Abstract:

Membrane-based organic solvent separations have the potential to reduce the energy and carbon footprint of existing and emerging processing industries. High-performance membranes will accelerate this transition especially if the membranes can be fabricated in a manner that matches the immense scales of the various processing industries. This work describes the successful fabrication of UiO-66/Matrimid® hollow fiber membranes, subsequent post-treatments, and successful application in organic solvent reverse osmosis separations. The work focused on creating a membrane with high selectivity and permeability through a facile and scalable fabrication process. A dry jet-wet quench method was employed to fabricate mixed matrix hollow fiber membranes containing UiO-66. The fibers were subsequently crosslinked and evaluated via separation of the alkylaromatic solvents toluene and triisopropylbenzene. The crosslinking dramatically improved the membrane stability over extended operating conditions, as expected, but certain modifications from traditional flat sheet crosslinking protocols were needed to realize success in the hollow fiber morphology. The addition of UiO-66 boosted the permeance of the hollow fiber membranes while retaining rejection of triisopropylbenzene. This work demonstrates the successful dual-layer fabrication and post-treatment of mixed matrix hollow fiber membranes for organic solvent reverse osmosis applications.

609047 A Generalized High-Viscosity Gel Conversion Method for Highly Selective Small-Pore and Medium-Pore Zeolitic Hollow Fiber Membranes

Abstract:

High-silica zeolite membranes made of small-pore CHA and medium-pore MFI zeolites have previously demonstrated excellent performance for important separations involving smaller molecules like CO_2 and hydrocarbons such as butanes and xylenes. To accelerate the scale-up and industrial application of these zeolite membranes, more versatile and scalable fabrication techniques are highly desired. Gel-based conversion methods for zeolite membrane formation have the advantages of being amenable to coating-based membrane deposition approaches and not requiring bulk hydrothermal solutions. However, gel conversion methods have not been studied and controlled in detail to obtain high-quality membranes. Here we discuss a versatile, high-viscosity gel conversion method for more generalized use in zeolite membrane fabrication, and demonstrate its application to both CHA and MFI-type zeolite membrane fabrication on ceramic hollow fiber substrates. The highly concentrated precursor gels - with viscosities high enough to overcome gravitational effects on uniformity of coatings - were coated on the hollow fibers and transformed into membranes in a water vapor atmosphere. Important parameters, such as H_2O/SiO_2 ratio, seeding and synthesis duration, were investigated in detail for better understanding of the membrane formation and optimization of synthesis conditions. The excellent separation properties of the membranes are discussed in detail. For example, CHA hollow fiber membranes with CO_2/propane separation factors > 10,000 and CO_2 permeance over 5000 GPU, and MFI hollow fiber membranes with *n*-butane permeances > 400 GPU and *n*-/*i*-butane separation factors ~40, are obtained. We discuss a path forward for larger-scale fabrication of these membranes.

597014 Fabrication of Highly Permeable Polyethersulfone Substrates with Bicontinuous Structure for CO_2-Selective Composite Membranes

Abstract:

The substrate pore structure has a significant effect on the CO_2 transport performance of a thin-film-composite (TFC) membrane. In this study, a novel solvent system comprising 2-pyrrolidone (2PD) and 2-methoxyethanol (2-ME) was used to prepare polyethersulfone (PES) substrates with bicontinuous structure by vapor-induced phase separation, followed by immersion in a nonsolvent, water. Compared with the common solvent, e.g., *N*-methyl-2-pyrrolidone (NMP), 2PD is a solvent with a lower solubility. However, the solubility was improved by the addition of 2-ME into 2PD. Moreover, the 2PD/2-ME system is more hydrophilic and can decrease the thermodynamic stability of the casting solution significantly. As a result, the phase separation during the membrane formation process was induced via the spinodal decomposition mechanism. The process parameters, including PES concentration, 2PD/2-ME weight ratio, water vapor exposure time, and water bath temperature, were optimized by characterizing the membrane morphologies and the CO_2 transport performance. Compared with the baseline PES substrate optimized previously with NMP, the new substrate showed a high surface porosity of 19.9%, indicating a 48.5% increase. Also, the new substrate exhibited a very high CO_2 permeance of 133,226 GPU at 57°C, which was 5 times more permeable than the baseline substrate. This high permeance was contributed mainly by the absence of a top dense layer and the presence of highly interconnected pores in the substrate. By using this substrate, the prepared TFC membrane containing amines showed a CO_2 permeance of 908 GPU, which was 48 GPU higher than that with the baseline substrate at 57°C. This improvement can be explained well using the resistance-in-series model. The improved CO_2 permeance was attributed to both the reduced lateral diffusion and substrate transport resistances. In addition, the scalability of the PES substrate with the bicontinuous structure was demonstrated by the successful roll-to-roll fabrication of a 50-foot long scale-up PES substrate with a width of 21″.

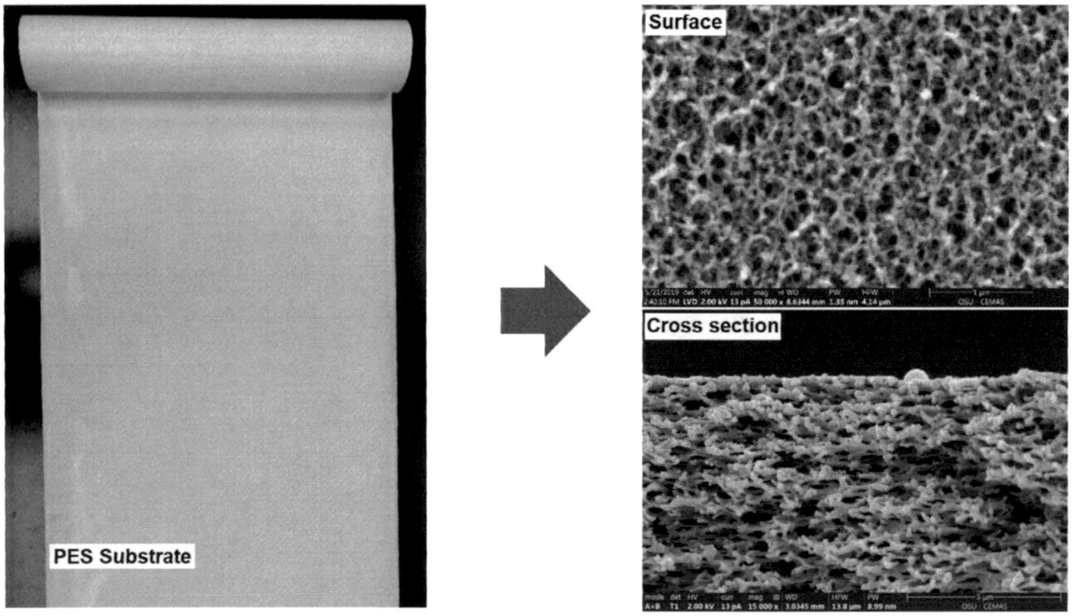

607697 New Carbon and Zeolite Structured Adsorbents Made from Expanded PTFE

Abstract:

The overall objective of this work is to develop novel structured adsorbents from expanded PTFE (e-PTFE) comprising activated carbon, carbon molecular sieve and zeolite adsorbents within the PTFE framework. So far, structured columns have been fabricated from 5A zeolite, Shirasagi molecular sieve carbon (MSC)-3R Type 162 and BPL activated carbon. A nearly identical column has also been fabricated with the corresponding commercial beaded, granular or pelleted counterparts for testing under identical conditions. Details about the fabrication technique will be discussed including converting sorbent-PTFE composite tapes into spiral wound adsorber columns with minimal bypass of the gas mixture input stream. Also, representative results from four different experimental tests will be presented. These experimental tests include single gas adsorption breakthrough curve experiments for CO_2 in N_2, volumetric equilibrium adsorption isotherm experiments for column adsorbent mass determination with N_2, 1-bed pressure swing adsorption (PSA) cycling experiments with CO_2 in N_2, single gas 10 Hz or 100 Hz volumetric frequency response experiments with CO_2, N_2 and possibly other gases, and column axial pressure drop experiments with air.

609404 3D-Printed Activated Carbon with Controlled Porosity for Removal of Toxic Chemicals from Contaminated Water

Abstract:

Removal of toxic chemicals from water is an attractive subject now, as the growing population and effects of climate change have rendered access to potable water more important than ever before.[1] Atrazine (ATZ) and carbamazepine (CBZ) are among the most frequently detected and top 10 contaminates in the water system.[1] In this study, we focused on removing ATZ and CBZ from water sources using a continuous flow system. To implement activated carbon into our continuous flow process, we developed 3D-printed monoliths which were optimized for aqueous stability. 3D printing was selected for monolith production because it allows for fine tuning of the monolith geometry, and physio/chemical/mechanical properties of monoliths.[2–4] To formulate these monoliths, we used bentonite clay as a binder and polyvinyl alcohol (PVA) as a co-binder. The 3D printing technology provides the ability to build activated carbon monoliths with various thicknesses and porosity in a fast and smooth operation. The 3D-prinetd carbon monolith precursors were pyrolyzed in nitrogen environment and then characterized. To study competitive adsorption, both dynamic and steady-state adsorption experiments were conducted by using different ratios of ATZ and CBZ. The effective diffusion coefficient of the 3D-prinetd carbon monolith adsorbates were increased with decease in monoliths internal patterns. The adsorption of both compounds onto powder and 3D-printed monoliths in dilute aqueous solutions (1-10 mg/L) were investigated for 6 h time-on-stream. All samples were analyzed by HPLC/ultraviolet equipped with a C-18 column. From these experiments, it was revealed that the monoliths were able to adsorb 27% of ATZ and 24% of CBZ from the system.

References

[1] M.J. Benotti, R.A. Trenholm, B.J. Vanderford, J.C. Holady, B.D. Stanford, S.A. Snyder, Environ. Sci. Technol. 43 (2009) 597–603.

[2] F. Magzoub, X. Li, J. Al-Darwish, F. Rezaei, A.A. Rownaghi, Appl. Catal. B Environ. 245 (2019) 486–495.

[3] S. Lawson, B. Adebayo, C. Robinson, Q. Al-Naddaf, A.A. Rownaghi, F. Rezaei, , Chem. Eng. Sci. 218 (2020) 115564.

[4] X. Li, F. Rezaei, A.A. Rownaghi, React. Chem. Eng. 3 (2018) 733–746.

597419 Novel Polymeric Fiber Sorbents Embedded with Porous Organic Cages

Abstract:

Focus on the synthesis and functionalization of porous organic cages (POCs) for separations have quickly grown over the last decade. However, to fully realize the potential of solid-phase POCs, they must be effectively packed into ordered structures for wide-scale use and implementation. Processes less destructive than pelletization are preferred and this necessitates alternate processing methods. This work shows the first fiber sorbent embedded with POCs, namely CC3R embedded throughout a cellulose acetate (CA) polymer matrix. CC3R was found to be stable after exposure to spinning solvents as confirmed by NMR, PXRD, and N_2 physisorption. CC3R-CA fibers were spun using the dry-jet wet-quench spinning method. Spun fibers retained the adsorptive properties of CC3R as confirmed by CO_2 and N_2 physisorption, reaching upwards of 60 wt% adsorbent loading. Addition of PVP led to significant leaching so it and other pore formers were excluded from the polymer dope. Fibers show slightly higher than expected CO_2 sorption at room temperature, believed to be due to a change in cage packing driven by non-solvent phase inversion. Fixed bed breakthrough experiments performed on both fibers and powder samples allowed for direct comparison of the sorption performance of the sorbent in different structures. The development of fiber sorbents embedded with POCs and displaying their adsorptive properties will help realize greater use of POCs as sorbents in other separation processes.

595319 Modeling and Optimization of Structured Contactor Adsorption Processes Based on Breakthrough Experiments

Abstract:

A structured adsorbent module for a gas separation process has been proposed with significant potential in terms of both its design diversity and operational flexibility. To characterize the module designs as for a given gas separation process, the pre-existing process model is modified with an additional parameter representing the structure designs. Based on our breakthrough experiments, the parameter, mass, and heat transfer coefficients are estimated in two different designs and varying gas flowrates. The model and parameters from this characterization are employed to search the best gas separation performances in different contactors and operations of a cyclic adsorption process using the structured contactors.

609013 CO_2 Capture Using 3D Printed PEI Adsorbents Supported By Carbon Nanostructures

Abstract:

Processes that utilizes solid adsorbents to capture CO_2 are promising alternatives to state-of-art technologies using water-based amine solutions as absorbents for capturing CO_2 from large point sources such as power plants. Although the energy needs of solid sorbent-based processes are low, the process footprint and consequently the capital cost connected to its implementation can be large due to the relatively long cycle times needed to get the required capture rate and purity of the CO_2 product. To overcome this challenge, processes having structured adsorbents like laminates, monoliths etc. are needed owing to their low pressure drop and better mass transfer characteristics[1]. In recent times, 3D printing of structured adsorbents has been gaining significant attention due the versatility in shapes of the adsorbents and good control over the channel dimensions[2].

Even though there are several published studies in the context of structured adsorbents, they are mostly restricted to characterization of the adsorbents with respect to equilibrium and mass transfer. Very few published studies exist on a process level demonstrating an improvement in the process performance in comparison with pellets[3-4]. It has been shown that, the true potential of any adsorbent only be obtained through rigorous optimization of a cyclic adsorption process. Secondly, most of the published studies in the context of adsorption-based carbon capture have neglected the presence of water in the flue gas. Supported amine sorbents are currently being studied as alternatives to zeolites due their moisture tolerating abilities but like 3D printed adsorbents, their performance in an actual adsorption process is not well understood[5].

Therefore, the aim of this work is evaluating the performance of a 3D printed adsorbent in the context of a vacuum swing adsorption process for CO_2 capture from a wet flue gas. In this work a series of novel adsorbents were prepared combining polyethylene-imine (PEI) with various nanomaterials such as multi-walled carbon nanotubes (CNT) in different ratios (Figure 1).These adsorbents were then printed in the form of monoliths by the 3D printing technique called robocasting. Breakthrough experiments were then carried out with a dry mixture of CO_2 and N_2 and a humid mixture containing 5% H_2O. Pressure drop tests were also carried out estimate the pressure drop across the 3D printed structure. Further, the breakthrough experiments were also used to estimate the adsorption kinetics. The information from the adsorbent characterization was then used to simulate and optimize a vacuum swing adsorption process in order to identify the operating conditions corresponding to 95% CO_2 purity and 90% CO_2 capture rate targets to capture CO_2 from a coal fired power plant containing 15% CO_2, 5% H_2O in N_2.

Keywords: Structured adsorbents, 3D printing, Vacuum Swing adsorption

References

1. Rezaei et al., *Eng. Chem. Res.* **2010,** *49,* 4832–4841
2. Couck et al., *Micropor Mesopor Mater 255 (2018) 185-191*
3. Nina Mohammadi, PhD thesis, 2017

4. Sharma et al., *Eng. Chem. Res.* **2020,** https://doi.org/10.1021/acs.iecr.9b05337
5. Gelles et al., *Adsorption* **2020,** 26:5–50

600726 Heat Intensification of TSA Processes with Packed Metal Foams: An Experimental and Modeling Study Applied to Post-Combustion CO_2 Capture

Abstract:

When performing a gas separation by Temperature Swing Adsorption (TSA), two key factors to increase process efficiency are the acceleration of the temperature swing kinetics, and the reduction of thermal gradients along the reactor cross section. To both aims, it is necessary to overcome the intrinsic heat transfer limitations associated with the operation of a packed bed, both in static and in flow conditions. Following an approach already suggested in the context of catalytic processes [1], it is possible to enhance heat transfer by introducing highly conductive open-cell metal foams filled with sorbent beads. Namely, the faster heat conduction through the metal adds onto the conventional heat conduction of the bed and the convection of the gas phase. This speeds up the reactor response to a change in temperature of the heating/cooling medium, and flattens temperature profiles transversally to the flow direction.

In the proposed configuration, the metal foam is placed in the reactor first. The void cells of the foam are then filled with the sorbent pellets. Keeping the average diameter of the particles less than half of the average diameters of the foam cells, filling can be simply performed by gravity. According to experimental measurements, this configuration also allows to house solid inventories (per unit reactor volume) much larger than those obtained with typical washcoated structured supports, and almost comparable to the ones of packed bed reactors. With this contribution, we propose the – to the best of our knowledge – first application of the packed metal foams as an alternative to the standard fixed bed configuration in adsorption based separations.

In particular, in this study, a post-combustion CO_2 capture process application has been investigated. In the context of CO_2 capture from flue gas, TSA processes are a promising alternative to the more established amine-wash[2]. Differently from absorption-based separations, a TSA cycle is not operated as a steady state process, but rather as a sequence of steps undergone by the gas-solid contactors. The simplest possible cycle configuration consists of: 1. an adsorption step during which the flue gas is fed, performed at low temperature to maximize CO_2 intake; 2. a desorption step, i.e. a heating ramp with the consequent release of the adsorbate, where the high-purity CO_2 stream is produced; 3 a final cooling step, which brings the system back to its initial state. The cooling step is performed in static conditions (i.e., both column ends are closed), whereas the other steps are typically performed at low velocities, well under the fluidization threshold, to contain the pressure decay.

In order to observe temperature gradients in both the axial and the radial direction within the reactor, a jacketed large cross-section (80 mm diameter and 200 mm length) cylindrical adsorption column has been adopted as experimental setup. It has been filled packing spherical zeolite 13X pellets (average diameter 1.8 mm) into an aluminum foam with a void fraction of 0.95 and a cell size of 5 mm. For this arrangement a total sorbent volumetric fraction of 0.50 has been measured, which represents a 25% reduction of the solid inventory compared to a standard packed column (at fixed reactor volume). A first series of steady-state and dynamic heat transfer tests has been completed by flowing inert gas (Argon) at different velocities through the column, and analyzing axial and radial temperature profiles across the packed foam. A detailed mathematical model for the description of the system dynamics in two spatial dimensions (axial and radial) has been validated by means of the experimental results.

Moreover, a second series of experiments has reproduced the operation of a basic TSA cycle (adsorption, heated regeneration, closed cooling) for the separation of a cry binary 12/88%v. CO_2/N_2 mixture. The cycle have been performed both in a packed bed and in a packed foam configuration. Pressure and temperature profiles within the column, as well as the composition of the outlet gas flow, have been monitored. The aforementioned reactor model has been extended to account for adsorption, also in view of its implementation into a process simulation tool, which could enable a systematic optimization of the process design. The final model has also been validated by means of the experimental results, for both packed bed and packed foam configurations.

Thus, in conclusion, experimental and modeling results demonstrate that heat transfer rates are increased by up to a factor two, as conjectured at the beginning of the study. This translates into a flatter temperature distribution along the reactor radius, and a faster transition toward the steady state when switching between heated and cooled steps.

As regards the specific application for CO_2 capture from flue gas, simulations have allowed, on the one hand, to quantify the productivity benefits obtained by cutting off the duration of the regeneration and cooling steps by a factor of three, and, on the other hand, to assess the disadvantages related to the lower packing densities, which limit the total adsorption capacity per unit volume.

References

[1] Visconti C.G., Groppi G. and Tronconi E., *Catal. Today.*, 2016, 273,178-186

[2] Joss L., Gazzani M., and Mazzotti M., *Chem. Eng. Sci.*, 2017, 158, 381-394

Acknowledgements

The authors thank Casale SA for the financial and technical support granted to the project.

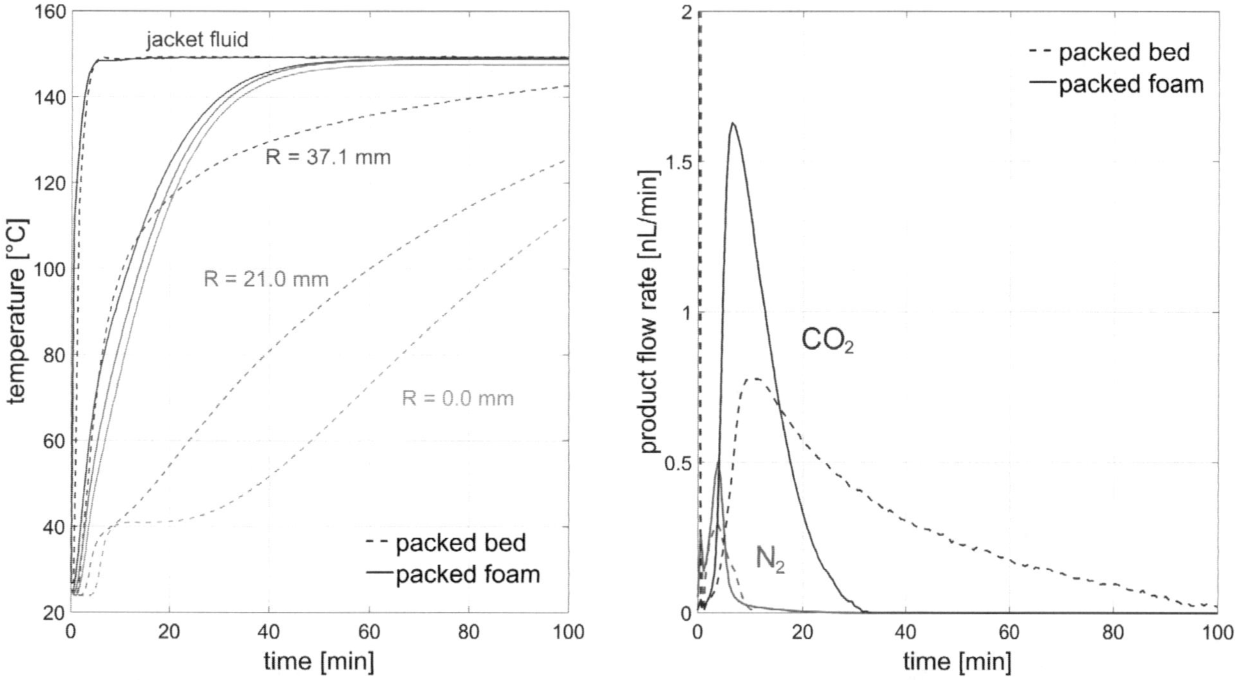

Figure 1: Experimental profiles in time for temperatures (at different radial positions) and outlet flow rates. Results for packed bed (dashed) and packed foam (continue) configurations during the regeneration step of a TSA cycle are shown. A hot fluid flowing in the jacket indirectly heats the reactor. The advantage of speeding up the transitory phases during the adsorption cycle is evident from the sharper CO2 flow peak at column outlet during the bed regeneration.

599297 Azeotropic Fluorocarbon Refrigerant Mixture Separations Using Porous Material

Abstract:

Fluorocarbon refrigerants have been evolving over the past three decades for the purpose of environmental responsibility, performance, and safety. The most widely used fluorocarbon refrigerants today are the hydrofluorocarbon refrigerants (HFCs). Many of these refrigerants are HFC mixtures and are azeotropic in nature. Although HFCs have a zero ozone depletion potential (ODP), some have relatively high global warming potentials (GWP). As a result, regulations such as the Kigali Agreement propose to regulate the future use of HFC refrigerants in certain applications. The refrigerant industry is therefore currently moving toward the production and marketing of hydrofluoroolefin (HFO) refrigerants and HFO/HFC refrigerant blends due to their lower GWPs.

Project EARTH (Environmentally Applied Research Towards Hydrofluorocarbons) focuses on the safe and sustainable recycling of azeotropic HFC refrigerant mixtures. Rather than venting or incinerating the millions of kilograms of these high GWP refrigerants currently in circulation, a more sustainable option would be to recycle them. In order to do this, however, these HFC mixtures must first be separated into their constituent species, a process made difficult due to their azeotropic nature. Project EARTH will investigate the separation of azeotropic HFCs beginning with R-410a, a near-azeotropic refrigerant mixture composed of 50 wt.% R-32 (CH_2F_2) and 50 wt.% R-125 (CHF_2CF_3). While R-410a itself has a high GWP (2088, relative to $CO_2 = 1.0$), R-32 has a more acceptable GWP of 675 and has the potential to be recycled and reused in refrigerant mixtures with HFOs. The GWP of R-125 (3500) is relatively high and must be separated from R-32 and in the future, R-125 can serve as a feedstock for producing lower GWP products. The separation of R-410a is an important and critically needed separation process that will have a significant impact on the environment and the heating, ventilation, air-conditioning, and refrigeration (HVACR) industry. This presentation will focus on the use of porous materials such as zeolites for the separation of R-32 and R-125.

607915 Nitrogen Rejection from Natural Gas Streams Using a Nitrogen Selective Metal Organic Framework

Abstract:

The adsorptive separation of CH_4 and N_2 is generally performed through a kinetic separation using a titanosilicate or carbon molecular sieve. However, a metal-organic organic framework (MOF) called $V_2Cl_{2.8}$(btdd) was recently discovered to separate CH_4 and N_2 through an equilibrium separation, with CH_4 as the light product [1]. To determine feasibility of the vanadium MOF for nitrogen removal, a series of simulated process studies were performed. Two cycles were considered: the Skarstrom cycle and a 6-step cycle with pressure equalization and light product pressurization steps. First, the Skarstrom cycle was considered with a wide range of operating conditions. Three inlet compositions (55/45, 80/20 and 92/8 mol% CH_4/N_2), three process temperatures (30, 40 and 50°C), three particle densities (600.0, 775.0 and 961.7 kg/m^3) and two high pressures (100 and 500 kPa) were considered. These conditions were chosen to span the range of possible landfill gas or natural gas well process conditions. The inlet conditions were used to generate at least 1000 samples for all possible combinations of these variables. These samples were trained using Gaussian process regression to perform process purity/recovery and energy/productivity optimizations. Artificial neural networks were used to predict axial bed profiles and to validate the machine learning predictions. From these simulations, it was determined that the material works best in vacuum swing adsorption cycles with the high pressure at 100 kPa. Due to the high isosteric heat of adsorption for N_2 on $V_2Cl_{2.8}$(btdd), the process performed better at higher temperatures. At 50°C, the best separations were achieved. The particle density did not seem to change the purity-recovery optimizations but did affect the energy-productivity optimizations. As the particle density decreased, the productivity also decreased. Most of these process conditions could yield the required CH_4 purity of 96 mol%, but often the recovery was poor. The maximum recovery at a purity of 96 mol% was about 80%. After the Skarstrom cycles process performance was determined, a more complicated 6-step cycle with pressure equalization and light product pressurization was used to improve the process recovery.

611574 Adsorption Separation of Methane and Carbon Dioxide for the Production of Renewable Natural Gas Using Activated Carbons

Abstract:

Activated carbons are a widely used in industry for different applications due to their low cost and low energy requirements. This study investigated commercially available activated carbons and carbon aerogel for CO_2 removal from biogas using adsorption, for the production of renewable natural gas. Pure component isotherms for methane were determined gravimetrically for temperatures of 10 to 90°C for pressures up to 6.5 atm. We also investigated the effect of some characteristics of the adsorbents such as pore size, surface, % oxidation, and ash content on this separation. It was observed that % oxidation had little effect on CH_4 adsorption, while ash content had a negative effect on CH_4 adsorption with these adsorbents. In the high-pressure region, pore size and surface area dominated the adsorption, whereas in the low pressure region, the heterogeneous nature of the carbon materials dominated adsorption, which has previously been observed for CO_2 adsorption with similar materials. Using previously obtained pure component CO_2 isotherms by the same materials, the Ideal Adsorbed Solution Theory (IAST) was used to predict binary isotherms for a CH_4-CO_2 mixture, and the Expected Working Capacity values (EWC) were calculated for different pressure swing adsorption applications. The most practical adsorbent used in this study for CO_2 separation from biogas was found to be OLC carbon.

603760 Pvsa and Tvsa Adsorbent Screening Tools for Post-Combustion and Industrial Carbon Capture

Abstract:

Metal-organic frameworks (MOFs) have received significant attention by the research community for CO_2 capture. However, there has been little focus on the industrial application of these materials in cyclic processes. This is unsurprising given that MOFs generally have low synthesis yields and comparatively higher costs, which can hamper technology development cycles leading to delayed commercialisation.

In this work, we aim to identify the MOF properties and process conditions that yield better performance for post-combustion and industrial carbon capture. The identification of these could better direct research efforts, leading to more efficient development process. To achieve this, we first developed a simplified pressure-vacuum swing adsorption (PVSA) model with integrated process economics [1]. The PVSA model is based on a one bed, three-step equilibrium cycle which reflects the non-isothermal nature of adsorption while still allowing rapid solutions [2]. While temperature swing adsorption (TSA) would traditionally have been disregarded for this style of application due to prohibitively long cycle times, rapid TSA technologies are gaining popularity and they may enable the deployment of TSA to these very large-scale applications. Therefore, an analogous study was carried out for TSA for which a simplified equilibrium temperature-vacuum swing adsorption (TVSA) model was developed. The separation performance output from these are then used to size and cost the process equipment required to process flue gas of a given flow rate. In addition to the traditional metrics of purity, recovery, working capacity, and specific energy, the cost of capture is a significant metric for technology selection and deployment and allows the identification of trade-offs not otherwise seen.

We surveyed the literature for MOFs with measured isotherms of CO_2 and N_2 at, at least, three temperatures. We estimated other required inputs such as density, porosity, and heat capacity by their respective methods. While the adsorption cycle used may not reflect the greatest attainable performance for each MOF, it does allow for rapid evaluation of the adsorbent-process ensemble.

We exploited this feature to determine the ideal MOF isotherms for a set of post-combustion and industrial flue gases. For the PVSA case, we found that improving selectivity by reducing N_2 adsorption, and improving CO_2 working capacity by having moderate enthalpies of adsorption yield the greatest process improvements both technically and economically. For TSA, there are a much wider range of isotherms which yield good performance. This is due to the fact that adsorbents which perform well for PVSA are those that do not show significant thermal effects, whereas for TSA, thermal effects are desired to take advantage of the temperature driving force.

The capture costs are dominated by process capital costs which are primarily comprised of vacuum requirements and adsorbent inventory due to low working capacities.

MOFs that displayed poor performance for PVSA such as Ni- and Mg-MOF-74, and HKUST-1, showed significantly improved separation performance in TSA due to their higher enthalpies of adsorption. Furthermore, by applying modest vacuum (30 – 50 kPa_a) the total energy consumed by the TVSA process was reduced compared to TSA while still maintaining product purity. The process economics for the TSA/TVSA case are currently in progress, however, initial results show that higher working capacities can be obtained which would lead to lower adsorbent inventories. Whether this increased working capacity can offset longer cycle times is to be determined.

The performance of published MOFs were evaluated for post-combustion and industrial carbon capture applications in PVSA, TSA, and TVSA processes. Ideal MOF isotherms were proposed for a selection of carbon capture scenarios, and areas to focus adsorbent development on were proposed considering both separation performance and process cost.

[1] – D. Danaci *et al.*, *Mol. Syst. Des. Eng.*, 2020, Advance Article

[2] – B.J. Maring, P.A. Webley, *Int. J. Greenh. Gas. Con.*, **15**, 16-31, 2013

607124 Modeling and Techno-Economic Optimization of a Rotary Packed Bed for Post-Combustion CO_2 Capture Using a Diamine-Appended Metal-Organic Framework

Abstract:

The amount of carbon dioxide in the atmosphere is currently rising globally by around six billion tons per year[1]. Thus, it is essential to separate and recover carbon dioxide from the flue gases released by power plants to avoid surplus CO_2 emissions. Carbon capture and sequestration (CCS) technologies will continue to play a vital role in CO_2 capture until substantial changes are made to the energy infrastructure[2]. Presently, carbon capture by amine absorption is the most widely used technology. Still, solvent regeneration is energy-intensive, and other significant drawbacks include solvent degradation and corrosion. Therefore, alternatives are being developed. The use of solid sorbents for post-combustion capture is considered one of the potential options, and design of the solids contactor plays a critical role for these sorbents to be economically viable. Presently, extensive research has been focused on fixed bed, moving bed, or fluidized-bed reactors. However, fixed beds exhibit difficulty processing large volumes of flue gas, and moving beds and fluidized beds suffer from particle attrition[3]. To that end, a rotary packed bed concept for carbon capture with integrated cooling and heating in between layers is designed and analyzed to eliminate the challenges of traditional contactor designs.

Metal-organic frameworks (MOF's) have been recognized as promising candidates for CO_2 capture due to desirable characteristics such as high pore volume, surface area, and remarkable tunability[4]. This work utilizes a novel class of diamine-appended MOF's that have strong potential in reducing the cost of carbon capture[5,6]. In this study, we model these diamine-appended MOF's to be impregnated on the surfaces provided by a rotary packed bed reactor. As the bed continuously rotates, CO_2 is adsorbed by exposing a portion of the bed to flue gas. In the regeneration portion of the bed, steam is used to regenerate the MOF and recover CO_2. Cooling water is used to decrease the temperature rise in the adsorption section, while regeneration occurs by direct injection of steam and indirect heating by an embedded heat exchanger.

The developed RPB model is a dynamic, pressure-driven, two-dimensional model. First-principles mass and energy balances coupled with internal and external mass transfer resistance and reaction kinetics are solved using a moving reference frame (MRF) to incorporate the effect of the rotational speed of the matrix and variation in the circumferential direction. Due to high heat of reaction for this specific MOF and due to the step-shaped isotherms coupled with a highly non-linear kinetics, there is considerable variation in the temperature and loading in the circumferential direction along with the axial direction motivating the development of the two-dimensional model. An increase in the rotation speed can improve the solids loading in the capture section for a given inlet lean solid loading. However, an increase in the rotation speed results in an increase in the lean solids loading from the regeneration section due to the lower residence time. Furthermore, as the rotation speed increases, the electric power requirement keeps increasing. The tradeoffs discussed above suggest that there is an optimum rotational speed, but would require to take into account both the capital and operating costs. Therefore, a techno-economic analysis is performed for obtaining the optimal operating conditions including the speed by minimizing the equivalent annual operating cost. Compared to the conventional MEA-based capture system, the optimal RPB using the diamine-appended MOF shows considerable decrease in the penalty of CO_2 capture.

References:

[1] Zhao, Z., et al. (2007). "Adsorption of carbon dioxide on alkali-modified zeolite 13X adsorbents." International Journal of Greenhouse Gas Control **1**(3): 355-359.

[2] Sumida, K., et al. (2012). "Carbon dioxide capture in metal–organic frameworks." Chemical reviews **112**(2): 724-781.

[3] Thiels, M., et al. (2016). "Modelling and design of carbon dioxide absorption in rotating packed bed and packed column." IFAC-PapersOnLine **49**(7): 895

[4] D'Alessandro, D. M., et al. (2010). "Carbon dioxide capture: prospects for new materials." Angewandte Chemie International Edition **49**(35): 6058-6082.

[5] McDonald, T. M., et al. (2015). "Cooperative insertion of CO 2 in diamine-appended metal-organic frameworks." Nature **519**(7543): 303.

[6] Milner, P. J., et al. (2017). "A diaminopropane-appended metal–organic framework enabling efficient CO2 capture from coal flue gas via a mixed adsorption mechanism." Journal of the American Chemical Society **139**(38): 13541-13553.

605431 An Equilibrium Adsorption Model to Estimate Gas-in-Place and CO_2 Storage Capacity of Shale Reservoirs

Abstract:

Advances in extraction technologies have propelled shale gas production; however, these processes can be quite inefficient, with about 90% of the original Gas-in-Place (GIP) remaining in the reservoir [1]. Fluid behaviour in these formations is not well understood, and the effect of confinement in the nanometric pores of shale presents additional complexity. The deployment of enhanced recovery using CO_2 injection can potentially improve production rates as well as lead to the permanent sequestration of CO_2 [2].

In shale, natural gas (composed mainly of methane and ethane) can be adsorbed in the pores of the organic matter and clay minerals and be held freely in larger pores and fractures [3]. The quantification of gas adsorption in these reservoirs is essential, as it dictates both GIP estimates and CO_2 storage potential of shale. Classical approaches for calculating the GIP are flawed, because they do not account for the reduction of the available pore space upon gas adsorption. In this work, we propose a method for estimating the GIP that utilises the so-called excess amount adsorbed, the sole quantity that can be directly accessed experimentally. As such, the method does not necessitate any assumptions of density or volume of the adsorbed phase. We further extend this formalism to evaluate the effect of CO_2 injection on recovery and storage in shale reservoirs.

The material balance equations using excess adsorption are presented for a typical shale reservoir and are applied to the Marcellus reservoir in the USA, which is considered as the base case. We first investigate primary recovery and obtain the GIP as a function of reservoir pressure (or P/Z, the ratio of reservoir pressure and gas compressibility). We also examine the effect of adsorption behaviour on the GIP and reservoir production by considering four different adsorption isotherms: Langmuir, Henry's Law, BET and Anti-Langmuir. We then use a cyclic production/injection process to investigate the effect of CO_2 injection on recovery. Our model consists of a Production stage, whereby the reservoir is depressurised to a certain production pressure, followed by an Injection stage, where a specific volume of CO_2 is injected into the reservoir. As mass transfer limitations are inherent in shale reservoirs, both the Production and Injection stages are not at equilibrium, i.e. there is no change in the adsorbed phase, and gas is added or removed only from the free gas phase. At the Soak stage, the reservoir is finally allowed to equilibrate, and gas is exchanged between the bulk phase and adsorbed phase. The cycle then continues, as the reservoir produces again. Competitive adsorption is incorporated in this model through the use of the binary Langmuir isotherm.

We observe that the unique features of the supercritical adsorption isotherms are also clearly evident in the GIP curves and that adsorption plays a dominant role in gas storage in shales. We also show that the conventional approach that does not use the excess amount adsorbed can lead to a significant overestimation of the GIP. Injection of CO_2 yields up to 70% more CH_4 as compared to primary recovery, and about 50% of the CO_2 remains permanently sequestered in the reservoir. We also compare this to a N_2 sweeping scenario, where the same volume of N_2 is injected instead of CO_2. We observe that although CH_4 recovery is similar to the CO_2 injection scenario, higher reservoir pressures are required to sustain this recovery. Although gas injection in general leads to more CH_4

from the reservoir, the main benefit of using CO_2 is its storage. We also test the sensitivity of the model to various factors such as reservoir pressure, porosity and the adsorption model parameters. The equilibrium adsorption model is a useful tool as it can provide a quick means of testing injection scenarios, evaluate production methodologies and produce accurate estimates of the GIP and CO_2 storage potential of shale reservoirs.

References

1. Jarvie, R. Hill, T. Ruble, R. Pollastro, AAPG Bulletin, 91 (2007) p.475-499
2. R Edwards, M. Celia, K. Bandilla, F. Doster, C. Kanno, Environmental Science and Technology, 49 (2015) p.9222-9229
3. Yu, K. Sepehrnoori, T. Patzek, SPE Journal, 21 (2016) p.589-600

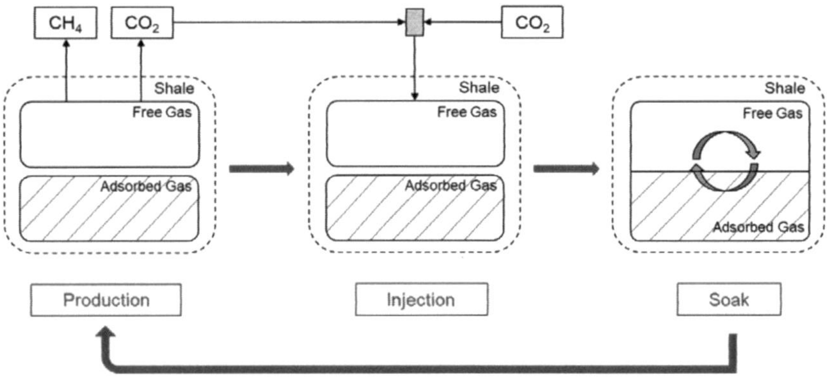

599752 Improved Synthesis of Sterically Hindered Polyvinylamine and Its Application in Facilitated Transport Membranes for CO_2 Capture from Flue Gas

Abstract:

Membrane separation technology requires highly CO_2-selective and permeable membranes for a cost-effective post-combustion CO_2 capture from flue gas. Amine-containing facilitated transport membranes are able to realize an increased permeability due to the reversible reaction between CO_2 and amino groups. Here, we have developed an improved method for the synthesis of a high-molecular-weight sterically hindered polyvinylamine (SH-PVAm) as the new fixed-site carrier in our ultra-thin CO_2 selective membranes.

Commercial polyvinylamine (PVAm) was mono-methylated into SH-PVAm using the stepwise reductive amination with a highly polar fluorinated alcohol as solvent to enhance the equilibrium shift to the imine formation. The method was found to prevent the over-alkylation, resulting in the increased yield of the target product. The synthesized SH-PVAm was characterized and confirmed by FTIR and ^1H NMR. The SH-PVAm solution retained a sufficiently high viscosity after incorporating the aminoacid salt, 2-(1-piperazinyl)ethylamine sarcosinate, as the mobile carrier for the membrane coating without any penetration issue. The SH-PVAm exhibited a significantly enhanced performance in the blended membrane of about 170 nm thickness over the unmodified PVAm for CO_2/N_2 separation at 57°C. The former membrane demonstrated a CO_2 permeance of 1071 GPU and a CO_2/N_2 selectivity of 183 whereas the latter showed 975 GPU CO_2 permeance and 163 selectivity. The improved membrane performance corresponded well with our computational results on the reaction chemistry of CO_2 between the sterically hindered vinylamine monomer (SH-VAm) and the pristine vinylamine monomer (VAm) by density functional theory (DFT). SH-VAm showed a preference for the more efficient bicarbonate formation pathway and contributed to a higher CO_2 loading capacity. Therefore, the steric hindrance effect of SH-PVAm enhanced the solubility of CO_2 in the polymer matrix and resulted in the higher CO_2 permeance.

600187 Polyoxymethylene-Based Amorphous Polymers with Superior CO_2/N_2 Separation Performance

Abstract:

Poly(ethylene oxide) (PEO) containing polymers are leading membrane materials for CO_2/N_2 separation with a balanced high CO_2 permeability and CO_2/N_2 selectivity because of the affinity of ether oxygen towards CO_2. Herein, we design polymers based on polyoxymethylene (POM) with higher ether oxygen content than PEO and thus more superior separation properties. The challenge is to disrupt the crystallization of these highly polar polymers and achieve high gas permeability. We first synthesize macromonomers of polyoxymethylene acrylate (POMA) by ring-opening polymerization of 1,3,5-trioxane (TX) following by capping with acetic anhydride. The POMA with short chain length can then be polymerized to obtain amorphous polymers (polyPOMA). The macromonomers were characterized using NMR, FTIR, and GPC for the chemical structure and chain length, and the polymers were characterized using DSC and XRD to confirm the amorphous nature. Pure- and mixed-gas CO_2/N_2 separation properties were determined, and the effect of POMA chain length on the gas separation performance was investigated. When the number of the repeating units of POM segments is 9, the membranes show the best performance with CO_2 permeability of ~200 Barrer and CO_2/N_2 selectivity of ~70 at 35 °C, above the 2008 Robeson's upper bound. When tested with simulated flue gas containing 15% CO_2 and 85% N_2, the POM-based membranes exhibit stable CO_2/N_2 separation performance in the presence of water vapor at temperatures up to 60 °C. We will also elucidate the structure/property relationship in this series of polymers with exciting separation performance.

598491 Tailoring Dual Anionic-Cationic Poly(ionic liquid) Composite Membranes for Advanced CO_2 Separation

Abstract:

Herein, we propose a strategy of developing novel dual anionic-cationic crosslinked poly(IL)-IL composite membranes via a photopolymerization method for enhanced CO_2 separations. Interestingly, these are the first examples of dually photopolymerized anionic-cationic poly(IL)-IL composite systems, in which the backbones of poly(IL)s feature both pendant anions and cations without any mobile counterions unlike poly(IL) reported so far in the literature. A new type of dual photopolymerizable anionic-cationic IL (DIL) monomer having methacrylate functional group tethered with highly delocalized sulfonimide anion ($–SO_2–N^{(-)}–SO_2–C_7H_7$) and a vinylimidazolium counterion ($[C_4vim]^+$) was successfully synthesized and photopolymerized with four distinct amounts of free IL with a structurally analogous cation ($[C_4mim][Tf_2N]$) and 20 wt% PEGDA crosslinker, offering novel composite matrices. All of the newly developed poly(IL)-IL composite membranes exhibited superior CO_2/CH_4 and CO_2/N_2 selectivities together with moderate CO_2/H_2 selectivity and reasonable CO_2 permeabilities. A "bottom-up" approach of our novel poly(IL) systems on their structure-property relationships as well as gas separation properties will be discussed in this presentation.

610478 Reverse-Selective PTMSP Supported Ionic Liquid Membranes for Light Hydrocarbon Recovery

Abstract:

Rich in hydrocarbons other than methane, growing unconventional gas production motivates new separations challenges for the growing supply of natural gas liquids (NGLs). Membranes are advantageous for pre-fractionating NGLs from methane in small-scale, decentralized shale operations. In this work we report a reverse-selective supported ionic liquid membrane (SILM) platform based on varying compositions of the polyacetylene PTMSP and the ionic liquid [hmim][Tf_2N]. The thermal stability, mechanical properties, and ionic liquid dispersion in these membranes is characterized in addition to their permeation properties. Gas permeability in SILMs is modest compared to those in the neat PTMSP. Ethane and propane permeability in SILMs increase nonlinearly with higher transmembrane pressure, which translates to highly pressure dependent pure-gas C_2H_6/CH_4 and C_3H_8/CH_4 selectivities. It is hypothesized that the ionic liquid first occupies the accessible free volume in the PTMSP support, thus exacerbating swelling due to highly condensable NGLs. Ionic liquid and NGL-induced swelling behavior is elucidated via density and ellipsometry measurements.

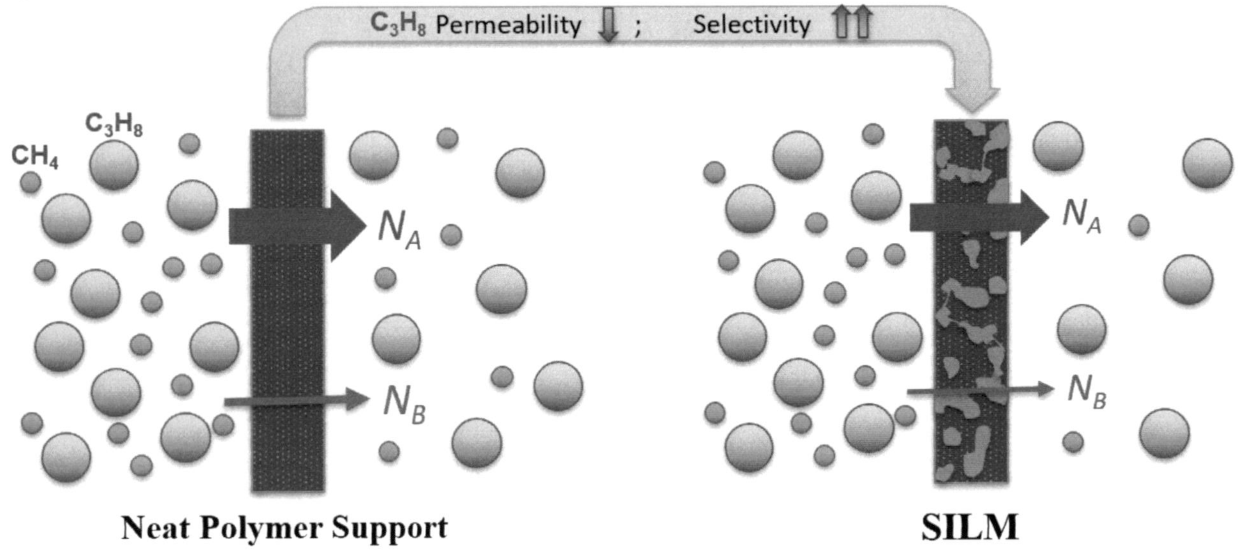

597465 Advancing Material Design Towards Enhanced Sour Gas Separations

Abstract:

The bulk removal of hydrogen sulfide (H_2S) from natural gas remains an operationally energy and cost intensive separation challenge. Polymeric membranes offer a compact and passive pretreatment step to reduce the burden of traditional amine absorption separation techniques. However, a lack of fundamental research on membrane materials for sour gas separations limits the industrial viability of such hybrid separation systems. Recent work at Aramco's sour gas testing facility has investigated novel membrane materials across 10 polymer families, tailored for harsh H_2S separations. This talk explores the fundamental trends observed in material transport properties and delves deeper into three particular regions of interest within the sour gas transport performance map. Specifically, simultaneously high H_2S/CH_4 and CO_2/CH_4 selectivities are achieved by novel poly(ethylene glycol)- and polyimide-derived polymers, ultrahigh H_2S/CH_4 selectivities are achieved by modified Pebax and PEG hydrogels, and high permeabilities and stability are achieved by alkoxysilanated polynorbornenes. A common strategy for improving sour gas separation performance across all studied polymer classes involves improving the polymer affinity for H_2S by incorporating polar moieties such as alkoxysilyl, amine, or ether content. Themes of plasticization, solubility versus diffusion controlled separations, and tradeoffs in transport properties are discussed throughout.

596349 Effect of Thermal Treatment on the Structure and Gas Transport Properties of a Triptycene-Based Polybenzoxazole Exhibiting Configurational Free Volume

Abstract:

Polymer membranes exhibiting superior stability at temperatures higher than 200°C would improve the thermal and economic efficiency of hydrogen separation from pre-combustion syngas in the Integrated Gasification Combustion Cycle (IGGC), as well as of H_2/CH_4 separation from refinery processes.

The effect of thermal treatment at 220°C for 10 days on the structure and transport properties of 30mm thick films of a triptycene-based polybenzoxazole (TPBO) was investigated experimentally and theoretically. In sharp contrast with the behavior of other rigid glassy polymers, such as PIMs and Matrimid polyimide, which lose most of their permeability upon exposure to high temperatures for a few hours, gas and water vapor diffusion and permeability coefficients in TPBO decrease by just 20%, while sorption is virtually unaffected by thermal treatment. Remarkably, the CO_2/CH_4 selectivity exhibit a negligible change. Fluorescence spectroscopy, WAXD and FTIR analysis indicate that, in contrast with typical behavior of glassy polymers, TPBO does not experience accelerated physical aging, and rule out formation of intermolecular charge transfer complexes upon thermal treatment. According with this physical picture, the diffusion coefficient of penetrant molecules sorbed in the Langmuir's mode, D_H, does not change after treatment. Small molecule diffusivity and permeability decline is caused by a decrease in polymer chain mobility, which makes more difficult opening gaps to allow penetrant diffusion jumps. According to this picture, the Henry's mode diffusion coefficient, D_D, substantially decreases upon thermal treatment. The higher stability exhibited by TPBO relative to other high Tg glassy polymers is ascribed to the presence of configurational free volume, which is not related to the non-equilibrium transient conformation, but to the molecular configuration and, as such, it is not relaxed upon protracted exposure to high temperatures.

601139 Influence of Aliphatic and Aromatic Fluorine Groups on the Gas Permeability and Morphology of Fluorinated Polyimides

Abstract:

Partially fluorinated polymers show exceptional separation performance for a large variety of gas pairs, in part due to inefficient packing caused by the comparatively large size of fluorine versus hydrogen atoms. While the incorporation of aliphatic fluorine groups in the polymer backbone is relatively commonplace (e.g., CF3 groups as a bulky side group), there are very few studies on the impact of aromatic fluorine groups on gas transport. Here, we present an analysis of two pairs of structurally-analogous, partially fluorinated polyimides – one pair altered via aliphatic fluorination, and the other via aromatic fluorination. Challenges related to the synthesis of highly fluorinated polyimides are discussed and the effects of differing fluorine group connectivity on solid-state morphology are related to observed changes in permeability and permselectivity, diffusion and diffusion selectivity, and their energetics. Developing an understanding of structure-property relationships related to fluorination can allow for improved rational design of high-performance polymers in the future.

CHARACTERIZATION OF FACTORS AFFECTING THE GAS WELL DELIVERABILITY OF THE JIAOSHIBA SHALE GAS FIELD BY MULTIVARIATE STATISTICAL ANALYSIS

Yizhong Zhang[1,2], Maolin Zhang[1], Bin Ju[3], Jincai Shen[4], Lan Ge[4], Fanhua Zeng[2]*

**Corresponding author. Email address: Fanhua.Zeng@uregina.ca.*

[1] Hubei Cooperative Innovation Center of Unconventional Oil and Gas, Yangtze University, Hubei, Wuhan, China, 430100

[2] University of Regina, Regina, SK, Canada, S4S 0A2

[3] School of Oil and Natural Gas Engineering, Southwest Petroleum University, Chengdu, Sichuan, China, 610500

[4] SINOPEC Chongqing Fuling Shale Gas Exploration and Development Co., Ltd., Chongqing, China, 408014

Abstract

The Longmaxi shale gas in the Jiaoshiba Shale Gas Field is the first and the largest successfully developed shale gas in China. The analysis of key factors of deliverability of gas wells plays a significant role in efficiently developing shale gas reservoirs. However, there are few methods analyzing the key factors of deliverability of shale gas wells by considering both geological characters and engineering parameters. In this study, the reservoir characteristics, drilling and completion data, fracturing data (including the well logging interpretation of horizontal section, the track and layer section that the well bore passed through, and fracturing parameters) and production data of 86 gas wells in the Wufeng-Longmaxi shale, Jiaoshiba Shale Gas Field in Eastern Sichuan Basin, China are statistically analyzed and ordered by multivariate statistical analysis. The gas well deliverability is described by absolute open flow potential and 9 factors (physical parameters of reservoir, including density of rock, porosity, total organic carbon, brittle mineral content of rock and drilling and fracturing engineering quality, including horizontal length of gas well, horizontal length ratio through different stages in the formation, number of fracturing stages, liquid volume, and sand amount) that may affect the deliverability of shale gas wells are considered. The results reveal that the absolute open flow potential has a close relationship with porosity, total organic carbon, brittle mineral content and total liquid volume. The multivariable nonlinear regression models of absolute open flow potential with different key factors considered are built, and the prediction results are reliable by goodness-of-fit test acceptable with real gas well deliverability testing data. The research methods and results in this study can provide a reference basis for reasonable production rate allocation, design, and adjustment of development plan for the shale gas reservoirs.

Keywords: Gas well deliverability, Longmaxi shale gas, Jiaoshiba Shale Gas Field, Multivariate statistical analysis, Multivariable nonlinear regression

1. Introduction

Following the increased demand of petroleum energy and the developed science and technology, the industry nowadays has focused on the unconventional resources. Shale gas reservoirs are one of the most popular kind of unconventional resources with extremely low permeability, tight porous structure, large specific area and high heterogeneity, which have attracted attentions of researchers worldwide [1-3]. Shale gas has self-generated, self-contained, organic-rich, nanoscale pore structure, and other unusual characteristics and it could be considered as a source as well as a reservoir [4-5]. In addition to free gas and small amount of dissolved gas, adsorbed gas is largely existed in the tight formation, which makes it difficult to produce owing to the complexity of the reservoir [6-8]. The Sichuan-Chongqing region in southwest China is one of the fastest shale gas production growing regions in the world. The annual shale gas deliverability of China has reached 2×10^{10} m^3 with the annual shale gas production of 1.5×10^{10} m^3 [9]. As one of the main contributors of natural gas production growth in China, the gas production of Jiaoshiba shale gas reservoir has exceeded the daily production of 1.67×10^{7} m^3/d since 2013 [10]. The large production of commercial shale gas flow of the Longmaxi Formation (e.g. more than 2×10^{5} m^3/d of JY1 well) has made the Jiaoshiba area one of the most important shale gas field in China.

Compared with typical shale gas reservoirs in North America, the shale gas in Sichuan Basin has lower TOC value, lower porosity, lower gas saturation, smaller thickness and higher buried depth, which makes its exploration more complicated [9]. In both the early test stage and current production stage, there are large differences among deliverability of the shale gas wells in the Jiaoshiba area. Therefore, the deliverability evaluation is the basis of reasonable shale gas production and prediction, and the research on deliverability evaluation and its related reservoir and engineering parameters of shale gas wells is particularly important. Previous studies have focused on the impacts of reservoir quality including sedimentary environment [11], structure [12-13], mineralogy and lithology [14-17] and engineering quality including drilling and completion techniques [12, 18] and multi-stage hydraulic fracturing performance [19-20] on the shale gas production. Their results revealed that both the reservoir quality and engineering quality will affect the produced gas rate of the shale gas wells. Most of researches did analysis based on the relationship between the interested variable and its single impacting factor, comprehensive multivariate statistical analysis with consideration of correlation and collinearity among factors needs to be further studied. Moreover, instead of the produced gas rate or gas bearing content, the absolute open flow (AOF) potential may be a better parameter to describe the deliverability of gas well. The AOF potential is a hypothetical gas flow rate that may be produced if the flowing pressure at the bottom of the well (BHFP) were equivalent to the atmospheric pressure, i.e. the maximum flow rate that the gas well may produce [21]. In this study, the well deliverability of Jiaoshiba shale gas is described by the AOF potential based on BHFP and the impacts of reservoir and engineering characteristics on the AOF potential will be comprehensively analyzed. The data of over 86 wells of Fuling Jiaoshiba shale gas field is analyzed and the impacts of the reservoir quality (i.e. density of rock, porosity, total organic carbon, brittle mineral content of rock) and the engineering quality (i.e. horizontal length of gas well, horizontal length ratio through different formation stages, number of fracturing stages, liquid volume, and sand amount) on the AOF are quantified. The most important factors affecting deliverability of the gas wells in the Jiaoshiba area will be obtained by hierarchical clustering analysis method. The regression correlation between the well deliverability and the important parameters will be built. The objective of this study is to find a new method to evaluate the key factors that affect the well deliverability of shale gas reservoir by multivariate statistical analysis with consideration of correlation and collinearity among factors and build multivariable nonlinear regression models based on these key factors, so that the well deliverability of shale gas reservoir could be predicted.

2. Shale gas reservoir characters

The Longmaxi shale gas in the Jiaoshiba area located in the Eastern Sichuan Basin in the southwest region of China is the first and the largest successfully developed shale gas reservoir in China [22]. The shale gas bearing formation is the Upper Ordovician Wufeng Formation to Lower Silurian Longmaxi Formation with a thickness of about 200–300 m, area of 272.44 km^2 and buried depth of less than 3500m. Comprehensively considering the lithology, electrical and physical properties, geochemical and gas-bearing characteristics, the gas bearing shale section in the Upper Ordovician Wufeng Formation to Lower Silurian Longmaxi Formation could be vertically divided into three parts with stable horizontal stratigraphic distributions for each member (i.e. Long1, Long2 and Long3 from bottom to top). Long1 member is rich in fossils (e.g., radiolarian, calthrop, and graptolite) and is composed of black mudstone and argillaceous siltstone. The Long2 member consists of argillaceous siltstone and fine-grained sandstone. The Long3 member is composed of dark-gray mudstone and gray shale with fewer fossils [5, 23].

Previous studies suggested that Long1 member consisted of one 3-ord sequence, and could be divided into three system tracts (i.e. the transgressive system tract (TST), early highstand system tract (EHST), and late highstand system tract (LHST)) [10, 15, 16]. As shown in **Fig.1** [5], this study is targeted in the Long1 member, which is abundant of organic matter with pores and could be divided into 3 stages (i.e. stage1, stage2 and stage3 from bottom to top) based on physical properties with gas bearing increasing from top to bottom. The Long1 member is of good brittleness, and the brittle mineral content gradually increases from top to bottom. The natural gas in the layer belongs to over-mature dry gas, in which methane is the main component without containing hydrogen sulfide, carbon monoxide and hydrogen.

3. Samples

The AOF potential is a hypothetical gas flow rate that may be produced if the flowing pressure at the bottom of the well (BHFP) were equivalent to the atmospheric pressure [21]. The AOF potential may be a better parameter to describe the deliverability of the gas wells as it gives the maximum flow rate that the gas well can produce. As shown in Fig.2, the AOF potential data of and 9 factors that may affect the deliverability of gas wells (i.e. horizontal length, number of fracture stages, total liquid volume, total amount of sand, rock density, porosity, TOC, brittle mineral content of rocks, horizontal length ratio through stage 1 and stage 3 of Long1 member) are collected (data is offered by SINOPEC Chongqing Fuling Shale Gas Exploration and Development Co., Ltd., Chongqing, China). Though **Fig.2** shows certain relationship between AOF potential and single impacting factor, the correlation and collinearity among different factors are not considered. According to the data, the average reservoir pressure of the Long1 member is 37.69MPa with pressure coefficient 1.55. The reservoir temperature is 85.99°C. The pay zone thickness is between 20m-200m and the most favorable thickness is about 38m. The kerogen type of the organic matter is mostly type II_1 to II_2. The total organic carbon (TOC) contents of the shale range from 1%-8% with average value of 5% and it is increasing from top to bottom. The vitrinite reflectance (R_o %) of organic matter ranges from 2.5%-3.7% with average value of 2.9%. The porosity of the shale is between 1.88% to 7.06%, with average value of 4.15% and the permeability of the shale is less than 0.001mD. The horizontal wells in the sample have lateral length between 1500m to 1800m, well spacing between 500m to 600m, and the distance between adjacent clusters between 20m to 25m. All the horizontal wells are multi-stage hydraulic fractured with number of fracturing stages between 15 to 29. The fluid volume per stage is between 1500m^3 to 1800m^3, and proppant volume per stage is between 50m^3-65m^3. The fractured half-lengths range from 300m to 350m, and propped half-lengths range from 250m to 270m. The fracture heights range from 40m to 50m, and main fracture width range from 5mm to 10mm.

4. Methods

4.1. Normality test

When working with multivariable regression problems, some assumptions should be followed. The regression process of independent variables and the target variable are assumed to follow Gauss–Markov theorem (i.e. the ordinary least squares (OLS) estimator has the lowest sampling variance within the class of linear unbiased estimators, if the independent variables have no perfect collinearity, and the errors in the regression model are uncorrelated with equal variances and expectation value of zero) [24]. Thus, the first step of the multivariate statistical analysis is to test whether the sample variables follow normal distribution. The Q–Q (quantile-quantile) plot is applied, which is a probability plot that shows a graphical comparison between the samples and normal distribution by plotting their quantiles against each other [25]. If the points in the Q–Q plot approximately lie on the 45-degree line (i.e. $y=x$), the distribution of the sample being compared are similar enough to normal distribution. If normality test of a variable behaves not well, certain transformation (i.e. logistic transformation, logarithm transformation, index transformation, power transformation, etc.) of the variable and nonlinear regression model should be considered.

4.2. Standardization

Before the key factor analysis, the calculation of correlations between each parameter is essential. Since the units, scales and ranges of the factors are different, the first step is to standardize all the parameters based on the normality test results. The method applied in the study is the Fisher Z-transformation method. The standard score (Z score) for each data point is its difference from sample mean divided by the sample standard deviation.

$$Z = \frac{X - \overline{X}}{s} \qquad (1)$$

where

X is the data point.
$\overline{X}$ is the sample mean.
s is the sample standard deviation.

4.3. Correlation analysis

After standardization, the correlations between each factor are calculated. This is to eliminate the repeated consideration of correlation and collinearity in the later multivariable regression. In other words, if two variables are strongly related, only the principle variable between them is considered in later analysis. The sample data has a number of n wells, and each well has a number of k variables including a dependent variable (X_k) which is Z scores for AOF potential and $k-1$ independent variables (X_1, X_2, ..., X_{k-1}) which are Z scores for the factors affecting AOF potential. The sample data could be shown in matrix form as follows

$$X = \begin{bmatrix} x_{1,1} & x_{1,2} & \cdots & x_{1,k} \\ x_{2,1} & x_{2,2} & \cdots & x_{2,k} \\ \cdots & \cdots & & \cdots \\ x_{n,1} & x_{n,2} & \cdots & x_{n,k} \end{bmatrix} \tag{2}$$

Then the correlation matrix for all variables would be

$$R = \begin{bmatrix} r_{11} & r_{12} & \cdots & r_{1k} \\ r_{21} & r_{22} & \cdots & r_{2k} \\ \cdots & \cdots & \cdots & \cdots \\ r_{k1} & r_{k2} & \cdots & r_{kk} \end{bmatrix} \tag{3}$$

$r_{ij}(i,j=1,2,...,k)$ is the correlation coefficient for $x_{ij}(i=1,2,...,k)$ and $x_{ij}(i=1,2,...,k)$, and the calculation equation is

$$r_{ij} = \frac{\sum_{m=1}^{k}(x_{im}-\overline{x_i})(x_{jm}-\overline{x_j})}{\sqrt{\sum_{m=1}^{k}(x_{im}-\overline{x_i})^2 \sum_{l=1}^{k}(x_{jl}-\overline{x_j})^2}} \tag{4}$$

The value of correlation coefficient should be between -1 and 1. If the correlation coefficient is close to 1, the involving two variables have a strong positive relationship, and if it is close to -1, the involving two variables have a strong negative relationship. If the correlation coefficient is close to 0, this suggests that the involving two variables may be independent to each other.

4.4. Hierarchical clustering analysis

As long as correlations among factors are obtained, clustering method could be applied to find the key factors while eliminate the factors with small effect on AOF potential or strong correlation with other factors. Hierarchical clustering analysis (HCA) is a method to find the hierarchy relationship among variables, so that the variables could be divided into clusters [26]. HCA is performed on the normalized sample data set of the 86 well data including AOF potentials and 9 different factors, using centroid linkage clustering (Eq.5) as linkage criteria. The method will try to link the two most related factors in the data set to form a new cluster at each step (**Fig.3** agglomerative hierarchical clustering). The new merged cluster is then considered as a variable in the relation map and agglomerative approach is again used. Finally, a tree-hierarchy structure with linkage lines that join all the factors would be built.

$$D_{uv}^2 = (\bar{X}_u - \bar{X}_v)'(\bar{X}_u - \bar{X}_v) \quad (5)$$

, where $\bar{X}_u$ is the centroid for cluster C_u. $\bar{X}_v$ is the centroid for cluster C_v.

4.5. Multivariable nonlinear regression

Once the main factors that affect the AOF potential are found, other factors are either unessential to AOF potential or have strong correlation to the main factors, which could be eliminated. The next step is to linearize the main factors, including logistic transformation, logarithm transformation, index transformation, power transformation, etc. Then, a set of multivariable linear regressions is carried out to find the optimal regression model. If there are k variables, the linear regression for the variables is

$$y_i = \alpha + \beta_1 X_{i1} + ... + \beta_k X_{ik} + \varepsilon_i = \alpha + \sum_{j=1}^{k} \beta_j X_{ij} + \varepsilon_i, i = 1, 2, ..., n \quad (6)$$

Least squares could be used to find the intercept a and coefficients β_j (j=1,2,...,k), which is to find the optimums for a and β_j (j=1,2,...,k) when the sum of squared residuals, *SSe*, is minimized. The squared residual r_i^2 is defined as the difference between real value and predicted value of the dependent variable.

$$r_i = y_i - \hat{y}_i = y_i - f(x_i, \alpha, \beta) \quad (7)$$

$$SSe = \sum_{i=1}^{n} r_i^2 \quad (8)$$

$\hat{y}_i$ is the estimation value of dependent variable y_i. $a, b_j (j = 1, 2, ..., k)$ are the estimation values of intercept and coefficients $\alpha, \beta_j (j = 1, 2, ..., k)$. Let the partial derivatives with respect to $a, b_j (j = 1, 2, ..., k)$ equals 0.

$$\begin{cases} \dfrac{\partial SS_e}{\partial a} = 0 \\ \dfrac{\partial SS_e}{\partial b_j} = 0 \end{cases} \quad j = 1, 2, \ldots, k \quad (9)$$

It could be also proved that $a, b_j (j = 1, 2, ..., k)$ is the unbiased estimation of $\alpha, \beta_j (j = 1, 2, ..., k)$.

5. Results and Discussions

5.1. Normality test

Fig.4 shows the Q–Q (quantile-quantile) plots for the 10 variables of the 86 wells. The results reveal that the points of most variables in the Q–Q plots approximately lie on the 45-degree line, which means they follow normal distribution well. The only variable may not behave with good normality is horizontal length ratio through stage 1 and stage 3 of Long1 member. This variable may need transformation (i.e. logistic transformation, log transformation, logistic transformation, etc.) in the later regression study. Thus, the multivariable nonlinear regression would be needed in this study.

5.2. Standardization and correlation of the variables

The matrix of correlation coefficients of all variables is shown in **Table 1**. **Table 1** suggests that there is no strong linear relation between the AOF potential and single one factor. Relatively, the AOF potential is positively related to the TOC, the brittle mineral content, and the total liquid volume. It is negatively related to rock density. Some factors show strong correlation with each other, and the number of fracture stages and total liquid volume have the highest correlation coefficient of 0.946. The drilling and fracturing engineering factors, including the number of fracturing stages, the length of horizontal well, the total liquid volume and the total sand amount, are highly linear related. Among the geological factors, the rock density is highly negative to TOC, brittle mineral content, and the TOC and the brittle mineral content of rock are highly linear related. The collinearity among some factors is very strong and may affect AOF potential modelling, which should be eliminated in the multivariate statistical analysis of AOF potential.

5.3. Hierarchical clustering analysis

According to **Table 1**, the number of fracturing stages and total liquid volume have the highest correlation coefficient, which is 0.946. These two factors could be firstly clustered together, and the central point of this cluster is calculated by centroid method. Then the correlation analysis is done again after applying the whole cluster as a variable. This progress is continued until all the variables converged to one cluster. The result of hierarchical clustering analysis with linkage lines is shown **Fig.5**, and **Table 2** gives the importance order of the effect of each factor on the AOF potential.

It is suggested in the hierarchical clustering analysis results that the order of the factors that affect AOF potential is: a. porosity, b. TOC, c. brittle mineral content of rock, d. horizontal length ratio through stage 1 and stage 3 of Long1 member, e. number of fracturing stages, f. total liquid volume, g. horizontal length, h. total sand amount and i. rock density. The AOF potential firstly has closest relationship to porosity. It then has closer relationship to the cluster of TOC, brittle mineral content of rock, horizontal length ratio through stage 1 and stage 3 of Long1 member. These four factors belong to the first cluster. The other four factors including number of fracturing stages, total liquid volume, horizontal length and total sand amount belong to the second cluster. Rock density belongs to the third cluster. Thus, the main factors that affect AOF potential are porosity, TOC, brittle mineral content of rock and number of fracturing stages.

5.4. Multivariable nonlinear regression

After the main clusters in the hierarchical clustering analysis are obtained, the key factors that affect the AOF potential are found. The independent variables in the multivariable nonlinear regression models could be chosen according to the cluster results. With consideration of the correlation and collinearity among factors, one factor in each cluster with smallest linkage distance is chosen in the regression model, and four models after testing (**Table 3**) are regressed in this study.

With trials of nonlinear transformation of the involving variables, the linear regression of the transformation is ideal, and all the coefficients have passed the student t test for significance. The four models are shown as follows.

Model 1:

$$Y' = \ln Y, X_1' = \ln X_1, X_2' = (\ln X_2)^8, X_3' = X_3 \tag{10}$$

$$Y' = -13.067 + 1.509X_1' + 0.012X_2' + 0.867X_3' \tag{11}$$

Model 2:

$$Y' = \ln Y, X_4' = \ln X_4, X_2' = (\ln X_2)^8, X_3' = \ln X_3 \tag{12}$$

$$Y' = -1.756 + 1.465X_4' + 0.011X_2' + 1.043X_2' \tag{13}$$

Model 3:

$$Y' = \ln Y, X_4' = \ln X_4, X_5' = X_5^{\frac{1}{100}}, X_2' = (\ln X_2)^8, X_3' = \ln X_3 \tag{14}$$

$$Y' = -1.603 + 1.434X_4' + 0.226X_5' + 0.011X_2' + 0.848X_3' \tag{15}$$

Model 4:

$$Y' = \ln Y, X_1' = \ln X_1^{\frac{1}{10}}, X_3' = X_3^{\frac{1}{100}}, X_2' = (\ln X_2)^8, X_6' = X_6 \tag{16}$$

$$Y' = -12.556 + 5.211X_1' + 0.149X_5' + 0.012X_2' + 0.035X_6 \tag{17}$$

It is shown from **Fig.6** that the total AOF potential of 86 wells is $5420.48\times10^4 m^3/d$ and the total AOF potential of 86 wells predicted by the 4 models range from $5110.24\times10^4 m^3/d$ to $5607.74\times10^4 m^3/d$. The average AOF potential of single well is $63.03\times10^4 m^3/d$ and the average AOF potential of single well predicted by the 4 models range from $59.42\times10^4 m^3/d$ to $65.21\times10^4 m^3/d$. The coefficient of determination (R^2) in the figure is the proportion of the variance in the dependent variable Y that is predictable from the independent variables $X_j (j = 1, 2, ..., k)$. The R^2 values show that all the 4 models could be relatively preferable models to predict the AOF potentials based on different factors. In total, the prediction value of model 1 and model 4 are closer to the real AOF potential value with higher coefficients of determination (R^2) and less prediction errors.

6. Conclusions

In this study, instead of the produced gas rate, the absolute open flow (AOF) potential is analyzed as a better parameter to describe the deliverability of gas well. The reservoir parameters, drilling and completion data, fracturing data (including the well logging interpretation of horizontal section, the track and layer section that the well bore passed through, fracturing scale and monitoring parameters) and sample production data of 86 gas wells in the Wufeng-Longmaxi shale, Jiaoshiba shale gas field in Eastern Sichuan Basin, China are considered in the multivariate statistical analysis. The hierarchical clustering analysis could give the orders of the key factors that affect gas well AOF potential as well as eliminate the collinearity among the factors, which gives the basis of multivariable nonlinear regression. The results suggest that the order of the factors affects AOF potential is: a. porosity, b. TOC, c. brittle mineral content of rock, d. horizontal length ratio through stage 1 and stage 3 of Long1 member, e. number of fracturing stages, f. total liquid volume, g. horizontal length, h. total sand amount and i. rock density. The estimation results of the AOF potential by the multivariable nonlinear regression models based on the key factors from hierarchical clustering analysis have small differences to that of the actual values of the sample with goodness-of-fit test acceptable. The predictions of AOF potential of the models are reliable, and the feasibility of this multivariate statistical analysis is verified in the study of shale gas well deliverability. The research results can provide a reference basis for design of drilling and completion, the reasonable production rate allocation, design and adjustment of development plan of the shale gas reservoirs.

Acknowledgement

The authors would like to acknowledge the support from National Natural Science Foundation of China (NSFC Grant Number: 520004032) and the Thirteen-Five National Science and Technology Major Projects of China: 1. Study on Shale Gas Fracturing and Gas Production Regularity in Wufeng Formation and Longmaxi Formation in Fuling Area (Sub-Issue Number: 2016ZX05060-026) and 2. Research on the efficient development technology and policy of inter-salt shale oil reservoir (Sub-Issue Number: 2016ZX05049005-010).
`

References

1. Zolfaghari, A., Dehghanpour, H., & Xu, M. (2017b). Water sorption behaviour of gas shales: II. Pore size distribution. *International Journal of Coal Geology, 179*, 187-195.
2. Zhang, Y., Yao, S., Zhang, M., Zhou, X., Mei, H., & Zeng, F. (2018). Prediction of adsorption isotherms of multicomponent gas mixtures in tight porous media by the oil–gas-adsorption three-phase vacancy solution model. Energy & Fuels, 32(12), 12166-12173.
3. Zhang, Y., Zhang, M., Mei, H., & Zeng, F. (2019). Study on salt precipitation induced by formation brine flow and its effect on a high-salinity tight gas reservoir. *Journal of Petroleum Science and Engineering, 183*, 106384.
4. Dong, T., Harris, N. B., Ayranci, K., Twemlow, C. E., & Nassichuk, B. R. (2017). The impact of composition on pore throat size and permeability in high maturity shales: Middle and Upper Devonian Horn River Group, northeastern British Columbia, Canada. *Marine and Petroleum Geology*, 81, 220-236.
5. Wang, C., Zhang, B., Hu, Q., Shu, Z., Sun, M., & Bao, H. (2019). Laminae characteristics and influence on shale gas reservoir quality of lower Silurian Longmaxi Formation in the Jiaoshiba area of the Sichuan Basin, China. *Marine and Petroleum Geology, 109*, 839-851.
6. Cao, T. T., Song, Z. G., Wang, S. B., & Jia, X. (2015). A comparative study of the specific surface area and pore structure of different shales and their kerogens. *Science China Earth Sciences, 58*(4), 510-522.
7. Chalmers, G. R. L., & Bustin, R. M. (2007). The organic matter distribution and methane capacity of the lower cretaceous strata of northeastern British Columbia, Canada. *International Journal of Coal Geology, 70*(1-3), 223-239.
8. Chalmers, G. R. L., & Bustin, R. M. (2008). Lower cretaceous gas shales in northeastern British Columbia, Part II: Evaluation of regional potential gas resources. *Bulletin of*

Canadian Petroleum Geology, 56(1), 22-61.
9. Ma, X., Li, X., Liang, F. , Wan, Y. , & Guo, W. . (2020). Dominating factors on well productivity and development strategies optimization in Weiyuan shale gas play, Sichuan Basin, SW China. *Petroleum Exploration and Development, 47*(3), 594-602.
10. Shu, Y., Lu, Y., Chen, L., Wang, C., & Zhang, B. (2020). Factors influencing shale gas accumulation in the lower Silurian Longmaxi formation between the north and South Jiaoshiba area, Southeast Sichuan Basin, China. *Marine and Petroleum Geology, 111*, 905-917.
11. Wang, Y. , Dong, D. , Li, X. , Huang, J. , Wang, S. , & Wu, W. . (2015). Stratigraphic sequence and sedimentary characteristics of Lower Silurian Longmaxi Formation in the Sichuan Basin and its peripheral areas. Natural Gas Industry B, 2(2–3), 222-232.
12. Hu, M., Huang, W., & Li, J. (2018). Effects of structural characteristics on the productivity of shale gas wells: A case study on the Jiaoshiba Block in the Fuling shale gasfield, Sichuan Basin. *Natural Gas Industry B, 5*(2), 139-147.
13. Zhang, H., Wang, J., & Zhang, H. (2016). Investigation of the main factors during shale-gas production using grey relational analysis. *The Open Petroleum Engineering Journal, 9*(1).
14. Wei, C., Wang, H., Sun, S., Xiao, Y., Zhu, Y., & Qin, G. (2012, January). Potential investigation of shale gas reservoirs, Southern China. In *SPE Canadian Unconventional Resources Conference*. Society of Petroleum Engineers.
15. Wang, C., Zhang, B., Lu, Y., Shu, Z., Lu, Y., Bao, H., ... & Chen, L. (2018). Lithofacies distribution characteristics and its controlling factors of shale in Wufeng Formation-Member 1 of Longmaxi Formation in the Jiaoshiba area. *Petroleum Research, 3*(4), 306-319.
16. Wang, C., Zhang, B. Q., Lu, Y. C., Shu, Z. G., Lu, Y. Q., Bao, H. Y., ... & Chen, L. (2018). Lithofacies distribution characteristics and main development controlling factors of shale in Wufeng Formation-Member 1 of Longmaxi Formation in Jiaoshiba area. *Acta Pet. Sin., 39*(6), 631-644.
17. Kim, G., Lee, H., Chen, Z., Athichanagorn, S., & Shin, H. (2019). Effect of reservoir characteristics on the productivity and production forecasting of the Montney shale gas in Canada. *Journal of Petroleum Science and Engineering, 182*, 106276.
18. Curtis, M. E., Cardott, B. J., Sondergeld, C. H., & Rai, C. S. (2012). Development of organic porosity in the Woodford Shale with increasing thermal maturity. *International Journal of Coal Geology, 103*, 26-31.
19. Cipolla CL, Lolon EP, Erdle JC, Rubin B. Reservoir modeling in shale-gas reservoirs. *SPE reservoir evaluation & engineering*. 2010 Aug 1;13(04):638-53.
20. Zhu W, Ma D, Qi Q, Song Z, Yue M. Multi-zone coupling productivity of horizontal well fracturing with complex fracture networks in shale gas reservoirs. Natural Gas Industry B. 2018 Feb 1;5(1):80-9.
21. Mohammed, K. R., Stigliano, H., Vicente, M. G., & Angulo, R. J. (2010, January). Comparison of Different Deliverability Testing Techniques in Low Permeability Gas Reservoir and Its Impact on the Absolute Open Flow (AOF) Estimation. In *SPE EUROPEC/EAGE Annual Conference and Exhibition*. Society of Petroleum Engineers.
22. Guo, X., Qin, Z., Yang, R., Dong, T., He, S., Hao, F., ... & Liu, K. (2019). Comparison of

pore systems of clay-rich and silica-rich gas shales in the lower Silurian Longmaxi formation from the Jiaoshiba area in the eastern Sichuan Basin, China. *Marine and Petroleum Geology*, *101*, 265-280.

23. Yang, R., He, S., Hu, Q., Hu, D., Zhang, S., & Yi, J. (2016). Pore characterization and methane sorption capacity of over-mature organic-rich Wufeng and Longmaxi shales in the southeast Sichuan Basin, China. Marine and Petroleum Geology, 77, 247-261.
24. Henderson, C. R. . (1975). Best linear unbiased estimation and prediction under a selection model. *Biometrics, 31*(2), 423-447.
25. Gnanadesikan, M. B. W. . (1968). Probability plotting methods for the analysis for the analysis of data. *Biometrika, 55*(1), 1-17.
26. Rokach, Lior, and Oded Maimon. (2005). "Clustering methods." Data mining and knowledge discovery handbook. Springer US, 321-352.

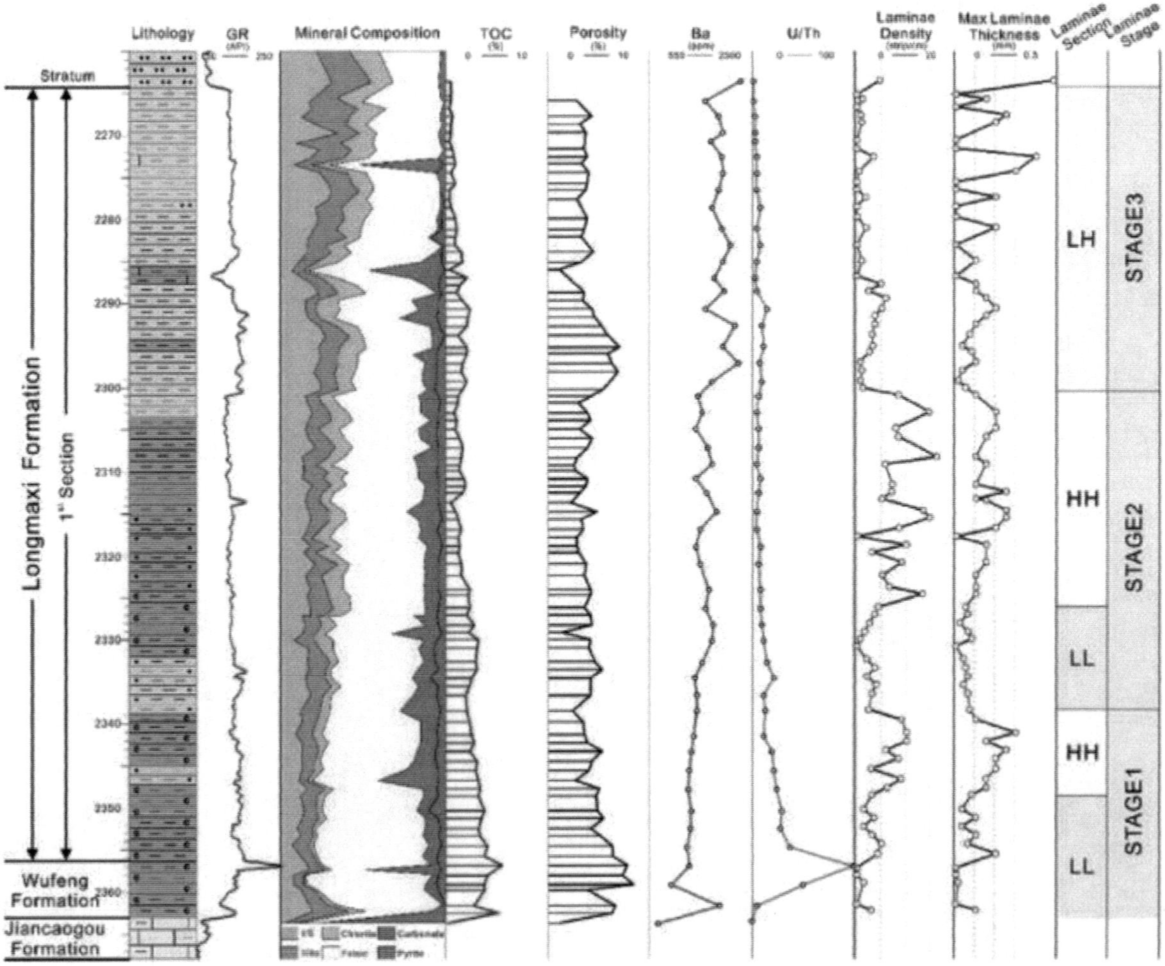

Fig.1 Lithofacies, gamma ray log, mineral content, TOC content, porosity, geochemical proxies, laminae density, and laminae stages of the Long1 member in well A (Wang et al., 2019).

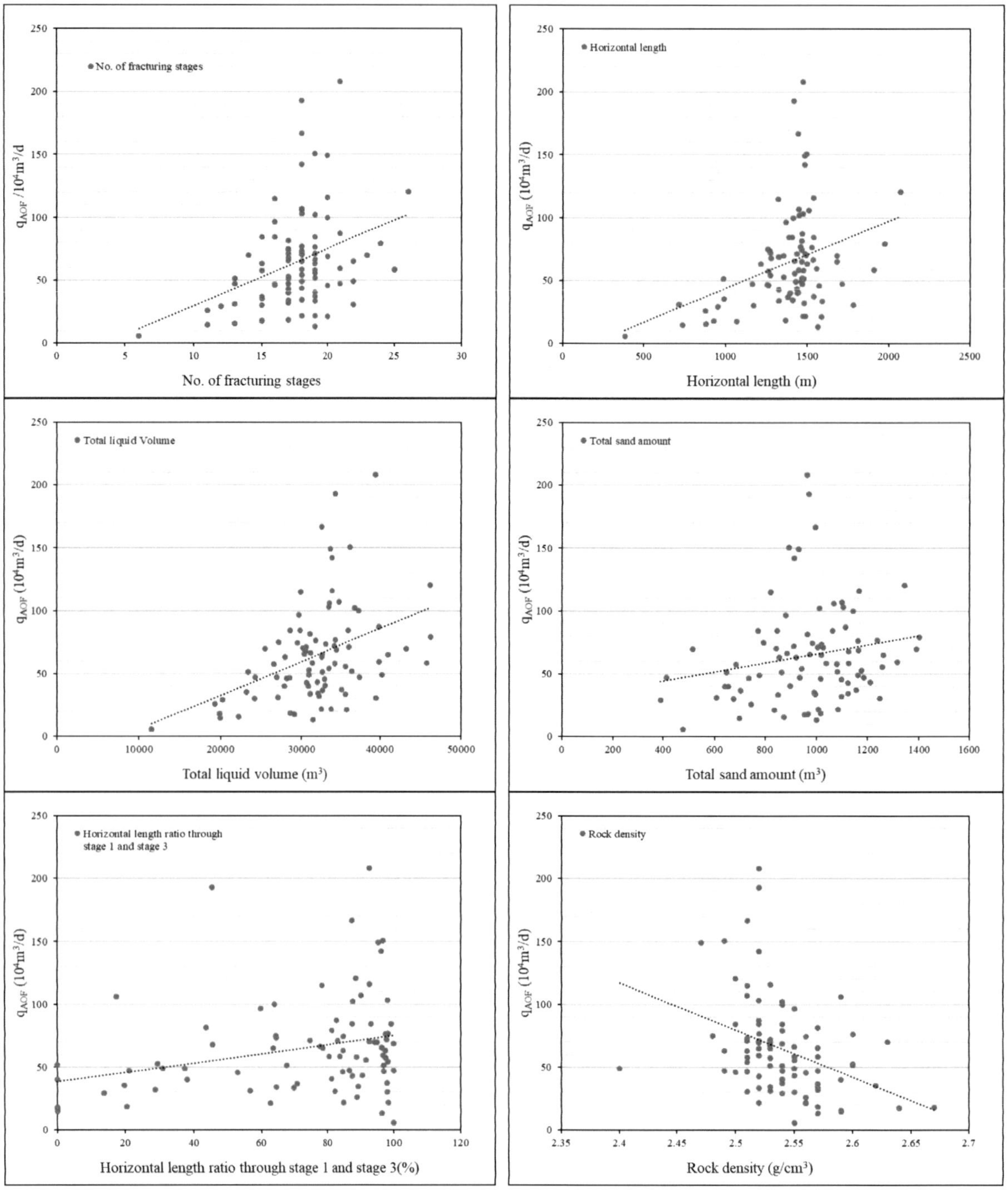
No. of fracturing stages
q_{AOF} /10^{4}m^3/d
No. of fracturing stages
Horizontal length
q_{AOF} (10^{4}m^3/d)
Horizontal length (m)
Total liquid Volume
q_{AOF} (10^{4}m^3/d)
Total liquid volume (m^3)
Total sand amount
q_{AOF} (10^{4}m^3/d)
Total sand amount (m^3)
Horizontal length ratio through stage 1 and stage 3
q_{AOF} (10^{4}m^3/d)
Horizontal length ratio through stage 1 and stage 3(%)
Rock density
q_{AOF} (10^{4}m^3/d)
Rock density (g/cm^3)

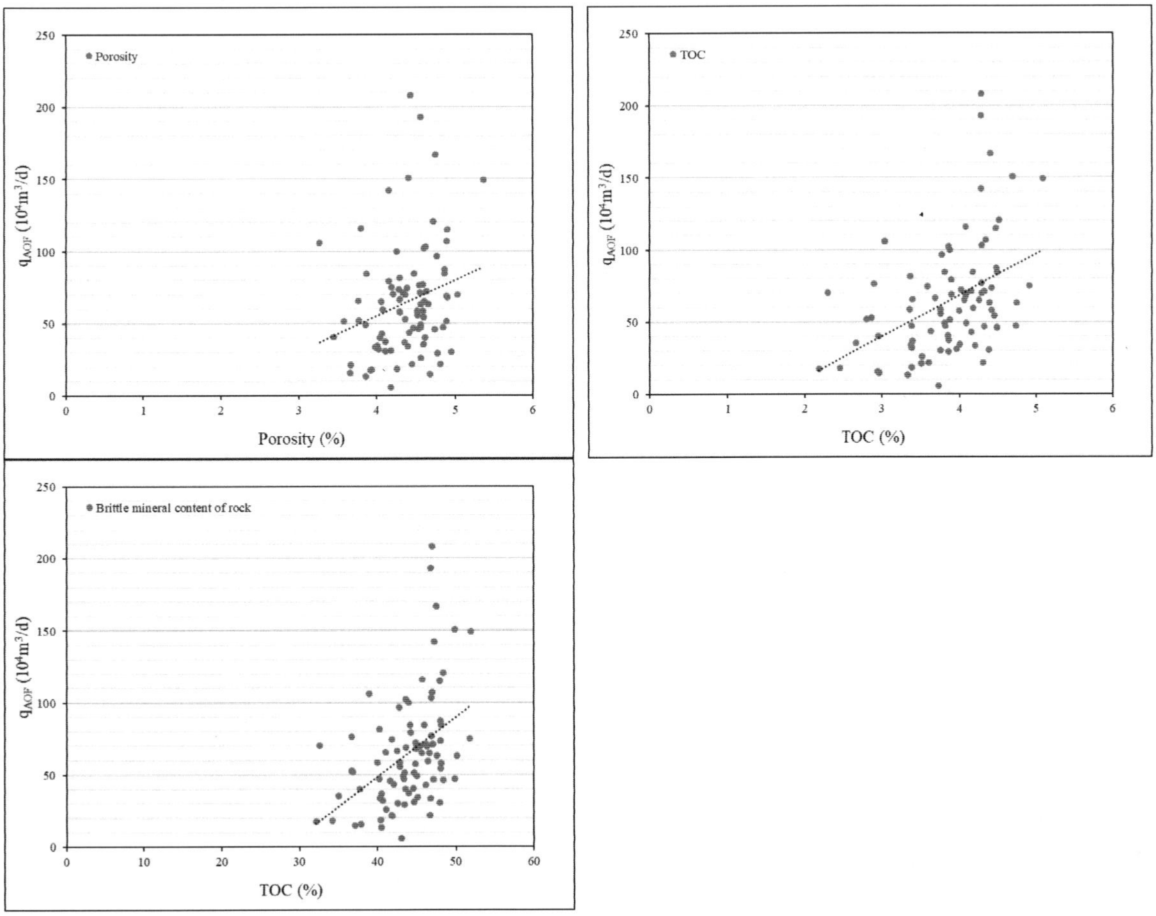

Fig.2 The scatter plots with single linear regression line of sample data of 86 gas wells in Jiaoshiba area with the AOF potential against 9 factors that may affect the deliverability of gas wells (i.e. number of fracture stages, horizontal length of well, total liquid volume, total sand amount, rock density, porosity, TOC, brittle mineral content of rocks, horizontal length ratio through stage 1 and stage 3 of Long1 member)

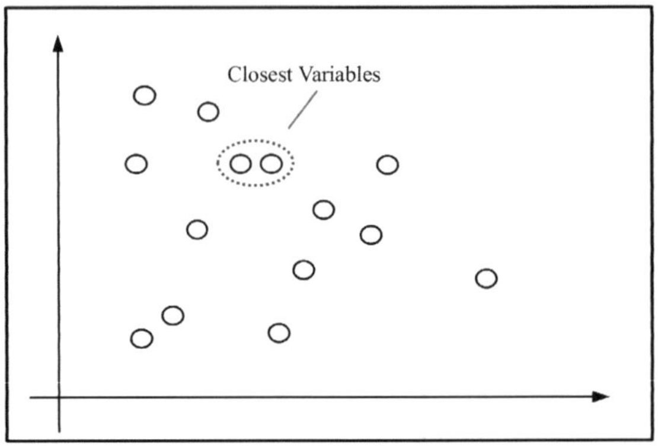

(a)

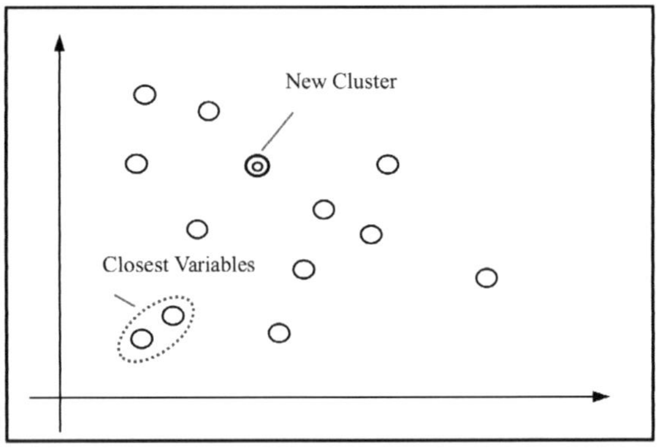

(b)

Fig.3 Hierarchical Clustering Analysis: (a) Clustering the variables by closest relation; (b) The center of new cluster is calculated by centroid method

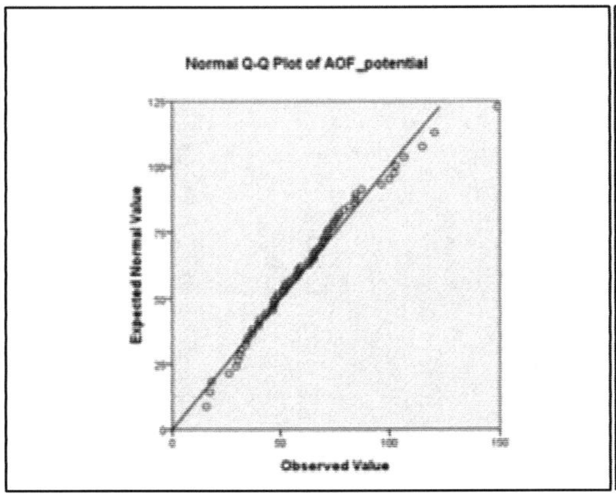
Normal Q-Q Plot of AOF_potential
Expected Normal Value
Observed Value

a.

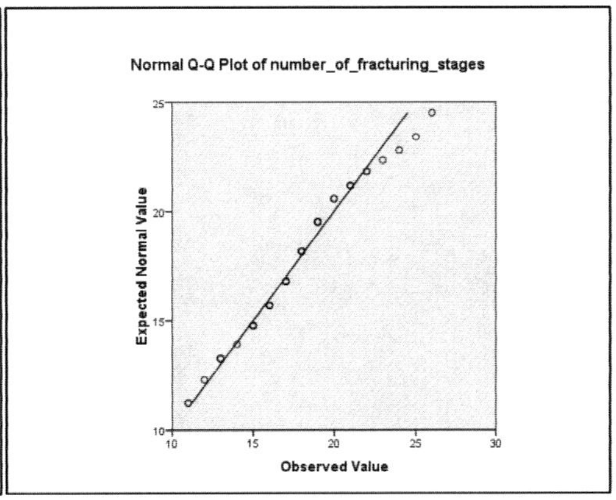
Normal Q-Q Plot of number_of_fracturing_stages
Expected Normal Value
25
20
15
10
10
15
20
25
30
Observed Value

b.

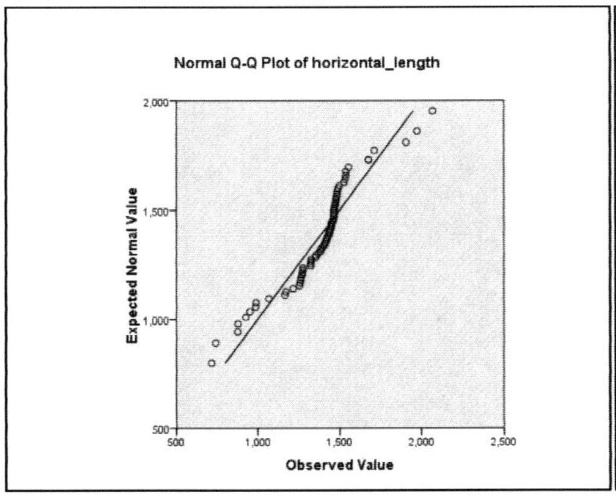
Normal Q-Q Plot of horizontal_length
Expected Normal Value
2,000
1,500
1,000
500
500
1,000
1,500
2,000
2,500
Observed Value

c.

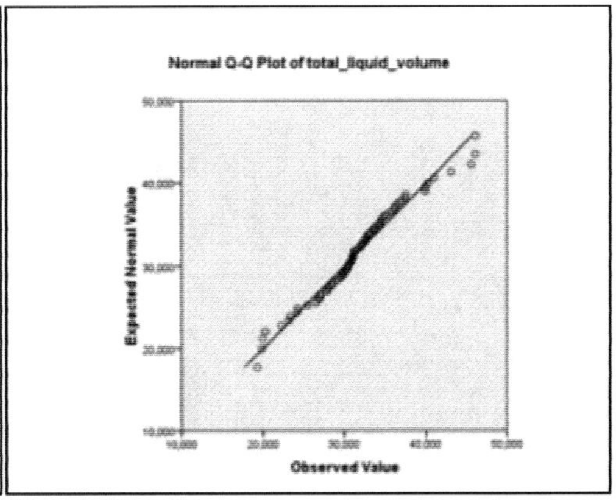
Normal Q-Q Plot of total_liquid_volume
Expected Normal Value
Observed Value

d.

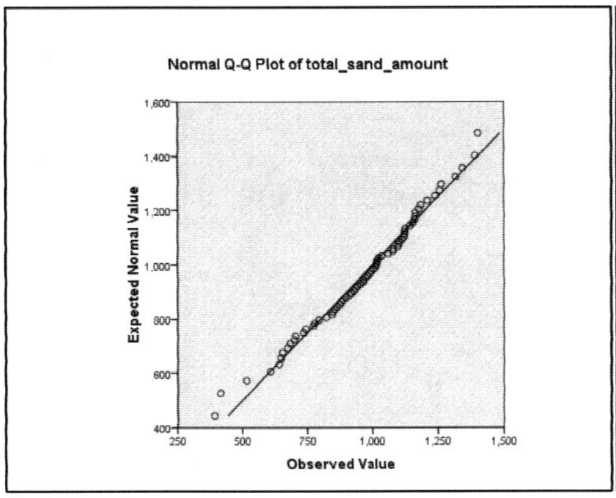
Normal Q-Q Plot of total_sand_amount
Expected Normal Value
1,600
1,400
1,200
1,000
800
600
400
250
500
750
1,000
1,250
1,500
Observed Value

e.

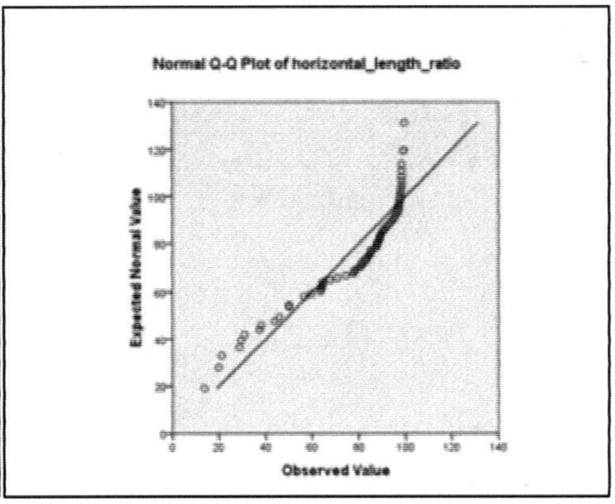
Normal Q-Q Plot of horizontal_length_ratio
Expected Normal Value
Observed Value

f.

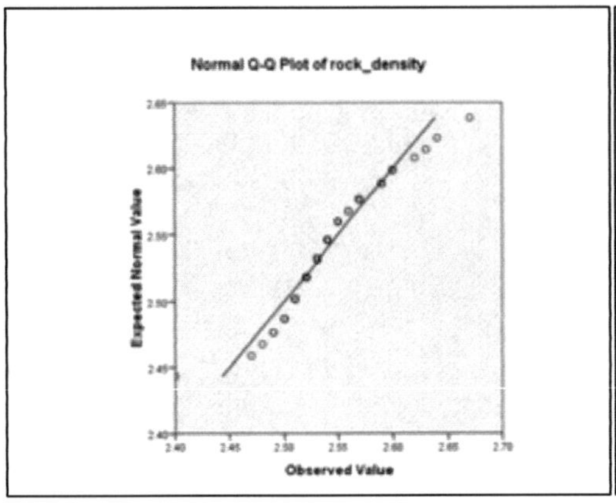

g.

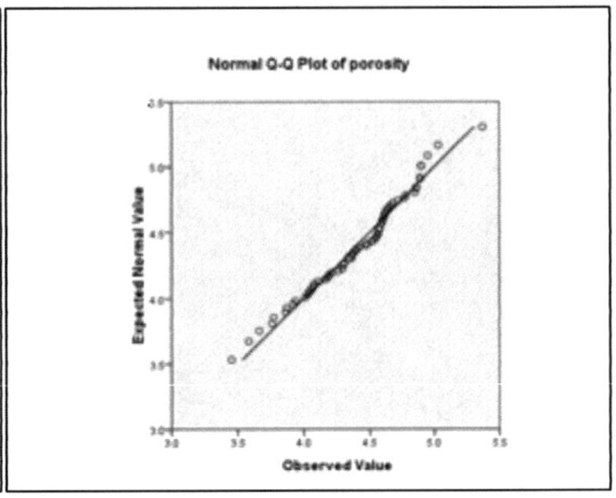

h.

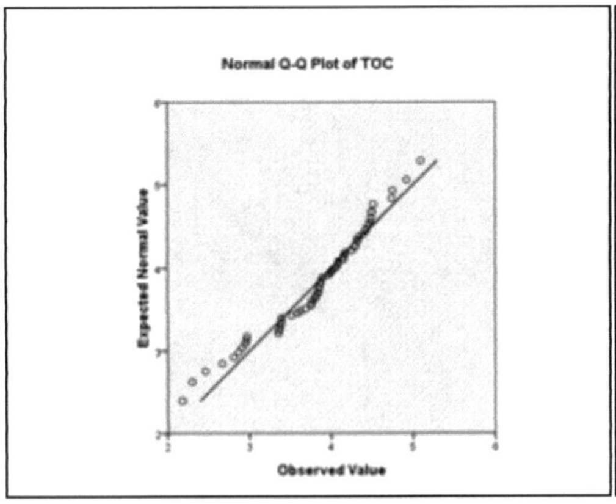

i.

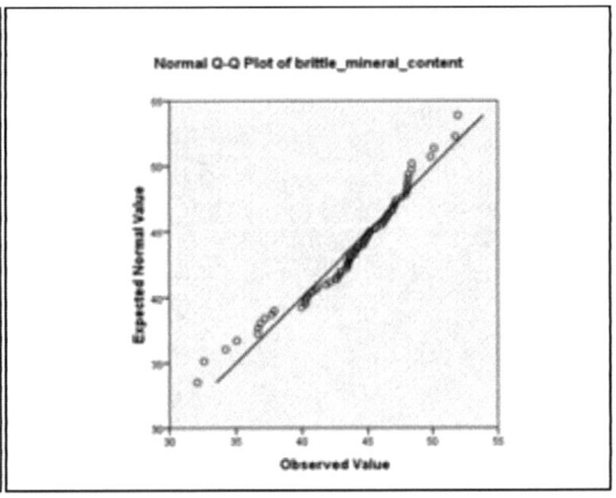

j.

Fig.4 The Q–Q (quantile-quantile) plots for the 10 variables of the 86 wells: a. AOF potential; b. number of fracturing stages; c. horizontal length of well; d. total liquid volume; e. total sand amount; f. horizontal length ratio through stage 1 and stage 3 of Long1 member; g. rock density; h. porosity; i. TOC; j. brittle mineral content of rock

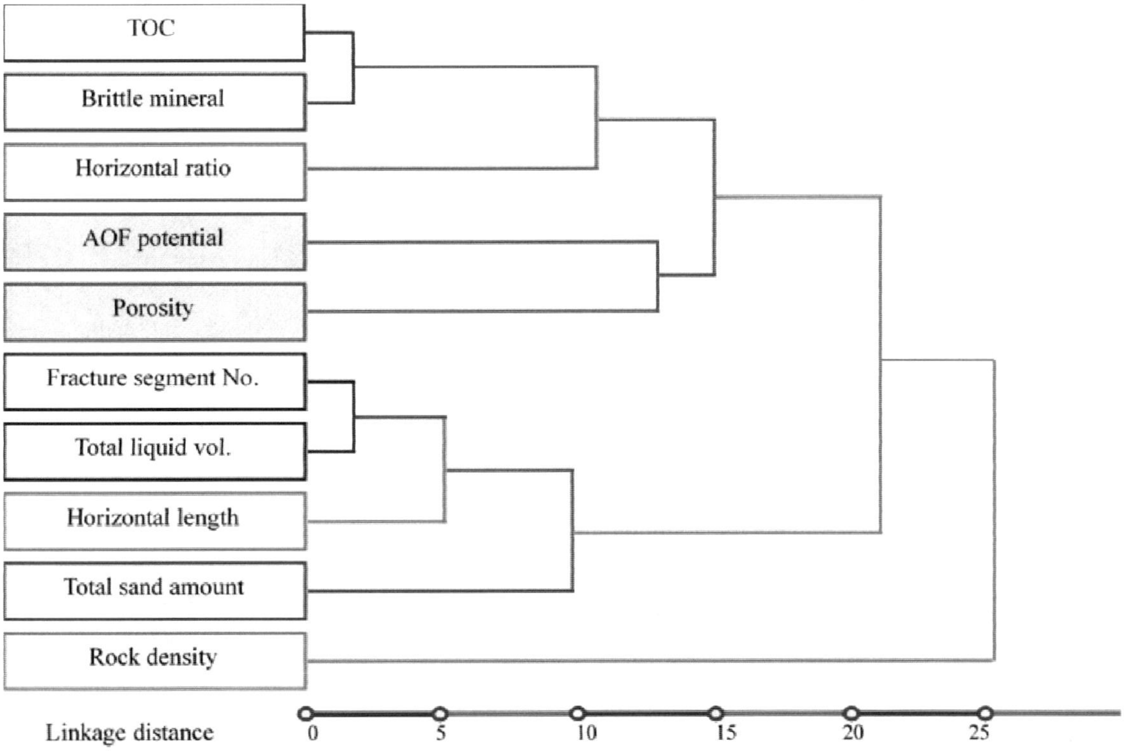

Fig.5 Hierarchical clustering analysis result of the sample data of 86 wells with linkage distance

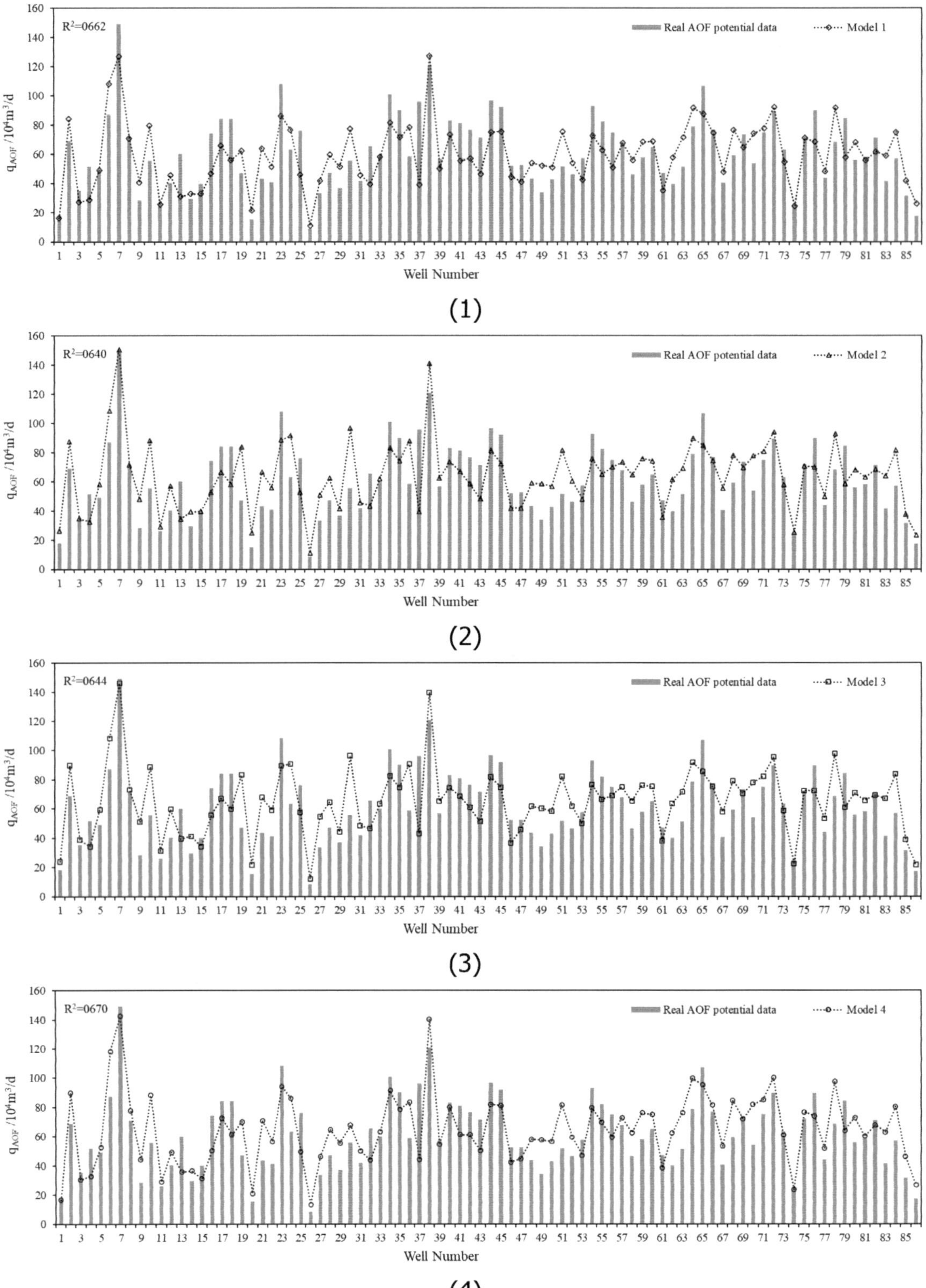

Fig.6 Comparison of actual and predicted AOF potential by the 4 models for the 86 wells with consideration of different key factors in Fuling Jiaoshiba shale gas field

Table 1 Correlation coefficients matrix of all variables in Jiaoshiba area

Factors	AOF potential	No. of fracturing stages	Horizontal length	Total liquid Volume	Total sand amount	Horizontal length ratio through stage 1 and stage 3	Rock density	Porosity	TOC	Brittle mineral content of rock
AOF potential	1	0.354	0.348	0.405	0.194	0.285	-0.373	0.246	0.433	0.438
No. of fracture stages	0.354	1	0.897	0.946	0.765	0.273	-0.203	-0.013	0.241	0.235
Horizontal length	0.348	0.897	1	0.863	0.633	0.331	-0.241	0.024	0.26	0.252
Total liquid Volume	0.405	0.946	0.863	1	0.736	0.302	-0.248	-0.01	0.274	0.282
Total sand amount	0.194	0.765	0.633	0.736	1	0.22	-0.002	-0.072	0.072	0.07
Ratio through stage 1 and stage 3	0.285	0.273	0.331	0.302	0.22	1	-0.492	0.287	0.568	0.578
Rock density	-0.373	-0.203	-0.241	-0.248	-0.002	-0.492	1	-0.289	-0.908	-0.917
Porosity	0.246	-0.013	0.024	-0.01	-0.072	0.287	-0.289	1	0.419	0.336
TOC	0.433	0.241	0.26	0.274	0.072	0.568	-0.908	0.419	1	0.982
Brittle mineral content of rock	0.438	0.235	0.252	0.282	0.07	0.578	-0.917	0.336	0.982	1

Table 2 Hierarchical clustering analysis result of factors in Jiaoshiba area

Parameter	Hierarchical clustering analysis result
AOF potential	1
Number of fracturing stages	2
Horizontal length	2
Total liquid volume	2
Total sand amount	2
Horizontal length ratio through stage 1 and stage 3	1
Rock density	3
Porosity	1
TOC	1
Brittle mineral content of rock	1

Table 3 Factors chosen in regression model according to hierarchical clustering analysis results

Regression model	Factors considered
1	a. Total liquid volume (X_1) b. Porosity (X_2) c. TOC (X_3)
2	a. Fracture segment No. (X_4) b. Porosity (X_2) c. TOC (X_3)
3	a. Fracture segment No. (X_4) b. Horizontal length ratio through stage 1 and stage 3 (X_5) c. TOC (X_3) d. Porosity (X_2)
4	a. Total liquid volume (X_1) b. Horizontal length ratio through stage 1 and stage 3 (X_5) c. Porosity (X_2) d. Brittle mineral content of rock (X_6)

601617 Solutions of C6F13- and C4F9-Based Polyfluoroacrylates in CO_2; Solubility, Viscosity, and Adsorption Onto Cement

Abstract:

High molecular weight poly(fluorinated acrylates) (PFAs) have drawn interest in the chemical and petroleum engineering fields due to their remarkable solubility in liquid and supercritical carbon dioxide (sCO_2) without the need for any co-solvent. This study does not include any examples of PFA based on monomers with the -C8F17 moiety because the ultimate hydrolysis degradation products of such polymers leads to the formation of PFOA, a compound deemed unsafe by the EPA. Instead, we have synthesized PFA based on fluoroacrylayte monomers containing the -C6F13 group (the functionality used in current commercial fluoroacrylate copolymers). These -C6F13-based polymers would degrade to form perfluorohexanoic acid (PFHxA) that, while significantly safer than PFOA, is also receiving more scrutiny from regulatory agencies in recent days. Therefore, we have also synthesized PFA based on monomers containing the -C4F9 functionality. The -C4F9-PFA ultimate degradation products include perfluorobutanoic acid (PFBA), which is significantly more environmentally benign than the -C6F13-based PFA. Therefore, one of the main objectives of this study was to determine if -C4F9-PFA retained the remarkable CO_2-solubility and CO_2-thickening capability as PFA based on the -C6F13 PFA and previously reported -C8F17-based PFA.

Many samples of PFA based on the -C6F13 and -C4F9-based fluoroacrylate monomers and varying ratios of initiator:monomer were bulk polymerized and then characterized for molecular weight and polydispersity. The -C4F9-PFA was shown to be nearly identical in CO_2-solubility as the -C6F16-PFA. Between 1-4 wt% PFA, the solubility of PFA in CO_2 was nearly independent of PFA molecular weight. Further, the cloud points loci for mixtures of CO_2-C4F9-PFA were nearly identical to the cloud point curves for the CO_2-C6F13-PFA at each temperature, with higher pressure required at higher temperature. This demonstrated that the -C4F9-PFA was just as CO2-soluble as the -C6F13-PFA. Further, both polymers were capable of thickening CO_2 in dilute concentration to a comparable degree (e.g. a 6-fold increase in viscosity with 4 wt% C4F9 PFA, 1.24×10^6 Da PFA in CO_2 at low shear rate).

PFA has previously been considered for use as a "CO_2 thickener" in enhanced oil recovery (EOR) operations. Unlike common chemical engineering applications where low viscosity is beneficial to reduce mass transfer resistance, low CO_2 viscosity is often problematic for EOR because it leads to viscous fingering (mobility control problems) and preferential loss of disproportionate amounts of CO_2 to high permeability "thief zones" that contain little recoverable oil (conformance control problems). However, in 2019 it was shown that PFA dissolved in CO_2 adsorbs so strongly onto sandstone or carbonate rock that the use of PFA in large-scale EOR projects to address mobility control would not be feasible, although near-wellbore conformance control applications (i.e. injecting the CO_2-PFA into isolated thief zones) may be viable.

In the current study, however, we are attempting to exploit the adsorption of PFA on solid surfaces for improved wellbore integrity. We are proposing to inject PFA-CO_2 solutions into small cracks within cement or between steel and cement (i.e. microannular cracks), thereby allowing a portion of the PFA to absorb and greatly reduce the crack permeability or completely seal the crack. Because PFA is a high molecular weight, amorphous, transparent, sticky, stretchy, oil-repellant, water-repellant polymer that is insoluble in water, oil or natural gas, it is a promising sealant for wellbore applications. Even though PFA is inherently expensive, this is a small volume application in which the cost of PFA would not exclude its consideration. Relative to other sealants (cement, resins, aqueous emulsions), PFA-CO_2 solutions are orders of magnitude less viscous, which would allow them deeper access into smaller cracks than other options. The PFA-CO_2 solutions would, however, be an inappropriate selection for the remediation of extremely large cracks or voids in cement.

In order to lay the groundwork for this proposed application, we have also determined the equilibrium adsorption of PFA onto Portland cement particles. This was accomplished by a direct measurement of the mass gain of a known quantity of sieved (22-30 mesh) Portland cement particles that was immersed in a PFA-CO2 fluid of specified original composition (e.g. 4 wt% PFA). The equilibrium concentration of PFA in the liquid CO_2 was determined via mass balance (original PFA placed in the cell – the amount of PFA found to adsorb onto the cement). This is, to the best of our knowledge, the first report of polymeric adsorption levels of solid surfaces immersed in CO_2-polymer solutions

611710 Competitive Sorption of CO2 in Shale Nanopores for Sequestration and Enhanced Gas Recovery Using Molecular Density Functional Theory

Abstract:

Enhanced gas recovery (EGR) by carbon dioxide (CO_2) injection to gas-bearing shale formations has been shown to be feasible and beneficial experimentally from pore scale to field scale. In addition to the improved recovery factor, CO_2 is trapped in the geological formation, which provides an opportunity to store CO_2. When supercritical CO_2 is injected, the preferential adsorption and dissolution of CO_2 in nanopores and organic matter enables the displacement of shale gas. This mechanism has been investigated by molecular simulation in graphite pores, but rarely for permeable pores by simulation or classical density functional theory (DFT). The compositional distribution of components from the organic matter source (kerogen) to nanoscale pores in the organic matter and to large fractures makes the estimation of fluid storage and the multicomponent compositional distributions in shale a challenging problem. We have introduced a density functional theory that accounts for molecular size and shape and kerogen maturity and swelling to predict the partitioning of components between the kerogen matrix and nano-pores in the kerogen [1].

In this work, a molecular thermodynamic model is presented that predicts both the adsorption of hydrocarbons in nanoscale pores and the dissolution of fluids in organic matter in equilibrium with the bulk kerogen phase. The swelling ratios of five kerogens of varying maturities and types in different solvents are well quantified using the new model. This kerogen matrix model is incorporated in interfacial statistical associating fluid theory (iSAFT) DFT to form a nanoporous kerogen composite model. The equilibrium partitioning of mixtures of fluids from the bulk phase to the nanoporous kerogen phase show enrichment of aromatic and cyclic molecules in kerogen, the preferential adsorption/absorption of high molecular weight molecules and carbon dioxide, and the significant amount of hydrocarbon storage both in pore space and kerogen matrix [1, 2].

Further, we apply molecular DFT to study the competitive sorption of CO_2 with shale gas in kerogen at various conditions. The DFT model is verified by grand canonical Monte Carlo (GCMC) simulation in graphite slit pores for pure and binary component systems at different temperatures, pressures, pore sizes and bulk compositions for methane/ethane/ CO_2. [2]. The model is utilized to predict the CO_2 adsorption selectivity in multicomponent systems which include two different shale gases. We show the selectivity of CO_2 decreases as temperature / pressure/ nanopore size/ shale gas average molecular weight increase. Subsequently, we model the CO_2/CH_4 adsorption in pores of more realistic conditions, in which we consider different levels of moisture in the pore and the gas dissolution in organic matrix. Interestingly, the presence of water inside a nanopore greatly reduces the CO2 adsorption but improves the CO_2 selectivity with respect to methane. The presence of organic matter kerogen also increases the CO_2 selectivity. In summary, a novel permeable pore DFT model enables prediction of partitioning and distribution of CO2 and hydrocarbons in shale related to enhanced gas recovery and CO_2 storage. [1, 2].

1. Jinlu Liu and Walter G. Chapman, "Thermodynamic Modeling of the Equilibrium Partitioning of Hydrocarbons in Nanoporous Kerogen Particles," Energy & Fuels, 33, 891-904 (2019).
2. Jinlu Liu, Shun Xi, and Walter G. Chapman, "Competitive Sorption of CO2 with Gas Mixtur in Nanoporous Shale for Enhanced Gas Recovery from Density Functional Theory," Langmuir, 35, 8144-8158 (2019).

609261 Spatial and Temporal Evolution of Natural Gas and Natural Gas Liquid Feedstocks in Shale Basins

Abstract:

Technological advancements in horizontal drilling and hydraulic fracturing have led to increasing production of light hydrocarbons from shale resources, including natural gas and natural gas liquids (NGLs). However, some basins that have insufficient gas gathering and pipeline networks (e.g. the Permian Basin in Texas) may flare significant quantities of gas. For example, in the Permian Basin, about 4.6% of gross gas production was estimated to be flared between May 2018 and March 2019 (Zhang et al., 2020). To mitigate flaring issues, hydrocarbon resources can be locally transformed to chemical or energy products. The National Science Foundation's Center for Innovative and Strategic Transformation of Alkane Resources (CISTAR) is developing modular processes to transform natural gas and NGLs to gasoline-like transportation fuels.

Determining the scale of operation of such modular chemical manufacturing requires a comprehensive understanding of the temporal and spatial availability of the hydrocarbon feedstocks. Preliminary work was performed in the Eagle Ford oil and gas producing region in south central Texas. With over 20,000 wells, the Eagle Ford produced 5.4 billion cubic feet natural gas per day and 1.2 million barrels of condensate and oil per day in 2019 (RRC, 2020). From the northwest to the southeast portion in the Eagle Ford Shale, well production varies from oil, to gas with condensate (wet gas), and to dry gas, as shown in Figure 1 (Gherabati et al, 2016). From oil to wet gas to dry gas regions, propane to ethane ratio varies from 1:1 to 1:3 (mass basis). In addition to varying gas compositions, the magnitude of available feedstocks also vary across the region. Ethane available at individual gas plants was estimated by assuming that produced gas from wells is sent to their nearest gas plant. Among 44 gas processing plants in the Eagle Ford, 60% of ethane production are represented by 43 plants with 200,000 – 750,000 kg of ethane per day, with one plant processing at ~3,000,000 kg/day, accounting for 40% of ethane production, shown in Figure 2.

Similar work has been performed in the Permian Basin in Texas. The Permian Basin is one of the largest oil-producing regions in the world and accounts for over 30% of oil production in the United States. Recent study (Zhang et al, 2020) shows that methane emissions in the Permian Basin are the highest among all the U.S. oil/gas producing regions. This presentation will provide detailed characterizations of spatial and temporal variability in light alkane production, and losses due to flaring for a case study region in the Permian Basin.

Collectively, the analyses in the Eagle Ford and the Permian Basin will provide two case studies of spatial and temporal characteristics of feedstock availability for modular light alkane manufacturing.

References

Gherabati, S. A., Browning, J., Male, F., Ikonnikova, S. A., & McDaid, G. (2016). The impact of pressure and fluid property variation on well performance of liquid-rich Eagle Ford shale. *Journal of Natural Gas Science and Engineering*, *33*, 1056-1068.

Railroad Commission of Texas (RRC) – Eagle Ford Shale Information. (2020). Available at: https://www.rrc.state.tx.us/oil-gas/major-oil-and-gas-formations/eagle-ford-shale-information/

Zhang, Y., Gautam, R., Pandey, S., Omara, M., Maasakkers, J. D., Sadavarte, P., ... & Zhang, R.

(2020). Quantifying methane emissions from the largest oil-producing basin in the United States from space. *Science Advances*, *6*(17), eaaz5120.

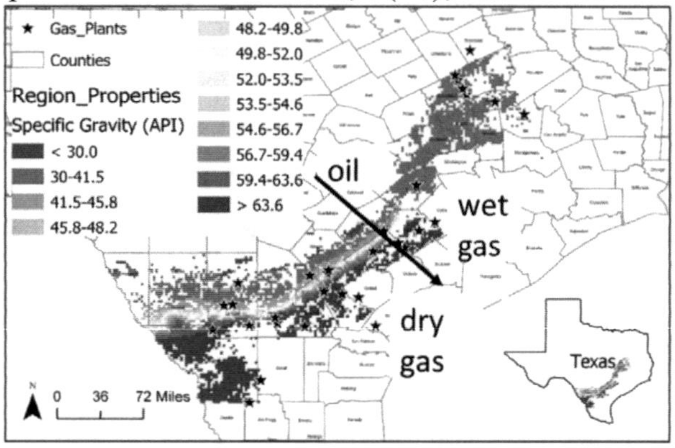

Figure 1. Spatial variabilities in well production in the Eagle Ford Shale.

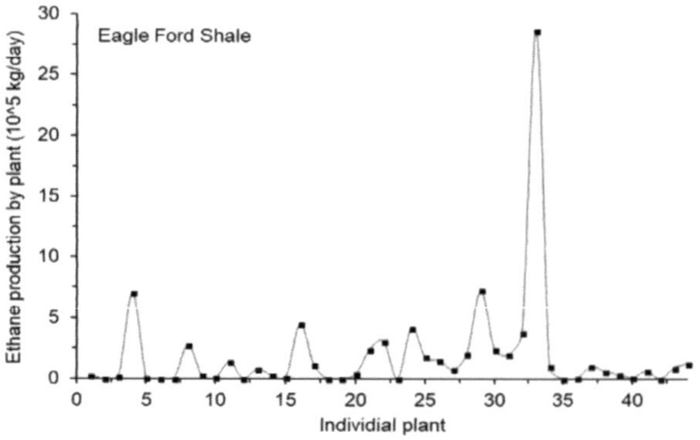

Figure 2. Ethane available at 44 gas processing plants in the Eagle Ford Shale

592706 A New Geology-Engineering Integration Solution for Shale Gas Exploration: A Case Study from Changning Block in China

Abstract:

Rich and quality shale gas reservoirs have been found in the Lower Silurian Longmaxi strata in the southern Sichuan Basin, which was identified as great development potential. However, shale gas reservoir is characterized by ultra-low permeability, which usually has no natural productivity and is called a typical “artificial gas reservoir”. We must pay attention to reservoir quality, drilling quality and completion quality in order to obtain the maximum EUR and the highest oil recovery. To tackle this challenge, the integrated Geoengineering solution was deployed in Changning Shale Gas Block, which consists of geological modeling, geomechanics, fracturing simulation and gas reservoir simulation. Geoengineering integration technology is used on the Changning A platform for field applications. The results show that: (1) Carrying out integrated research on geological engineering can realize the organic combination of geology and engineering. The geological model provides technical support for engineering optimization. The engineering practice constantly corrects the geological model and improves the precision; (2) Through the integration of geological engineering, we can create "transparent" shale gas reservoir, which can improve the predictive accuracy of geological and engineering sweet spot mapping; (3) Hydraulic fractures can be fine characterization by using geological engineering integration technology, which is useful for quantitative description and analysis of key parameters such as geometrical shape, net pressure, proppant distribution and conductivity of hydraulic fractures; (4) the refined unstructured-grid (<10m) based 3D numerical simulation model was built for hydraulic fractured horizontal wells in shale gas reservoir, the history match for shale gas wells has been hugely improved: gas production rate was matched to more than 98%; the wellhead pressure was matched to more than 80%; the cumulative liquid production rate was matched to more than 90%. With the implementation of the Geoengineering study, the shale gas production in Changning area has been increasing rapidly in the past 2 years.

593227 New Technology of EOR of Ultra Deep and Complex Condensate Gas Reservoirs in Tarim Basin， China

Abstract:

A number of condensate gas reservoirs (both sandstone and carbonate) with depth of over 5000m have been developed in Tarim Basin, China. Those reservoirs are featured with thick reservoirs pay zones (over 100 meters) with ,strong reservoir heterogeneities in both rock properties and fluid properties. Gas channeling in the process of gas injection development and retrograde condensation in the process of depletion development have seriously affected the recovery of condensate. In order to enhance the condensate recovery in those condensate gas reservoirs, a series of new technologies have been developed during the last decade.

Laboratory experiment analysis and development practice discovered that three major mechanisms controlling the gas migration process in the reservoirs: displacement, gravity and diffusion for cyclic gas injection. It is proposed that gravity differentiation is the main factor affecting the development effect of cyclic gas injection, and the fluid distribution law of underground condensate gas reservoir is re-recognized. For the sandstone condensate gas reservoir developed with large well spacing (over 1000 meters), the displacement and diffusion will be replaced by gravity differentiation gradually in the process of cyclic gas injection, and the fluid will be redistributed vertically according to its density, which makes different phase behaviors. Similarly, the reservoir space of depleted carbonate condensate gas reservoir is mainly dissolution cave and fracture. Under the action of gravity, the reverse condensate oil usually accumulates at the bottom of fracture cave system. On this basis of this understanding, the technology of gravity assisted three-dimensional displacement gas injection for sandstone condensate gas reservoirs and horizontal well placement technology for carbonate condensate gas reservoirs were developed to optimize the well placement. The technologies have been successfully applied to the developed condensate gas field in Tarim Basin, and the recovery rate of condensate oil has been increased by over 19%.

Production capacity evaluation method of condensate gas reservoir considering the influence of stress sensitivity and condensate oil

DU Chunxiao, LIU Xiaohong, ZHANG Fengyi, XU Jing, GENG Zhigang

(Tianjin Branch of CNOOC Ltd., Tianjin 300459)

Abstract: Accurate evaluation of production capacity is an important guarantee for efficient development of condensate gas reservoir. For the deep condensate gas reservoir with low permeability and low pressure difference between the initial formation pressure and the dew point, the influence of reservoir stress sensitivity and condensate oil on the production capacity of the gas reservoir cannot be ignored. Nevertheless, at present, the widely used method to calculate the gas reservoir production capacity is based on single-phase gas, without considering the influence of condensate oil and stress sensitivity on reservoir permeability. On the basis of researching a large number of literatures, combined with laboratory experiments and numerical simulation, the production capacity evaluation model of condensate gas reservoir under the comprehensive influence of stress sensitivity and condensate oil is derived by using the seepage flow mechanics theory, and the research results are applied to the Bohai Z gas reservoir. The results show that the production capacity evaluation model can more accurately reflect the production status of this kind of gas reservoir in different development stages, and can be widely used in the productivity evaluation of similar condensate gas reservoir.

Key words: deep condensate gas reservoir; stress sensitivity; condensate oil; evaluation of production capacity; low permeability

Gas reservoir production capacity evaluation is an important research content of gas reservoir engineering research, which provides key guidance for the design of gas reservoir development plan. For condensate gas reservoir, the fluid flow law and phase change are more complex, and the reverse condensation phenomenon has a serious impact on gas well productivity. According to theoretical and experimental studies, many scholars have confirmed that low permeability reservoirs have strong stress sensitivity. At present, the widely used method to calculate the gas reservoir production capacity is based on single-phase gas, without considering the influence of condensate oil and stress sensitivity on reservoir permeability. Many scholars have carried out research on the influence of reservoir stress sensitivity or retrograde condensation on gas well productivity, but lack of comprehensive consideration of both. Therefore, a new productivity evaluation method of condensate gas reservoir considering stress sensitivity and retrograde condensation is derived by combining laboratory experiments and numerical simulation methods, and is applied to BZ gas reservoir in Bohai oilfield.

1. Influence of stress sensitivity on reservoir permeability

The relationship between rock permeability and effective confining pressure has been studied deeply by many scholars using a large number of experiments. The relationship between permeability and effective confining pressure can be divided into three types: power-law relationship, exponential relationship and polynomial relationship.

BZ gas reservoir in Bohai oilfield is a low permeability gas reservoir. The core stress sensitivity test results of BZ-1 well are shown in Table 1 and figure 1. It can be seen that the dimensionless permeability of core decreases with the increase of effective stress, and BZ gas reservoir has strong stress sensitivity.

Table 1 Data of core stress sensitivity test

Net pressure	Sample 1	Sample 2	Sample 3	Sample 4
MPa	K_n/K_i	K_n/K_i	K_n/K_i	K_n/K_i
3.5	0.748	0.792	0.774	0.781
5.0	0.637	0.670	0.642	0.643
7.0	0.476	0.521	0.503	0.524
9.0	0.391	0.394	0.402	0.414
11.0	0.313	0.329	0.308	0.369
15.0	0.208	0.238	0.239	0.248
20.0	0.152	0.170	0.180	0.170
25.0	0.117	0.121	0.141	0.131
30.0	0.095	0.102	0.122	0.112
35.0	0.081	0.087	0.097	0.083

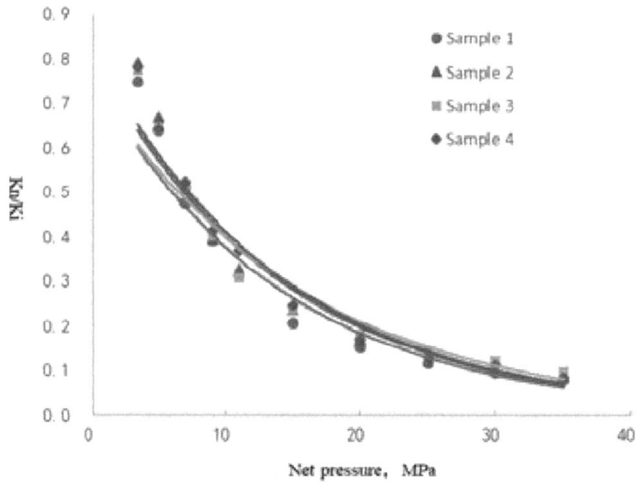

Fig 1 Stress sensitivity curves of the cores

By fitting the relationship between rock permeability and effective confining pressure, it is found that there is a good exponential relationship between them（Table 2）. It can be expressed as：

$$k_n = \mathrm{N}k_i e^{-\alpha(p_i - p)} \qquad (1)$$

Table 2 Stress sensitivity index fitting of core permeability

Core number	Exponential fitting expression	R^2
1	$y = 0.7721e^{-0.072x}$	0.9553

2	$y = 0.8186e^{-0.071x}$	0.9595
3	$y = 0.7609e^{-0.064x}$	0.9489
4	$y = 0.8331e^{-0.07x}$	0.9727

According to the principle of seepage mechanics, the square distribution expression of stable seepage pressure of gas reservoir is as follows.

$$p^2 = {p_e}^2 - \frac{{p_e}^2 - {p_{wf}}^2}{\ln(\frac{r_e}{r_w})} ln\frac{r_e}{r} \quad (2)$$

Formula (1) is deformed.

$$k_n = Nk_i e^{-\alpha\left\{p_i - \left[{p_e}^2 - \frac{{p_e}^2 - {p_{wf}}^2}{ln(\frac{r_e}{r_w})} ln\frac{r_e}{r}\right]^{\frac{1}{2}}\right\}} \quad (3)$$

2 Influence of condensate on gas phase relative permeability

The influence of reverse condensate oil is studied by numerical simulation method. By using single well numerical simulation, the variation law of condensate oil saturation at different distances from well during gas field development is obtained.

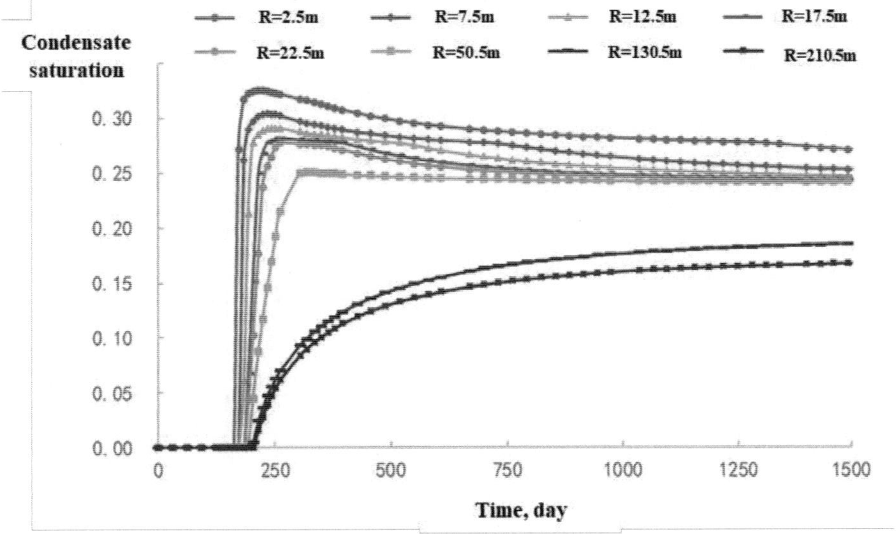

Fig.3 variation law of condensate oil saturation

It can be seen from Figure 3 that the condensate saturation decreases with the increase of radial distance, and tends to be stable after increasing with the extension of production time. The results show that the condensate saturation changes with time in the near well zone, and the distribution function of condensate saturation with radial distance $s_o(\mathrm{r})$ can be obtained at any time.

According to the relative permeability curve, the regression relationship between gas relative permeability and oil saturation can be obtained. The relative permeability curve of BZ gas reservoir in Bohai oilfield is shown in Fig. 4, the regression expression is:

$$k_{rg}(\mathrm{r}) = 4.7367s_o(r)^3 - 2.0497s_o(r)^2 - 2.1398s_o(r) + 1.0015 \quad (4)$$

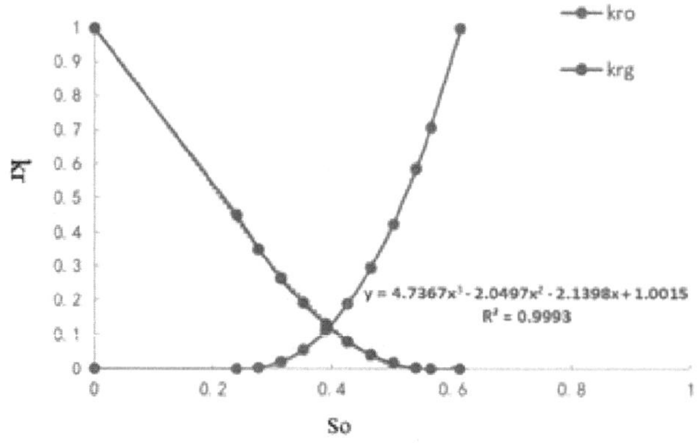

Fig.4 Relative permeability curve of Z gas reservoir

3 Improved gas well productivity evaluation model

The expression of effective permeability is:

$$k_g = k_{\mathrm{n}} \cdot k_{rg} \tag{5}$$

Replace formula (3) and formula (4) into formula (5):

$$k_g = Nk_i[4.7367s_o(r)^3 - 2.0497s_o(r)^2 - 2.1398s_o(r) + 1.0015]e^{-\alpha\{p_i - [p_e^2 - \frac{p_e^2 - p_{wf}^2}{\ln(\frac{r_e}{r_w})} ln\frac{r_e}{r}]^{\frac{1}{2}}\}} \tag{6}$$

The condensate saturation beyond a certain distance from the well is lower than the critical flow saturation, the distance is defined as the effective radial distance and expressed as r_o.

The whole stratum can be divided into two regions. The effective permeability can be expressed as:

$$k_g(\mathrm{r}, p_{wf}) = \begin{cases} Nk_a[4.7367s_o(r)^3 - 2.0497s_o(r)^2 - 2.1398s_o(r) + 1.0015]e^{-\alpha\{p_i - [p_e^2 - \frac{p_e^2 - p_{wf}^2}{\ln(\frac{r_e}{r_w})} ln\frac{r_e}{r}]^{\frac{1}{2}}\}}, & r_w < r < r_o \\ Nk_a e^{-\alpha\{p_i - [p_e^2 - \frac{p_e^2 - p_{wf}^2}{\ln(\frac{r_e}{r_w})} ln\frac{r_e}{r}]^{\frac{1}{2}}\}}, & r_o \le r \le r_e \end{cases} \tag{7}$$

① $r_w < r < r_o$, average permeability can be expressed as:

$$k_1(p_{wf}) = \frac{\int_{r_w}^{r_o} k_g(r, p_{wf}) \cdot 2\pi r dr}{\int_{r_w}^{r_o} 2\pi r dr} \tag{8}$$

② $r_o \le r \le r_e$, average permeability can be expressed as:

$$k_2(p_{wf}) = \frac{\int_{r_o}^{r_e} k_g(r, p_{wf}) \cdot 2\pi r dr}{\int_{r_o}^{r_e} 2\pi r dr} \tag{9}$$

According to the principle of seepage mechanics, the average gas permeability of the formation can be obtained:

$$k(p_{wf}) = \frac{k_1(p_{wf}) \cdot k_2(p_{wf}) \cdot \ln(\frac{r_e}{r_w})}{k_2(p_{wf}) \cdot \ln(\frac{r_o}{r_w}) + k_1(p_{wf}) \cdot \ln(\frac{r_e}{r_o})} \tag{10}$$

Based on the above research, the productivity evaluation formula of gas well considering stress sensitivity and condensate oil is obtained.

$$p_e^2 - p_{wf}^2 = \frac{1.273\times10^{-3}T\mu Z}{k(p_{wf})h}\ln\left(\frac{r_e}{r_w}\right) q_{sc} + \frac{2.162\times10^{-10}\gamma_g ZT}{r_w h^2 k(p_{wf})^{1.5}} q_{sc}^{\;2} \qquad (11)$$

4 Application

BZ is a rare condensate gas reservoir with complex reservoir and fluid characteristics. The main performance is low pressure difference between the initial formation pressure and the dew point (0.21-7.13MPa), the maximum amount of condensate is 40%, the reservoir permeability is 1.00~6.80mD, strong stress sensitivity.

The evaluation equation is applied to BZ gas field. The IPR curves of three cases are calculated, under the comprehensive influence of condensate oil and stress sensitivity, only considering the influence of condensate oil or stress sensitivity. The results are compared with those of the traditional binomial formula.

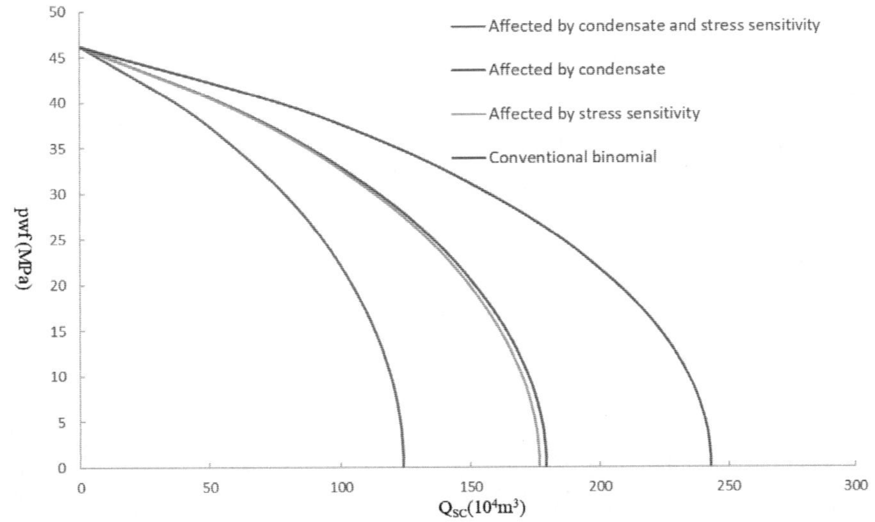

It can be seen that reservoir stress sensitivity and retrograde condensation have great influence on gas reservoir productivity. Under the combined effect of both, the productivity of gas reservoir is reduced by 50%. The IPR curves of gas wells under different formation pressures under the combined influence of stress sensitivity and retrograde condensation are further obtained.

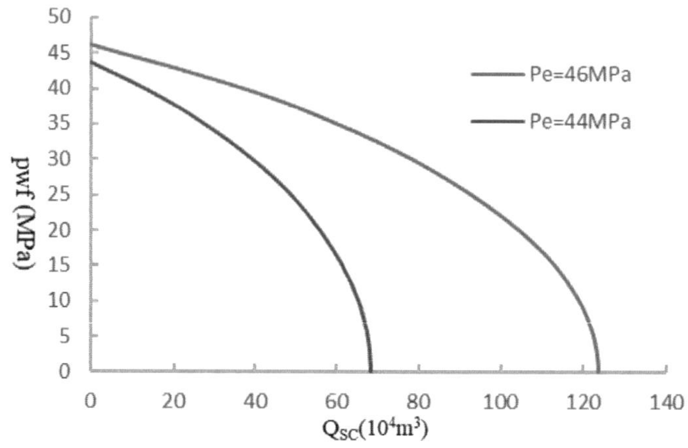

It can be seen that the formation pressure decreases, stress sensitivity and retrograde condensation increase, the productivity of gas wells also decreases obviously. The formation pressure drops by 2MPa, and the open flow rate of gas wells decreases by 57%. Therefore, in order to ensure the gas well productivity, it is necessary to adopt the development mode of gas injection and pressure maintaining, and the gas well production allocation should not be too high, in order to slow down the adverse impact of pressure drop.

6 Conclusion

(1) Stress sensitivity and condensate oil have great influence on the productivity of low permeability condensate gas reservoirs with low pressure difference between the initial formation pressure and the dew point. Most studies lack of comprehensive consideration of the two factors.

(2) Combined with laboratory experiment and numerical simulation, the formation permeability considering the effect of reservoir stress sensitivity and retrograde condensation is derived by using the principle of seepage mechanics, and then the evaluation method which can reflect the real productivity of gas reservoir is obtained.

(3) The research results show the necessity of gas injection to maintain reservoir pressure, and put forward suggestions for reasonable production of gas wells.

Symbol description

k_i-Original permeability, $10^{-3}\mu m^2$

k_n-Absolute permeability, $10^{-3}\mu m^2$

k_g-Effective permeability, $10^{-3}\mu m^2$

k_{rg}-Gas phase relative permeability, $10^{-3}\mu m^2$

p_i-Original formation pressure, MPa

p-Formation pressure, MPa

N-Stress sensitivity coefficient

α-Stress sensitivity index

p_e-Formation pressure at supply radius, MPa

p_{wf}-Bottom hole pressure, MPa

r_e-Supply radius, m

r_w-Wellbore radius, m

s_o-Condensate saturation near wellbore

r-Radial distance from well, m

T-Formation temperature, K

μ-Gas viscosity, mPa·s

Z-Gas deviation factor

h-Thickness of gas reservoir, m

β-Characteristic parameters of pore structure affecting turbulence and inertia resistance

γ_g-Relative density of natural gas

q_{sc}-Gas well production under standard conditions, m^3/d

References

[1] Fussel D.D. Single well Performance for gas condensate reservoir[J]. JPT, 1973, 25(12): 860-876.

[2] Jones J.R. and Raghavan R. Interpretation of Pressure buildup reservoirs in gas condensate well[J]. SPEFE, 1989.

[3] FAN Xueping, XU Xiangrong. Experimental and mechanism research about permeability damage with the change of stress[J]. PETROLEUM EXPLORATION AND DEVELOPMENT, 2002

[4] XUE Zhan, LIU Pengcheng. A New Method of Correction of Gas Condensate Well Productivity Considering Capillary Number[J]. Journal of Chongqing University of science and Technology (NATURAL SCIENCE EDITION), 2014.

[5] LUO Ruilan, CHENG Linsong. A new method of determining relationship between permeability and effective overburden pressure for low-permeability reservoirs[J]. Journal of China University of Petroleum, 2007.

Numerical simulation study of a complex naturally-fractured gas reservoir with extra-high condensate in Bohai Oilfield

GAO Zhennan, SU Yanchun, GENG Zhigang, LIU Weilin, HUANG Baogang

(CNOOC China Limited Tianjin Branch, Tianjin 300459, China)

Abstract:

The BZ condensate gas field is the largest natural gas field in the history of the Bohai bay basin, with proven reserves of more than 100 billion cubic meters of natural gas and 100 million cubic meters of condensate oil. This gas field has unique features, such as deep burial, fracture development, strong stress sensitivity, small pressure difference between formation pressure and dew point pressure, high condensate content, etc. There is no similar condensate-gas field development cases in the world and proper reservoir numerical simulation study is crucial for efficiently developing such a reservoir.

This paper develops a numerical simulation flowchart for this kind of complex naturally fractured condensate gas reservoirs. First, based on logging, core and well testing data, the reservoir of BZ gas field is divided into three different types of deposits in the vertical direction: glutenite, semi-weathered zone and inner buried hill. The fluid flow among the three different types of deposits is calculated based on the equivalent fluid flow conduction between adjacent grids. Secondly, the correlation of permeability changes with stresses is established based on experimental data and implemented in the reservoir simulation based on the imbedded programming module in the simulator. Thirdly, through PVTi numerical simulation, EOS are tuned based on 7 pseudo-components. Finally, a dual-porosity compositional model is established, and the numerical simulation study is carried out to optimize the development plan. The developed flowchart will be a very useful tool for modeling naturally-fractured condensate gas reservoirs.

The simulation results indicated that: 1) considering the strong stress sensitivity, gas injection can effectively increase the permeability of the reservoir and control the condensate precipitation; 2) considering the structural characteristics of the reservoir, gas injection in the high position can form top-gas drive and expand the swept volume; 3) considering the influence of fracture, injection- production line perpendicular to fracture trend can effectively delay the breakthrough of injected gas and expand the injection capacity; 4) considering the characteristics of condensate, keeping formation pressure higher than dew point pressure can minimize the loss of condensate oil; 5) considering the mechanism of gas injection displacement, slug gas injection can form pressure pulse, expand the sweep volume and improve the recovery of condensate oil. Those findings can also be used to guide the development of other gas condensate reservoirs.

Key words: buried-hill fracture; condensate gas reservoir; stress sensitive characterization; reservoir numerical simulation; cyclic gas injection development

● 0 Introduction

BZ condensate gas field is the largest natural gas field in Bohai Sea, with proven natural gas reserves of over 100 billion cubic meters and proven condensate reserves of more than 100 million cubic meters. The gas field has the following geological characteristics: buried deep(-4400~-5500m), complex seepage (fracture development), stress sensitivity (permeability loss 89.2%), and the fluid characteristics are as follows: low pressure difference between formation pressure and dew point pressure (1.07MPa), ultra-high condensate (711g / m3), large amount of reverse condensate (32.3%). The efficient development of BZ condensate gas field is a world problem! It is found that there is no similar gas field development case at home and abroad, so it is impossible to accurately evaluate the development effect of this gas field. Therefore, it is urgent to carry out reservoir numerical simulation research to provide support for early development plan and subsequent efficient development[1-3].

● 1 Research method

According to the reservoir classification, the complex seepage system gas reservoir model is established. Based on the stress sensitive experiment, the variation law of seepage capacity is regressed. Based on the matching of fluid phase, the dual medium component model is established. And the development mode and gas injection development strategy are studied through reservoir numerical simulation.

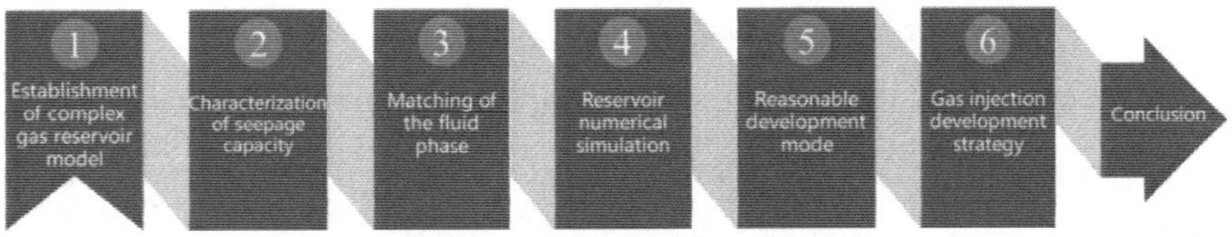

Fig 1. Technology Roadmap

● 2 Gas reservoir model

2.1 Reservoir zonation

Combined with the analysis of logging, core and testing data, BZ gas field has good reservoir connectivity, and it can be divided into sandy conglomerate zone, weathered zone and buried hill inner zone from top to bottom[4].

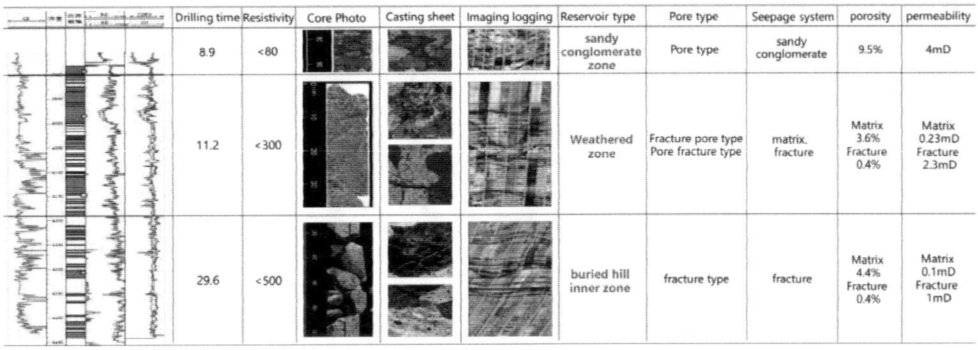

Fig 2. Vertical development characteristics of BZ condensate gas field reservoir

2.2 Establishment of complex gas reservoir model

Establish a structural model based on the understanding of reservoir zoning; and using lithofacies modeling and discrete fracture network modeling to generate matrix and fracture porosity, permeability and interporosity flow coefficient. By zeroing the matrix conductivity inside the buried hill, the coexistence of three types of complex seepage models in the BZ gas field can be realized.

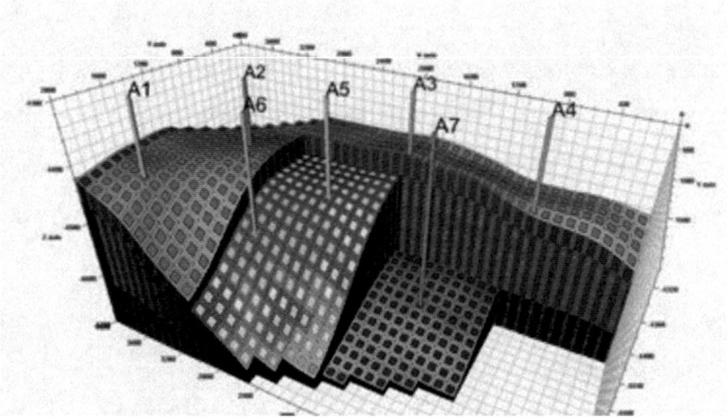

Fig 3. Gas reservoir model of complex seepage system

3 Stress sensitive characterization

3.1 Stress sensitivity analysis

According to the strong stress sensitivity of BZ gas field, the relationship between core permeability residual coefficient and net stress is fitted, and there is a good power function relationship between them; with the increase of net stress (reservoir pressure decreases), the rate of permeability loss gradually decreases[5-11]. In early stage, the net stress increased to 10MPa, Kr reduced by 60%; in mid stage, the net stress increased to 20MPa, Kr reduced by 20%; when reach the abandonment pressure(the net stress increased to 30MPa), the remaining permeability is about 10%.

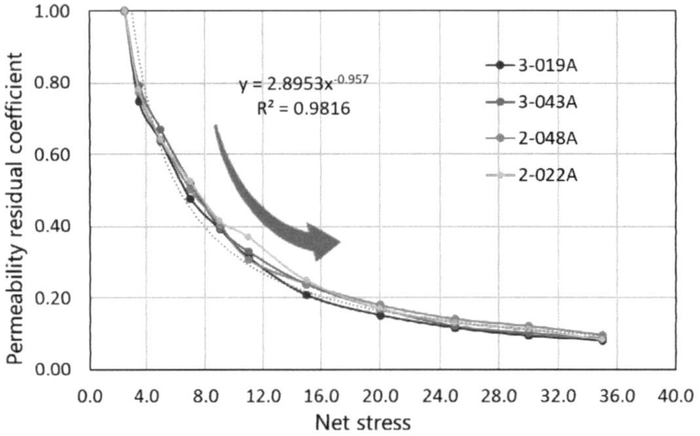

Fig 4. Permeability versus stress curve

$$K_r = 2.8953\ (P - P_0)^{-0.957}$$

K_r: Ratio of permeability to initial permeability, zero dimension;

P: Reservoir pressure at any time, MPa;

P_0: Initial formation pressure, MPa。

3.2 Stress sensitive characterization

According to the regression relationship, the characterization code of reservoir physical properties with time is compiled, and the sensitivity analysis is carried out by reservoir numerical simulation.

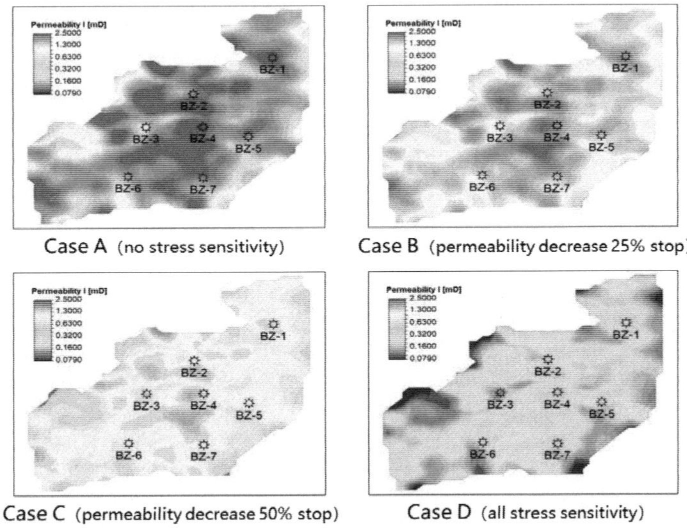

Fig 5. Comparison of permeability of different stress sensitive cases

● 4 Characterization of fluid characteristics

4.1 Fluid phase analysis

The BZ gas field is a condensate gas reservoir with high saturation and ultra-high condensate content through the study of fluid phase behavior characteristics, and the maximum condensate oil saturation is as high as 41%. When the pressure drops about 15MPa, the maximum saturation can be reached. Condensate oil precipitation seriously pollutes the reservoir, increases gas phase seepage resistance, and reduces gas well productivity and recovery[12-13].

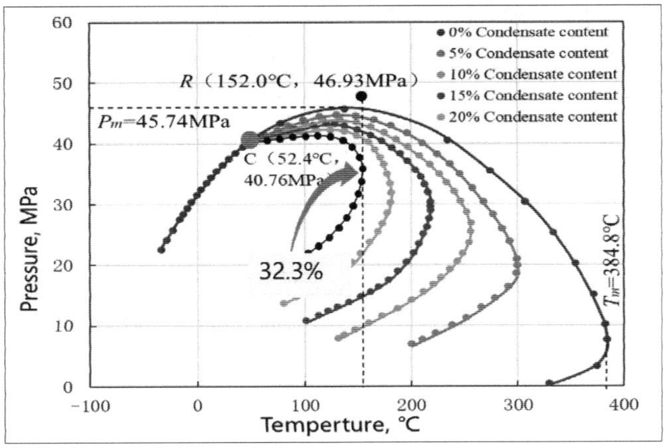

Fig 6. Phase diagram of fluid temperature and pressure in BZ gas field

4.2 Fluid phase matching

Considering the accuracy and speed of numerical simulation, 4 cases of component splitting combinations were set according to the principle of similarity combination. Combined with the calculation time and results of Numerical simulation of component model, the 7-component splitting plan is determined as the best plan.

11 components：C1、N2、C2、CO2、C3、C4、C5、 C6、C7-C10、C15+C22、C23+

7 components：C1+N2、C2+CO2、C3-C4、C5-C6、C7-C10、C15+C22、C23+

5 components：N2、CO2、C1、C2- C6、C7+

3 components：C1+N2、C2-C6+CO2、C7+

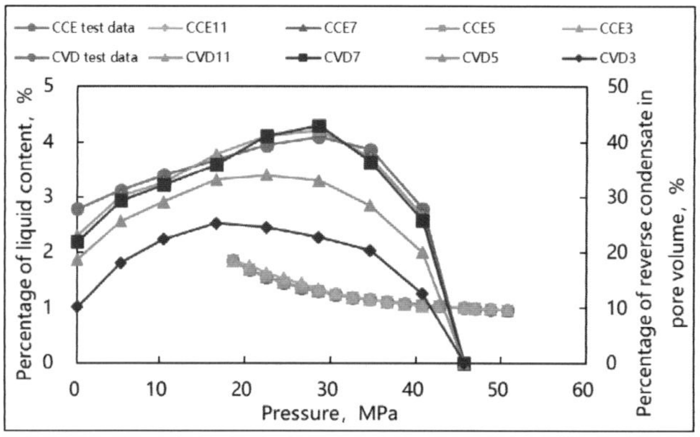

Fig 7. Comparison of experimental fitting accuracy

● 5 Study on gas reservoir numerical simulation

5.1 Optimization of development mode

As oil and gas fluids are produced, the formation pressure decreases, the physical properties become worse, the seepage ability decreases, and the recovery of gas field decreases. In depletion exploitation mode, the formation pressure drops rapidly, the condensate oil quickly reaches the maximum condensate velocity, and lost in the formation, so the condensate oil recovery rate is low. In gas injection development mode, the formation pressure is maintained through gas injection, and the retrograde condensate rate is controlled, and the condensate oil recovery is high. Considering reservoir stress sensitivity and retrograde condensate characteristics, from the perspective of improving oil and gas recovery and reservoir protection, it is recommended that the BZ condensate gas field development adopts first cycle gas injection and then depletion development[14-17].

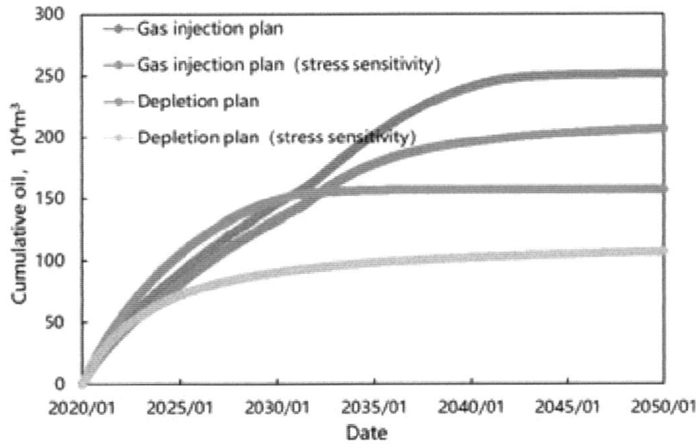

Fig 8. Comparison of different development mode

5.2 Cycle gas injection development strategy

Condensate gas reservoirs are usually developed with gas injection. The main purpose is to supplement formation energy, reduce condensate oil precipitation, and improve condensate oil recovery. When evaluating the effect of development, gas injection sweep volume and formation pressure maintenance level are two very important factors.

5.2.1 Optimization of injection and production

During gas injection development, the gas injection well position is directly related to the dry gas sweep volume. The structural and the fracture trend are two important factors.

A. Structural factor

The sandy conglomerate zone in structural high area of the BZ gas field has good physical properties. Priority is given to production in the early stage, and gas injection at the low area. In later stage, the condensate oil is enriched in the structural low area by gravity, and the top gas injection forms secondary gas and top drive. Vertical displacement, effectively expanding the gas sweep volume.

Tab 1. Development effect of structural factor

Plan	Cumulative gas 10^8m^3	Gas recovery %	Cumulative oil 10^4m^3	Oil recovery %
Inject in low area 10 years	36	53	181.3	26.6
Inject in low area 5 years,And in high area 5 years	36.6	53.3	192.4	28.3
Inject in high area 10 years	36.1	53.1	189.2	27.8

B. Fracture trend factor

When the injection-production line is parallel to the fracture trend, it is easy to form gas channeling, which reduce oil and gas recovery. When the injection-production line is vertical to the fracture trend, it can maximize the sweep volume, and effectively maintain formation energy and the seepage ability, and inhibit condensate oil to separate out.

Tab 2. Development effect of fracture trend factor

Plan	Cumulative gas 10^8m^3	Gas recovery %	Cumulative oil 10^4m^3	Oil recovery %
The injection-production line isparallel to the fracture	35.1	51.1	168.6	25.1
The injection-production line isperpendicular to the fracture	36.6	53.3	192.4	28.3
The injection-production line isoblique to the fracture	36	53	186.1	27.3

5.2.2 Reasonable pressure maintenance level

The most important thing in the development of condensate gas reservoirs is control the formation pressure to slow down the precipitation of condensate oil. The timing of gas injection and the rate of reinjection are two key factors.

A. Timing of gas injection

The timing of gas injection determines the degree of condensate oil loss, seepage capacity, and condensate precipitation rate during gas injection development. Since the difference between the formation pressure and the dew point pressure of the BZ gas field is very small, gas injection at the beginning of production can effectively delay the retrograde condensation time and reduce pollution, which is the optimal solution.

Tab 3. Development effect of timing of gas injection

Plan	Cumulative gas 10^8m^3	Gas recovery %	Cumulative oil 10^4m^3	Oil recovery %
Formation pressure 30MPa (4 years later)	26.4	48	127	23.4
Formation pressure 40MPa (2 years later)	26.3	47.9	132	24.4
Formation pressure 50MPa (0years later)	26.3	47.9	135.9	25.1

B. Re-injection rate

The higher the gas re-injection rate, the more stable the formation pressure remains. Before the formation pressure is higher than the dew point pressure, the 100% re-injection rate can always maintain the formation pressure higher than the dew point pressure, minimize the loss of condensate oil, and increase condensate oil recovery rate.

Tab 4. Development effect of Re-injection rate

Plan	Cumulative gas 10^8m^3	Gas recovery %	Cumulative oil 10^4m^3	Oil recovery %
100% re-injection rate	36.2	52.7	198.5	29.2
80% re-injection rate	36.4	53	196.1	28.8
50% re-injection rate	36.6	53.3	192.4	28.3

5.2.3 Optimization of gas injection mode

Considering the mechanism of gas injection and the demand for gas supply in the Beijing-Tianjin-Hebei region in winter, a comparative study of 3 plans: cyclic gas injection, slug gas injection, and Inject gas and soak. Circulating gas injection is prone to gas channeling, thereby reducing the condensate oil and gas recovery rate. The plan of inject gas and soak delays the gas channeling time on the basis of maintaining formation pressure, so the recovery rate is higher. In slug gas injection plan, the pressure pulse formed by the gas injection and gas production cycle can effectively expand the swept volume. And when injection stopped, the condensate in matrix will flow into the fracture due to the gravity. Therefore, the slug gas injection plan has the highest oil and gas recovery.

Tab 5. Development effect of Optimization of gas injection mode

Plan	Cumulative gas 10^8m^3	Gas recovery %	Cumulative oil 10^4m^3	Oil recovery %
Cyclic gas injection	35.3	50.2	188.4	26.8
Slug gas injection	37	53.9	198.4	29.1
Inject gas and soak	36.5	52.3	194.7	28.2

• 6 Conclusion

(1) The BZ gas field are divided into sandy conglomerate zone, weathered zone and Buried hill inner zone. The gas reservoir model of complex seepage system is established by the zeroing matrix conductivity inside the buried hill.

(2) The BZ gas field has strong stress sensitivity, and Nearly 90% loss of seepage capacity under abandonment pressure. Therefore, it is necessary to carry out numerical simulation research through time-varying reservoir physical properties.

(3) BZ gas field is a gas field with high saturation and high condensate content. Through PVTi matching and numerical simulation sensitivity analysis, the 7-component splitting plan is determined as the best plan.

(4) The numerical simulation study of the BZ gas field shows that: firstly inject gas in low area of the structure and then in the high area, inject gas at the start of production, the injection rate is 1.0, and seasonal slug gas injection is the optimal plan. This research provides strong support for

ODP program preparation and subsequent efficient development.

Reference:

[1] Fang Na, Jiang Guanghong, Li Yunpeng, et al. Development characteristics and potential tapping direction of buried-hill fractured oil reservoirs[J]. Special Oil & Gas Reservoirs, 2017,24(3):90-94.

[2] Xu Chunqiang, Zhang Zhen, Zhang Xintao, et al. Tectonic evolution and hydrocarbon accumulation model of western sag in western Bohai sea[J]. Journal of Northeast Petroleum University, 2018, 42（1）: 68-76.

[3] Zhang Feifei. Characterization and evaluation of pore-throat structures in ultra-low permeability and tight sandstones and their impacts on petrophysical properties: a case study of the lower Cretaceous in northern Songliao Basin[J]. Journal of Northeast Petroleum University, 2018,42(5):54-62.

[4] Tong Kaijun, Zhao Chunming, Lv Zuobin, et al. Petroleum exploration and development[J]. Petroleum Exploration and Development, 2012,39(1):56-63.

[5] Lin Guangrong, Liu Xuegang, Lu Yan, et al. Impact of stress change on percolation pattern of two-phase in low permeability reservoir[J]. Special Oil & Gas Reservoirs, 2008,15(1):43-45.

[6] Yang Bin, Hou Dali, You Gang, et al. Research and application of stress sensitive experiment on low permeable reservior[J]. Reservoir Evaluation and Development, 2012,5(2):41-45.

[7] Cao nai, Lei Gang. Stress sensitivity of tight reservoirs during pressure loading and unloading process[J]. Petroleum Exploration and Development, 2019,46(1):132-138.

[8] Jiang Ruizhong, Huang Lei, Cui Yongzheng, et al. Transit pressure analysis of a horizontal well in fractured composite reservoirs with stress-sensitivity effect[J]. Journal of Northeast Petroleum University,2018,42（3）: 84-91.

[9] Fang Yang. Research on relationship between deliverability of gas reservoir and the stress-sensitiveness of the reservoir[J]. Well Testing,2009,18（6）:18-21.

[10] Fu Lanqing. The effect of fractures on stress sensitivity and seepage characteristics in sandstone reservoirs[J]. Journal of Yangtze University（Natural Science Edition）,2016,13（20）: 14-18.

[11] Zhang Rui, Ning Zhengfu, Zhang Haishan, et al. New insights and discussions on stress sensitivity of fractured tight reservoir[J]. Natural Gas Geoscience, 2016,27(5):918-923.

[12] Wang Lezhi, Qi Zhilin, Jiang Yiwei. Study on retrograde condensate characteristics of deep condensate gas reservoirs with low permeability [J]. Natural Gas Technology, 2007, 1（1）: 52-54.

[13] Xiao Lixian, Du Jianfen, Guo Ping, et al. Pollution caused by retrograde condensate in condensate gas reservoir[J]. Fault-block Oil & Gas Field, 2009, 16（4）: 102-104.

[14] Cheng Liang, Zhang Xiaodong, Wang Juan, et al. Study on periodic gas injection mechanism of condensate gas reservoirs with high to high-super content of condensate oil in central Tarim by numerical simulation[J]. Journal of Xi'an shiyou University (Natural Science Edition),2016,31(2): 57-64.

[15] Wang Jing, Liu Huiqing, Liu Renjing, et al. Numerical simulation for low-permeability and extra-low permeability reservoirs with considering starting pressure and stress sensitivity effects[J]. Chinese Journal of Rock Mechanics and Engineering, 2013,32(2):3317-3327.

[16] Li Yong, Zhang Jing, Wu Xueyong, et al. A new reservoir simulation approach for fractured gas-condensate reservoirs[J]. Petroleum Exploration and Development, 2010,37(5):592-595.

[17] Jiang Hai，Yang Zhaozhong，Hu Yuehua，et al. Optimization of the cyclic gas injection parameters in condensate gas reservoir[J]. Journal of Xi'an shiyou University (Natural Science Edition),2009,24（2）:54-56.

602060 Monitoring and Detection of Byproduct Crystallization Using in-Situ Microscopy

Abstract:

In-situ monitoring of solid-phase composition during crystallization is often accomplished by Raman spectroscopy, but not all compounds in the system of interest are strongly Raman active. We are developing an image-based analysis to detect a second undesired solid phase with in-situ microscopy before other PAT tools detect the appearance of the solid. Our system involves the reactive crystallization of beta-lactam antibiotics, wherein a byproduct of the enzyme-catalyzed reaction can crystallize within the antibiotic slurry.

The current target of our research is cephalexin, a critical antibiotic, with a byproduct of phenylglycine; however, several other beta-lactam antibiotics exhibit the same behavior, including ampicillin, cefaclor, and cefazolin. The byproduct phenylglycine is present in solution at concentrations on the order of tens of millimoles per liter with a supersaturation less than 1.5; detecting a concentration decrease of about 10 millimolar (indicating initiation of phenylglycine crystallization) within a mixture of several components at >100 millimolar concentration proved difficult with in-situ liquid-phase tools like ATR-FTIR. Phenylglycine shares many functional groups with the desired product and reactants, making its spectroscopic fingerprint similar to the product and reactants. With in-situ Raman spectroscopy, we were unable to detect low loadings of solid phenylglycine in a cephalexin slurry, either because of overlap with the complex solution of similar functional groups or because of low Raman activity compared to cephalexin. Changes in focused beam reflectance measurement (FBRM) counts and chord length distributions did not show up at low phenylglycine solids loadings, nor did changes in turbidity. The solution pH exhibited the strongest change upon formation of phenylglycine solids, but pH is not a robust measure of the appearance of phenylglycine as many (often benign) system perturbations give similar pH responses. As crystals of beta-lactam antibiotics are (predominately) needle-shaped and phenylglycine crystals are plate- or column-shaped, the difference in shape, as well as backscatter intensity, can be used to differentiate the two classes of particles. Additionally, phenylglycine crystals are known to partition to air-water interfaces, a fact that can be exploited as the crystals tend to clump on bubble surfaces making them easier to identify in images (Hoeben et al. 2009).

Image analysis enables features and statistics of in-situ images are extracted and recorded in real time, and we have shown that phenylglycine can be detected from in-situ images. First, objects within the image are identified by edge detection. Then properties of the objects such as size, aspect ratio, pixel intensity, and more, as well as distributions of these properties for every object in each image are used to detect phenylglycine. By tracking these features and statistics, a classifier can be constructed to identify phenylglycine from image data and can be implemented in real-time for monitoring and detection. The statistics can also be recorded as a historical record, which is more feasible than storing the experiment as complete images.

Hoeben, M. A., et al. (2009). "Design of a Counter-Current Interfacial Partitioning Process for the Separation of Ampicillin and Phenylglycine." Industrial & Engineering Chemistry Research **48**(16): 7753-7766.

598054 Integration of In-Silico Tools and High-Throughput Automation for Enhanced Crystallization Development

Abstract:

The presentation will discuss a case study, where multiple in-silico solubility prediction tools and high throughput automation were utilized to accelerate solubility and morphology screening in binary and ternary solvent systems. Regression-based prediction tools using single solvent solubility data and first-principle models were used to identify potential solvent systems to improve crystallization performance. Over 85 solvent combinations were prepared using an automated platform for solubility measurement and morphology screening, and several synergistic solubility behaviors were observed. Experimental solubility trends conformed well with in-silico predictions, indicating these tools can be a beneficial first screen at identifying synergistic solvents to explore. The presentation will delve into the utility of an integrated approach comprising of the solubility modeling and automated experimental screening platform in the efficient data generation to enable timely crystallization process development.

All authors are employees of AbbVie and may own AbbVie stock. AbbVie sponsored and funded the study; contributed to the design; participated in the collection, analysis, and interpretation of data, and in writing, reviewing, and approval of the final publication.

606112 Utilization of Quality-By-Control for Rapid Process Design of Agrochemical Crystallization

Abstract:

Crystallization is a key separation process that is used in the agrochemical industry to separate the agrochemical actives from impure solutions. A poorly designed crystallization step can lead to poor yield, low purity, and long filtration time. Traditionally, a series of design of experiments led by the Quality-by-Design (QbD) approach is used to optimize operating profile of the crystallization process (i.e. temperature profile, solvent/antisolvent addition profile, pH profile).[1] An exhaustive list of experiments covering all factors is needed to explore the whole design space experimentally. To minimize the number of experiments and personnel exposure to toxic chemicals, Quality-by-Control (QbC) utilizes control strategies implemented for the target critical quality attributes (CQAs) to determine the operating profile of the process.[2] For rapid process design, direct design or model free approaches can be used to quickly determine an operating profile of a process leading the system to the desired CQAs. Two direct design approaches, direct nucleation control (DNC) and supersaturation control (SSC), are used in this work to control the crystallization of a model agrochemical compound. Both DNC and SSC utilize process analytical technology (PAT) tools to acquire data and control the process. DNC utilizes closed-loop feedback control approach with particle measurements from focused beam reflectance measurement (FBRM) to generate temperature cycles.[3] On the other hand, SSC uses concentration measurement via UV/Vis detector in a closed-loop feedback control approach to control the concentration by manipulating temperature.[4]

In this work, both direct design approaches were implemented to improve the particle shape, length, and filtration time of needle shaped particles. Preliminary results have indicated significant improvement in the overall crystallization-filtration process performance. Using a DNC-based direct design approach, the filtration time was reduced by a factor of four compared to the standard recipe. The improved procedure not only reduces the unit operation cycle times, but also improves the particle shape for better downstream operations (i.e. drying, transport). The analysis of the product crystals using Ultra Performance Liquid Chromatography (UPLC) indicates that the impurity profile of the agrochemical compound was not compromised with the thermocycles. With SSC-based direct design approach, the concentration of the system was controlled via a set point of the absolute supersaturation. The system was able to perform a cooling profile and generate uniform needle particles. The QbC-based direct design approaches were also evaluated in larger scale (5 L) crystallization systems indicating that the feedback control based rapid design approach can lead to fast and robust scale-up of agrochemical crystallization processes.

References:

(1) Bondi, R. W.; Drennen, J. K. *Quality by Design and the Importance of PAT in QbD*; Academic Press, 2011; Vol. 10.

(2) Su, Q.; Ganesh, S.; Moreno, M.; Bommireddy, Y.; Gonzalez, M.; Reklaitis, G. V.; Nagy, Z. K. A Perspective on Quality-by-Control (QbC) in Pharmaceutical Continuous Manufacturing. *Comput. Chem. Eng.* **2019**, *125*, 216–231.

(3) Bakar, M. R. A.; Nagy, Z. K.; Saleemi, A. N.; Rielly, C. D. The Impact of Direct Nucleation Control on Crystal Size Distribution in Pharmaceutical Crystallization Processes. *Cryst. Growth Des.* **2009**, *9* (3), 1378–1384.

(4) Saleemi, A. N.; Rielly, C. D.; Nagy, Z. K. Comparative Investigation of Supersaturation and Automated Direct Nucleation Control of Crystal Size Distributions Using ATR-UV/Vis Spectroscopy and FBRM. *Cryst. Growth Des.* **2012**, *12* (4), 1792–1807.

609295 Dual Chamber and Membrane Embedded Continuous Microfluidic Device for MOFs Screening

Abstract:

Metal Organic Frameworks (MOFs) are a class of three-dimensional materials comprised of organic and inorganic building unit cells. The self-assembly of metal ions and organic linker through coordinate bonds results in synthesis of infinite and diverse micro-porous structures. High porosity and ultra-high surface to volume ratio of MOFs along with their fascinating chemical and design flexibility have prompt to numerous applications. Despite of the ubiquitous applications of MOFs, great challenges still exist for developing a relatively cheap, safe and readily scalable process that is sought after by industries. Here we have developed a continuous-flow synthesis of MOFs by modifying the design of the previously developed microfluidic mixer to accommodate for the separated nucleation and growth chambers to induce synthesis of the uniform nanosized particles at constant supersaturation. Both chambers have identical designs, yet the growth chamber has a membrane that separates inlets and outlets. Nucleation chamber is initially filled with premixed solutions of the aqueous solutions of the metal ligand and organic ligand while keeping its outlet open. After running the system for few minutes to ensure all the air has been moved out and system is filled with the solution, the outlet is closed, and circulation is started. Using an optical system, the nucleation of the crystals can be observed within the nucleation chamber. By that time, outlet is opened, and nucleated crystals will move to the growth chamber. Crystals growing in the growth chamber are kept inside and extracted for the further analysis. Using this microfluidic platform, parametric screening is performed and the effect of the ratio of metal ligand to the organic linker on the morphology, growth rate and yield of the synthesized MOFs are measured.

598093 In Situ Polymorph Transformation of a Pharmaceutical Intermediate

Abstract:

We discovered that the polymorph of a pharmaceutical intermediate transformed *in situ* during crystallization from solution. Initially, we noted an anomaly in the signal recorded during a crystallization experiment using a Mettler-Toledo Particle Track G400 instrument, an unexplained decrease followed by an increase in the average chord length of the particles. At roughly the same time, we discovered that there is a clear change in the rheology of the suspension. Rapid crystallization early in the process produces a suspension with a yield-stress that flows and mixes poorly. This gradually transforms during subsequent antisolvent addition to a free-flowing suspension without yield stress. We determined that the phenomenon causing both the observed change in particle size and rheology is crystal polymorph transformation accompanied by a dramatic change in habit. Our first indication of the transformation was obtained observing suspension sampled early in the crystallization under an optical microscope. We observed a clear habit change with time. We then further explored this transformation using simultaneous *in situ* Raman spectroscopy (Kaiser RXN1-785) and microscopy (Mettler-Toledo PVM and EasyViewer). We verified that a polymorph transformation occurs during the antisolvent addition phase of crystallization, accompanied by a change in crystal habit. The initial form grows rapidly as acicular crystals, followed by a gradual transformation to a second, more stable form characterized by roughly equant crystal aggregates. This transformation is responsible for the change in suspension rheology and the observed change in particle size. The dependence of the transformation on process conditions and seeding, as well as the kinetics of the transformation, have been determined. Knowledge of this transformation allowed us to understand the experimental observations and enabled confident scale-up.

608146 Detection of Morphology and Modality Changes during Crystallization Using in Situ Chord Length Monitoring and Inference System Model.

Abstract:

A new framework has been recently introduced to correlate in situ chord length distribution (CLD) measurements and offline particle size distribution (PSD) data (Irizarry et al, 2017) and applied to different particle morphologies (Schoell et al, 2019). In this work we are expanding this modelling scheme to consider multimodal distribution measurements (Irizarry et al, 2020). This work also applies the proposed approach to predict if the particle morphology or distribution of different APIs can be predicted from a CLD signal. To accomplish this a data-driven inference system and a more robust CLD2PSD model was built for the studied pharmaceutical compounds since the differences in optical properties and particle morphology require a unique model for each individual substance. The inference system is trained on PSDs annotated by modality and morphology, and the individual models have been combined with online solute concentration data to allow for the estimation of crystal growth, secondary nucleation and agglomeration kinetics, depending on the supersaturation level encountered during the process. The validity of this approach is studied in terms of classification and prediction error rates on test sets. The main advantage of the proposed approach is that all data is generated online and in situ, thus avoiding large sampling errors. Also, it can operate in areas of high concentrations and on small particle sizes where PVM pictures are more difficult to resolve.

References

- Irizarry, R. et al., 2017. Data-driven model and model paradigm to predict 1D and 2D particle size distribution from measured chord-length distribution. Chemical Engineering Science, 164, pp. 202-218.
- Schoell, J. et al., 2019. Determining particle-size distributions from chord length measurements for different particle morphologies. AIChE Journal, 65, e16560
- Irizarry, R. et al., 2020. CLD-to-PSD model to predict bimodal distributions and changes in modality and particle morphology. Submitted to Chemical Engineering Science.

611603 Industrial Scale Downstream Process Development for Low Concentration Amino Acid Product Using PAT

Abstract:

This work utilizes PAT to develop a downstream purification process on an industrial scale for a low yielding synthetic amino acid from a dilute chemical reaction product. Cooling crystallization optimized by use of inline IR measurement was used to remove 90% of the highest impurity paving the way for efficient use of continuous chromatography to remove lower concentration impurities. The steps leading up to the final design of the Downstream Process (DSP) process are outlined in this work.

The food grade product, an amino acid (AA) is used globally for animal feed as well as human consumption. Production is targeted to be on several MT/year scale. The process of synthesis of the AA involves several reactions involving organic and inorganic compounds.

The resulting reaction product is dilute, over 50% water and contains a higher molar ratio of impurities in the form of organic and inorganics compounds than the product AA. The salts in the reaction product are metal sulfate and sulfite, both at concentrations higher than the AA. Ion exclusion chromatography pulse test results were promising with AA eluting after all impurities. However, scale up to continuous chromatographic separation using Simulated Moving Bed (SMB) chromatography failed due to pressure build up in the columns with the sulfates precipitating out and high ionic load causing osmotic pressure to build up, shattering the resins.

To allow the use of SMB chromatography, precipitation of sulfates prior to chromatography was imperative. Various methods of sulfate salt crystallization including evaporative, cooling and antisolvent crystallization were studied. Evaporative crystallization lead to formation of anhydrous sulfates while cooling crystallization forms decahydrate sulftates, removing considerable amounts of solvent in the process. In line ReactIR® was used to determine the end point for cooling crystallization when the sulfate concentration in the ML plateaued. Given the translucent nature of decahydrate crystals Inline particle track FBRM® was not capable of accurate crystal count measurement. Nucleation and growth kinetics were estimated by off line microscopy as well as real time solution concentration measurement from ReactIR. Use of PAT allowed for faster optimization of the salt precipitation step with 90% of sulfates removed prior to SMB. Salt precipitation lead to some product loss which required an additional cake wash step to recover AA and recycle back into the process with the feed. Permeability of decahydrates crystals were considerably better than anhydrous crystals allowing for better wash and product loss reduction in salt precipitation step.

A second round of crystallization process is required for purification of the AA in the SMB extract from 95% purity to commercial grade product. Inline particle track FBRM(R) was utilized for kinetic measurements in that process.

599917 Highly Selective Carbon Molecular Sieve Membranes Derived from Polybenzimidazole Doped with Pyrophosphoric Acid for H_2/CO_2 Separation

Abstract:

Membrane technology with superior H_2/CO_2 separation properties at high temperatures has attracted significant interest for pre-combustion CO_2 capture. Polybenzimidazole (PBI) is a leading material for this separation with H_2 permeability of 27 and H_2/CO_2 selectivity of 16 at 150 °C because of the strong size-sieving ability. We demonstrate that the PBI can be first cross-linked with pyrophosphoric acid (PPA) and then carbonized to form carbon molecular sieves (CMS) membranes achieving more superior H_2/CO_2 separation properties. The effects of the doping level (defined as the ratio of the PPA to the PBI repeating units) and carbonization temperature on the separation properties are thoroughly characterized. For example, at a doping level of 0.34, the blend (PBI-$PPA_{0.34}$) exhibits H_2/CO_2 selectivity of 72 with H_2 permeability of 32 Barrer at 150 °C. However, a further increase in the doping level to 0.42 decreases H_2 permeability to 18 Barrer and H_2/CO_2 selectivity to 42. When carbonized at 600 °C, the obtained CMS of PBI-$PPA_{0.42}$ exhibits H_2 permeability of 460 Barrer and H_2/CO_2 selectivity of 37 at 150 °C, surpassing the Robeson's upper bounds. By comparison, the CMS derived from PBI at 600 °C show a similar H_2 permeability (~450 Barrer) but much lower H_2/CO_2 selectivity of 5.3 at 150 °C. Moreover, we have optimized the carbonization to be 600 °C or below so that the PBI can still retain its mechanical properties, enabling excellent processability for the CMS membranes. We will also present the long-term testing results with simulated syngas in the presence of water vapor. The structure/property relationship in these CMS materials will also be elucidated.

599953 Novel Facilitated Transport Membrane and Process for Post-Combustion Carbon Capture

Abstract:

Large-scale application of membrane in post-combustion carbon capture has been limited by the trade-off between CO_2 permeance and CO_2/N_2 selectivity of most polymeric membrane materials. In order to overcome this limitation, research efforts on facilitated transport membrane (FTM) have been devised with the objectives of (1) developing carriers with high CO_2 loading capacity and reactive diffusivity, and (2) designing membrane processes that can capitalize on the outstanding selectivity shown by FTM. In this presentation, a novel FTM was synthesized in a composite membrane configuration with a 170-nm selective layer coated on a nanoporous support. In the selective layer, polyvinylamine was used as fixed-site carrier and an amino acid salt, synthesized by deprotonating sarcosine with 2-(1-piperazinyl)ethylamine, was blended as mobile carrier. The membrane was used to fabricate 1.4-m^2 spiral-wound modules, which exhibited a CO_2 permeance of 1450 GPU and a CO_2/N_2 selectivity of 185 at 67°C with actual flue gas at the National Carbon Capture Center in Wilsonville, AL, USA.

In addition, for the range of CO_2 partial pressure relevant to carbon capture from coal-derived flue gas, significant increase in CO_2 permeance can be achieved upon the bulk removal of CO_2. For instance, at 67°C, the CO_2 permeance can increase from 1464 to 1918 GPU when the CO_2 partial pressure reduces from 74.1 to 3.9 kPa. Such a carrier saturation phenomenon has been modeled and incorporated into a two-stage membrane process featuring partial retentate recycle. It has been demonstrated that the bulk CO_2 removal reduces the CO_2 partial pressure gradually in the membrane module. This feature mitigates the carrier saturation and results in an uprising CO_2 permeance upon the CO_2 removal. For the membrane performance at 67°C, an attractive capture cost of \$41.75/tonne can be achieved for 90% CO_2 capture in addition to offering the reduced system footprint.

602659 A Hybrid Membrane/Adsorptive Reactor Process for Pre-Combustion CO_2 Capture. Experimental, Modelling and TEA Studies

Abstract:

The growing demand for energy worldwide combined with concerns about global warming due to CO_2 emissions from fossil fuel-burning power plants motivates advanced approaches that can produce electricity from fossil fuels without harmful CO_2 emissions. One popular approach is the Integrated Gasification Combined Cycle (IGCC) power generation process integrated with carbon capture and storage (CCS) technologies. In IGCC plants, the fossil fuel (e.g., biomass/coal) is first gasified to produce a syngas mixture which is then first cooled-down and treated in a cold-gas-clean-up unit (CGCU) to remove trace contaminants (e.g., COS, H_2S, HCI, Hg, etc.), and is then heated-up to the appropriate temperature to further convert the CO into H_2 and CO_2 via the water gas shift (WGS) reaction. In its place this group has proposed a hybrid membrane/adsorptive reactor (MR-AR) process with enhanced yield and selectivity due to the simultaneous removal from the reaction zone of both H_2 and CO_2, to help overcome the equilibrium limitations of the WGS reaction. This technology provides added flexibility for the IGCC application for which, in addition to efficient high-purity H_2 production, CO_2 recovery and purity are also key drivers.

We study here a hybrid system combining a MR and an AR in tandem, with the AR following the MR, and the MR's reject stream serving as the AR's feed. We apply this system to the water gas shift (WGS) reaction in the context of its application in the Integrated Gas Combined Cycle (IGCC) process for H_2 generation and simultaneous CO_2 capture from coal and biomass. The MR-AR hybrid system under study attains a high WGS reaction conversion exceeding equilibrium, produces an ultra-high purity H_2 product for power generation, and delivers a high-pressure CO_2 stream ready for sequestration. A carbon molecular sieve membrane (CMSM – previously field-tested by this team) and a sour shift WGS catalyst ($Co/Mo/Al_2O_3$) is used in the MR and a hydrotalcite adsorbent is used in the AR. Lab experiments were carried-out to investigate the membrane characteristics, and the performance of the individual MR and AR components and combined MR-AR system under IGCC-relevant conditions. Multi-cycle experiments were carried-out to validate the MR-AR system's performance with respect to CO conversion, H_2 and CO_2 purity and recovery, and to monitor whether membrane, adsorbent, and catalyst are stable during each run, and to compare results with those from a lab-scale packed-bed reactor (PBR) under same operating conditions. As part of this work, we have also carried-out parametric studies for optimization of the operation of the MR-AR hybrid system by investigating various operating conditions for both the MR and the AR, and a detailed model for the MR-AR system was also developed and validated using the experimental data. Using this model, a technical and economic analysis (TEA) was generated for process design/optimization and economic evaluation of the WGS-MR-AR system for a broad range of operating conditions and design parameters. The experimental and model findings manifest the ability of the WGS-MR-AR process to operate under the desired conditions and to improve the efficiency of the WGS reaction and demonstrate the potential of the system to be achieved in H_2 and CO_2 separation with the CMS membranes.

During the experiments involving ~ 750 hr of syngas exposure, the CMSM, adsorbent and the catalyst all displayed robust and stable performance. The MR-AR multi-cycle runs demonstrated performance superior to a conventional PBR with attaining conversions >>95%, while the AR is functional, by producing a high-purity hydrogen product, which can be directly usable in a hydrogen turbine for power generation, and a pure high-pressure CO_2 stream ready for sequestration. By employing a steam sweep in the permeate side of the MR and appropriately adjusting the permeate-side pressure >2 bar, we were able to protect the catalyst in the AR from deactivating, avoid carbon loss and reduce the energy penalty of IGCC with CCS. The CMSM-based MR-AR system is, therefore, deemed a good candidate technology for environmentally benign power generation. We are currently constructing a pilot-scale system for field demonstration of the technology to commence in late 2020.

602710 Carbon Capture Performance of Crosslinked Polyphosphazene Blend Membranes

Abstract:

Thiol-ene photopolymerization was a highly effective technique to cross-link gummy polyphosphazene elastomers containing methoxyethoxyethoxy (MEE) side groups. The solid material has a pure gas CO_2 permeability of 610 barrer and CO_2/N_2 selectivity of 35 at 40°C. The immediate benefit of this facile cross-linking technique was enabling the material to be applied as a solid selective coating on a porous membrane support. In this talk, we will demonstrate recent exciting advances we have achieved with this membrane. The material was successfully tested under coal-derived flue gas at the National Carbon Capture Center, demonstrating stable performance over 500 hours of continuous testing. The material performance has been characterized with mixed gases containing various levels of humidity: gas permeability decreased with increasing humidity, and was fully reversible, with selectivity remained barely changed. Finally, we successfully fabricated a thin film composite membrane with CO_2 permeance of 1200 gpu with CO_2/N_2 selectivity of 31. The membrane performance, monitored over 2000 hours, did not exhibit reduction due to physical aging because of the rubbery nature of the coating material.

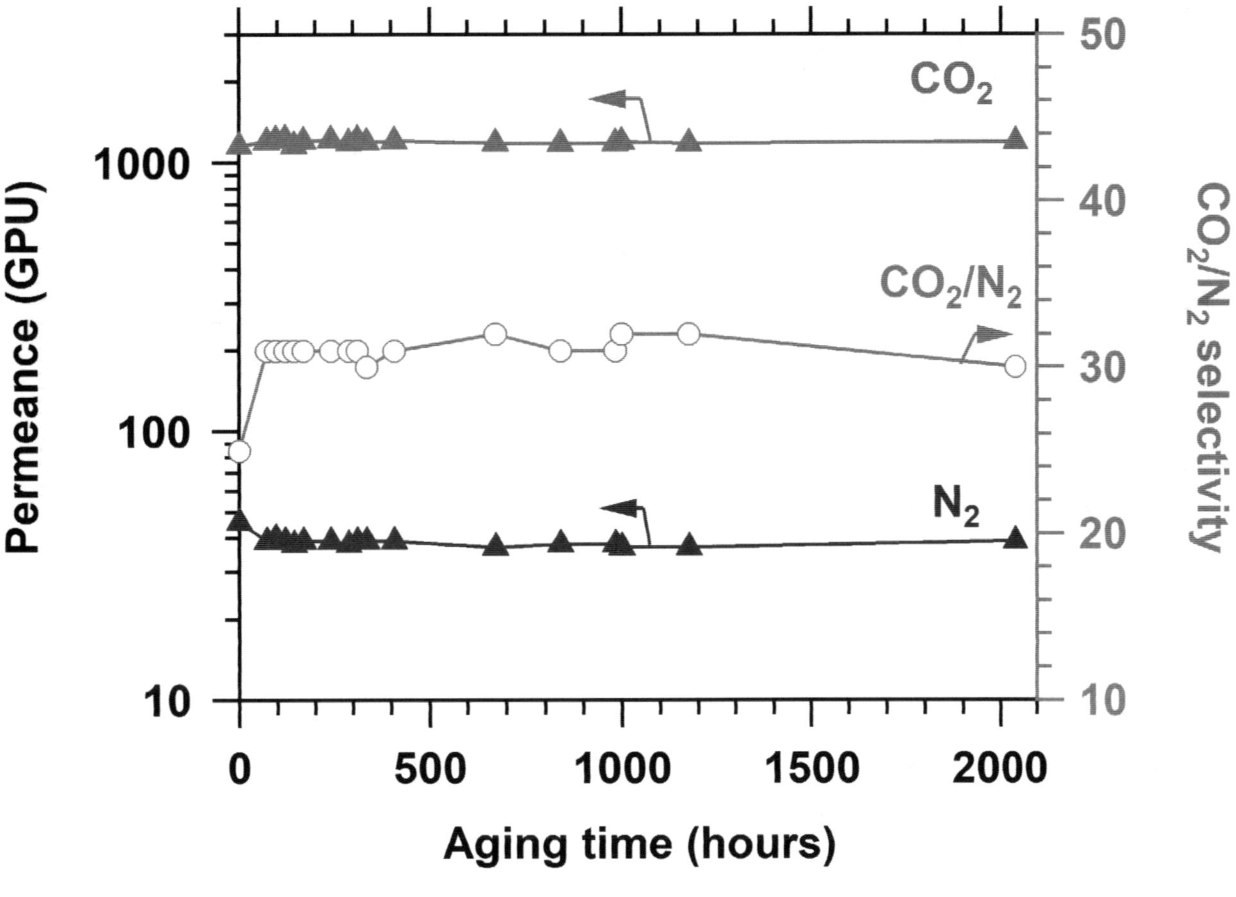

605717 Leveraging Free Volume Manipulation to Improve Membrane Separation Performance of Amine-Functionalized PIM-1

Abstract:

Diffusion-selective microporous polymers have gained increased attention as energy-efficient alternatives for membrane-based gas separations. The archetypal polymer of intrinsic microporosity, PIM-1 has stood out due to its excellent performance and facile post-synthetic modification through functionalization and cross-linking. Here, we take advantage of this post-synthetic versatility to tune the performance of amine-functionalized PIM-1 (PIM-NH_2). In previous work, PIM-NH_2 displayed high CO_2 sorption compared to PIM-1, resulting in reduced CO_2-based diffusion selectivity. At the same time, PIM¬-NH¬¬2 had moderately improved H_2-based separation performance through hydrogen-bonded mediated size-sieving. We have developed a post-synthetic method to address sorption and diffusion mediated effects in PIM-NH_2 and achieve optimal CO_2- and H_2-based separation performance. Using protection and de-protection of the amine group through thermal and solution-based methods, we tuned the porosity and sorption capacity of the films. Protection and de-protection of a bulky group enabled discrete tuning of the film's porosity to achieve size-sieving through narrowing of the pore size distribution or increased permeation through enlarged pores. Through de-protection, we also introduced cross-links which reduced CO¬2 sorption and enhanced CO_2 diffusion. The thermally processed sample displayed significant enhancements in CO_2-based selectivity and H_2-based selectivity, ultimately crossing the 2008 Robeson upper-bound for H¬2-based pairs.

607347 Synergistic CO_2-Sieving from Polymer with Intrinsic Microporosity Masking Nanoporous Single-Layer Graphene Membrane

Abstract:

A synergistic improvement of the molecular transport properties of a composite film is highly attractive for advances in the energy-efficient membranes. Herein, in the context of carbon capture, we demonstrate a remarkable enhancement in the CO_2 selectivity by uniquely masking nanoporous single-layer graphene with polymer with intrinsic microporosity (PIM-1). In the process, we overcome a major bottleneck of the state-of-the-art pore-incorporation techniques in graphene where in addition to the molecular sieving nanopores, larger nonselective nanopores are also incorporated, which so far, has restricted the realization of CO_2-sieving from the single-layer graphene membranes. [1-2] By masking nanoporous graphene (NG) with PIM-1, the direct gas-phase transport from the larger nanopores is eliminated, benefitting the transport of CO_2 over N_2. Overall, we achieve much higher CO_2/N_2 selectivity (33) from the composite film than that from the standalone NG (10) and the PIM-1 membranes (15), and high CO_2 permeance up to 2500 GPU. The selectivity enhancement is explained by an analytical gas transport model for nanoporous graphene, which shows that the transport of the stronger-adsorbing CO_2 is dominated by the adsorbed phase transport pathway whereas the transport of N_2 benefits significantly from the direct gas-phase transport pathway. Further, slow positron annihilation Doppler broadening spectroscopy reveals that the interactions with graphene reduce the free volume of interfacial PIM-1 chains which is expected to contribute to the selectivity. Masking of graphene with nanoporous polymer is doubly attractive because the polymer film also acts as the mechanical reinforcement for graphene, enabling the fabrication of graphene membranes without cracks.[3]

References:

[1] He, G., Huang, S., Villalobos, L.F., Zhao, J., Mensi, M., Oveisi, E., Rezaei, M. and Agrawal, K.V.* High-Permeance polymer-Functionalized single-Layer graphene membranes that surpass the postcombustion carbon capture target. *Energy & Environmental Science* 2019, *12*(11), 3305-3312.

[2] Zhao, J., He, G., Huang, S., Villalobos, L.F., Dakhchoune, M., Bassas, H. and Agrawal, K.V.* Etching gas-sieving nanopores in single-layer graphene with an angstrom precision for high-performance gas mixture separation. *Science advances* 2019, 5, eaav1851.

[3] Huang, S., Dakhchoune, M., Luo, W., Oveisi, E., He, G., Rezaei, M., Zhao, J., Alexander, D.T., Züttel, A., Strano, M.S. and Agrawal, K.V.* Single-layer graphene membranes by crack-free transfer for gas mixture separation. *Nature communications* 2018, *9*(1), 1-11.

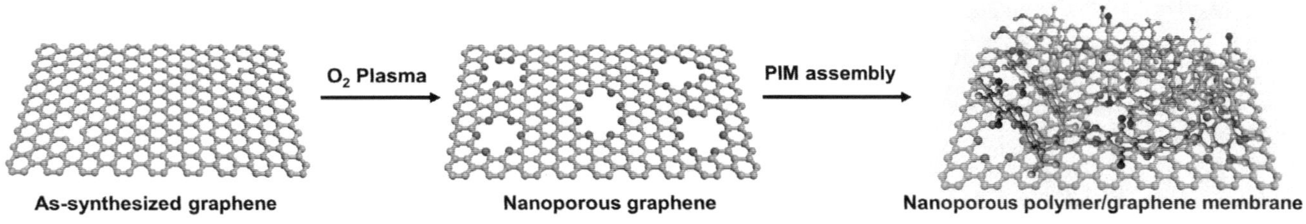

608851 Discovering Unique Behaviour of Molecules Diffusing through the Tight Spaces of Amorphous Silica and Zeolite Membranes

Abstract:

Gases and hydrated ions exhibit unique behaviour inside inorganic materials such as amorphous microporous silica and zeolites, where the pore spaces are only slightly larger than they are. Prof Lin was one of the earliest researchers to master successful fabrications of defect-free microporous inorganic membranes for molecular scale separations of gases and liquids. This inspiring early work led to a collaboration starting 12 years ago to explore molecular sieving of gases through tailored pore size amorphous microporous silica, and an even newer topic on the selective behaviour of hydrated ions in zeolite cage structures.

The pores of amorphous microporous silica are unique in their ability to be tailored to selectively permeate small molecules like He and H_2, while reject only slightly larger CO_2, N_2, CH_4 and CO, implying these materials have a pore size of around 0.3nm [1]. This size is challenging for most analytical techniques to measure, but with positron annihilation lifetime spectroscopy (PALS), the hierarchical structure having pore sizes peaking at 0.3nm, 0.8nm and 1.2nm was revealed for the first time, together with confirmation of the proposed silica network assembly mechanisms upon thermal treatment [2].

More recently, a study was extended to hydrated ions in zeolite structures. The first known studies on this commenced with MFI-type zeolite membranes made by in-situ crystallization, demonstrating a unique aqueous ion sieving effect, for example Na^+ rejection of 77% whereas Mg^{2+} rejection was reported at 88% [3]. Despite being significantly more expensive than traditional well established desalination technologies, MFI-type zeolites interact uniquely and specifically to hydrated ions yielding potentially more useful effects. For example, monovalent ions such as Na^+ and K^+ more easily enter the zeolite intrinsic pores, altering the unit cell dimensions, while divalent ions such as Ca^{2+} and Mg^{2+} mostly interacted within the intercrystalline pores [4]. Further tailoring of the ion behavior was achieved through Si/Al ratio of the zeolite [5], while increasing temperature from 20°C to 90°C saw a unique ability to reversibly trigger selective passage of K^+ [6]. Meanwhile, larger multivalent ions Al^{3+} and Fe^{3+} showed strong binding uniquely into the intercrystalline pores, offering the ability to block grain boundary defects and increase rejection of smaller ions such as Na^+ [7]. Although zeolite membranes perform well as extremely robust desalination membranes [8], their higher cost implies viable outcomes in niche areas such as sensors or targeted high value ion separations.

References

1. Diniz da Costa, J.C., Lu, G.Q., Rudolph, V., and Lin, Y.S., *Novel molecular sieve silica (MSS) membranes: characterisation and permeation of single-step and two-step sol-gel membranes.* Journal of Membrane Science, 2002. **198**(1): p. 9-21.
2. Duke, M.C., Pas, S.J., Hill, A.J., Lin, Y.S., and Diniz da Costa, J.C., *Exposing the Molecular Sieving Architecture of Amorphous Silica Using Positron Annihilation Spectroscopy.* Advanced Functional Materials, 2008. **18**(23): p. 3818-3826.

3. Li, L., Dong, J., Nenoff, T.M., and Lee, R., *Desalination by reverse osmosis using MFI zeolite membranes.* Journal of Membrane Science, 2004. **243**(1): p. 401-404.
4. Zhu, B., Zou, L., Doherty, C.M., Hill, A.J., Lin, Y.S., Hue, X., Wang, H., and Duke, M., *Investigation of the effects of ion and water interaction on structure and chemistry of silicalite MFI type zeolite for its potential use as a seawater desalination membrane.* Journal of Materials Chemistry, 2010. **20**: p. 4675-4683.
5. Zhu, B., Doherty, C.M., Hu, X., Hill, A.J., Zou, L., Lin, Y.S., and Duke, M., *Designing hierarchical porous features of ZSM-5 zeolites via Si/Al ratio and their dynamic behavior in seawater ion complexes.* Microporous and Mesoporous Materials, 2013. **173**(0): p. 78-85.
6. Zhu, B., Hong, Z., Milne, N., Doherty, C.M., Zou, L., Lin, Y.S., Hill, A.J., Gu, X., and Duke, M., *Desalination of seawater ion complexes by MFI-type zeolite membranes: Temperature and long term stability.* Journal of Membrane Science, 2014. **453**(0): p. 126-135.
7. Zhu, B., Hu, X., Shin, J.-W., Moon, I.-S., Muraki, Y., Morris, G., Gray, S., and Duke, M., *A method for defect repair of MFI-type zeolite membranes by multivalent ion infiltration.* Microporous and Mesoporous Materials, 2017. **237**: p. 140-150.
8. Zhu, B., Myat, D.T., Shin, J.-W., Na, Y.-H., Moon, I.-S., Connor, G., Maeda, S., Morris, G., Gray, S., and Duke, M., *Application of robust MFI-type zeolite membrane for desalination of saline wastewater.* Journal of Membrane Science, 2014. **475**: p. 167-174.

599114 Influence of Seeds on the Micro-Structure and the Separation Properties of FAU Zeolite Membranes

Abstract:

About half of the total energy at the chemical industries is consumed by separation processes [1]. Membranes are recognized as a potential candidate for an alternative energy-efficient separation process. In the last few decades, micro-porous inorganic membranes have been attracting a significant attention, due to their superior resistance compared to conventional polymeric membranes in addition to their unique separation abilities as summarized in various reviews [2,3].

Zeolite membranes are the first micro-porous inorganic membranes commercialized. Since 1997, more than 300 units have been applied industrially to dry solvents. Research interests are extended to apply these hydrophilic zeolite membranes as membrane reactors in esterification, dimethyl ether synthesis, hydrogenation of carbon dioxide to methanol and other reactions [4-6]. In these new applications, membrane stability is often discussed besides their permselectivity.

To understand the membrane properties, we have been interested in the contribution of inter-crystalline permeation pathways existing in the poly-crystalline zeolite membranes. TEM observation suggested assembly of small crystals with sub-micrometer size in the early stage of the densification of a FAU membrane, followed by a growth at the twin boundary [7]. In this study, FAU membranes were prepared with different seeds. Formed membranes were tested by pervaporation with 10% water -90% ethanol (EtOH). 10wt% Methanol (MeOH) - 90% methyl methacrylate (MMA) mixtures, and pure components. While the dehydration selectivity was almost the same to the seed type, membranes prepared with smaller seeds had lower MeOH/MMA selectivity, showing larger contribution of the inter-crystalline permeation.

Acknowledgement: Authors acknowledge the financial supports from various funding agencies, in particular, the JST-eAsia JRP project and the JST SICORP Grant Number JPMJSC18C5, Japan.

1. D.S. Sholl, R.P. Lively, *Nature*, *532*(7600), 435 (2016).
2. Y.S. Lin, *et al.*, *Separation and Purification Methods*, 31(2), 229 (2002)
3. N. Rangnekar, *et al.*, *Chemical Society Reviews*, *44*(20), 7128 (2015)
4. C. Zhou, *et al.*, *Angewandte Chemie International Edition*, 55(41), 12678 (2016)
5. H. Li, *et al.*, *Science*, *367*(6478), 667 (2020)
6. R. Raso, *et al.*, *Catalysis Today* (2020)
7. I. Kumakiri, *et al.*, *Microporous and Mesoporous Materials*, 272, 53 (2018)

600407 ZSM-5 Zeolite Nanosheet-Nafion Composite Membrane As an Ion Separator for Redox Flow Batteries

Abstract:

Zeolite nanosheets with nanometer-scale thickness and very large aspect ratios offer new opportunities to control crystal orientation and minimize thickness of zeolite-composite membranes for achieving enhanced selectivity and flux in molecular and ion separations. Here, we report the synthesis of a zeolite nanosheet laminated membrane (ZNLM) on polymer substrate and evaluation of this ZNLM-polymer composite membrane as ion separator for the vanadium redox flow battery (RFB). To enable the fabrication of ZNLM on polymer substrates, a room temperature UV-irradiation method has been established for template removal from the zeolite nanosheets that avoids the difficulty of zeolite activation after the composite membrane formation. The composite membrane fabrication is accomplished by synthesis of a ~350 nm-thick free-standing UV-activated ZSM-5 ZNLM, which is subsequently sandwiched in between two Nafion thing films to form a layered structure with an overall thickness of ~120 μm. This zeolite nanosheet-Nafion composite membrane exhibited drastically improved proton selectivity over the vanadium ions and reduced proton transport resistance as compared to the common benchmark Nafion-117 membrane. The composite membrane also exhibited higher Coulombic and voltage efficiencies in the all-vanadium RFB than the Nafion-117 membrane.

601725 Postsynthetic Modification of ZIF-8 Membranes Via Membrane Surface Ligand Exchange for Light Hydrocarbon Gas Separation Enhancement

Abstract:

The ability to tailor the pore structure of metal–organic framework (MOF) membranes enables synthesis of new or modified MOF membranes with enhanced separation characteristics. This work employs a modified version of solvent-assisted ligand exchange, termed membrane surface ligand exchange (MSLE), to modify the pore structure of zeolitic imidazolate framework-8 (ZIF-8) membranes. This paper is the first to perform a time-based, ex situ characterization and gas permeation study of ZIF-8 MSLE with 5,6-DBIM (DBIM, dimethylbenzimidazole) to effectively narrow the ZIF-8 pores, enhance light hydrocarbon gas-phase separations and give insight into the exchange mechanism with respect to time and temperature. The results show that relatively fast exchange kinetics occur mainly at the outer surface of the ZIF-8 membrane during the initial 30 min of exchange and enables significant (40–70%) increases in propylene/propane selectivity with minimal (10–20%) propylene permeance losses for the modified ZIF-8 membranes. We postulate as the reaction time proceeds, the ligand-exchange rate slows as the DBIM linker diffuses into the ZIF-8 membrane beyond the external surface, exchanges with the original linker, disrupts the original framework's crystallinity, and then increases long-range order/crystallinity as the reaction proceeds. The H_2/C_2 separation factor increases with increased 5,6-DBIM content in the ZIF-8 framework which is facilitated by increased MSLE time and reaction temperature.

614072 The Role and Challenges of Industrial Scale Distillation Research

Abstract:

It is crucial in industrial distillation column designs to predict the liquid/vapor distribution, tray/packing hydraulics, and concentration gradients, etc., which affect the performance of column internals. However, all of these phenomena cannot be studied in small laboratory columns due to the significant difference in the 'shape of a theoretical stage' between laboratory and industrial scale distillation research. Fractionation Research, Inc (FRI) has been embarking on industrial scale distillation studies since 1952. In this presentation, the role of the large scale distillation studies at FRI will be discussed. The challenges and difficulties associated with the large scale research will be described and discussed. The measures and solutions to address those challenges and difficulties will be presented in detail.

614047 Molecularly Engineering Polymeric Membranes for Pre-Combustion CO_2 Capture at 150 °C and Above

Abstract:

Membrane materials with high H_2 permeability and H_2/CO_2 selectivity at 150–300 °C hold promise for pre-combustion CO_2 capture. However, polymers with high permeability tend to exhibit weak size-sieving abilities and thus low selectivity. Polybenzimidazole (PBI) is one of the leading membrane materials for H_2/CO_2 separation for H_2 purification and pre-combustion CO_2 capture. For example, PBI exhibits H_2 permeability of 27 Barrers and H_2/CO_2 selectivity of 14 at 150 °C, which is on the Robeson's upper bound. Herein we demonstrate that PBI can be engineered to further improve the H_2/CO_2 separation performance. For example, PBI can be doped by polyprotic acid to increase the size-sieving ability and thus H_2/CO_2 selectivity to as high as 130; PBI can be carbonized to obtain "hourglass" free volumes and thus improve the H_2 permeability to 54 Barrer and H_2/CO_2 selectivity to 80; and PBI can be doped with Palladium nanoparticles to increase H_2 solubility and thus achieve H_2 permeability of 48 Barrer and H_2/CO_2 selectivity of 43. We determined the mixed-gas separation properties of these materials, demonstrating their stability for long-term operation in the presence of water vapor. The structure/property relationship in these materials will also be elucidated.

614049 Distillation Tray Efficiency

Abstract:

Distillation tray efficiency is an important performance indicator for trayed towers. Tray efficiency converts stages calculated in a process simulator to actual trays to be installed. Tray efficiency is a measure of whether an existing tower is operating to specifications. Tray efficiency provides context to tower capacity – with definitions such as useful capacity. Tray efficiency is central to the very objective of a distillation tower.

However, the measurement, prediction, and application of tray efficiency has been one of the most challenging aspects of tray performance - because a large number of variables effect efficiency – and moreover, the effects are correlated, varied, and sometimes counter-acting. Physical properties, liquid and vapor flow rates, and the geometry of the various elements of a tray all play a role in determining the tray efficiency.

In this talk, I will present advances in measurement and prediction of distillation tray efficiency with references to work performed at Oklahoma State University and Fractionation Research, Inc.

614710 Finding the Right Balance in Separations

Abstract:

The Separations Division is more than an organization—it is a community, and the areas within this community compete with each other in a collegial manner. This fact notwithstanding all areas face the need to optimize separation efficacy and cost factors—including energy, CO_2 footprints and equipment. Award winners presenting in this session have contributed broadly and deeply to finding the right balance of the above factors within their subareas. Our division members are fortunate to have this crosscutting session to see how highly different areas address this issue. There is no single, unchanging answer within the whole field of separations, since society imposes an evolving set of needs on our Division. *Innovation is key*, and seeking ways to achieve revolutionary change by collaboration and evolutionary development is well suited to the large-scale separations field. This presentation will provide a framework with some examples to illustrate how such collaboration and revolutionary changes can be achieved using materials, devices and systems—with some specific examples.

612300 Separation Challenges in Membrane-Based Water Applications

Abstract:

Various separation processes are used extensively in dilute aqueous applications ranging from valuable chemical/pharmaceutical recovery to high value metals recovery from mining effluents to water purification and detoxification. Technologies include adsorption, affinity separations, ion exchange, solvent extraction, membrane separations, etc. Membrane separation technologies (such as, RO, NF, MF, UF, ED, CDI, etc,) are expanding at a rapid rate with applications ranging from complete desalination to selective protein and low MW organics recovery to water reuse and water detoxification. In addition, by integrating life sciences and nanotechnology one can overcome a variety of separation challenges through bioinspired and responsive nanostructured membranes. Integrating the principles of other separation technologies, such as, adsorption/ion exchange/chromatography to membranes, one can reduce the effect of diffusional mass transfer issues and high level sorption site utilization issues. Various membranes and module designs also suffer from use in elevated temperature (even 90 ^{o}C) and corrosive condition applications. Selective separations are often needed where one needs very low (may be less than 5%) NaCl separation while rejection divalent or trivalent ions to eliminate osmotic pressure issues and allowing water reuse. The presentation will include: research needs, challenging selective separation examples, hybrid systems, responsive and bioinspired membranes for selectivity, need for surface functionalization, and discussions on flux vs. selectivity.

599009 Development of Silica-Derived Molecular Sieving Membranes Via Controlling Microporous Structure

Abstract:

Silica is one of the most attractive materials for hydrogen separation because of its amorphous structures, through which small molecules such as helium and hydrogen can permeate, in a wide temperature range. At the end of the 1980s, sol-gel derived amorphous silica membranes were found to have excellent separation performance for hydrogen or helium at high temperatures.[1] Accordingly, many research groups have focused on the development of amorphous silica membranes for gas separation and pervaporation.[1,2]

However, two critical issues remain to be solved in the practical application of amorphous silica membranes in the separation and purification process: the hydrothermal stability of amorphous silica networks, and the difficulty of controlling the silica network size. In the present study, pore size controllability via "spacer" method and fluorine doping was introduced based on the results of gas permeation properties.

Pore size control and gas permeation properties of organosilica membranes

The selection of Si precursors made it possible to control the network size of silica structure.[3-7] Organosilica membranes were fabricated using bridged organoalkoxysilanes (bis(triethoxysilyl)methane (BTESM), bis(triethoxysilyl)ethane (BTESE), bis(triethoxysilyl)propane (BTESP), bis(trimethoxysilyl)hexane (BTMSH), bis(triethoxysilyl)benzene (BTESB), and bis(triethoxysilyl)octane (BTESO)) to produce highly permeable molecular sieving membranes. Organosilica membranes showed H2/N2 and H2/CH4 permeance ratios that ranged from 10-150. H2 selectivity was decreased with an increase in the carbon number between 2 Si atoms, indicating that the network pore size could be successfully controlled even by utilizing the flexible organic bridges (Si-C3-Si, Si-C6-Si, and Si-C8-Si).

Pore size control and improved stability via fluorine doping

The utilization of triethoxyfluorosilane (TEFS) containing Si-F bonds, the addition of NH4F and alkali fluorides, and HF-catalyzed reactions can be accepted as a common method to eliminate the Si-OH density. The incorporation of Si-F bonds in an amorphous structure affected the Si-O-Si bond angle, and this change in the SiO2 structure made it possible to form looser and more-uniform structures.[8-10] Highly permeable CO2 separation membranes (CO2 permeance: 4.1 x 10-7 mol m-2 s-1 Pa-1, CO2/CH4 permeance ratio: 300 at 35 oC) were successfully developed via fluorine doping of the SiO2 matrix.[8] Recently, a molecular sieving membrane was fabricated using TEFS, and the densification of the network structure was apparently suppressed by comparison with a TEOS-derived structure, due to the presence of fewer Si-OH groups.[10]

References

[1] Ockwig, N. W., Nenoff, T. M., Chem. Rev., 107, 4078 (2007). [2] Lin, Y. S. et al., Sep. Purif. Method., 31, 229 (2002). [3] Kanezashi, M. et al., J. Am. Chem. Soc., 131, 414 (2009). [4] Kanezashi, M. et al., Ind. Eng. Chem. Res., 51, 944 (2012). [5] Kanezashi, M. et al., AIChE J., 58, 1733 (2012). [6] Kanezashi, M. et al., J. Membr. Sci., 415-416, 478 (2012). [7] Kanezashi, M. et al., AIChE J., 63, 4491 (2017). [8] Kanezashi, M. et al., ChemNanoMat, 2, 264 (2016). [9] Kanezashi, M. et al., ACS Appl. Mater. Interfaces, 9, 24625 (2017). [10] Kanezashi, M. et al., J. Membr. Sci., 549, 111 (2018).

608859 Addressing the Dual Challenge through Advanced Separations: From Materials to Process Scale-up

Abstract:

The dual challenge of supplying energy to a growing middle class while reducing emissions is shaping the future of energy globally. Large-scale hydrocarbon separation technologies account for roughly 45% of energy consumption in gas processing, refining and petrochemicals manufacturing. In response to these challenges, ExxonMobil has been developing advanced separation concepts focused on some of the highest impact streams such as olefin/paraffin and xylene isomer separations natural gas processing and carbon capture. Development of novel materials, fundamental and data driven process models and lab-bench-pilot scale evaluation are often simultaneously required to take these applications from a concept to technical demonstration. The talk discusses the research portfolio of the separations and process chemistry group at ExxonMobil's Corporate Strategic Research highlighting the potential to dramatically reduce the energy/carbon and physical footprint of critical operations not feasible via conventional processing.

607299 Microporous Polymers to Microporous Carbons: Linking Structure to Guest Transport and Sorption

Abstract:

“Polymers of intrinsic microporosity” (PIMs) are an emerging class of microporous materials that have the unique capability to be solution processed into adsorbents, membranes, and other fluid-solid contactors. Moreover, it has been recently discovered that PIM materials can be pyrolytically converted into carbon molecular sieves (CMS) with well-defined micropores in the range of 5-8 Å, which is ideal for the separation of similarly-sized molecules. From an applications perspective, these materials harness the scalability of the polymer processing platform while providing flexibility in materials design for targeted mass transfer problems; i.e., will a microporous polymer or a microporous carbon be better suited for a particular task? A series of lingering scientific gaps include missing links between the chemical structures of the polymer and the resulting polymer-derived carbon as well as the linkage between these structures and the void space inherent in both materials. To address these gaps, the textural properties of the polymer precursor materials are typically characterized based on cryogenic sorption techniques developed for “rigid” microporous solids (e.g., zeolites). These approaches can lead to poor predictions of guest sorption and transport within the polymer, as the free volume regions present in PIMs are not persistent as in traditional rigid solids, but may change in time due to both chain motions and penetrant-induced dilation. However, the cryogenic sorption experiment still provides important information regarding the current free volume of the polymer and its ability to dilate. After pyrolysis of the PIM material, the textural properties of the rigid CMS are easily determined via cryogenic sorption techniques, but relating the void space microstructure to the structure of the carbon itself is more difficult. Neutron diffraction techniques provide this insight and an example will be given that compares the pore size from cryogenic sorption and diffraction. Finally, a relationship between void space microstructure and small molecule separation capabilities will be provided to begin creating a tentative design pathway from polymer structure to carbon-based separation device.

602281 Beyond Power: CO_2 Capture from Industrial Sources with Polaris^{TM} Membranes

Abstract:

To mitigate the harmful effects of global climate change, CO_2 from large point sources must be captured and either sequestered or utilized in processes such as enhanced oil recovery. Membrane technology is an attractive CO_2 capture option because of advantages such as simple passive operation, tolerance to acid gases and oxygen, no hazardous chemical emissions or handling issues, minimal water requirements, and near instantaneous response to process conditions.

Working with the US Department of Energy, MTR has developed new membranes and process designs to recover CO_2 from power plant flue gas. MTR Polaris membranes have CO_2 permeances at least ten times higher than standard commercial membranes, which greatly reduces the cost of a membrane capture system. For high capture rates, these membranes are combined with a novel, patented process design that uses incoming combustion air to sweep membranes and recycle CO_2 to the boiler. Since 2009, MTR has performed bench scale or small pilot field tests capturing up to 20 tons per day of CO_2 (equivalent to 1 MW_e) at a natural gas combined cycle plant, coal-fired power plants, and a boiler manufacturer research facility. These field tests have aided in further development of the Polaris membrane, refinement of the process design, and the design of a novel membrane module specifically for high-volume, low-pressure, and low-pressure drop applications. Current field tests include a 1 MW_e Gen-2 Polaris membrane system and 1 MW_e membrane-sorbent hybrid system to be tested at the Technology Centre Mongstad in Norway. MTR is also conducting a FEED study in preparation for building and operating a 10 MW_e membrane system at the Wyoming Integrated Test Center outside of Gillette, WY.

Cement and steel plants are also large point sources of CO_2 emissions and are potentially easier targets for a CO_2 capture process because the CO_2 in their flue gas tends to be more concentrated compared to power plants. In this talk, a detailed analysis of membrane-based systems for the capture of the CO_2 emitted from cement or steel production processes will be presented. Process designs, process costs and economics, and the effect of the CO_2 capture rate on the CO_2 capture cost will be discussed.

614711 Molecular Separations in Organic Liquid Systems: The Selectivity Challenge

Abstract:

Organic liquids are ubiquitous in chemical-science-based industries, which range in scale from refining to pharmaceutical production. It is generally accepted that 40-70% of capital and operating costs in these industries are dedicated to separations. Membrane technology has the potential to provide game changing alternatives to conventional concentration and purification technologies such as evaporation, liquid extraction, adsorption, crystallisation and chromatography. To achieve this, membranes must offer resilience in organic environments, and have good permeance. Crucially, in many applications to molecular separations in liquids a membrane system should also offer an outstanding selectivity between solutes to provide an efficient process. This paper will explore materials science and engineering based approaches to improving selectivity, and illustrate the effect of selectivity on overall process complexity and feasibility.

Materials science approaches to improving selectivity between solutes revolve around designing materials for the membrane separating layers which can discriminate cleanly between two solutes. Approaches include incorporating pre-structured materials with microporosity, such as MOFs, COFs, and POCs into the separating layers of membranes, or designing membranes with increasingly rigid polymer networks which build-in desired microporosities.

An engineering-based approach to improving system selectivity is to use multiple separation stages. In these systems, the permeate from one stage is typically fed into a subsequent stage for further solute fractionation. This idea has been explored through simulation and a limited number of experimental demonstrations – these latter requiring the control of pressures and flows.

The impact of these approaches will be considered in the context of high value molecules from pharmaceutical contexts, and the potential for each approach to achieve desired outcomes will be assessed.

603839 Membrane Filtration with Liquids: A Global Approach with Prior Successes, New Developments and Unresolved Challenges

Abstract:

After 70 years, modern pressure-driven polymer membrane processes with liquids are mature and accepted in many industries due to their good performance, ease of scale-up, low energy consumption, modular compact construction, and low operating costs compared with thermal systems. Successful isothermal operation of synthetic membranes with liquids requires consideration of three critical aspects or "legs" in order of relevance: selectivity, capacity (i.e. permeation flow rate per unit area) and transport of mass and momentum comprising concentration polarization (CP) and fouling (F). Major challenges remain with respect to increasing selectivity and controlling mass transport in, to and away from membranes. Thus, prediction and control of membrane morphology and a deep understanding of the mechanism of dissolved and suspended solute transport near and in the membrane (i.e. diffusional and convective mass transport) is essential. Here, we focus on materials development to address the relatively poor selectivity of liquid membrane filtration with polymers and discuss the critical aspects of transport limitations. Relevance to nano-channels such as synthetic, biomimetic (imidazole assembly) and biological (aquaporin) will be addressed. Machine learning could help optimize membrane structure-by-design and transport conditions for improved membrane filtration performance. (REF: Belfort, G., Membrane Filtration with Liquids: A Global Approach with Prior Successes, New Developments and Unresolved Challenges. *Angew. Chem. Int. Ed.* **2018,** *57*, 2 – 13).

607766 Process Scaling and Design for Large Scale CO_2 Capture By PSA

Abstract:

Extremely large gas flow rates, like that associated with flue gas produced from a 550 MW coal fired power plant, can exceed 25,000,000 SLPM. Even larger flue gas flow rates can be produced from steel mills. The flue gas produced from a 550 MW coal fired power plant consists primarily of 15 vol% CO_2, 65 vol% N_2, 8 vol% O_2, 12 vol% H_2O and ppm levels of SO_2 and NO_x at a pressure slightly above about 100 kPa and a temperature between ambient and 125 ^oC. According to the Department of Energy (DOE), the goal is to capture and concentrate the CO_2 and produce it in a stream containing at least 95 vol% CO_2 with at least 90 % recovery and with less than 10 ppm of O_2. Pressure and vacuum swing adsorption processes are being touted for this application. However, an issue arises with these low pressure flue gas streams when fed to a pressure swing adsorption (PSA) process.

The issue is that the feed cannot be compressed very much because of the cost of compression. Hence, the feed must be compressed only slightly with a blower to about 120 kPa, thereby limiting the axial pressure drop along the bed to be less than about 20 kPa atm differential. This pressure drop limit/issue does not exist for PSA processes that operate with a feed pressure well above atmospheric pressure, like from 80 psia to 800 psia, the higher the better. Moreover, because the feed pressure is so low, to effectively regenerate the beds a vacuum must be applied, typically down to absolute pressures of around 5 kPa to 10 kPa. This is commonly referred to as a vacuum swing adsorption (VSA) process. To make matters even worse, the regeneration steps, typically consisting of countercurrent depressurization (CnD) and light reflux (LR) steps, also have a limit on the magnitude of the axial pressure drop, for if it is too large, then the beds are not effectively regenerated because the light end of the bed never experiences the lowest vacuum pressure experienced by the heavy or feed end of the bed. These limits on the axial pressure drop during the feed and regeneration steps of a VSA cycle are dictated by the interstitial velocity in the bed which varies with time and position.

These axial pressure drop limits give rise to two constraints: 1) the longer the bed the higher the axial pressure drop and 2) the higher the interstitial velocity the higher the axial pressure drop. When limited to bed aspect ratios of no less than one, these constraints limit both the height of the bed and the diameter of the bed, which, in turn, limits the amount of gas that can be processed in a bed during the feed step. A unique scaling procedure was developed to meet these constraints while minimizing the total number of beds required in the VSA system by operating one or more identical VSA units in parallel. Each of these VSA units contains a minimum number of identical beds. Many of these trains of multibed VSA units meet the required process performance, including the pressure drop constraints, but not the limit on the O_2 concentration in the CO_2 product. The latest results of this study PSA scaling and design will be presented.

598959 Designing Efficient Vacuum Pressure-Swing Adsorption (VPSA) Processes for Air Separation Applying an Oxygen-Binding Adsorbent

Abstract:

Air separation is currently one of the most important gas operations in the chemical industry [1,2]. It is performed typically at large-scale by cryogenic distillation to produce high-purity product streams. Adsorption processes have the potential to substitute the energy- and capital-intensive distillation operations with the right adsorbent and process technology for several applications. Currently, commercial adsorption technologies of different production scales, i.e., PSA & VPSA, exist to realize this separation to generate oxygen-rich streams as well as nitrogen-rich streams, which are useful in a variety of industries [1,2]. Traditionally, adsorption-based air separation employs materials that selectively adsorb nitrogen in order to recover high-purity oxygen as light-product, e.g., commercial LiX zeolites [2]. Nitrogen-rich product streams may also be generated by applying carbon molecular sieves (CMS), modifying cycle configuration & process operating conditions [1]. In the last decades, novel classes of materials have been developed & investigated to alternatively reverse the adsorptive selectivity, i.e., adsorbing preferentially the oxygen species, thus enabling its recovery as heavy-product stream of pressure-swing operations, whilst rejecting nitrogen as light-product—see e.g., Hutson & Yang [3] among others. In this contribution, we investigate vacuum pressure-swing adsorption (VPSA) cycle configurations via process modeling; these cycle designs target specifically high-purity recovery of oxygen. We parametrize single component adsorption equilibria from experimental measurements for nitrogen and oxygen species, adsorbed by a tailored oxygen-binding adsorbent. We feed a full-order VPSA process solver with this equilibrium information, as well as appropriate mass & heat transfer approximations. We apply a state-of-the-art multi-objective optimization strategy developed recently [4], in order to determine optimal trade-offs between performance variables relevant to these cyclic processes. These modeling efforts allow us to better evaluate the potential of oxygen-binding adsorbents, as well as assisting in the design of efficient adsorption-based air separation processes for oxygen production.

References

[1] Böcker, N.; Grahl, M.; Tota, A.; Häussinger, P.; Leitgeb, P.; Schmücker, B. *Ullmann's Encyclopedia of Industrial Chemistry*; Wiley-VCH Verlag GmbH & Co. KGaA, **2013**; pp. 1–27.

[2] Kirschner, M. J.; Alekseev, A.; Dowy, S.; Grahl, M.; Jansson, L.; Keil, P.; Lauermann, G.; Meilinger, M.; Schmehl, W.; Weckler, H.; Windmeier, C. *Ullmann's Encyclopedia of Industrial Chemistry*; Wiley-VCH Verlag GmbH & Co. KGaA, **2017**; pp. 1–32.

[3] Hutson, N. D.; Yang, R. T. Synthesis and Characterization of the Sorption Properties of Oxygen-Binding Cobalt Complexes Immobilized in Nanoporous Materials. *Industrial & Engineering Chemistry Research* **2000**, 39, 2252–2259.

[4] Rubiera Landa, H. O.; Kawajiri, Y.; Realff, M. J. Efficient evaluation of vacuum pressure-swing cycle performance using surrogate-based, multi-objective optimization algorithm. *Proceedings of the 30th European Symposium on Computer Aided Process Engineering (ESCAPE30)*, **2020** (accepted).

606702 Multi-Scale Modeling of a Partial Pressure Temperature Swing Adsorptive Reactor (PPTSAR) Process for the Water Gas Shift Reaction

Abstract:

In this work, a novel Partial Pressure Temperature Swing Adsorptive Reactor (PPTSAR) Process for the Water Gas Shift Reaction (WGSR) is proposed. The process is realized using a dynamic, multiscale CFD model. It couples the reactor and catalyst/adsorbent pellet domains at their respective scales to accurately model heat and mass transport phenomena occurring within each domain. The resulting multiscale model elucidates the dynamic behavior of the PPTSAR during start-up, adsorption/desorption cycles, as well as eventual long-term, periodic behavior. The use of two or more PPTSARs allows continuous modes of operation for the WGSR, enabling complete CO2 capture with simultaneous CO conversion greater than 95%.

607786 Design and Scale-up of Tvsa Cycle for CO_2 Removal from Spacecraft Cabins Using a Structured Adsorbent

Abstract:

The overall goal of this project is to design and scale up a temperature vacuum swing adsorption (TVSA) process that utilizes an IntraMicron (IM) structured adsorbent that consists of a copper microfibrous media with entrapped ground and sieved 13X adsorbent particles. This TSA cycle design would be perhaps a drop-in replacement for the TVSA unit on board the ISS or it may be a grass roots design for future applications. This design is based on simulations being carried out with the USC dynamic adsorption process simulator (DAPS) and validated experimentally using the USC single bed apparatus in a TVSA mode.

An extensive bench scale experimental study is being carried out around the base case separation that meets or exceeds the required CO_2 removal rate for four astronauts. This removal rate is 1 kg of CO_2 per day per astronaut. This corresponds to processing around 740 SLPM of a feed steam containing 2,700 ppm of CO_2 and removing 68.5 % of this CO_2. The effects of the feed flow rate, CO_2 concentration, cycle time, regeneration temperature and low vacuum pressure on the process performance are being investigated. DAPS will utilize these bench scale experimental results to design and scale up a new column utilizing the IM structured adsorbent. The latest results of this project will be presented.

608178 Artificial Neural Network-Based Surrogate Models for Rapid Simulation, Optimization of Pressure Swing Adsorption

Abstract:

Adsorption based processes such as Pressure Swing Adsorption (PSAs), due to its cyclic nature and the presence of multiple design variables are notoriously challenging to simulate and optimize. Using detailed models that are based on first principles require hundreds of core hours of computation to accurately simulate and optimize a given process for a given material. In this talk we introduce MAPLE (Machine-assisted Adsorption Process Learner and Emulator) a framework that is capable of learning from a training-set obtained from detailed modelling of thoroughly sampled operational space and with the potential of evaluating PSA performance at any other arbitrary operating condition. As an example CO_2 capture from post-combustion flue gas is considered and two different process cycles are evaluated. For each of these cases, the trade-off between training effort and accuracy and the model's ability to predict PSA performance are evaluated. The MAPLE framework is then employed to optimize PSA performance for a variety of objective functions. The talk will also discuss how the framework can be suitably adopted for reliably screening large databases of adsorbents.

601230 New Adsorptive PSA Separation Process for Production of Isopentane and Di-Methyl Branched Paraffins from C5/C6 Light Straight Naphtha Feeds

Abstract:

The conventional Total Isomerization Process (TIP) for improving the octane rating of light straight-run naphtha (LSRN) (C5/C6 range), is one of the first and most successful processes in the application of adsorption phenomena to industrial processes. The process typically isomerizes straight-chain hydrocarbons (pentanes (C5) and hexanes (C6)) with low research octane number (RON) into their more valuable mono- and di-branched isomers in a thermodynamically controlled reaction. Afterward, linear hydrocarbons non-isomerized are separated (for recycling) from the output stream of the catalytic reactor by adsorption using the molecular sieve zeolite 5A that completely separate linear from branched paraffins. In terms of performance, actual TIP processes can produce a final isomerate mixture with an average RON around 87-89. However, it suffers from the disadvantage that in the final stream, there are still 30% of molecules with low RON, such as the mono-branched isomers of hexane.[1,2]

In this work, we report a new advanced recycling technology[3] (Figure 1) developed to fractionate C5/C6 isomers mixtures into low RON (LRON) and high RON (HRON) fractions by also recycling the mono-branched C6 isomers in TIP processes. For that, a new adsorber unit is used through a combination of two adsorbents: i) a benchmark Zeolite and ii) a Metal-Organic Framework (MOF). The Zeolite separates the linear n-pentane (nC5, RON:61.7) and n-hexane (nC6, RON:24.8) from the branched paraffins. In contrast, the MOF, separates low RON mono-branched hexane isomers (2-methylpentane (2MP, RON:73.4), 3-methylpentane (3MP, RON:74.5)) from the high RON pentane and hexane di-branched isomers (isopentane (iC5, RON:93), 2,3-dimethylbutane (23DMB, RON: 101.7), and 2,2-dimethylbutane (22DMB, RON:91.8)). As a result, the LRON linear and mono-branched C6 isomers are separated through a synergy between size selectivity (zeolite) and host-guest interactions (MOF), and recycled to the catalytic reactor providing a final HRON isomerate product having values higher than 92 RON (Figure 1).

To develop this new technology, and previously to cyclic Pressure Swing Adsorption tests (PSA), several fixed bed breakthrough experiments with equimolar C5/C6 isomers mixtures were carried out on a mixed bed of Zeolite [beads (1.0 to 1.8 mm)] and MOF [shaped beads (1.0 to 1.8 mm)], to select the best operating conditions. The breakthroughs were measured at the temperature range of 373 - 473 K and total hydrocarbon pressure up to 50 kPa. The example shown in Figure 2, is for a breakthrough curve performed at 423 K and 50 kPa, where a clear separation between the HRON (22DMB, 23DMB and iC5) and the LRON isomers (nC6, nC5, 2MP, 3MP) is obtained, resulting in a selectivity (hierarchy) elution order: nC6 >> nC5 >> 2MP ≈ 3MP >> 23DMB > iC5 ≈ 22DMB. For this experiment the calculated productivity of the mixed bed, is 0.59 $mol.kg_{ads}^{-1}$ for a RON product of 92.3.

Figure 3 shows the steady-state effluent concentration profiles of the separation of C5/C6 isomers in a 4 step PSA experiment: I – Pressurization with feed (adsorption); II – High-Pressure feed (adsorption); III – Countercurrent blowdown (desorption) and IV – Countercurrent Low-pressure purge (desorption). As shown in Figure 3, it is clear that at the end of the adsorption steps (I and II) the mass transfer front of the HRON isomers iC5, 22DMB and 23DMB has left the column as the final product. Regarding the other LRON isomers, the majority of their respective mass transfer fronts remains inside the column, with only a very small concentration leaving the column. It is also clear from Figure 3 that at the end of the desorption steps (III and IV), the concentration of the HRON isomers at the effluent of the column is very small which means that the bed is efficiently cleaned. This experiment gives a final product in the end of the adsorption step with a RON around 93. A mathematical model was also developed to simulate the fixed bed and PSA experiments.

As a final conclusion, this works shows the remarkable capacity of a mixed bed of a Zeolite/MOF adsorptive system to separate all C5/C6 isomers by cyclic PSA adsorption processes into distinct HRON and LRON fractions. This result is of significance importance to upgrade actual existing TIP processes.

References:

1 - Holcombe, T. C.; Sager, T. C.; Volles, W. K.; Zarchy, A S. Isomerization Process U.S. Patent 4,929,799, 1990.

2 - Minkkinen, A., Mank, L., Jullian, S., Process for the isomerization o C5/C6 normal paraffins with recycling of normal paraffins, U.S. Patent 5,233,120, 1993.

3 - Serre, C.; Nouar, F.; Rodrigues, A.; Maurin, G; Brântuas, P.; Henrique, A.; Silva, J., Unpublished results, 2020.

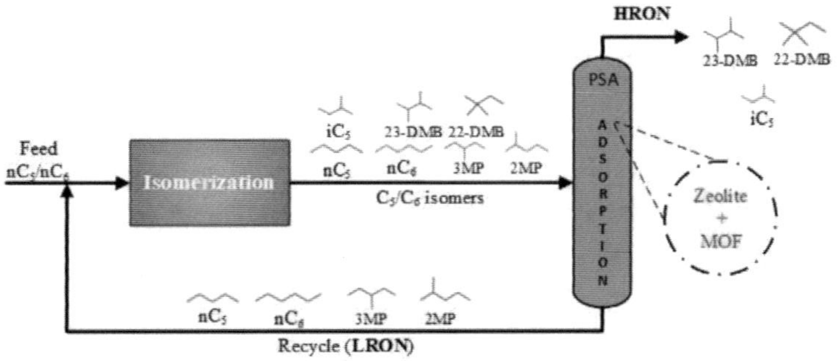

Figure 1. Upgraded TIP process using a mixed adsorption bed of a Zeolite/MOF operating under a PSA cycle to separate C5/C6 isomers into HRON and LRON fractions.

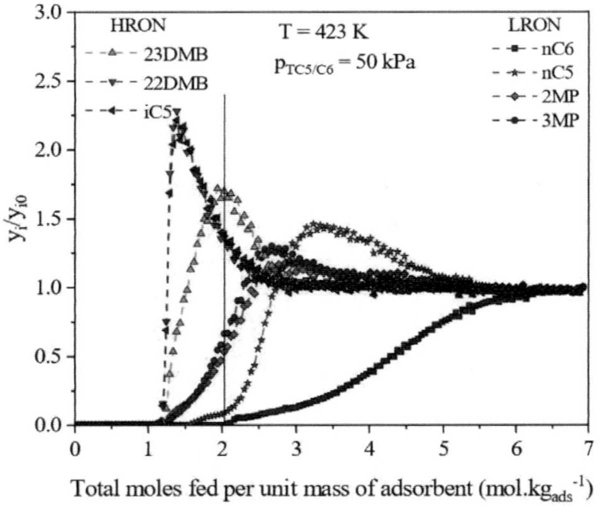

Figure 2. Breakthrough experiment with a mixture of C5/C6 isomers in a mixed bed of Zeolite/MOF adsorbents plotted in terms of normalized feed as a function of total moles fed per unit mass of adsorbent. The vertical line remarks the time of the experiment where the productivity of the mixed bed is 0.59 mol.kg_{ads}^{-1} for a RON product of 92.

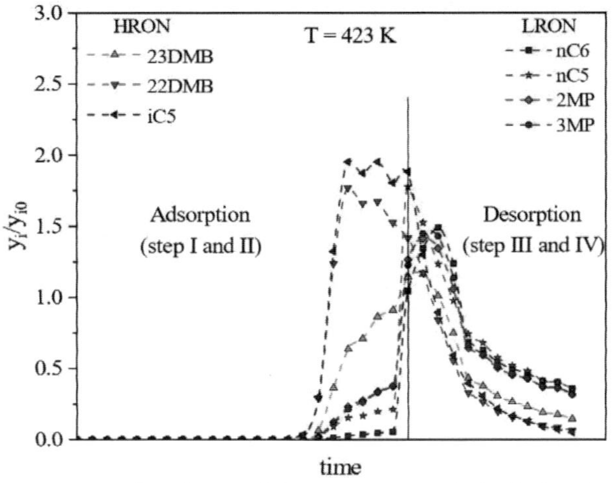

Figure 3. Steady-state effluent concentration for a four-step PSA cycle separation of C5/C6 alkane isomers into HRON and LRON isomer fractions, in a mixed bed of Zeolite/MOF adsorbents.

597429 Fundamental Study of Water Transport in Polyvinylidene Fluoride (PVDF) Derived Carbon Molecular Sieve (CMS)

Abstract:

The molecular transport in porous spaces is an essential area of research in industrial separations, catalysis, and geology. Transport of water-organic mixtures under high-loading conditions is often encountered in industrial separation applications such as purification of produced water, solvent recovery from water-organic mixtures, and various biorefinery separations. Carbon molecular sieves (CMS) are a useful and promising membrane material for studying the fundamentals of molecular transport, as these materials are chemically and thermally stable in addition to having a tunable pore size distribution. The study of organic solvent transport in carbon molecular sieve membranes has been previously investigated. However, there is a dearth of data on water transport in these small pore CMS systems. In this work, we investigated the sorption, diffusion, and permeation behavior of water in polyvinylidene fluoride (PVDF) derived CMS (PVDF-CMS) of various pore size distribution. The water sorption and diffusion coefficients are compared with an exemplary organic compound, p-xylene. The comparison shows how hydrophilicity and molecular size of permeate (kinetic diameter of water and p-xylene are 2.65Å and 5.85Å, respectively) affect the transport behavior in CMS with its ultra-micropore size in a range of 5Å – 7Å. Understanding the fundamentals of how pure water and organic solvent transport in CMS will further be used for exploring how water-organic mixtures transport in CMS membrane.

611656 Resilient Hollow Fiber Nanofiltration Membranes Fabricated from Crosslinkable Phase-Separated Copolymers

Abstract:

As membrane technologies become more critical to meeting the growing demand for energy efficient separations, a need has emerged for membrane platforms that can be tailored to accommodate the highly varied applications, feed compositions, and treatment demands of wastewater treatment, desalination, and pharmaceutical product purification. Nanofiltration (NF) membranes based on copolymer materials are a promising separations platform because they can be engineered at the molecular scale to address an array of process needs. Here, for example, a resilient NF membrane is developed through the molecular design of a poly(trifluoroethyl methacrylate-*co*-oligo(ethylene glycol) methyl ether methacrylate-*co*-glycidyl methacrylate) [P(TFEMA-OEGMA-GMA)] copolymer that can be dip-coated onto hollow fiber supports. By exploiting the microphase separation of the oligomeric ethylene glycol side chains from the copolymer backbone and by elucidating the processing-structure-property relationships for the dip-coating process, membranes with pores 2 nm in diameter that exhibit a hydraulic permeability of 15.6 L m^{-2} h^{-1} bar^{-1} were generated. The GMA repeat units were functionalized post-coating with hexamethylene diamine to incorporate positively-charged moieties along the pore walls. This post-coating functionalization yielded membranes that rejected 98% of the $MgCl_2$ from a 1 mM feed solution. Moreover, the reaction with the diamine-crosslinked the copolymer such that the membranes could operate stably in ethanol, an organic solvent that otherwise damaged the unreacted parent membranes. Finally, the chemical stability of the P(TFEMA-OEGMA-GMA) copolymer resulted in membranes that could operate continuously in aqueous solutions containing 500 ppm chlorine for a 24-hour period without exhibiting deterioration in the membrane nanostructure. These results demonstrate the ability to develop resilient, valence-selective NF membranes through the molecular engineering of the copolymer. The stability of these membranes has led to investigations of how they can withstand exposure to various organic solvents. By combining the versatility of molecularly designed copolymers with functionalities that can be utilized to increase the integrity of the membrane can further the use of membrane technologies into emerging applications.

607744 RO Treatment of Small Community Wellhead Water for Contaminant Removal and Salinity Reduction

Abstract:

The rising salinity and nitrate levels in groundwater water supplies in agricultural regions pose a threat to potable water supplies of many rural communities who are without suitable and immediate alternative water supplies. Salinity and nitrate reduction can be effectively achieved via RO membrane treatment of such impaired water supplies. The challenge in small remote communities, however, is in (a) attaining high recovery while avoiding mineral scaling, (b) providing effective nitrate monitoring to comply with regulatory requirements, (c) effective low-cost operation, affordable by the communities, and (d) community septic system capable of handling the combined volume of discharged domestic wastewater and RO concentrate stream. Accordingly, the present study reports on the technical, operational and economic feasibility of a water treatment system, based on RO treatment train (i.e., RO, MF, media filtration, activated carbon adsorption, and remineralization), designed, constructed and deployed in a small agricultural community in Northern California. RO system operation was implemented at high recovery (up to ~90%) allowing a sufficient retention time in the septic tank, while ensuring the target nitrate removal and salinity reduction. The RO system featured a unique approach to recovery enhancement via partial concentrate recycle and product water quality monitoring (with respect to nitrate and salinity) via online nitrate and conductivity sensors, in addition to establishment of a correlation between membrane salt and nitrate passage. Field tests demonstrated the potential for deployment of a novel RO system, integrated with a pressure intensifier, in which the RO feed pump outlet pressure requirement was significantly reduced, thereby enabling utilization of renewable energy sources (e.g., solar or wind). Field studies have demonstrated the ability to mitigate mineral scaling with the combination of periodic triggering a freshwater flush and optimal antiscalant dosing based on the direct membrane surface monitoring. It was also shown that operating costs could be significantly reduced by implementing a unique remote supervisory control and decision support system.

606711 Adsorption of Organic Micropollutants to Polymer Surfaces Probed By Second Harmonic Scattering Laser Spectroscopy

Abstract:

Modern thin-film composite reverse osmosis membranes suffer from poor rejection of organic micropollutants, including pharmaceuticals, endocrine disruptors, and disinfection byproducts. The physico-chemical interactions between organic solutes and the polymer membrane surfaces are poorly understood, but may play a significant role in determining membrane rejection.

The purpose of this study was to quantitatively measure the adsorption of several organic micropollutants and surrogate molecules to a model polymer surface. To begin developing a quantitative understanding of the interactions between organic micropollutants and polymer membrane surfaces, angle-resolved second harmonic scattering laser spectroscopy was used to probe the adsorption of resonant aqueous cationic and anionic dye molecules to polystyrene surfaces. Calculation of the Gibbs free energy for adsorption of malachite green to negatively charged polystyrene and naphthol yellow S to positively charged polystyrene revealed highly favorable interactions. The Gibbs free energy of adsorption for the adsorption of malachite green to negatively charged polystyrene was in agreement with other independent measurements.

For non-resonant organic solutes, including the micropollutants caffeine and pentoxifylline, a competitive displacement methodology was employed to obtain values for the Gibbs free energy of adsorption. The mechanism of adsorption for these organic micropollutants appears to be dominated by van der Waals interactions, with surface charge playing a relatively minor role, even for ionic adsorbers. Preliminary results on the adsorption of organic solutes to polyamide surfaces will also be presented.

This experimental technique provides a straightforward way of quantifying the adsorption of critical organic micropollutants to polymer membrane surfaces. These measurements could inform the design of membranes with improved rejection of organic micropollutants, reducing the need for costly secondary treatment of purified wastewater.

598973 Synchrotron-Based X-Ray Spectroscopy of Foulants on Reverse Osmosis Membranes

Abstract:

Membrane reverse osmosis is the gold standard for water desalination and purification. Its key limitation is membrane fouling. Efforts to reduce membrane fouling require a mechanistic understanding of the reactions involved, but this understanding has been difficult to achieve. One reason is that operating conditions involve multiple types of solutes, such as dissolved salts and natural organic matter. These species have large concentration gradients across individual membrane modules. In addition, different modules in RO systems, such as the lead module and the tail module, may have vastly different feed stream conditions. These interactions combine to form a complex fouling layer that may vary across the length of the RO system.

In this study, we investigated the nature and distribution of fouling on membrane surfaces to inform the design of anti-fouling methods. Equipped with a set of synchrotron-based X-ray spectroscopic tools, we studied the surface chemistry of a fouling layer on membranes harvested from the Orange County Water District (OCWD) pilot plant, which converts wastewater effluent into potable water with a filtrate flow rate of 12 gallons per minute. Although the plant uses microfiltration pretreatment and an anti-scalant, membrane fouling still occurs. Our X-ray tools enabled elemental mapping of the fouling layer for particle composition, size, and spatial distribution. They also provided definitive chemical speciation such that inorganic and organic materials could be identified.

We probed membranes from various sections of the pilot plant for multiple elements using micron-resolution X-ray spectroscopic imaging. We determined that the tail module had a higher particle count, with a larger particle size and size distribution, than the lead module. Out of a large range of elements found in both modules, calcium concentrations were the highest. With this insight, we employed X-ray absorption spectroscopic characterization of the calcium speciation on the membrane, noting that calcium contributes to the formation of inorganic scales, organic complexes, and biofouling. We demonstrated that limited inorganic, crystalline scales form, but we have confirmed the presence of asymmetric calcium-organic complexes, corresponding to a calcium-ligand coordination number of 7 or 8. We hypothesize that calcium can connect organic ligands to negatively-charged functional groups on the polyamide membrane. These findings provide the foundation for a mechanistic understanding of the multiple chemical reactions leading to membrane fouling, and this understanding is critical to design effective anti-fouling methods.

605231 Graphene Oxide Nanofiltration Membranes for Desalination at Realistic Conditions

Abstract:

A new generation of nanofiltration membranes is essential to address the rapidly increasing difficulties in sustainable production and recycle of industrial water. Graphene oxide (GO)-based membranes have shown potential to revolutionize nanofiltration, but sustaining solute rejections at realistic concentrations has been a main challenge. We show that a series of membranes consisting of GO bound with polycyclic π-conjugated cations display dramatically enhanced rejections for salts and neutral solutes over a wide concentration range. Furthermore, we find that the counter-intuitive solute rejection behaviors observed in these π-intercalated GO membranes can be explained by a dual mechanism of interlayer spacing modulation and creation of diffusion barriers in the two dimensional interlayer galleries. These membranes are easily scalable, and display high chemical and mechanical robustness in desalination of a multicomponent industrial stream at elevated pH, temperature, stream velocity, and solids content.

611465 High-Perfromance Nanofiltration Hollow Fiber Membranes for Wastewater Treatment

Abstract:

Due to the well-defined pore size (i.e., 0.5–2.0 nm in diameter) and rationally designed surface charge characteristics, nanofiltration (NF) membranes can effectively remove a variety of inorganic and organic contaminants from wastewater. In order to simplify the fabrication process of conventional polyamide thin film composite (TFC) NF membranes and leverage the advantages of hollow fiber configurations, NF hollow fiber membranes with ultrahigh water permeability and well-defined nanopores and surface charge characteristics were developed via a single-step spinning process. The as-spun NF hollow fiber possesses a high pure water permeability (PWP) of 80 $L \cdot m^{-2} \cdot h^{-1} \cdot bar^{-1}$ with a small pore size of 1.0 nm in diameter and a MWCO of 1000 Da. Surface modification by hyperbranched polyethylenimine (PEI) can further lower the pore diameter to 0.85 nm and MWCO to 680 Da. In addition, the NF hollow fibers exhibit outstanding performance stability, low fouling tendency, and great fouling reversibility in textile wastewater treatment. This study provides important insights for the design and fabrication of wholly integral asymmetric hollow fiber NF membranes with ultrahigh flux and good rejection for wastewater treatment and water purification.

609149 Measurement and Modeling of Water Isotherms on Zeolite Crystals

Abstract:

A systematic study of water adsorption isotherms of zeolite crystals over a pressure range of 10^{-5} to 10^{-1} bar and temperatures from 25 to 250 °C was carried out. This work focuses on water isotherms of 3A and 4A zeolite crystals, not pelletized zeolites containing binders. Most water adsorption data reported in the literature result from the use of zeolite pellets or beads forms. To exclude the effects of binders on water adsorption, one must measure adsorption on zeolite crystals without any clays. This is especially important for the design of nontraditional beds with structured contactors, in which different formulations can be used compared to commercial pellets or beads used in packed beds.

Multi-temperature isotherms of 3A crystals are analyzed to determine the isotherm model that offers the best representation of data. Both Toth and dual-site Langmuir model provide reasonable description, but fail in the region of high water loadings over 11 mol/kg. A simple triple-site Langmuir model has been proposed and is proven to satisfactorily model multi-temperature isotherms on 3A and 4A crystals as well as literature data on 3A and 4A pellets. The accuracy of the three sites model (triple-site Langmuir) to describe water adsorption is consistent with and related to the three cation sites (inside α-cage. β-cage, and sodalite cage) in the zeolite LTA structure. Thus, the water molecules have different bonding energy to the cations depending on the locations within the zeolite. These results provide fundamental water isotherms and a simple model to evaluate 3A and 4A in deep dehydration with temperature-swing, or partial pressure-swing adsorption processes, or their combination using non-traditional monolith contactors.

599330 Binary Adsorption Equilibria Measurement for Biogas Separation Using a Zero Length Column Method

Abstract:

Biogas is a promising alternative to natural gas as the global energy sources shift from fossil fuels to renewables. The main components in biogas are CH_4 (35-70%), CO_2 (15-60%), and N_2 (15-50%), with trace impurities such as NH_3, H_2O, and H_2S. Release of biogas into the atmosphere causes rising levels of greenhouse gases, and flaring biogas produces SO_X due to the H_2S present in biogas. It's critical to remove H_2S and CO_2 in biogas as H_2S causes equipment corrosion and poisons catalysts for downstream processes, and CO_2 reduces the heating value. Purifying biogas for electricity generation or biomethane production has gained interest among researchers. Many are exploring less energy-intensive separation processes compared to traditional methods involving liquid scrubbing. Adsorption using microporous materials such as zeolites, metal-organic frameworks (MOFs), and porous organic cages (POCs) is a promising technology as it provides tuned selectivity and requires less regeneration energy. However, the majority of research is on single-component adsorption, which does not well predict the competitive interaction in gas mixtures. This work focuses on the design of a Zero Length Column (ZLC) for rapid binary adsorption equilibria measurement. The binary adsorption equilibria are studied in both pristine and acid-gas exposed MOFs and zeolites such as MIL-125-NH_2, ZIF-8, mmen-Mg_2(dobpdc), and zeolite 13X. With a better understanding of competitive adsorption behavior, sorption materials for complex gas separation can be screened more rapidly, and processes such as pressure swing adsorption (PSA) can be modeled with better accuracy.

611672 Adsorption Equilibria of CO_2/CH4/N2 Binary Mixtures on Flexible Zeolitic Imidazolate Frameworks (ZIF-7)

Abstract:

In conjunction with a high surface area and pore volume, structural transformation in flexible Metal-Organic Framework (MOFs), which generally attributed to the alterations between narrow-pore or non-porous (np) and large-pore or porous (lp) states of the adsorbent framework, makes these materials one of the most promising in the field of gas separation. Although the structural transition of these materials in the presence of pure gases and their adsorption capacity has been studied widely in the literature, limited research investigated the adsorption behaviour and gas separation performance of these hybrid adsorbents in the presence of gas mixtures experimentally. Experiments with binary mixtures provide more accurate and direct measurements of an adsorbent's equilibrium selectivity for the component gases than obtained from estimates based only on pure fluid capacity measurements.

Here, a dynamic column breakthrough (DCB) apparatus was first used to measure the adsorption capacities of pure CO_2, CH_4 and N_2 on flexible ZIF-7 at temperatures in the range 263.15–303.15 K and pressures up to 900 kPa. Structural changes in ZIF-7 were observed for CO2, and CH4. Equilibrium adsorption capacities of the component gases in the binary mixtures of $CO_2 + N_2$, $CO_2 + CH_4$ and $CH_4 + N_2$ were then determined by these dynamic experiments. In addition, we have investigated the application of a new adsorption isotherm model, which was successfully regressed to the pure capacity data over the ranges of temperature and pressure measured; this model avoids the need for piecewise definitions or discontinuities in the definition of the adsorption isotherm. The IAST selectivity estimated from the pure fluid adsorption data were compared with the measured selectivities in the gas mixtures; reasons for the discrepancies observed will be discussed.

612063 Combining Novel Methodologies Based on NMR Relaxometry and Gas Adsorption for a Reliable Surface Area Analysis of Nanoporous Materials

Abstract:

In the past, major progress has been made concerning the textural characterization of porous materials in the gas/dry state. The main drawback of these methods stems from their inability to test wet materials (i.e., in presence of the liquid/mobile phase), thus providing no direct information concerning the state of the nanoporous materials immersed a liquid phase, which is important for many applications (e.g. chromatographic operations). This is particularly important for functionalised nanoporous materials, where the interaction of surface functionals with the chosen liquid may affect the effective surface area and pore characteristics for nanoparticle separation. Hence, for the optimization of processes in the liquid phase (e.g. liquid chromatography) characterization only based on gas adsorption and liquid intrusion techniques (e.g., mercury porosimetry) is not sufficient. However, far fewer advances have been made in assessing textural properties in the liquid phase. One of the important challenges in liquid phase characterisation is a reliable and fast assessment of surface area. For this, the application of NMR relaxation measurements has recently received a lot of attention. The NMR relaxation measurement method is based on the fact that the liquid in contact with a surface behaves differently from that of the bulk phase (or "free" liquid) in response to a changing magnetic field, i.e., a material with a high surface area has a shorter relaxation time, as more fluid is bound to the surface. Despite the fact that the application of NMR relaxometry for textural characterization of various material types (including silicas and titania, carbons and even metal organic framework materials (MOFs)) has been reported in the literature, a rigorous and systematic validation of the method utilizing materials with true surface area benchmark data has not yet been reported. One problem is that many of the materials used in the reported studies contain a significant amount of microporosity (e,g,, MOFs) and here the Brunauer-Emmett-Teller (BET) method cannot be applied in a straightforward way for determining reliable, probe accessible surface area data. Additional uncertainties result from the choice of nitrogen as the probe molecule, leading to errors in assessing surfaces up to 20-30 % ; due to its quadrupole moment, the orientation of a nitrogen molecule is very much dependent on the surface chemistry of the adsorbent.

In this work the surface area of nonporous and mesoporous silica model adsorbent materials immersed in a liquid phase is investigated by methodologies based on NMR relaxometry and compared with benchmark adsorption data (by applying procedures based on argon adsorption). We further investigate the effect of confinement and pore network characteristic on NMR relaxation and consequently on a reliable assessment of surface area, but also explore the potential of NMR relaxation for pore size analysis.

601785 Surface Area Prediction of Nanoporous Materials Using Machine Learning Methods

Abstract:

The surface area of a porous material plays a critical role in a broad range of science and engineering applications, as many chemical and biological events occur near or on the surface. Surface areas of porous materials, such as metal-organic frameworks (MOFs), are commonly characterized using the Brunauer-Emmett-Teller (BET) method. The BET areas depend significantly on the linear region (the region of the isotherm) selected for the BET analysis, which is chosen *via* four consistency criteria as per the state-of-the-art. However, it has been shown that even with the fulfillment of these criteria, the BET method does not always provide an accurate measure of the monolayer area, often overestimating it significantly, especially for high-surface-area MOFs. In this work, we extend this argument to systematically show that even if we use the best possible consistency criteria, the BET method cannot always predict the monolayer areas accurately. To this end, we propose, for the first time, a data-driven approach to accurately predict the surface areas of MOFs. Machine learning (ML) is employed to train models based on adsorption isotherm features of over 300 diverse MOF structures to predict a benchmark measure of the surface area known as the true monolayer area. The true monolayer area is computed by counting the number of adsorbate molecules present in the first layer from the framework in molecular simulation snapshots. We demonstrate that the ML-based methods can consistently predict true monolayer areas significantly better than the BET method, especially for high surface area materials, showing great promise as a more accurate alternative to the BET method in the structural characterization of porous materials.

600352 Pore Size Distribution of Silica Colloidal Crystals from Nitrogen Adsorption Isotherms

Abstract:

Silica colloidal crystals, often referred to as opals, are face-centered closely packed cubic crystals comprised of silica spheres. The voids between the spheres form a network of octahedral and tetrahedral pores. These pores have sizes that could vary in the range of tens to hundreds of nanometers and can be probed by nitrogen adsorption porosimetry.

Here, we used sphere diameters 93 nm, 106 nm, and 565 nm to prepare two mesoporous samples and a macroporous sample and then measured nitrogen adsorption and desorption isotherms for further characterization [1]. Furthermore, we modified the surface of the samples with organics and measured new adsorption isotherms [2]. The IUPAC recommendation for the calculation of pore-size distribution (PSD) from a nitrogen adsorption isotherm is a combination of the density functional theory (DFT) with the integral adsorption equation. Although the DFT works for different pore geometries, the libraries available with commercial software are rather limited and the DFT predictions for nitrogen adsorption on a silica surface are based on one reference silica material. Therefore, the application of this model for deriving the PSD of silica materials with modified surfaces can lead to incorrect PSD. Furthermore, often available kernels of DFT isotherms for spherical pore geometry in the libraries are limited to ca. 40 nm, [3] which is not sufficient to describe the characteristic pores of silica colloidal crystals samples, estimated to be ca. 60 nm [1].

To resolve this problem we proposed a straightforward procedure for calculation of the PSD of silica colloidal crystals from nitrogen adsorption isotherms requiring only three parameters to fit. The procedure is based on the adsorption integral equation solution with a kernel of theoretical isotherms, consistent with the procedure used for many other porous materials [4]. The solution is carried out using the non-negative least squares (NNLS) regression with Tikhonov regularization. The kernel of mesoporous isotherms is built on the basis of the macroscopic Derjaguin–Broekhoff–de Boer theory of capillary condensation considering the voids as a network of spheres. Application of our procedure for the analysis of the adsorption branches of experimental isotherms resulted in bimodal distributions, where the modes matched well with the sizes of the voids in the colloidal crystals face-centered cubic structure: the main mode corresponds to the octahedral voids and the second mode to the tetrahedral voids. The resulting pore-size distribution for the samples with the modified surface matched the original one quite closely. It demonstrates the procedure as a simple and efficient technique to estimate the pore-size distribution and justifies the spherical shape approximation for the voids in the silica colloidal crystals.

References:

[1] A. Galukhin, D. Bolmatenkov, A. Emelianova, I. Zharov, G. Y. Gor, Langmuir, 35, 2230 (2019).
[2] M. A. Maximov, A. Galukhin, G. Y. Gor, Langmuir, 35, 14975 (2019).
[3] Quantachrome NOVAtouch brochure.
https://www.quantachrome.com/pdf_brochures/novatouch_rev1.pdf
[4] J. Landers, G. Y. Gor, A. V. Neimark, Coll. Surf. A., 437, 3, (2013).

597022 Characterization of Metal Sites on Reverse Phase Chromatography Column

Abstract:

In liquid chromatography, separations are ideally governed by analyte interactions with active sites. However, the presence of metal contaminants can result in metal-analyte interactions that can have detrimental effects on the performance of chromatography processes. Metal contamination is not always avoidable, and can be introduced as impurities in the media manufacturing process, as impurities in the feed stream, and from equipment used. Therefore understanding the metals contamination in a chromatography column, and its impact on separation performance is of interest. In this work, we characterize the adsorption behavior of the 4, 4-bipyridyl and 2, 2-bipyridyl on the metal sites of reverse phase chromatography columns. Because of its geometry, 4, 4-bipyridyl does not readily form a chelate with the metal sites, while 2, 2-bipyridyl does. In this study, we compared the adsorption strength of the compounds on metal sites vs. on the hydrophobic sites. To prevent the results from being complicated by exposed silanol sites, we compared the results with and without sulfinic acid.

Through a bi-Langmuir adsorption model, we were able to determine the population of the metal sites and estimated the adsorption strength. The simulation and parameter estimation were conducted using the CADET model. A few different metal contaminations were studied: Zn^{2+}, Fe^{3+}, Cu^{2+}. Furthermore, we explored strategies to control the effect of metal through use of EDTA as an additive. The mathematical model also revealed whether EDTA removes the metals from the stationary phase or simply masks the metal-analyte interaction. The results of this study can be used to characterize the extent of metal contamination in a column, and therefore troubleshoot column performance issues on both analytical and preparative chromatography. Based on the data obtained through this study, we proposed a strategy to characterize column life.

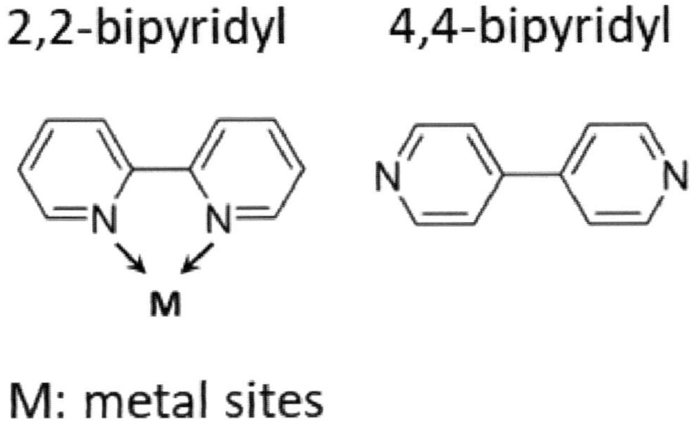

606261 Determination of Orientation and Conformation of Lysozyme at the Air-Water Interface Using an Integrated MD/SFG Approach

Abstract:

Proteins have markedly complex and diverse structures due to a high variability in size, shape, amphiphilicity, and charge needed to carry out their biological function. The interfacial behavior of adsorbed proteins govern properties such as foaming and stabilization, relevant to applications in pharmacy, biotechnology, and the food industry.[1-2] By studying proteins at the air water interface (AWI), which occurs naturally in many biological processes, we can explore how conformation and orientation of these complex biomolecules lead to self-assembly at these interfaces. To date, the interfacial structure of even one of the most widely studied proteins, lysozyme, remains unresolved.

In this talk we describe our approach to uncovering this interfacial structure by combining molecular dynamics simulation, vibrational sum frequency generation (SFG) spectroscopy, and spectral calculations to determine the conformation and orientation of lysozyme at the AWI. Computationally, two force fields are used to simulate lysozyme. With this approach we determine agreement in a single interfacial pose at high atomistic resolution. We validate the proposed structure of lysozyme by comparing signals of experimentally derived structures, to the spectra calculated from simulation, showing strong agreement by the pose predicted by MD. Lastly, we provide additional atomistic insight, discussing how pH may lead to an orientation change from head on to side-on at the interface, explaining previous discrepancies in the literature with regards to the thickness of the experimentally derived monolayers. This work provides a template for future studies of proteins at interfaces to make maximum use of integrated computational and experimental approaches.

[1] Dickinson, E. Proteins at Interfaces and in Emulsions Stability, Rheology and Interactions. *J. Chem. Soc. Faraday Trans.* **1998**, *94* (12), 1657–1669.

[2] McClements, D. J. Protein-Stabilized Emulsions. *Curr. Opin. Colloid Interface Sci.* **2004**, *9* (5), 305–313.

600717 Density Functional Theory Model for Adsorption-Induced Deformation of Materials with Non-Convex Pore Geometry

Abstract:

Adsorption of fluids in nanoporous media causes mechanical stresses and results in deformation [1]. This phenomenon is ubiquitous, and its magnitude depends on the pore size and geometry. Adsorption and adsorption-induced deformation is typically modeled in convex pores (cylindrical or spherical pores) or planar geometry (slit pores). However, many porous materials are composed of spherical grains, so that the pores are formed by the intergranular spaces between the convex surfaces, thus becoming non-convex. Here we present a first theoretical study of adsorption-induced deformation of materials with non-convex pores. We take the templated mesoporous carbons (CMK-3) which have an ordered hexagonal structure of parallel cylindrical rods [2], and model adsorption of nitrogen in this system using the classical density functional theory (DFT) within the local density approximation. Adsorbed fluid density profile in CMK-3 geometry can be reduced to two dimensions, but unlike slit or cylindrical pore model it cannot be made one dimensional. The fluid-fluid interaction parameters are fit to experimental nitrogen binodal, and reproduce other thermodynamic properties (such as surface tension and vapor pressure). The solid-fluid interaction parameters are fit to reproduce the nitrogen adsorption at 77 K on a reference carbon surface. Using the resulting parameters we predict the adsorption isotherms and strain isotherms for nitrogen adsorption in CMK-3 carbons. The shape of adsorption isotherm matches the shape of experimental isotherm. The predicted strain isotherms are qualitatively different from the strain isotherms in cylindrical pores [3]. We attribute this difference to formation of liquid bridges between the adjacent rods [4]. Our results suggest that adsorption-induced deformation in materials with non-convex pore walls cannot be predicted within the existing models. These results may shed some light on understanding adsorption-induced deformation of consolidated granular media.

References:

1. Gor, G. Y.; Huber, P.; Bernstein, N., 2017. Adsorption-induced deformation of nanoporous materials - a review. Applied Physics Reviews, 4(1), p.011303.
2. Jun, S.; Joo, S. H.; Ryoo, R.; Kruk, M.; Jaroniec, M.; Liu, Z.; Ohsuna, T.; Terasaki, O., 2000. Synthesis of new, nanoporous carbon with hexagonally ordered mesostructure. Journal of the American Chemical Society, 122(43), pp.10712-10713.
3. Gor, G. Y.; Neimark, A. V., 2011. Adsorption-induced deformation of mesoporous solids: Macroscopic approach and density functional theory. Langmuir, 27(11), pp.6926-6931.
4. Ludescher, L.; Braxmeier, S.; Balzer, C.; Reichenauer, G.; Putz, F.; Hüsing, N.; Gor, G. Y.; Paris, O., 2020. Capillary Bridge Formation between Hexagonally Ordered Carbon Nanorods, Adsorption, DOI: 10.1007/s10450-020-00215-6

604204 Optimal Packing of CO at High Coverage on Pt(100) and Pt(111) Surfaces

Abstract:

High coverage structures for CO on Pt(111) and Pt(100) surfaces are studied by density functional theory modeling and compared to high pressure scanning tunneling microscopy experiments. Semilocal exchange correlation functionals are known to provide incorrect adsorption site and overestimated adsorption energy for CO on Pt. We develop a simple first-principles correction for the adsorption energy of CO on Pt(111) and Pt(100) using the bond length of adsorbed CO as a descriptor. The developed correction which decreases in the order top > bridge > hollow site is used to derive the surface stability diagram for CO adsorbed on 111 and 100 facets of Pt showing the thermodynamically stable CO configurations on the surface as a function of temperature and pressure. High coverage (θ>0.5) configurations of CO on Pt(111) lead to formation of superimposed hexagonal/quasi-hexagonal lattice of CO on hexagonal Pt(111) layer from a systematic exploration of such structures. Non-hexagonal CO structures seen in vacuum and low temperature conditions - c($\sqrt{3}$x3)rect-4CO, c($\sqrt{3}$x5)rect-3CO, and c($\sqrt{3}$x7)rect-5CO are however only 5 meV/Å^2 less stable than the hexagonal lattice at similar coverage. For Pt(100) at θ>0.75, the CO molecules adopt a one dimensional coincidence lattice and we observe the formation of (nx2) unit cells (n=4,6,8) with (2n-2) CO molecules in each cell on top/quasi-top and bridge/quasi-bridge positions creating a skewed hexagonal lattice to reduce CO-CO repulsion with increasing coverage. The computational results agree with the available experimental observations for Pt(111) and Pt(100). The integrated theoretical simulation and experimental observation provide structural data for the study of catalytic reactivity on Pt surfaces in reactions involving high CO pressures and suggest an approach for understanding roles of CO molecules on other metal catalyst surfaces in high pressure CO and during catalysis CO participates.

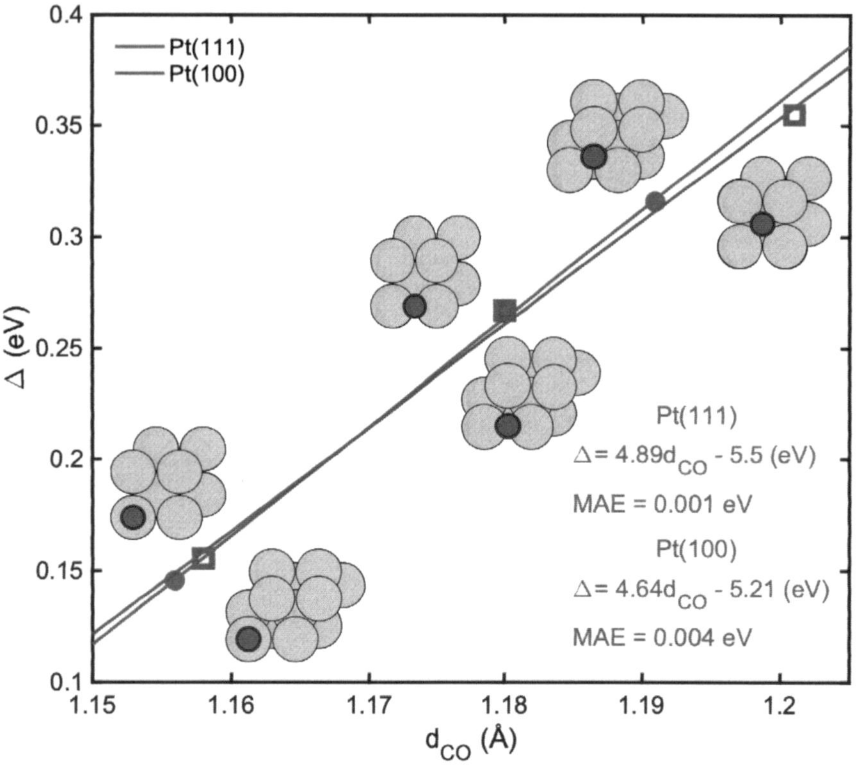

609577 Quantitatively Accurate Theory to Predict Adsorbed Configurations of Surfactants at Polar Surfaces

Abstract:

Surfactant molecules adsorb on polar surfaces in various morphologies, such as planar, spherical and cylindrical. So far, there is no encompassing theoretical framework that related molecular properties to the adsorbed morphologies. We have developed a theoretical model that is able to predict adsorbed morphologies of surfactant molecules on polar surfaces given their geometry and strength of interactions with the surface. We show, via coarse-grained simulations, that our theoretical model is quantitatively accurate in predicting the adsorbed configurations. Specifically, our model accurately predicts that linear-shaped surfactants adsorb either as a self-assembled monolayer or as stripes on the surfaces, depending on the relative interaction strengths of the polar head group and the alkyl tails with the surface. Our model correctly predicts the formation of different cylindrical and spherical configurations of the adsorbed asymmetrically shaped surfactant molecules.

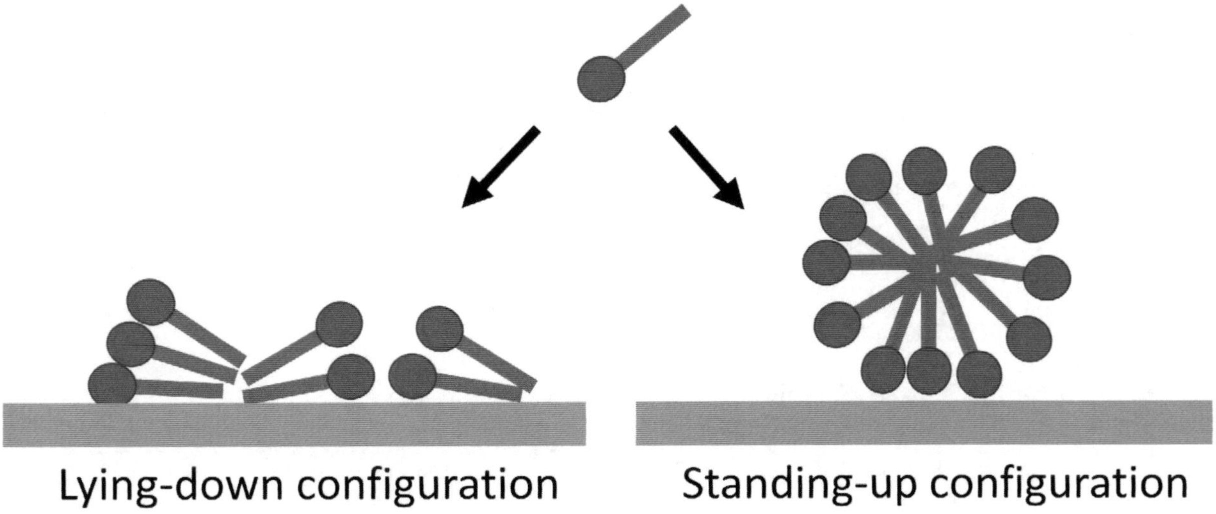

601439 2D Phase Transitions of Fluid Adsorbed in Cylindrical Mesopores: Influence of Adsorbent-Adsorbate Interactions

Abstract:

Recently, molecular simulations of methane adsorption in mesoporous cylindrical silica identified a two-dimensional phase transition associated with non-trivial densification of the adsorbed liquid, a high-density local maximum in the isosteric heat of adsorption, and hexagonal rearrangement of annular layers of adsorbed fluid [1]. This observed rearrangement of adsorbed methane was proposed as a plausible explanation for the densification of methane adsorbed in MCM-41 and SBA-15 that was observed in small-angle neutron scattering (SANS) experiments [2]. The proposed phase transition was, however, unique to the methane-silica adsorption systems, not being observed in, for example, nitrogen-silica, argon-silica, or argon-carbon systems. To better understand this unique phase transition and why it occurs for the methane-silica systems in contrast to others, we perform additional Monte Carlo molecular simulations to identify key characteristics of the methane-silica system that give rise to its appearance. Specifically, we measure adsorption isotherms, heats of adsorption, and bond-orientational order parameters for the benchmark systems that do not exhibit such phase transitions, and then smoothly transform those benchmark systems into the methane-silica system to clearly identify the boundaries in adsorbent-adsorbate parameter space that trigger this type of phase transition. Lastly, we discuss engineering consequences of the observed phase transition and how it may impact other thermophysical properties of the adsorbed fluid.

[1] D.W. Siderius, W.P. Krekelberg, W-S. Chiang, V.K. Shen, and Y. Liu, *Langmuir*, 33:14252 (2017).
[2] W-S. Chiang et al., *Langmuir*, 32:8849 (2016).

597618 Impact of Solvated Water Layer on Adsorption of Selenium Oxo-Anion on <012> Hematite Surface

Abstract:

Selenium is highly soluble in water as selenate and selenite and is toxic to human and aquatic life even at very low concentration (TWA 0.2 mg/m^3). Increased usage of selenium in the electronic and glass industry threatens to further increase the concentrations of these species in our water resources. Adsorption is a promising remediation strategy, particularly in low population densities area and communities with self-run wells, because of its potentially lower cost than the construction of chemical treatment plants. The adsorption of oxo-anions on existing sorbents is , however, not selective; this is exacerbated by the fact that Selenium oxoanions occur at much lower concentrations than similar S ions. This leads to increased financial burden for water treatment. Recent adsorption experiments of selenium oxoanion of different hematite surface found that selenate and selenite have higher selectivity on [012] and [110] hematite facet respectively. Here we use first principle calculations to study the facet dependent behavior of hematite based adsorption. Adsorption energies are calculated for varying oxo-anion configuration including inner and outer sphere complexes on [012] surface. In this work, we delineate the most important effects controlling adsorption, including surface configuration, pH, solvation, etc. We find the solvation configuration directed by the surface controls the adsorption behavior. In addition to shedding light on this particular adsorption species, this work highlights the criticality of correctly considering solvation, and outer sphere adsorption, in first principles analysis of oxoanion adsorption.

600557 Microkinetic Modeling Strategy Considering Topologic Aspect of a Cu Catalytic Surface Based on Kinetic Monte-Carlo

Abstract:

Introduction

It is obvious that surface reactions occur between "neighboring" reactants adsorbed on the surface, but conventional microkinetic modeling has not been conducted majorly in this respect. Traditional microkinetic modeling methods, also in our previous study too, consist only of surface coverage, which is the ratio how much a reactant is adsorbed on the surface, -oriented rate equations. Moreover, the reaction of the gaseous reactants is usually calculated by equations based on chemical equilibrium, and the surface reactions are usually calculated by equations based on reaction kinetics. We thought that this inconsistency should be improved so that we used kinetic Monte-Carlo method on microkinetic modeling.

Incorporating kMC simulation into a microkinetic model provides three advantages:

1. The model can take it account into that the topologic effects of surface reactions.
2. All reactions, including gas phase reactions, can be modeled based on reaction kinetics not on equilibriums.
3. The diffusion reactions on the catalytic surface which is not possible with conventional microkinetic plug-flow reactor (PFR) 1D modeling, can be considered.

On the other hand, there are two major disadvantages of the method of simulating the catalytic surface with kMC:

1. The computational load is greatly affected by the size of the catalytic surface.
2. The calculation time is much long compared to the general microkinetic model, which is disadvantageous for parameter estimation.

In order to solve the first problem, this study used periodic boundary unit cell structure to overcome the finite size effect problem.[2,3] Also, to solve the second problem, we derived the correlation between kinetic constants in two different scales of which are molecular level based on kMC, -number oriented form and continuum level based on general microkinetic model, -concentration oriented form. To drive that correlation, we used Gaussian Process (GP) based on Bayesian Optimization using Deep Neural Network (DNN) and performed parameter estimation using microkinetic model using the rate constant oriented from kMC simulations. [4]

Materials and Methods

Our system is Cu catalytic system consists of syngas which is a composition of CH4, CO and CO2 gases. The Cu catalytic surface of our system is based on Cu (111), and the lattice structures of unit cell are varied with 10-by-10, 20-by-20 and 50-by-50. Experimental data of adsorption / desorption reactions on Cu surface at equilibrium will be used to estimate kMC parameter, especially on desorption kinetics.

The reactor type is PFR and it is modeled by a series of Continuous Stirred Tank Reactors (CSTRs). With no assumption of rate determinate step (RDS), the model will be a microkinetic model with total

26 reactions which are taken from the work of C. V. Ovesen. *et al.* consist with five adsorption/desorption reactions and eight reversible, forward and backward surface reactions. [5] Kinetic constants, pre-exponential factors, activation energies and binding energies are taken from our previous study and Ovesen, C.V., *et al.*'s work which are tuned with experimental data. [1,5]

The tools we used are total two. One for DNN is TensorFlow® library of python and another for kMC simulation and optimization of microkinetic model with kinetics is MATLAB®.

Significance

In point of science, it has its novelty at taking topological reaction into account microkinetic modeling. In addition, we can present the adequate strategy for modeling a desorption kinetics from continuum scale experimental data which has been unclear yet.

601219 Application of Derjaguin-Broekhoff-De Boer Theory and Molecular Simulation for Calculation of Solvation Pressure in Spherical Mesopores

Abstract:

Adsorption of fluids in porous structure takes place due to the intermolecular interactions between solid surface and adsorbate. During this process, the adsorbate confined in pore exhibits high "solvation" pressure on the pore walls and causes the deformation of porous materials [1]. The high pressure in the pores have additional effects other than mechanical strains, in particular, some of the high-pressure chemical reactions take place in the pores at the moderate gas pressures [2,3], or the compressibility of fluids in pores changes [4]. Here we focus on solvation pressure in model silica pores of spherical shape induced by the adsorption of nitrogen.

We use the thermodynamic model for calculating solvation pressure from adsorption isotherms, applied before for various microporous materials [5,6] and mesoporous materials with cylindrical pores [7]. We apply it for spherical mesopores using the adsorption isotherms predicted from the macroscopic Derjaguin-Broekhoff-de Boer (DBdB) theory [8-10] and grand canonical molecular simulations. We show that both approaches give consistent results, which are also typical for the solvation pressure in mesoporous materials, yet noticeably differ from the results for cylindrical pore geometry [7]. Furthermore, we show that the dependence of the solvation pressure on the reciprocal pore size can be used for the calculation of the solid-liquid surface energy. Finally, we relate the calculated solvation pressures to the compressibility of liquid nitrogen in spherical confinement and found that the departure of the compressibility from the bulk value is consistent with the calculated values of solvation pressures.

[1] Gor, G. Y., Huber, P., & Bernstein, N. (2017). Appl. Phys. Rev., 4(1), 011303.
[2] Kaneko, K., Fukuzaki, N., Kakei, K., Suzuki, T., & Ozeki, S. (1989). Langmuir, 5(4), 960.
[3] Byl, O., Kondratyuk, P., & Yates, J. T. (2003). J. Phys. Chem. B, 107(18), 4277.
[4] Gor, G. Y., Siderius, D. W., Shen, V. K., & Bernstein, N. (2016). J. Chem. Phys., 145(16), 164505.
[5] Ravikovitch, P. I. & Neimark, A. V. (2006). Langmuir, 22(26), 10864.
[6] Kowalczyk, P., Ciach, A., & Neimark, A. V. (2008). Langmuir, 24(13), 6603.
[7] Gor, G. Y. & Neimark, A. V. (2010). Langmuir, 26(16), 13021.
[8] Derjaguin, B. V. (1940). Acta Physicochim. URSS, 12, 181.
[9] Broekhoff, J. C. P. & De Boer, J. H. (1967). J. Catal., 9(1), 8.
[10] Broekhoff, J. C. P. & De Boer, J. H. (1968). J. Catal., 10(4), 153.

603493 A Systematic Study of State-of-the-Art Methods in Crystal Structure Prediction for Organic Hydrates

- Download/View Extended Abstract

Abstract:

Hydrates are co-crystalline materials containing water as one of the molecules in the crystal lattice. The incorporation of water into the crystal lattice produces a unit cell different from that of the anhydrate and, consequently, the physical properties of the hydrate can differ significantly from those of the anhydrate. The existence and stability of hydrates is an important consideration in the development of pharmaceutical products: the prevalence of water during manufacturing and storage can mean that neat forms of an active pharmaceutical ingredient can undergo a phase transition to hydrate form, impacting the effectiveness of the drug. Crystal structure prediction (CSP) methods can in principle be useful in identifying likely hydrates, by undertaking searches for all polymorphs of water and one or more given compounds for a given co-crystal stoichiometry. Minimal information is needed, typically just the chemical connectivity diagram[1], to search for the low lattice energy arrangements of the constituent atoms in space.

Applications of CSP to hydrates have resulted in mixed success so far. In the fifth blind test[2] organised by Cambridge Crystallographic Data Centre, one of the targets was a hydrate but none of the 10 groups that attempted to predict its structure put forward the correct structure within their shortlist. In the sixth blind test[3], only 8 groups submitted predicted structures for the hydrate target, and only one group generated the experimental structure within their shortlist.

In order to gain a better understanding of the challenges that make CSP for hydrates difficult, we present a systematic evaluation of a CSP state-of-the-art method for organic hydrates, in which the lattice energy is partitioned into intramolecular and intermolecular contributions. Intramolecular interactions are modelled via quantum mechanical calculations[4], and intermolecular interactions are divided into electrostatics, modelled using *ab initio* derived distributed multipoles[5,6,7], and repulsion/dispersion interactions modelled using a semi empirical potential. A total number of 107 hydrates extracted from the Crystal Structure Database are minimized locally using the *CrystalOptimizer* algorithm[8] with six models of different levels of sophistication (functional, basis set, use of continuum polarizability). The geometric differences between experimental structures and the corresponding minimization outputs are compared in terms of root mean-squared deviation and a CPU time required. Five of the six models are found to give a good degree of accuracy in more than 95% of cases, but with varying computational costs. A further assessment of the proposed models is undertaken by determining relative energy rankings, which are critical in generating reliable polymorphic energy landscapes.

References:

(1) Pantelides, C. C.; Adjiman, C. S.; Kazantsev, A. V. General Computational Algorithms for Ab Initio Crystal Structure Prediction for Organic Molecules. In *Prediction and Calculation of Crystal Structures Methods and Applications. Topics in Current Chemistry*; Atahan-Evrenk, S.; Aspuru-Guzik, A., Eds.; Springer, **2014**; Vol. 345, pp. 25–58.

(2) Bardwell, David A., Claire S. Adjiman, Yelena A. Arnautova, Ekaterina Bartashevich, Stephan XM Boerrigter, Doris E. Braun, Aurora J. Cruz-Cabeza et al. Towards crystal structure prediction of complex organic compounds–a report on the fifth blind test. *Acta Crystallographica Sect. B: Struct. Sci. Cryst. Eng. Mater.* **2011**, 535-551.

(3) Reilly, A. M.; Cooper, R. I.; Adjiman, C. S.; Bhattacharya, S.; Boese, A. D.; Brandenburg, J. G.; Bygrave, P. J.; Bylsma, R.; Campbell, J. E.; Car, R.; *et al.* Report on the Sixth Blind Test of Organic Crystal Structure Prediction Methods. *Acta Crystallogr. Sect. B Struct. Sci. Cryst. Eng. Mater.* **2016**, *72*, 439–459.

(4) Gaussian 09, Revision A.02, M. J. Frisch, G. W. Trucks, H. B. Schlegel, G. E. Scuseria, M. A. Robb, J. R. Cheeseman, G. Scalmani, V. Barone, G. A. Petersson, H. Nakatsuji, X. Li, M. Caricato, A. Marenich, J. Bloino, B. G. Janesko, R. Gomperts, B. Mennucci, H. P. Hratchian, J. V. Ortiz, A. F. Izmaylov, J. L. Sonnenberg, D. Williams-Young, F. Ding, F. Lipparini, F. Egidi, J. Goings, B. Peng, A. Petrone, T. Henderson, D. Ranasinghe, V. G. Zakrzewski, J. Gao, N. Rega, G. Zheng, W. Liang, M. Hada, M. Ehara, K. Toyota, R. Fukuda, J. Hasegawa, M. Ishida, T. Nakajima, Y. Honda, O. Kitao, H. Nakai, T. Vreven, K. Throssell, J. A. Montgomery, Jr., J. E. Peralta, F. Ogliaro, M. Bearpark, J. J. Heyd, E. Brothers, K. N. Kudin, V. N. Staroverov, T. Keith, R. Kobayashi, J. Normand, K. Raghavachari, A. Rendell, J. C. Burant, S. S. Iyengar, J. Tomasi, M. Cossi, J. M. Millam, M. Klene, C. Adamo, R. Cammi, J. W. Ochterski, R. L. Martin, K. Morokuma, O. Farkas, J. B. Foresman, and D. J. Fox, Gaussian, Inc., Wallingford CT, **2016**.

(5) Nyman, J.; Day, G. M. Static and Lattice Vibrational Energy Differences between Polymorphs. *CrystEngComm* **2015**, *17*, 5154–5165.

(6) Price, S. L.; Leslie, M.; Welch, G. W. A.; Habgood, M.; Price, L. S.; Karamertzanis, P. G.; Day, G. M. Modelling Organic Crystal Structures Using Distributed Multipole and Polarizability-Based Model Intermolecular Potentials. *Phys. Chem. Chem. Phys.* **2010**, *12*, 8466.

(7) Stone, A. J. Distributed Multipole Analysis, or How to Describe a Molecular Charge Distribution. *Chem. Phys. Lett.* **1981**, *83*, 233–239.

(8) Kazantsev, A. V.; P. G. Karamertzanis; C. S. Adjiman; C. C. Pantelides. Efficient handling of molecular flexibility in lattice energy minimization of organic crystals. *J. Chem. Theory Comput.* **2011**, *7*, 1998-2016.

597238 Effects of Inorganic Salts and Volume Reduction on Glycine Polymorphism

Abstract:

Studies have demonstrated that inorganic salts can drive polymorphic selectivity of glycine from the metastable α-polymorph to the stable γ-polymorph. Here, we explore the effect of NaCl and $CaCl_2$ together with volume reduction on glycine nucleation. The salts were added to supersaturated aqueous solutions of glycine at a 2:1 Gly:NaCl/$CaCl_2$ ratio at volumes from 1000 µL to 50 µL. Analysis showed that a primary γ-glycine nucleation event occurs for Gly:NaCl systems, with observed increase in induction times as volume was decreased. However, β-glycine primary nucleation was also detected for Gly:NaCl systems. Interestingly, it was observed that β-glycine nucleation is preferred over the salt-directed γ-glycine nucleation at a certain volume. The increase in induction times combined with a decrease in probability for γ-glycine nucleation may have allowed preferred nucleation of the β-polymorph. On the other hand, no nucleation events were observed for Gly:$CaCl_2$ systems at a time span of 48 hours. These results provide insight on how polymorph modifiers act in confined environments.

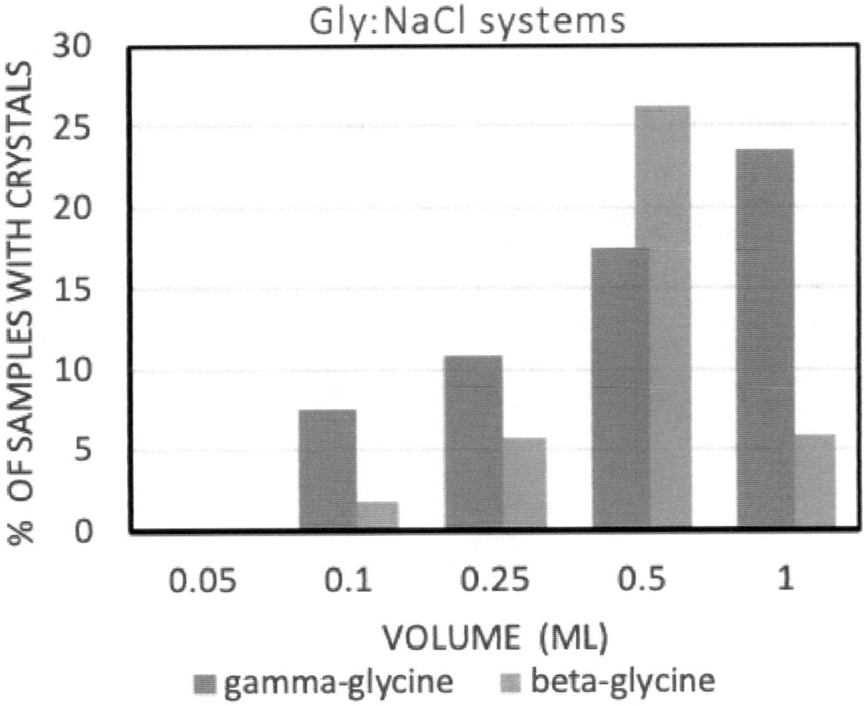

609279 Using Fluid Dynamics and Interface Engineering to Control Polymorphism and Morphology of Organic Molecules.

Abstract:

Crystallization of organic materials as thin films is an area of interest in multiple fields, such as in the pharmaceutical industry, organic electronics and for the multiple applications afforded by metal organic frameworks (MOFs). Over the past decade, significant work has been accomplished in terms of understanding how polymorphism of organic molecules can be enhanced through crystallization as a thin film, through innovative techniques such as confinement, polymer additives, and control of deposition conditions such as the use of electric fields. A deeper understanding of how these polymorphs are formed is critically missing from literature.

In this work, we use a mix of experimental and computational work to study the formation of thin film polymorphs of pharmaceutical compounds and organic semiconductors. The use of interface engineering can influence both the morphology and the polymorphism of acetaminophen, a pharmaceutical compound, and TIPS-pentacene, an organic semiconductor. We also show that changing the flow coating parameters can tune the nucleation and growth rates of organic semiconductors, which then impact the molecule polymorph, crystal size, and crystal orientation. In-situ X-ray diffraction data shows the changes in crystal growth conditions due to changing the fluid dynamics conditions and provides valuable insight in the mechanism of crystal nucleation. Controlling the interface precisely to obtain the desired polymorphs and crystal morphologies will be discussed as well.

609384 Acetaminophen Polymorphism and Texture Control Using Novel Coating Regimes

Abstract:

Spray coating and antisolvent addition have emerged as viable techniques that for use in the continuous processing of pharmaceutical molecules. Through controlling supersaturation during crystallization, these techniques are able to create particle distributions. In this work, meniscus guided flow coating is explored as an alternative to spray coating that creates high surface area crystals for use in pharmaceutical formulations. We have observed the formation of film morphology similar to that obtained using a spray coating technique, and unique from what has been previously described using a flow coating method. Acetaminophen is used as a model compound in this study, where films were fabricated and characterized using polarized optical microscopy and grazing incidence x-ray diffraction (GIXD) to determine particle sizes and crystal orientation on the substrate. Results suggest that the coating regime deviates from traditional evaporative or Landau-Levich coating regimes described in literature. Here we present our current hypothesis for the mechanism of crystallization that produces the observed particulate morphology.

594770 Ternary Phase Diagram and Phase Transformation Behaviors of Various Creatine Phosphate Sodium Hydrates

Abstract:

The thermodynamic stability of creatine phosphate sodium (CPS) in water–methanol system was investigated by the determination of CPS–water–methanol ternary phase diagram. The involved transformation behaviors were explored in solid-state, solutions, and vapor environment by VT-PXRD, Raman, FTIR spectroscopy and DVS. Results show that CPS can exist in three hydrate forms, namely, heptahydrates, tetrahydrates, and dihydrates, in the ternary phase diagram. These three crystal forms become the stable form in turn as the water content of the solvent mixture increases, with no anhydrate form appearing, which was seldom found in current reports. The transition points move towards higher water activity as the temperature increases. Besides, solution-mediated phase transformation could happen between tetrahydrates and dihydrates, which is influenced by solvent composition, temperature, and solid loading. While solid-state phase transformation of them couldn't occur when heating. Heptahydrates and tetrahydrates could transform into each other through adsorption or desorption of the water vapor. The study provides a guide for the control of CPS hydrate forms in theory and in the production to obtain the commercial crystal form with high purity.

610809 Poly(homopiperazine-amide) Thin Film Composite Nanofiltration Membrane for Heavy Metal Ions Removal

Abstract:

The development of membrane-based technologies for the treatment of wastewater streams and resources containing heavy metal ions is at high demand. Among various technologies, nanofiltration (NF) membranes are attractive choices, and the continuous development of novel materials to improve the state-of-the-art NF membranes is highly desired. Here, we report on the synthesis of poly(homopiperazine-amide) thin-film composite (HTFC)-NF membranes, using homopiperazine (HP) as a monomer. The surface charge, hydrophilicity, morphology, crosslinking density, water permeation, solute rejection, and antifouling properties of the fabricated NF membranes were evaluated. The fabricated HTFC-NF membranes demonstrated water permeability of 7.0 ± 0.3 L/m^2h bar and rejected Na_2SO_4, $MgSO_4$, and NaCl with a rejection value of 97.0 ± 0.6%, 97.4 ± 0.5%, and 23.3 ± 0.6%, respectively. The membranes exhibit high rejection values of 98.1 ± 0.3% and 96.3 ± 0.4% for Pb^{2+} and Cd^{2+} ions, respectively. The fouling experiment with humic acid followed by crossflow washing of the membranes indicates that a flux recovery ratio (FRR) of 96.9 ± 0.4% can be obtained.

612346 Molecular Thermodynamics for Scaling Prediction in Membrane Distillation

Abstract:

The scaling phenomenon is a major challenge in membrane distillation (MD) because MD is typically applied for treatment of hypersaline wastewater. Scaling not only decreases the vapor flux through the membrane but also reduces the lifetime of the membrane. The scaling-induced pore wetting caused by precipitation of the minerals at the membrane-water interface leads to water breakthrough in the membrane and eventually failure in the MD performance. Theoretically, the necessary conditions for a dissolve salt to precipitate can be found from its solubility constant and activity of the constituents in the brine. The composition and concentration of the dissolved solids can be used to determine the onset of precipitation in MD. A comprehensive thermodynamic model is, thereby, needed to estimate the electrolyte behavior and to predict the onset of precipitation at different brine compositions.

The MD has shown potential for treating hypersaline brine streams, but it is significantly impaired by the scaling of minerals. In this presentation, we will discuss the thermodynamic model that we have developed based on the eNRTL model (Song and Chen, *I&ECR,* 2009, 48(16), 7788-97) - a state-of-the-art electrolyte model - to investigate the fluid phase equilibria of brine at different compositions. We chose produced water (PW) resulted from hydraulic fracturing of the unconventional oil and gas as the model hypersaline brine for our thermodynamic modeling. The eNRTL model includes all the major ionic species that are present in PW, which could be incorporated into custom modeled simulations for more accurate prediction of the thermodynamic properties of the brine in the membrane-based separations. With fully-parameterized eNRTL model, we predicted salt precipitation in the treatment of PW samples with MD. Our thermodynamic modeling findings is in agreement with the experimental results from our scaling MD tests. This model is a tool that can help forecasting the precipitation of salts for any brine composition at different concentrations and temperatures. This approach could be further expanded to other desalination processes that are impaired by scaling of the dissolved electrolytes.

607767 Development of Polymeric Membranes for Pervaporation Application

Abstract:

A critical challenge facing all water desalination, regardless of feedwater type, is management of the produced concentrated waste solutions. Reverse osmosis (RO) is the most commonly used desalination process because it has the lowest energy consumption compared to alternative thermal technologies and operated near its theoretical thermodynamic floor. RO typically achieves 40-90% recovery of the feed water (dependent on concentration and composition), leaving up to 30% of the water as concentrated waste for disposal. For brackish feedwaters (total dissolved solids < 5 g/L) the maximum recovery is typically limited by sparingly soluble salts or minerals (e.g., calcium sulfate or silica, respectively) and, ultimately, depends on the feed water composition. However, for high salinity waters, such as seawater (total dissolved solids ~ 32g/L) or produced waters from hydraulic fracturing (total dissolved solids 30-200 g/L) the maximum recovery is limited when the osmotic pressure of water exceeds the maximum achievable pressure within the system. Pervaporation is a membrane process that operates based on a vapor pressure difference across a dense, selective membrane. Pervaporation is more energy intensive than RO, therefore ultimately it is best suited for applications that exceed the limitations of RO (such as concentrate management or treatment of waters with TDS > 60-80 g/L). However, it shows potential to greatly reduce the volume of concentrate in processes such as RO.

Our team has been working collaborative with researchers from at Rice University and University of Texas El Paso and we have developed a facile solution casting method for freestanding pervaporation membranes. These polymeric membranes show incredible promise for use in pervaporation desalination due to ease of casting and excellent performance parameters, with permeance of 135.5 ± 29 kg m^{-2} hr^{-1} bar^{-1} and salt rejection consistently greater than or equal to 99.5%. This reliably outperforms commercial pervaporation membranes and shows promise for continued desalination and produced water applications.

607774 Operational Improvements of Ultrafiltration Treatment of RO Feedwater Driven By Neural Network Models of UF Fouling and Backwash

Abstract:

Ultrafiltration treatment of reverse osmosis (RO) feedwater has been widely practiced in both inland and seawater desalination. However, UF requires periodic cleaning via backwash and eventually chemical in-place (CIP). Effective control and UF operation to enhance process performance and decrease frequency of CIP requires accurate system self-learning (for reasonable short-term forecasting and causal analysis) of the progression of fouling and backwash efficacy for the source water being treated. Admittedly, membrane fouling is a complex phenomenon governed by UF system design, operating conditions, water quality and temperature which can be temporally variable. In order to quantify UF operation and fouling progression under field conditions that are typically vary temporally, various fouling indicators have been proposed and evaluated with respect to quantification of UF backwash efficacy, UF resistance and post-backwash resistance. In the present work, UF system behavior is modeled via machine learning approaches, based on extensive data generated from field studies of UF-RO seawater desalination. UF operational models were developed based on artificial neural networks (ANN) and Alopex-based evolutionary algorithm (AEA) to predict fouling behavior of ultrafiltration (based on inside-out multichannel elements). The models were trained with data collected from a field study (at Port Hueneme, California) with an integrated UF-RO system (capacity of up to ~18,000 gallons/day) that utilized 8-inch RO elements and three inside-out polyethersulfone (PES) multi-bore hollow fiber UF membranes (0.02 μm pore size each module having an area of 50 m^2). The neural network was trained considering various input variables including, for example, pressure, flow rate, temperature, and feed turbidity and chlorophyll concentration. Initially, an ANN model with AEA and Bayesian theory was developed to predict UF resistance, including both filtration and backwash modes for each operational cycle, with the outcome being the filtration and backwash performance for different water quality conditions. The analysis also included the impact of coagulant dose to the UF modules. An additional ANN model that includes AEA was also developed to predict the UF post-backwash resistance based on training of data collected during a storm event and where the coagulant dose was adjusted via a unique coagulant dose controller developed previously at UCLA. The ANN model with AEA demonstrated excellent performance (RMSE~0.0099, R^2~0.947) in being able to properly account for backwash efficiency and prediction of the UF post-backwash resistance. The above modeling approach, as illustrated in the present work, provides a foundation for implementing improved self-adaptive operation of UF operation under variable field conditions.

597515 Nanofiltration of per- and Polyfluoroalkyl Substances (PFAS): Rejection Mechanisms

Abstract:

Perfluoroalkyl and Polyfluoroalkyl substances (PFAS) are an emerging and extremely pervasive contaminant with a potentially high health risk. They are currently being found throughout the United States in ground and surface waters above the EPA's limit of 70 ppt. Membrane separation is a potentially effective process to remove PFAS from water sources. Past research has shown that nanofiltration (NF) membranes are capable of removing PFAS from water, however the rejection mechanism for removal is unknown. Dead-end filtration experiments were conducted to infer the rejection mechanism used by nanofiltration membranes to remove PFAS from water. Nanofiltration membrane NF270 was found to intrinsically reject the highest risk PFAS molecules: PFOA (Perfluorooctanoic acid), PFOS (Perfluorooctanesulfonic acid), PFNA (Perfluorononanoic acid) and PFHxS (Perfluorohexanesulfonic acid). Rejections of 95-99.8% were observed at a neutral pH. A direct relationship was found between both molecular weight and radius and rejection, demonstrating the presence of the size exclusion mechanism. The electrostatic interaction rejection mechanism was investigated by changing testing solution pH. Changing pH was found to significantly impact PFAS rejection, with highest rejections at a basic pH of 10 (99.8%) and lowest rejections at an acidic pH of 3 (94.2%). At basic pH the zeta potential, and thus the electric charge, of the NF270 membrane surface is negative repelling the negatively charged PFAS molecules. These findings indicate that both electrostatic interactions and size exclusion are important for nanofiltration rejection of PFAS.

601559 Ultrafiltration Membranes and Modules for Microalgae Dewatering and Fouling Mitigation

Abstract:

Microalgae culturing is an effective way to utilize CO_2 and mitigate the CO_2 emissions to the atmosphere, and membrane technology has emerged as a key technology for microalgae dewatering, the most energy-intensive step in algae harvesting. However, the dewatering performance is subject to fouling by many components in the culturing solutions. Current studies regarding fouling are mostly conducted using a single model foulant, and there lacks an understanding of membrane fouling by the real algae products. In this work, we evaluated the dewatering performance of different commercial membranes and modules in dead-end cells and a cross-flow system. The effect of membrane pore size, algae concentration, membrane cleaning, and surface modification on the dewatering performance of the membrane coupons and modules were thoroughly investigated. Interestingly, ultrafiltration (UF) membranes exhibit greater steady-state dewatering permeance than microfiltration (MF) membranes despite lower pure-water permeance in the UF membranes. Increasing the algae concentration decreases the steady-state permeance of PSf-100. Nevertheless, when challenged by a feed containing 20 g/L algae, the PSf-100 still exhibited a water permeance of 687 LMH/bar. In multiple-cycle tests of fouling and cleaning, the polysulfone UF membrane (PSF-100) was cleaned by DI water or NaOH solution, and the cleaning efficiency of both methods reach 97%. Commerical membranes and modules were also modified by surface-grafting of superhydrophilic zwitterions to improve antifouling properties. The modified module shows water permeance 18% higher than the unmodified one. This study shows that membranes can achieve excellent dewatering performance and the surface modification can mitigate the fouling for the culture of microalgae.

603851 Pilot Scale Osmotic Concentration Process for Reducing Injected Wastewater Volumes in Gas Field Operations

Abstract:

Over the past 10-15 years, there has been increasing attention in the development of forward osmosis (FO) technology as a low-energy technical solution to wastewater treatment. The significant energy benefit arises in applications where direct recovery of the permeate product from the draw solution (DS) is obviated such as osmotic concentration (OC) process.
In the current work, a flexible pilot-scale unit, based on FO process, was designed and constructed for the treatment produced and process water from Qatar's gas operations. The FO pilot unit, without the DS recovery step, was applied as an option for wastewater volume reduction for gas field operations in Qatar. The pilot unit uses seawater (40 g/L TDS) as a DS and wastewater generated during gas operations as a feed. This feed water is of comparatively low conductivity (~ 2 g/L TDS), making it unusually suited to treatment by OC. The diluted DS could be directly discharged into the ocean; so the energy-intensive step of DS recovery is entirely eliminated.
Two hollow fiber FO membranes (Toyobo and NTU) were tested to assess their performance and fouling propensity on both synthetic and real wastewaters. Results demonstrated that the membrane can achieve feed water recoveries up to 90% without any scaling issues. Flux ranges from 2.25 to 12 LMH for feed recoveries between 60 and 90% using a constant 4X dilution rate of the draw solution. Furthermore, the pilot unit maintained a stable water flux of 10 and 1.92 LMH at 75% feed recovery using NTU and Toyobo respectively. The robustness of the pilot unit was investigated for over 48 h of continuous operation. This work will present the operational data of the pilot unit for both membrane modules under synthetic and real wastewaters

606906 Hydraulics and Operation Performance of Gas Dispersed Liquid-Liquid Extraction in a Multiphase Taylor-Couette Disc Contactor

Abstract:

Industrial separation and reaction steps often require more than one phase. Distinct phase contact of solids (S), liquids (L), and gases (G) forms the physical backbone of state of the art process intensification and optimization. Novel reactor and contactor design for liquid-liquid operations serve in optimizing the contact of two phases as for example in bubble columns or liquid-liquid extraction columns [1].

The novel and simple Taylor-Couette Disc Contactor (TCDC) design is based on the rotating disc contactor but without stator discs and with an increased rotor shaft diameter. The optimized TCDC design avoids any formation of hydraulic deadzones [2,3]. Previous investigations showed that stable continuous three phase flow (LLS) in the TCDC offers access to combinations of heterogeneously catalyzed chemical reactions with phase transfer of products in one single apparatus. Adding a fourth gaseous phase to the TCDC column opens a completely new fields of application. Continuous four phase contact (GLLS) enables to perform heterogeneously catalyzed carbonylation reactions with simultaneous phase transfer of the reaction product in one single column. The Taylor-Couette Disc Contactor perfectly complies with this requirement and provides access to continuously contact various combinations of solid, liquid and gaseous phases.

First investigation of an additional gas phase showed that the gas dispersed liquid-liquid extraction column allows to overcome the operation limits given by the flooding point of the column. Overcoming this operation limit comes with a with improved mixing and mass transfer in the TCDC column. This work is focused on the gas phase behavior in the Taylor-Couette Disc Contactor. Perforated discs allow the gas phase to continuously rise up in the column. A gaseous phase can be used as inert auxiliary power input to improve column performance or even as reactive component in the multiphase flow system. The Adjustable rotational speed, and airflow rate provide defined and uniform volumetric gas holdup up to 23% along the column height. The simple TCDC design allows easy scale up in height and diameter to ensure the required throughput and defined residence time. The influence of different perforation pattern regarding the gas phase holdup was observed in a TCDC column with 50mm inner diameter and 700mm active height. For the hydrodynamic test system air was used as gaseous phase, deionized water as continuous liquid phase, ShellSol-T as dispersed liquid phase and Amberlyst 15 as solid phase. In addition, the hydraulics of continuous GLS and GLL and even GLLS flow in terms of droplet size distribution, residence time and holdup of the gas phase, the dispersed liquid phase and the solid phase were investigated in reliance to the interaction of rotational, speed and specific phase flow rates. Flow regime maps for the gas phase show the changing flow pattern according to changing flowrates and rotational speeds. The first results of this investigations impressively show the unique and flexible field of application for industrial use of the TCDC column.

[1] Aloisiyus Y. Widianto, et al., Catalysis Today, https://doi.org/10.1016/j.cattod.2019.02.064

[2] E. Aksamija, C. Weinländer, R. Sarzio, M. Siebenhofer, *Sep. Sci. Technol.* 2015, 50 (18), 2844–2852. DOI: 10.1080/01496395.2015.1085406

[3] A. Grafschafter, G. Rudelsdorfer, M. Siebenhofer,(2018). Hydraulics and Operation Performance of TCDC-Extractors. Chemie Ingenieur Technik. 90. 10.1002/cite.201800031.

604437 Phase Separation Kinetics and Settler Design

Abstract:

Besides appropriate design of mass transfer equipment the successful design of phase separation equipment is an important issue in liquid-liquid extraction. Based on the balance of settling processes the design of batch settlers and the design of continuous settlers should be a simple task. Actually the rate of settling w is not constant but changes with height (w = w(h)).

Several investigations have focused on the development of design algorithms, addressing the phenomena which affect the rate of hindered settling of the dispersed phase. The review of Kamp et al [1] presents a very detailed discussion of the state of research, experimental techniques and modeling approaches. In this review a paper on the "Prediction of Steady State Dispersion Height from Batch Settling Data, published by Jeelani and Hartland in 1985 [2] is outlined. In the discussion of the theoretical background Jeelani and Hartland addressed proposals of Stonner and Wohler [3] for determining the rate of sedimentation of dispersed solvents from aqueous carrier phase. They mentioned that the settling height as well as coalescence height may show sigmoidal shape because of initial turbulences. Latter approach was used by Bol [4] and Rajcharak [5], who investigated several liquid-liquid dispersion systems. Bol [4] compared the monitoring of phase separation of liquid-liquid dispersions of MIBK dispersions in an optical cell and in an ultrasonic scanner. The system properties density, viscosity of the aqueous carrier phase and the interface tension were adjusted by adding PEG 4000 and sodium chloride to the aqueous carrier phase.

From a kinetics point of view droplet sedimentation of solvent dispersions in water may follow a zero order rate law for constant rate of sedimentation, a mixed order rate law (MOR), combining zero order and first order rate of sedimentation, or a consecutive step first order rate law (FOCR). As already suggested by Jeelani and Hartland [6] initial turbulences may induce pronounced retardation of phase separation, best modeled by a first order consecutive step rate law with sigmoidal shape of the sedimentation height versus sedimentation time.

The mixed order rate law and the consecutive step first order rate law were adapted to modeling droplet sedimentation from solvent dispersions in aqueous carrier, resulting in two easy to apply algorithms for highly accurate specification of the sedimentation rate (w(h)). Application of the mixed order approach results in a two parameter function (k_1 and k_2) of the dispersion height h and the maximum height of the clear liquid aqueous phase h_{max}, with either correlating the height with the independent variable time t for batch settling or the residence time for continuous settling. The advantage of this mixed order sedimentation rate model is the simple development of the height dependent settling rate w(h). When applying the consecutive step first order rate law the correlation of the height dependent rate of settling again follows a two parameter algorithm.

The rate models were compared with settling experiments. In both models the fit parameters k_1 and k_2 compare well with the physical properties of the systems and the preparation method of the corresponding dispersion.

Exemplarily the rate constants k_1 and k_2 of the FOCR model compare well with the physical properties viscosity and interfacial tension and the system property droplet size dp, as derived from the sedimentation rate for rigid spheres. For validation a series of 16 experiments with different PEG 4000 content and different NaCl concentration of the aqueous phase and the dispersed solvent MIBK was investigated.

From the MOR model the initial rate of settling can easily be determined, giving quick access to the initial mean drop size. The initial rate of settling was compared with the Henschke algorithm [7] and the proposal of Jeelani and Hartland [8]. While latter algorithm compares well with the data derived from experiments, the Henschke approach is about 25 % lower throughout.

From the comparison of the actual rate of sedimentation with the initial rate of sedimentation rate retardation can easily be determined, providing access to the volumetric amount of dispersed phase in the dense packed zone.

Conclusions

Rate based sedimentation models, as already suggested by Jeelani and Hartland [2] provide simple algorithms for quantifying the sedimentation process of droplets from solvent systems dispersed in aqueous carrier with simple and quick determination of the height dependent rate of sedimentation, as needed for successful settler design.

[1] Kamp, J., Vilwock, J., and Kraume, M., 2017: Drop coalescence in technical liquid/liquid applications: a review on experimental techniques and modeling approaches, Rev. Chem. Eng. 33, 1–47

[2] Jeelani, S.A.K., Hartland S., 1985: Prediction of steady state dispersion height from batch settling data., AIChE J., 31: 711–720.

[3] Stonner, H. M., and Wohler, F., 1975: An Engineer's Approach to a Solvent Extraction Problem, Inst. Chem. Engrs. Symp. Ser., 42,14.1

[4] Bol, P., 2015: Investigation of drop dispersion in batch settling processes, Dissertation, TU Graz.

[5] Rajcharak, B., 2016: Modeling of droplet sedimentation in liquid – liquid phase separation, Master Thesis, TU Graz.

[6] Jeelani, S.A.K., Panoussopoulos, K., and Hartland, S., 1999: Effect of Turbulence on the Separation of Liquid-Liquid Dispersions in Batch Settlers of Different Geometries, Ind. Eng. Chem. Res., 38, 493-501.

[7] Henschke, M., Schlieper, L. H. and Pfennig, A., 2002: Determination of a coalescence parameter from batch-settling experiments, Chemical Engineering Vol. 85. - pp. 369-378.

[8] Jeelani, S.A.K., and Hartland,S., 1998, Effect of Dispersion Properties on the Separation of Batch Liquid-Liquid Dispersions, Ind. Eng. Chem. Res. 1998, 37, 547-554

610836 Isolating Large Quantities of Liquid Crystalline Pyrene Trimer Via Continuous Supercritical Extraction

Abstract:

Semi-continuous, supercritical extraction (SCE) with toluene was used to fractionate polycyclic aromatic hydrocarbon (PAH) pitches into more favorable, highly concentrated materials. In particular, the starting pitch was produced by the catalytic polymerization of pure pyrene using $AlCl_3$, or by thermal polymerization, with both being done at temperatures from 300-500 °C. The trimer of pyrene has been shown to be 100% mesophase, making it an ideal precursor for advanced carbon materials. Unfortunately, yields for the trimer are quite low, typically ranging from 5-10% of the pitch at best. Our previous work in studying the kinetics of the reaction has allowed us to optimize the conditions for producing more trimer-rich pitches, approaching 15% by weight. With this, we have also coupled our 2-stage continuous SCE technique to further concentrate the trimer in the pitch and enable us to produce large quantities of trimer-rich material that exhibit liquid crystalline properties. Concentrations of the samples were measured by GPC-RI and MALDI-ToF, and using this technique, we are able to produce up to 100's of grams of materials in a single run. Fractions created were shown to be 50% trimer and higher, with the remaining weight being comprised of dimer or tetramer.

605975 Advanced Process Designs and Operating Modes of the TAPPIR®-Technology for Continuous Aqueous Two-Phase Extraction

Abstract:

Abstract Withheld

596930 Development from Batch to Continuous Solid Liquid Extraction- Artemisinin Extraction from Artemisia Annua Leaves

Abstract:

Artemisinin (ARTE) is recommended as the first-line treatment of malaria. It can be produced by chemically synthesis, semi-synthetically or by extraction from plants. Due to the complexity and expenses related to the synthetic processes, the isolation of artemisinin from sweet wormwood (*Artemisia annua*) is recommended to provide the larger quantities required on the international market [1].

Batch-wise extraction of valuable target components from plants is the state of the art. Less systematic activities have been devoted to continuous extraction techniques, in particular for simultaneous co-extraction of equally valuable precursors also present in the plants.

Our research is focused on studying continuous countercurrent ARTE extraction from Artemisia annua with joint isolation of dihydroartemisinic acid (DHAA), which is known to be an important precursor for the biosynthesis of ARTE. Significant amounts of DHAA were reported to be present in different extracts of *Artemisia annua*. In [2] it was shown that DHAA could be oxidized photocatalytically to ARTE, which allows increasing the overall ARTE yield of the extraction process. For the mentioned photoreaction the co-extraction of chlorophyll is also of large interest since it acts as an efficient green photosensitizer [3].

We will present results regarding the distribution equilibria of the components mentioned. This information is incorporated into a multi-stage equilibrium stage model describing a countercurrent extraction process. The model allows identifying suitable operating conditions, which maximize the sum of the amounts of ARTE and DHAA extracted. Finally, results of continuous extraction carried out in pilot plant will be presented to demonstrate the efficiency of the process.

[1] E.A.Brisibe; African Journal of Biotechnology, 7(25), 2008; 4884-4896

[2] F.Levesque, P.H.Seeberger, Continuous-flow synthesis of the anti-malaria drug Artemisinin, 51(7), 2014, 1706-1709

[3] S.Triemer, K.Gilmore, G.T.Vu, P.H.Seeberger, A.Seidel-Morgenstern, Angewandte, 57(19), 2018, 5525-5528

605571 Linker Doping As an Effective Strategies to Tune in-Situ Formed Zeolitic-Imidazolate Framework-8 Fillers in Mixed-Matrix Membranes for Propylene/Propane Separation

Abstract:

Zeolitic-imidazole framework-8 (ZIF-8) has been extensively studied as a molecular sieving filler in mixed-matrix membranes (MMMs) for propylene/propane separation. The propylene/propane separation performance of ZIF-8-containing MMMs has yet to meet the commercial viability criteria. Doping or mixing linkers in ZIF-8 framework has been found effective in tuning the molecular sieving properties of ZIF-8. Recently, these doping/mixing strategies have been successfully applied to polycrystalline ZIF membranes. However, there have been no reports on applying the same strategies in tuning the molecular sieving properties of ZIF fillers in MMMs. In this presentation, we plan to present tuning of the molecular sieving properties of ZIF-8 fillers in MMMs by doping with 2-ethylimidazole (eIm) linkers in order to improve their propylene/propane separation performance. MMMs were prepared by in-situ forming eIm-doped ZIF-8 fillers in polyimide using our recently reported polymer-modification-enabled in-situ MOF formation (PMMOF) process. As eIm linkers were doped into ZIF-8 filler crystals, the propylene/propane separation factor of the resulting MMMs was dramatically enhanced. Lastly, doped-linker ZIF-8 based mixed-matrix hollow fiber membranes were successfully prepared, demonstrating potential of their scalable practical applications.

605942 Photoresponsive Azo-Uio-66 Mixed Matrix Membranes for CO_2/N_2 Separation

Abstract:

Novel responsive molecular separation materials are interesting from a variety of perspectives. For example, materials that can be easily switched between adsorption mode and desorption mode could find new applications in sensors, environmental remediation and controlled release (including as *in vivo* controlled drug release materials). In another sense, materials that have existing applications with conventional release mechanisms, such as gas sorbents which rely on temperature or pressure swing, could be used in quite radically new applications if they could be operated with new stimuli that is faster acting, cheaper, or more energy efficient.

Our research group has worked with international collaborators for several years to develop photo-responsive gas sorbent materials, which can be rapidly and repeatedly switched between adsorption and desorption mode by illumination with light (usually UV light, but other wavelengths are possible). We showed in small-scale laboratory tests that it was possible to produce a range of different materials that exploit different strategies to impart photo-responsive nature to metal organic framework adsorbents, including through the use of guest molecules, photo-responsive groups pendant to ligand, and directly photo-responsive ligands.

In this most recent work, we have successful produced a highly photoresponsive sorbent based on the very stable UiO-66 MOF topology, through the synthesis and systematic incorporation of a modified ligand with azobenzene functional group (which adsorbs strongly in the UV region, leading to a *cis-trans* photoisomerization and subsequent transformation of the localized gas sorption environment). We denote this material Azo(X)-UiO-66 where X represents the proportion of the ligands which have been substituted with the modified Azobenzene-functionalised ligand. The sorbents were comprehensively characterized for textural properties using nitrogen sorption at 77K, composition (including quantifying ligand substitution) using 1H-NMR spectroscopy, crystallinity and structure using PXRD, chemical properties using FTIR and Raman spectroscopy, and thermal stability using TGA. The static and dynamic CO_2 sorption was evaluated in a custom-designed adsorption cell which allowed for *in-situ* illumination with UV light, while carefully controlling sample temperature to ensure local heating effects were minimized.

As shown in the Figure, samples with Azo content ranging from 16.7% (A), through to 33.3% (B), 66.7% (C) and 100% (D) all demonstrated some degree of dynamic photo-switching capability. The optimum was obtained for 66.7% Azo-content, with the samples at higher loadings typically displaying reduced overall CO2 sorption capacity, likely due to the steric hindrance imparted by the relatively bulky pendant azobenzene groups. For all samples, the isoteric heat of adsorption was relatively constant, although slightly higher for the 100% Azo-content sample.

The performance of mixed matrix membranes using Matrimid and PIM-1 as the continuous phase and Azo(X)-UiO-66 as the dispersed phase, was evaluated for CO2/N2 separations. PIM-1 membranes had CO_2 permeability around 7500 Barrer (similar to other reports in the literature), and incorporating Azo(100)- and Azo(66.7)-UiO-66 increased it to around 11 000 Barrer. Azo(33.3)- and Azo(16.7)-UiO-66 increased the permeability further to around 13 000 Barrer, indicating that all filler particles introduced additional free volume, but the samples with higher CO_2 sorption capacity (16.7 and 33.3) led to a higher mixed-matrix membrane permeability, consistent with standard models for gas transport in mixed-matrix membranes. Both single gas studies and mixed-gas experiments were used to evaluate the selectivity of the mixed-matrix membranes, revealing that the mixed-ligand filler particles could increase selectivity for CO_2 over N_2 from around 14 for pristine PIM-1 to 19 for mixed matrix membranes with Azo(100)-UiO-66.

The photo-response properties of mixed-matrix membranes containing photo-responsive filler particles is more subtle than the powder properties, but at a first approximation follows the trend that would be expected assuming mixed-properties response. Our initial results show that it is possible to prepare mixed-matrix membranes that demonstrate photo-responsive behavior that can be dynamically switched with application of external light, however the effect is modest (while still being statistically significant). This presentation will highlight the latest developments in our research in this area, including the development of a new test station with robust capabilities to evaluation photo-responsive membranes (not just mixed-matrix membranes), and new collaborations with materials scientists to test a wide range of novel membrane materials. We also present a collaboration with data scientists to incorporate exceptional levels of data storage, analysis and open data sharing through the adoption of the Chemotion electronic lab notebook and data repository.

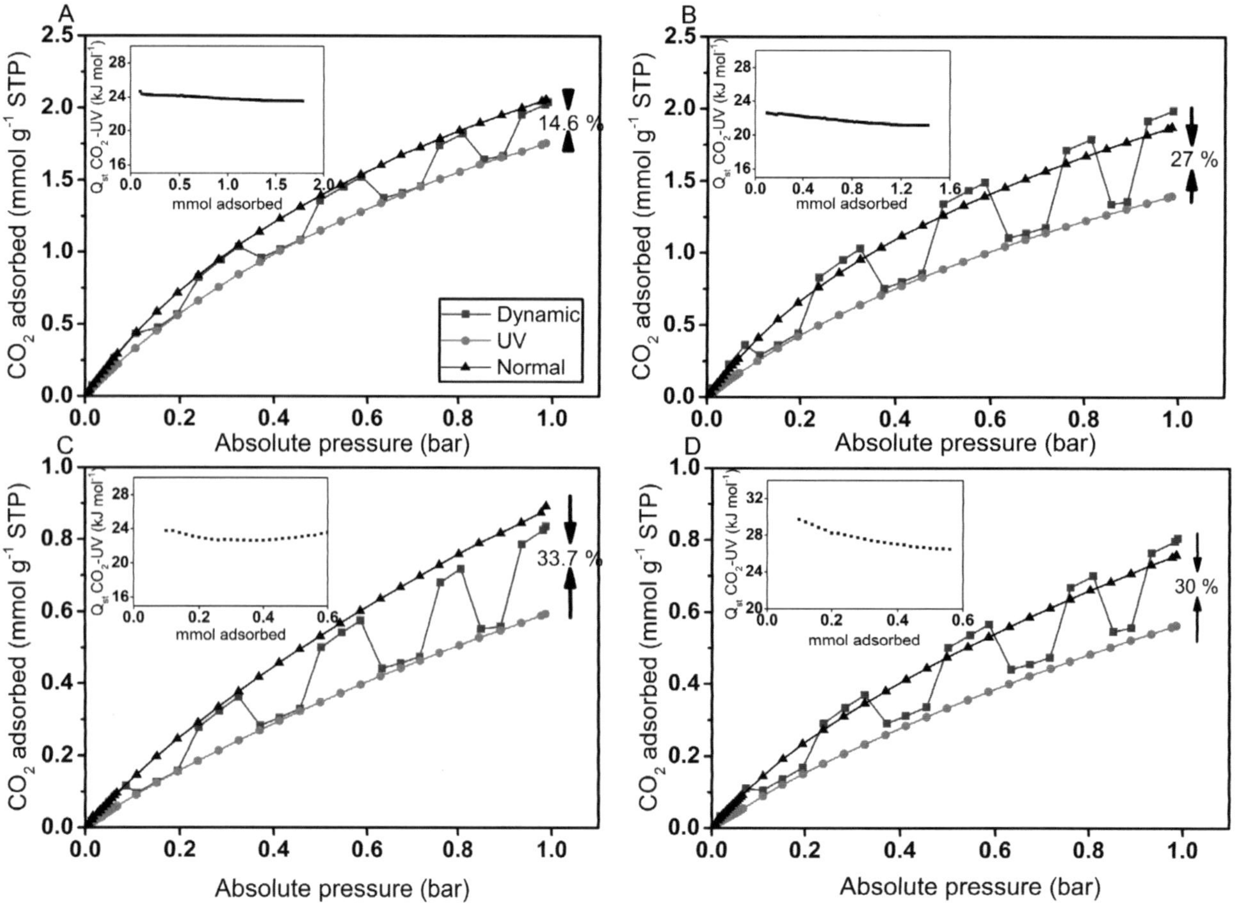

611548 Rational Formulation Design of Multicomponent Membranes for Carbon Dioxide Separation

Abstract:

The discovery of a membrane that can effectively remove carbon dioxide from coal-burning power plant emissions would help lower carbon dioxide emissions worldwide. To make membrane-based carbon capture economically viable, new formulations must be identified with high selectivity and high CO_2 permeance. We demonstrated herein the ability to break the permeability/selectivity trade-off by using multicomponent mixed-matrix membranes (McMMMs) with two, three or four components wherein each constituent has a specific function and have been designed for compatibility and high gas separation performance. The chemical interaction between the metal-organic framework (MOF) nanoparticles and the polymers is the key factor for optimizing the MOF-polymer interfacial compatibility. Compared with the neat polymer material, the CO_2 permeability and CO_2/N_2 selectivity were significantly enhanced, while the predicted cost of carbon capture was reduced. The permeability/selectivity values surpass the Robeson upper bound and suggest the potential of these membranes for practical CO_2 separations.

611816 Ab Initio Study of Polymer/MOF Interfaces

Abstract:

Mixed matrix membranes (MMMs) have received great attention due to their potential for improved separation performance compared to conventional membrane materials. Conventional membranes suffer from trade-offs between permeability and selectivity and are famously limited by the so-called “Robeson upper bound”. While MMMs have a similar trade-off, it has been shown that such materials can perform beyond the Robeson upper bound. Metal-organic frameworks (MOFs), crystalline nanoporous materials comprised of metal nodes coordinated by organic ligands, have been demonstrated as a promising filler for MMMs owing to their tunable porosity and functionalities. Most work to-date regarding separation performance of MOFs, however, has focused on the crystalline bulk of the MOF without regard for the external surface area and chemical environment. In fact, chemical and structural differences at surfaces as compared to the bulk can determine MOFs’ potentials to interface with other materials, as in MMMs. Additionally, interactions available on the MOF surfaces play a key role in adsorption properties of these materials.[1] It is therefore critically important that MOF surfaces and their interactions with polymer materials be well understood in order to facilitate development of optimal MOF/polymer pairs for optimization of MMMs for target applications. Thusly motivated, we study interactions of MOF-2 surfaces (“M-BDC”, with M = Zn, Co, and Cu and BDC = benzene-1,4-dicarboxylate) with multiple polymers (e.g. Polyetherimide, Kapton). Interactions of molecules with MOFs, and especially with undercoordinated metals which are likely to exist on MOF surfaces, are poorly described by conventional force fields, motivating our use of density functional theory (DFT) for such a study. In order to model the interactions of large polymer molecules, which are inaccessible with DFT due to scaling, we must develop accessible but electronically appropriate models for these materials. To this end, we pursue a cluster treatment of polymer molecules, which may involve decomposing a polymer or its monomers into fragments. We pursue an approach wherein these polymer fragments are terminated with various electron donating/withdrawing groups to capture the electronic environment of the larger monomers or polymer systems to appropriately model polymer—MOF interactions. We evaluate these models using charge partitioning analysis based on the Density Derived Electrostatic and Chemical (DDEC) charges approach.[2,3] We then employ DFT to study interactions of relevant chemical features of polymers to understand the fundamental interactions at the polymer/MOF interface in such MMMs. This approach is general, and may be of use in extension to other comparable systems where fundamental interactions are poorly described by established theories of material interfaces.

References

1) Howe, J. D.; Liu, Y.; Flores, L.; Dixon, D. A.; Sholl, D. S. Acid Gas Adsorption on Metal-Organic Framework Nanosheets as a Model of an "All-Surface". J. Chem. Theory Comput. **2017**, 13, 1341−1350.

2) Manz, T. A.; Limas, N. G. Introducing DDEC6 Atomic Population Analysis: Part 1. Charge Partitioning Theory and Methodology. RSC Adv. **2016**, 6, 47771−47801.

3) Limas, N. G.; Manz, T. A. Introducing DDEC6 Atomic Population Analysis: Part 2. Computed Results for a Wide Range of Periodic and Nonperiodic Materials. RSC Adv. **2016**, 6, 45727−45747.

611822 Carbon Dioxide Separation from Syngas Using Polymeric and Mixed Matrix Membranes

Abstract:

Greenhouse gases are major contributors to climate change and CO_2 emissions are responsible for huge percentage of these greenhouse gas emissions. Fuel combustion has been shown to be the major cause of increase in CO_2 emissions into the atmosphere. This has increased interest in developing means of decreasing the net CO_2 emissions. These technologies include carbon capture and storage (CCS), as well as carbon recycle into fuels and valuable chemicals.

This study centers around post combustion strategy whereby stored CO_2 after combustion is used to produce hydrocarbons through Reverse Water Gas Shift (RWGS) reaction and Fischer-Tropsch reaction. It is important to separate CO_2 from CO and H_2 (syngas) in between these reactions. This can be done using membranes, which is the focus of this study.

Poly Vinyl alcohol (PVA) membranes were chosen to be studied due to their good film forming ability and their chemical and thermal stabilities. Formaldehyde cross-linker is chosen due to the previously promising results observed with their use. Mixed matrix membranes were synthesized with the addition of carbon dots. The synthesized membranes were supported by Polyacrylonitrile (PAN) support and 2wt% of PVA is used with 10wt% of carbon dots with some of these membranes to produce mixed matrix membranes (MMM). The permeability of CO_2 has been shown to decrease with increase in pressure difference and that of CO and H_2 has been shown to increase. Also, the addition of formaldehyde and other mobile cross-linkers increase the permeability of the membrane for CO_2 while not changing the permeability for CO and H_2, thus increasing the selectivity of the membrane for CO_2 over CO and H_2. This effect was furthered with the addition of carbon dots into the matrix of the PVA membrane with much higher selectivity values observed for CO_2 over CO and H_2.

601658 Effect of Metal Ions in Polymer Electrolytes on Gas Transport Characteristics

Abstract:

Metal-organic frameworks (MOFs) have been dispersed in polymers to improve gas separation performance, and favorable interactions between the MOFs (usually metal centers) and polymers are preferred to improve interfacial compatibility and eliminate any non-selective voids. However, there lacks an understanding of the effect of polymer-metal ion interactions on gas separation properties. In this study, we investigate the effect of metal ions dissociated in amorphous cross-linked poly(ethylene oxide) (XLPEO) on physical properties and gas transport characteristics. Polymer electrolytes were prepared by photo-polymerization of solutions containing poly(ethylene glycol) diacrylate (PEGDA), poly(ethylene glycol) methyl ether acrylate (PEGMEA), and different salts such as $LiClO_4$, $Ni(BF_4)_2$, and $Cu(BF_4)_2$. The salts can be fully dissociated, and the Li^+, Ni^{2+}, and Cu^{2+} ions form conjugations with the ethylene oxide in XLPEO, as validated by FTIR and XRD. Increasing the salt content increases the density (following the additive model) and glass transition temperature (following the Gordon-Taylor equation) except for $Cu(BF_4)_2$. For instance, adding 17 wt.% $Ni(BF_4)_2$ to PEGDA-co-PEGMEA increases the Tg from -65 °C in pure polymer to -33 °C. Gas permeation experiments at 35 °C reveal that gas permeability and selectivity decrease by adding more salt to the polymer network. As an example, adding 17 wt.% $LiClO_4$ to XLPEO results in decreasing CO_2 permeability from 720 to 20 Barrer and CO_2/N_2 selectivity from 48 to 31 because of the competing effect for the polar ether to interact with metal ions, instead of CO_2. The effect of the salt loading on the gas permeability can be described using the Vogel-Fulcher-Tammann (VFT) equation. The effect of the salt loading on gas solubility and diffusivity is investigated. The elucidation of the effect of the polymer and metal ion interactions on gas transport properties may provide guidance in designing mixed matrix materials containing MOFs.

595628 Accelerating CO_2 Capture of Highly Permeable Polymer through Incorporating Highly Selective Hollow ZIF

Abstract:

The current large-scale CO_2 capture and storage (CCS) technology based on liquid phase adsorption tower is extremely energy intensive. Membrane separation is widely regarded as an alternative to conventional absorption, however the gas fluxes through commercially selective membranes are so low that hundreds of thousands of square meters of membrane area are required. The key to future membrane technology is to significantly decrease the total membrane area, which requires high flux membrane materials that retain high selectivity.

A facile strategy is via mixed matrix membranes (MMMs), which could combine the attractive properties of polymers with molecular-sieve materials. PIM-1 is an ideal candidate for large scale CO_2 separation MMMs because of its intrinsically high permeability. The tightly regulated pore structure of zeolite imidazolate frameworks (ZIFs) and the compatibility with the polymer matrix can be finely-tuned through rationally matching metal ions and organic ligands. In this study, we prepared MMMs that are both highly permeable and selective through the synergistic combination of PIM-1 with an imidazole-based ZIF-67 hollow nanoparticles (ZIF-HNPs). The molecular sieving shell of ZIF-HNPs can be efficiently utilized, improving both selectivity and permeability of the PIM-1 matrix.

High monodispersity of ZIF-HNPs with mean diameter of 330 nm were achieved. As a control, ZIF-67 solid nanoparticles (ZIF-SNPs) with mean diameter of 370 nm were prepared. A high magnification of ZIF-HNPs revealed that ZIF-HNPs with a cavity diameter of 220 nm consist of fused ZIF-67 nanoparticles. The measured powder XRD patterns of ZIF-HNPs matched well with the simulated XRD pattern of ZIF-67, indicating the intact crystalline structure of ZIF-HNPs. The N_2 sorption test for ZIF-HNPs revealed a type I isotherm with BET surface area of 1170 m^2/g and micro-pore volume of 0.501 cm^3/g, comparable to those of ZIF-67 nanoparticles. Also, ZIF-HNPs are well dispersed in PIM-1 matrix.

Single-component permeation measurements of CH_4 and CO_2 were performed on membranes with different loading of ZIF-HNPs. When less than 28 vol% ZIF-HNPs were embedded into PIM-1 matrix, the CO_2 permeability significantly increased with the loading, whereas the CH_4 permeability nearly remained unchanged, leading to the enhancement of selectivity by 31.4%. The improved selectivity strongly confirmed that the defect-free MMMs were successfully prepared. Nevertheless, as the loading further increased to 36 vol%, the selectivity for CO_2/CH_4 mixture dropped to 15.1, while the CO_2 permeability kept increasing, resulting from the non-selective interfacial defects occurred around the agglomeration of ZIF-HNPs. We also compared the separation performance between PIM/ZIF-HNPs and PIM/ZIF-SNPs. By replacing ZIF-SNPs with ZIF-HNPs at the same volume loading, the CO_2 permeability further increased by factor of 57.7% without sacrificing the selectivity enhancement.

The membranes loaded with ZIF-SNPs shows well matched permeability with Maxwell Model. However, the permeability of membranes loaded with ZIF-HNPs are much higher than the Maxwell model predicted values. This behavior indicates the non-ideal nature of the PIM/ZIF-HNPs originated from the inner cavity. Therefore, a modified Maxwell model was developed in this study to describe the gas permeability of membranes loaded with ZIF-HNPs. In our modified Maxwell model, the membrane was assumed to be a pseudo-two-phase with the first phase of the filler's cavity and shell. The second phase consists of the first phase with the polymer matrix. The effect of cavity size and ZIFs loading on gas permeability were quantified by the modified Maxwell model. It should be noted that the cavity size of the hollow fillers should not be too large regarding the overall gas permeation performance. According to the modified Maxwell model, an optimal permeability ratio of continuous polymer matrix to the dispersed ZIF-HNPs ($R_{c/d}$) equaling to 0.2-0.3 and cavity size ranging from 200 to 240 nm are preferred.

In conclusion, ZIF-67 hollow nanoparticle was synthesized and embedded as fillers in PIM membranes to prepare mixed matrix membranes. Compared with pristine PIM-1 membrane, the optimal CO_2 permeability and CO_2/CH_4 selectivity of mixed matrix membranes increased by 57.7% and 31.4%, respectively. Compared with membranes incorporated with the same volume loading of ZIF-67 solid nanoparticles, the gas permeability of mixed matrix membranes increased by 37%. These increments can be explained by the enhanced gas diffusivity coefficient resulting from the hollow cavity, which creates extra fractional free volume for fast gas diffusion. Moreover, a modified Maxwell model was developed to analyze the effects of cavity size and filler loading on gas permeability. The predicted results well match with the experimental gas permeability of membrane loaded with ZIF-HNPs (with optimal permeability ratio $R_{c/d}$ equals to 0.2-0.3), indicating the validity of this modified model. The optimized permeation results for CO_2/CH_4 gas pairs surpass their state-of-the-art upper bounds, manifesting that the strategy of combining PIMs and ZIF hollow nanoparticles is an attractive approach to prepare highly permeable membranes for a number of applications such as carbon capture and hydrogen purification.

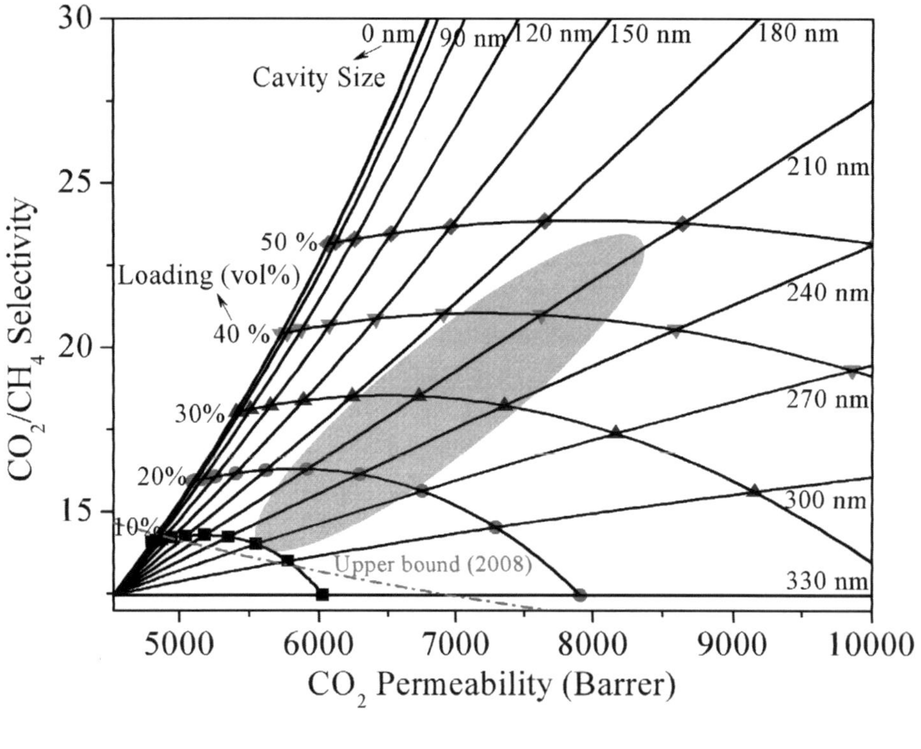

611012 Development of Membranes with Special Wettability for a Solar-Driven Water Desalination Scheme

Abstract:

Creating reentrant structures on flexible and porous substrates poses a significant challenge to the scalable fabrication of omniphobic membranes. Designing such membranes requires control over the surface topography and chemistry of the interfacial domains. Here, we present a semi-continuous bottom-up polymerization approach–based on initiated chemical vapor deposition–that allows for creating interfaces with the desired chemical and structural properties. By adjusting the free radical polymerization rate and controlling the dynamics and growth mechanism in our process, we assembled desired reentrant structures on a variety of porous substrates, making their surface superhydrophobic. Subsequently, by depositing a conformal thin film of low surface-energy polymer on the assembled structures, we further improve the non-wetting properties of the membranes, rendering them omniphobic. The developed method was applied to a variety of substrates, and variations of the method were developed to create Janus hydrophilic/omniphobic membranes. Here, we demonstrate the versatility of the developed scheme and create a vapor gap on a microfluidic channel, constructing a water separation module. The fabricated modules were tested for desalination of a saline feed water and the feasibility of constructing a stand-alone solar-driven desalination module was evaluated. Additionally, we developed a representative two dimensional model of the system. The result of our simulations indicates that the module can provide a specific water production rate of 10 L/m^2 per hour at a specific energy consumption of 3 Kwh/m^3.

605083 Electrospun Pretreatment Membranes

Abstract:

Approximately 96.5% of Earth's water is found in the seas and oceans, 1.7% in the ice caps, 0.8% considered as fresh waters and the remaining percentage consists of brackish water, slightly salty water found as groundwater in aquifers and as surface water in estuaries.[1] Water shortages due to growing industrialization and urbanization has existed for a long time, and there arises a dire need to convert saltwater into potable water. Today, the demand for potable water exceeds the conventional available water resources and its production has become a worldwide concern. Recently, researchers have underlined reverse osmosis (RO) as the most optimized technology for gaining access to clean water and meet global demands.[2] RO is an energy intensive process and often RO membranes are susceptible to fouling and scaling that drives up operational cost and hinder the efficiency. To increase the performance of RO membranes the feed water is pretreated to remove pollutants before desalination. This work aims to fabricate pretreatment membranes to prevent the effects of fouling and scaling by introducing hydrophilic character to membrane surface.

This work explores electrospinning, a cost-effective and scalable technique, to blend two polymers into a nonwoven membrane comprised of fibers ~100 nm – 100 μm in diameter. A rotary drum collector holding the mat simultaneously collects the hydrophobic poly(vinyl chloride) and hydrophilic poly(vinyl alcohol) (PVA) fibers being electrospun from two separate solutions. The work aims to tune the hydrophilicity of the resulting membrane by controlling the relative deposition rate of PVA onto the mat. After electrospinning, the mats are cross-linked with a poly(ethylene glycol) diacid to impart mechanical strength and tune the porosity. It was found that the pore size of the membrane mats varied from 500 nm to 80 nm when the concentration of the crosslinker was reduced. Fiber morphologies were characterized by scanning electron microscopy and the elemental composition across the membrane was mapped by energy dispersive X-ray spectroscopy. Additionally, membrane characterization tests (e.g., differential scanning calorimetry, thermogravimetric analysis, and Fourier-transform infrared spectroscopy) confirmed the presence of both polymers on the mat. A rigorous analysis method was developed combining the above results to estimate mass deposition rate, relative concentration of both polymers on the mat, and correlate PVA concentration with hydrophilic character.

References:

1. Greenlee, L. F., Lawler, D. F., Freeman, B. D., Marrot, B., Moulin, P., & Ce, P. (2009). Reverse osmosis desalination: Water sources, technology, and today's challenges. *Water Research*, *43*(9), 2317–2348. https://doi.org/10.1016/j.watres.2009.03.010
2. Lee, K. P., Arnot, T. C., & Mattia, D. (2011). A review of reverse osmosis membrane materials for desalination-Development to date and future potential. Journal of Membrane Science, 370(1–2), 1–22. https://doi.org/10.1016/j.memsci.2010.12.036

604867 Impact of Organic Compounds in Produced Water on Desalination By Membrane Distillation

Abstract:

In recent years, hydraulic fracturing has been widely employed by U.S. oil and gas industry for the extraction of oil and gas from unconventional reservoirs. This extraction process is intrinsically associated with production of water (i.e., produced water) that may contain chemicals that were added to the fracturing fluid as well as salts and hydrocarbons from the formation. Global production of this wastewater is increasing and it is currently estimated at 250 million barrels per day. Increasing cost for disposal of produced water in the injection wells emphasises the need for treatment and reused to control the disposal cost and protect the environment. Produced water is characterized by extremely high salinity that can range from 30-300 g/L. It also contains bacteria, sand or mud, oil and grease, and Naturally Occurring Radioactive Material (NORM). As such, it is not amenable to treatment using conventional wastewater treatment processes.

Membrane Distillation (MD) is an emerging potential solution for desalination of produced water. Unlike pressure-driven membrane desalination processes, the main driving force for membrane distillation is the vapor pressure difference across the membrane, which is not significantly affected by salinity. Therefore, MD is uniquely suitable for treatment of high salinity produced water. The presence of high concentration of organic compounds in produced water may be challenging for MD technology because they can cause membrane fouling and wetting and reduce the selectivity of this process. In particular, surfactants can promote membrane wetting by adsorbing onto the membrane surface to alter its hydrophobicity. If that happens, it would be necessary to remove surfactants in a pre-treatment step to achieve high water recovery without membrane fouling to maintain high permeate flux.

In this study, raw produced water from Permian Basin, Texas was obtained and tested in a laboratory-scale membrane distillation system. Increase in permeate salinity observed when treating this produced water was attributed to membrane wetting was observed in this system. It was found that the total organic carbon (TOC) of the produced water can be reduced from 120 mg/L to 20 mg/L by biological treatment and further reduced to 2 mg/L by activated carbon adsorption. However, this reduction in TOC did not eliminate apparent membrane wetting. These preliminary results suggest that membrane wetting by this produced water is primarily due to low surface tension caused by non-adsorbable polar compounds or due to volatilization of organic or inorganic salts present in the produced water.

Several organic compounds and surfactants, including 'Nonyl phenol ethoxylate', were detected in both raw water and the permeate. Polymeric form of this surfactant 'Nonylphenol polyethylene glycol' was tested in synthetic produced water and membrane wetting was observed. However, membrane wetting due to 'Nonylphenol polyethylene glycol' was irreversible while the wetting caused by the actual produced water can be reversed by simple washing with DI water. Several synthetic solutions were tested for their ability to pass through the membrane and it was found that volatilization of organic and inorganic salts is the primary the observed increase in permeate conductivity. Therefore, the usual practice of using the permeate conductivity to monitor membrane wetting may not be appropriate when treating complex wastewaters.

605409 Broader Wastewater Reclamation with Mixed Matrix Membranes

Abstract:

The global water crisis demands new technologies to support reuse of impaired water sources. However, wastewater reclamation remains limited by the poor ability of even the most selective technologies to sequester certain small neutral solutes. We hypothesize that the development of a membrane with more rigid void spaces will enhance selectivity toward emerging contaminants. Carbon molecular sieves (CMS) were incorporated in reverse osmosis (RO) thin film composite membranes. Carbon derived from polyamide powders was made by emulsifying aqueous and organic solutions traditionally used in interfacial polymerization and subsequent pyrolysis with up to 50 ppm of oxygen. Porosity measurements of char indicate a minimum pore size of 3 Å that increases to 10 Å as soak temperature ranges from 500 °C to 900 °C. Permeation testing in a bench-scale RO system under cross-flow conditions at 400 PSI show control membranes, without CMS, had a water permeability of 0.76 ± 0.12 L m^{-2} h^{-1} bar^{-1} and NaCl rejection of 99.03 ± 0.32 %. Mixed matrix membranes had a water permeability of 0.78 ± 0.11 L m^{-2} h^{-1} bar^{-1} and NaCl rejection of 90.39 ± 5.78 %, due to defects introduced from sieves. Although initial results are modest, they suggest defect-free incorporation of CMS from highly cross-linked polymers could increase selectivity toward contaminants in aqueous separations.

605441 Evaluation of Integrated Electrocoagulation-Microfiltration and Direct Contact Membrane Distillation Processes for Produced Water Treatment

Abstract:

Treating the wastewater generated from hydraulic fracturing operations is challenging because of the present of high salinity (total dissolved solids (TDS)), toxic organic & inorganic compounds, oil & grease, and naturally occurring radioactive materials. Membrane distillation (MD) is an emerging technology of interest for the treatment of high salinity brines. In this study, microfiltration (MF) was used to remove the flocs formed during the electrocoagulation (EC) pretreatment, and the MF permeate was treated further by direct contact membrane distillation (DCMD). Different types of membranes were investigated for the DCMD. In this work, we aimed to demonstrate the feasibility of an integrated EC-MF-DCMD system for water recovery from high salinity produced waters obtained from shale gas extraction facilities. During EC, a high removal efficiency of total organic carbon (48.4%) was obtained by using 6 aluminum plates with bipolar configuration at 9.5 A for 20 min. Keeping the feed tank temperature at 40 ^{o}C, while the temperature inside the model at 60 ^{o}C can effectively suppress crystallization on the membrane surface, which can significantly decrease scaling. 50% water recovery was obtained using commercial membrane over two DCMD cycles. However, organic fouling was still a big problem, especially surfactants due to forming micelles and bilayers, which can significantly decrease the flux and eventually block the membrane.

612363 Performance of Multi-Stage Vacuum Membrane Distillation Integrated with Mechanical Vapor Compression

Abstract:

Treatment of hypersaline (>100g/L) wastewater is challenging, and only distillation-based technologies can treat such streams. Among distillation-based technologies, membrane distillation (MD) has shown great potential for treating highly concentrated brine. Different MD configurations have been utilized for desalination of hypersaline brine; among those, vacuum membrane distillation (VMD) is an ideal candidate where the driving force is increased by reducing the permeate pressure, thereby high fluxes are usually achieved. Recovering the heat of condensation from the permeate vapor product is crucial for improving the energy efficiency of the system. Recent studies showed that a multi-stage vacuum membrane distillation (MSVMD) enables the consecutive heat recovery between the feed and the permeate vapor via a series of heat exchangers.[1] Results on analysis gained output ratio (GOR) - a representative of the energy efficiency in MD - indicate that the multi-stage configuration enhances energy efficiency. More enhancement in energy efficiency and water recovery could be achieved when VMD is integrated with other processes such as mechanical vapor compression (MVC).

In this presentation, we will discuss the strategies to integrate MSVMD with MVC to enhance the heat/water recovery. In this effort, our goal was to optimize MSVMD-MVD to achieve higher energy efficiency. The VMD model was established based on the mass and heat balances involved in the VMD process in Aspen Custom Modeler (ACM) V10 [2]. The pre-validated thermodynamic model (eNRTL-RK) for thermo-physical properties of produced water was incorporated in Aspen Properties for the first time. The VMD unit was first validated against the lab-scale experimental results in terms of water vapor flux at different feed temperatures, feed flow rates, and permeate pressures.[3] Process simulation on MSVMD- MVC was implemented in Aspen Plus V10 by importing the validated VMD model from ACM into the MVC-VMD flowsheet. The effects of operating conditions were investigated with respect to production and gain output ration (GOR).

References

(1) Chung, H. W.; Swaminathan, J.; Warsinger, D. M.; Lienhard V, J. H. Multistage Vacuum Membrane Distillation (MSVMD) Systems for High Salinity Applications. *J. Memb. Sci.* **2016**, *497*, 128–141. https://doi.org/10.1016/j.memsci.2015.09.009.

(2) Mengual, J. I.; Khayet, M.; Godino, M. P. Heat and Mass Transfer in Vacuum Membrane Distillation. *Int. J. Heat Mass Transf.* **2004**, *47* (4), 865–875. https://doi.org/10.1016/j.ijheatmasstransfer.2002.09.001.

(3) Malmali, M.; Fyfe, P.; Lincicome, D.; Sardari, K.; Wickramasinghe, S. R. Selecting Membranes for Treating Hydraulic Fracturing Produced Waters by Membrane Distillation. *Sep. Sci. Technol.* **2017**, *52* (2), 266–275. https://doi.org/10.1080/01496395.2016.1244550.

601032 High Flux CO_2-Selective Single-Layer Graphene Membranes: Synthesis and Scale-up

Abstract:

Nanoporous single-layer graphene (N-SLG), prepared by incorporating subnanometer vacancy defects in the graphene lattice, is highly promising for high flux gas separation because the resistance to diffuse is controlled by a single transition state at the nanopore [1–3]. Molecular sieving resolution (MSR), defined as the difference in the kinetic diameters of molecules to be separated, of 0.3-0.5 Å is needed for CO_2/N_2 and CO_2/CH_4 separation, respectively [4-5].

In this presentation, I will share recent developments in our laboratory on advances on N-SLG membranes including their scale-up to practical dimensions [6-8]. Strategies including controlled etching of graphene lattice, functionalization of graphene lattice as well as hybrid graphene-polymer membranes will be discussed. Specifically, I will discuss a novel etching chemistry, allowing controlled nucleation and expansion of vacancy defects, and yielding MSR of 0.2 Å. We show that molecular cutoff can be adjusted by 0.1 Å by a slow expansion of nanopores, making O_2/N_2 separation possible from atom-thick membranes. Large CO_2 permeances (above 10000 GPU) combined with attractive CO_2/N_2 separation factor (above 20), makes N-SLG membranes highly promising for postcombustion carbon capture. We show that molecular sieving (selectivities > 20) can be obtained from large graphene coupons reaching 1 cm^2 in size.

References

- Jiang et al., *Nano Lett.* 9, 4019–24 (2009).
- Celebi et al,. *Science.* 344, 289–92 (2014).
- Li et al., *Science.* 342, 95–98 (2013).
- Yuan et al., *ACS Nano.* 11, 7974–7987 (2017).
- Koenig et al., *Nanotechnol.* 7, 728–732 (2012).
- He et al., *Energy Environ. Sci.* 12, 3305–3312 (2019).
- Zhao et al., Sci. Adv. 5, eaav1851 (2019).
- Huang et al., Nat. Commun. 9, 2632 (2018).

597905 Impact of Post-Synthetic Modification Routes on Filler Structure and Performance in Metal–Organic Framework Based Mixed-Matrix Membranes

Abstract:

This work systematically investigates an exemplary mixed-matrix membrane (MMM) system comprised of a UiO-66-NH_2 metal–organic framework (MOF) and a 6FDA-Durene polyimide. The UiO-66-NH_2 MOF is post-synthetically modified with 6FDA-Durene oligomers through either chemical or thermal imidization routes. The thermal(T)-PSM approach does not require reactive small molecules for synthesis, thereby allowing the cross-linking reaction between the MOF and the polymer to occur only on the MOF surface. Conversely, the chemical(C)-PSM approach requires reactive small molecules for synthesis, thereby altering internal amine functionality of UiO-66-NH_2. Upon embedding these PSM-MOFs in high molecular weight 6FDA-Durene polymers, CO_2 and CH_4 pure-gas permeation tests reveal that the T-PSM-MOF has synergistic improvements in both permeability and selectivity due to enhanced interfacial compatibility without significant modification to the internal MOF framework. Conversely, MMMs formed with the C-PSM-MOF experienced a significant decrease in permeability due to reactions that occurred inside of the MOF pore structure. These findings suggest that there is a critical need to consider reaction conditions and their corresponding effect on MOF structure when pursuing a PSM approach to form MMMs with desired performance enhancements.

595629 2D Layered Double Hydroxide Membranes with Inherent CO_2-Selective Transport Channels for Efficient Carbon Capture

Abstract:

2D membranes have shown great potential for precise molecular separation owing to the atomic-scale thickness and well-defined interlayer nanochannels. Rational design of channel's size and surface chemistry is critical for the selective transfer of target molecules. The currently reported 2D material membranes for efficient CO_2 separation are primarily built by introducing cross-linking agents or intercalating agents. In this study, we demonstrate a novel 2D material membrane with inherent CO_2-selective transport channels based on layered double hydroxide (LDH) without any additional crosslinking or intercalating steps. The intrinsic breathing effect toward CO_2 of LDH enables CO_2 molecules transfer through the interlayer nanochannels by CO_3^{2-} ions form. Moreover, the membrane nanochannel size is tuned by the CO_3^{2-} ions from the initial 0.7nm to 0.3nm, which exhibit a distinct sieving property for CH_4 molecules (0.38nm) while not for CO_3^{2-} (ionic radius of 0.136 nm). The unique breathing effect and size-sieving effect jointly contribute to the inherent CO_2-favorable nanochannels within LDH membranes. The resulted ultrathin 2D LDH membrane with thickness of 30nm exhibit excellent CO_2/CH_4 separation performance with CO_2 permeance of 150 GPU and CO_2/CH_4 selectivity of 33. This study is anticipated to extend the materials and strategies for design of the CO_2 preferential transport channels in membranes.

602046 Few-Nanometer Organosilica Membranes for High-Temperature H_2/CO_2 Separation

Abstract:

Traditional silica membranes exhibit excellent H_2/CO_2 separation properties for H_2 purification and CO_2 capture, but they are brittle and faced with challenges for low-cost and large-scale production. Herein, we demonstrate the fabrication of ultrathin silica layer on polymeric membranes via plasma treatment to achieve superior H_2/CO_2 separation properties and excellent scalability of membrane production. Specifically, silica membranes were produced by applying oxygen plasma on polydimethylsiloxane (PDMS) thin films supported by microporous polymer substrates. The plasma treatment successfully converted the PDMS surface layer to a silica ultrathin film, indicated by the formation of SiO_2 as thin as 10 nm. We thoroughly investigated the silica formation mechanism and the effect of the plasma treatment conditions on physical properties and H_2/CO_2 separation properties of the silica membranes. With a plasma exposure time of 120 seconds, the membrane comprising 1.0 μm PDMS exhibits H_2 permeance of 370 GPU (1 GPU = 10^{-6} cm^3(STP)/cm^3 s cmHg) and H_2/CO_2 selectivity of 51 at 200 ºC, which surpasses the Robeson's upper bound. The membrane shows stable mixed-gas separation performance in the presence of water vapor at 200 °C for 47 h. When the PDMS layer thickness was decreased to 200 nm, the 120-second plasma treatment increases the H_2 permeance to 930 GPU and H_2/CO_2 selectivity decreases to 32, which is still above the upper bound. The robust H_2/CO_2 separation properties coupled with the facile fabrication demonstrate the potential of these organosilica membranes for practical H_2 purification and CO_2 capture.

611599 MOF@MOF Composite Membranes for Carbon Dioxide Capture and Separation

Abstract:

There are already over 60,000 reported MOFs, and many of them are promising adsorbents and membranes for many applications. Therefore, another synthetic route for the development of task-specific materials rather than designing new MOFs is to fabricate composites based on these well-characterized MOFs that combine the advantages of their constituents. Here, we present the first examples of pure inorganic MOF@MOF membranes constructed from different MOFs with lattice mismatch: SIFSIX-3-Ni@SIFSIX-1-Cu and SIFSIX-3-Ni@HKUST-1. The CO_2/N_2 selectivities of the HKUST-1 and SIFSIX-1-Cu membranes were improved from 5.7 to 13.4 and 2.3 to 7.4, respectively, after the growth of a selective layer of SIFSIX-3-Ni onto the surface of those materials. The integration of multiple layers into a single composite membrane allows for improved control over the membrane properties including gas selectivity, permeability, stability and mechanical properties—factors that must be tuned to meet the requirements for practical membrane-based separations.

PECTIN EXTRACTION FROM WATERMELON PEEL (*Citrillus Lanataus*) BY ACID HYDROLYSIS USING HYDRCHLORIC ACID SOLUTIONS.

Karina A. Gómez[1], Lorena P. González[1], José A. Pérez[1*]*

[1], *Chemical engineering department, Universidad del Atlántico, Km 7, Puerto Colombia, Atlántico, Colombia*

*corresponding authors: jandrespmen@gmail.com; Tel: +57 3002517865

kalejandragomez@mail.uniatlantico.edu.co; Tel: +57 3205768683.

Abstract

Watermelon peel was used for the pectin extraction using hydrochloric acid. The effects of pH (2.0, 2.5 and 3.0), extraction time (45, 60 and 75 minutes) and liquid-solid ratio (1:10, 1:25, 1:40 g/ml) on the yield of pectin extraction were investigated. A Box-Behnken experimental design and surface response methodology were used for the estimation of the optimal of the pectin yield. The extraction time was the most influential variable and the optimal conditions were found to be at pH of 2.0, extraction time of 75 minutes, and a liquid-solid ratio of 1:25 g/ml with a pectin yield of 7.70 %. The pectin was characterized by Fourier transform infrared spectrophotometry (FTIR) to confirm the presence of its characteristic functional groups. FTIR analysis showed that the extracted pectin has properties that are similar to the industrial pectin and may be classified as rapid set pectin.

Key words: *Watermelon peel, pectin, acid hydrolysis.*

Introduction

Watermelon is characterized by large and juicy fruits protected by a hard rind. It is known worldwide for its health benefits because is a source of folic acid, A vitamin, phosphorus, magnesium and potassium [1]. The composition of watermelon is divided into the pulp or pericarp, which is the edible part and represents the 90 to 95% of the water content of its size, the seeds are distributed along with the pulp and finally the rind or epicarp, that represents between a 30 and a 40% of the fruit. The last one is considered a waste and it is used mostly for pig feeding and in lesser proportion as an input for obtaining flour in bakeries, dairy products, and the rest is discarded generating contamination [2].

Watermelon is used as a raw material in the agro-industry to produce juices. This process generates a considerable amount of waste such as rind, pulp, and seeds. The waste generated by this process can be used to obtain other pectin products to increase its value, reduce environmental impact and use it in different industries where pectin has an application, such as food and pharmaceuticals industries [3].

The focus of this study is to evaluate the effect of the variables pH, extraction time and liquid-solid ratio, on the yield of the pectin extraction process from the watermelon rind using a Surface Response Methodology (SRM) and Box Behnken Central Composite Design (BBD). Besides characterize the product obtained by infrared spectrum analysis (IR) and determine some variables such as the percentage of ash, methoxyl content and degree of esterification of the pectin obtained.

Pectin

Pectin is a complex carbohydrate, which contains at least 65% of the galacturonic acid units. Its chains are formed by rings of this acid: each ring of the chain has a carboxyl group (-COOH) that can be esterified with methanol to produce methyl esters (-COOCH3) or can be neutralized by a base (-OH) [4].

Pectin sources and applications

The traditional and most important sources of industrial pectin are apple waste and citrus peels, which are available as a waste of the production of juices and contain high amounts of pectin substances. On the other hand, one of the unconventional sources of pectin that generate more interest are the waste of products obtained from bio-industries, to minimize environmental issues. These sources are banana, mango, and passion fruit peels, and others [5].

In the food industry, pectin is the working ingredient and source of dietary fiber. Due to its ability to form aqueous gels, it is important for modifying or creating the texture of certain foods such as compotes, jellies, salads, sweets, among others [6].

In the Pharmaceutical industry, it is used for preparations of different medications such as antidiarrheals, detoxicants, and others that are encapsulated with a pectin film [7].

Pectin properties and classification

The properties that characterize and evaluate the pectin performance are pH, viscosity, ash content, esterification degree, and purity, which is a measure that goes according to the percentage of galacturonic acid [8].

The main criteria to classify pectin is the esterification degree because the gelling properties are defined by it. Esterification Degree is defined by the ratio of methyl esterified galacturonic acid residues to the total number of galacturonic acid units present in the sample [9].

High methoxyl pectins have more than 50% of carboxyl groups esterified. They are also classified as fast gelling, with a gelification time less than 5 minutes and an esterification degree between 68 and 75%, and slow gelling with a gelification time of more than 5 minutes and an esterification degree between 60 and 68%. In contrast, we have Low methoxyl pectins, where less than 50% of the carboxyl groups are esterified. Differences between High and Low methoxyl pectins are shown in Figure 1 [10].

Figure 1. Pectin classification

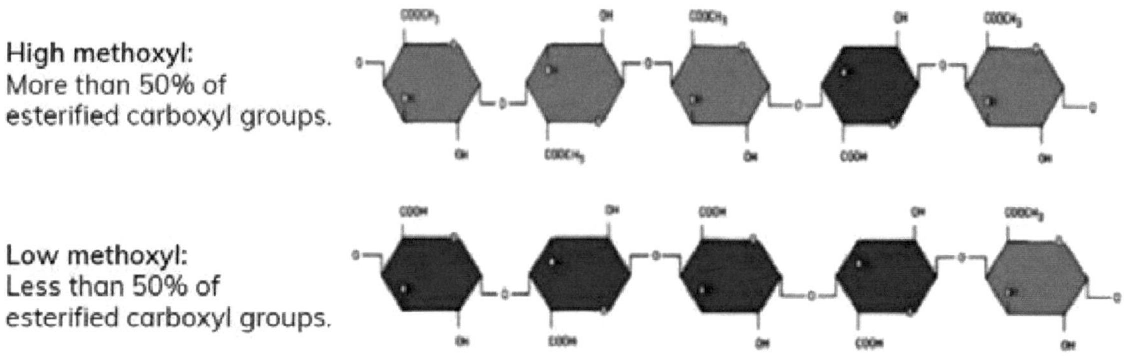

Pectin extraction

The extraction process consisted of a pre-treatment process of the raw material, followed by extraction of the pectin using acid hydrolysis. Then was carried the clarification process of the pectin, ending with the drying and characterization of the pectin obtained. The methodology used to carry out this project is summarized in the flow diagram of Figure 2.

Figure 2. Methodology.

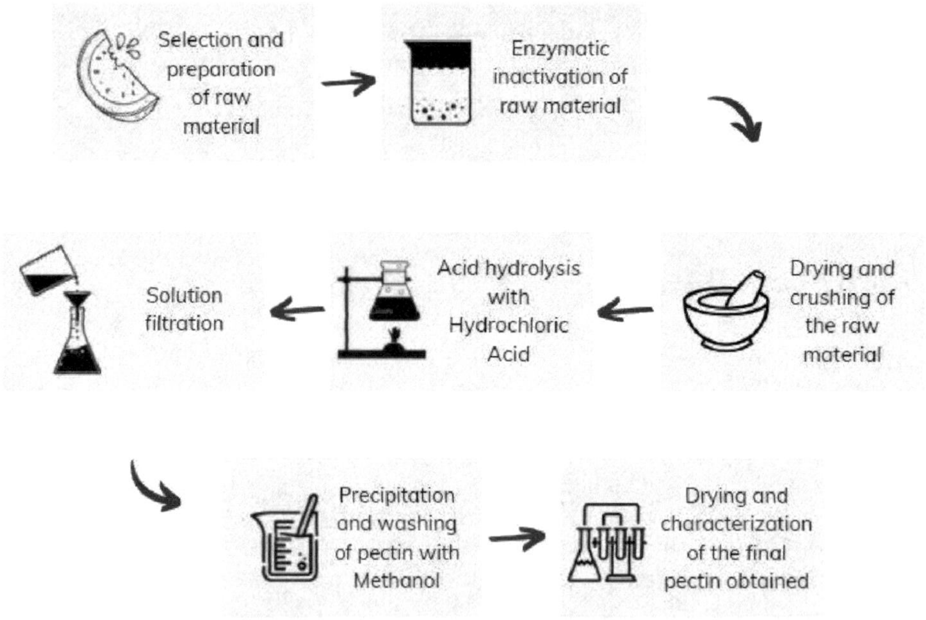

The variables that are relevant during extraction and that determine the characteristics of pectin in extraction yield are pH, temperature, and extraction time [4].

Experimental Design

To evaluate and optimize the effects of the process variables on the response variable, the surface response methodology, and a Box Benhen central composite design were used. As independent variables, we have the pH, extraction time and the liquid-solid ratio, the experimental design box Behnken (Table 1) established the best combination of variables that influenced the results.

Table 1. Box Behnken experimental design.

Temperature (°C)	Extraction solvent	Level	pH	Extraction Time (minutes)	Liquid-Solid Ratio (g/ml)
80°	Hydrochloric Acid	-1	2.0	45	1:10
		0	2.5	60	1:25
		1	3.0	75	1:40

Fifteen experimental runs were conducted in duplicate and data analysis was performed using R programming software. The results were studied statistically using the analysis of variance (ANOVA).

Discussions and results

Pectin Yield

The results of this investigation, a maximum pectin yield of 7.7% was obtained in conditions of pH 2, extraction time of 75 minutes, and a liquid-solid ratio of 1:25 g/ml as is showed in **Table 2.**

Table 2. Conditions for maximum extraction yield.

pH	2.0
Time (minutes)	75
LSR (g/ml)	1:25
Maximum extraction yield (%)	7.70

Pectin yield was favored by low pH levels. Due the depolymerization of pectin and others biopolymers during the extraction, the low pH dissociates the ionic bonds that hold the pectin in the watermelon rind and hydrolyzes the glycoside and ester bonds, to which the pectin could be esterified [11].

On the other hand, by increasing the extraction time, the pectin extraction yield increased. This can be explained by the composition of pectin where the alpha units of galacturonic acid require more time to weak its structure [12].

Finally, for the liquid-solid ratio, the extraction yield was favored by increasing it. This is due to the increase of the driving force for the mass transfer of the polysaccharides [13]. However, a slight decrease is evident when increasing the liquid-solid ratio in extreme conditions, because this leads to saturation that prevents mass transfer [14].

Statistical Analysis

The statistical analysis showed that were 3 factors and interactions that had a significant effect, which were pH, time, and the quadratic value of the liquid-solid ratio.

The evaluation of the effect of the variables we can observe in Figure 3, indicates that the yield was favored with longer extraction times and increased when the pH decreased. Also, a curvature for the liquid-solid ratio was present to maximize the yield

Figure 3. Effects of the variables during the extraction of pectin.

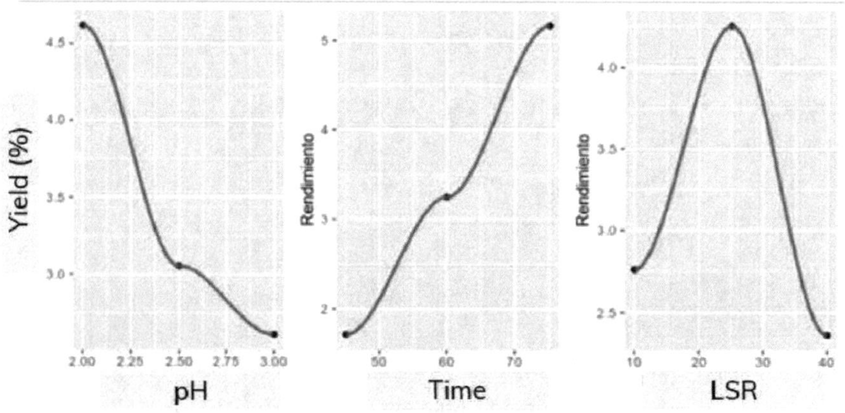

A second order statistical model was obtained (Equation 1) from the experimental data, including linear, interactive and quadratic terms and the conditions for reaching the optimal extraction yield are expressed on the table 3. It is observed that at a pH of 2, extraction time of 75 minutes, and a liquid-solid ratio of 22.1 g/ml, a theoretical maximum yield of 7.10% was obtained.

$$\textbf{Equation 1}.\ Pectin\ Yield\ (\%) = 7.98 - 11.70X_1 + 0.073X_2 + 0.607X_3 + 1.801{X_1}^2 + 0.024X_1X_2 - 0.031X_1X_3 + 0.00046{X_2}^2 - 0.0029X_2X_3 - 0.0073{X_3}^2$$

Table 3. Optimization of the model

Factor	Low	High	Optimum
X_1	2.0	3.0	2.0
X_2	45	75	75
X_3	10	40	22.1
Optimum extraction Yield: **7.10%**			

Surface Response Methodology

The surface response methodology (**Figure 4**) allowed to corroborate the results obtained previously. Where, at lower pH levels and prolonged extraction times, the yield is favored. In the interaction between pH and liquid-solid ratio, a maximum is reached at lower pH levels and ratio at its medium level. Finally, in the interaction between time and liquid-solid ratio, it is observed that yield is favored in prolonged times and ratio in its medium level.

Figure 4. Surface response Methodology for pectin extraction.

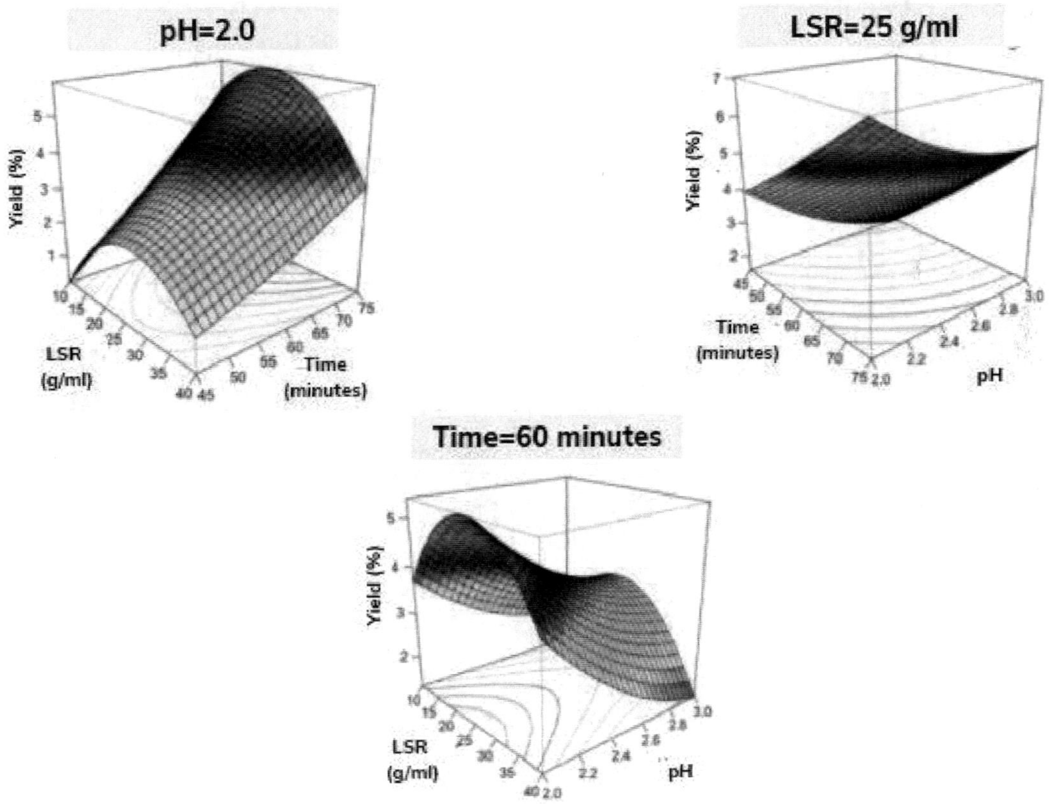

Characterization

The comparison of the characterization of the extracted pectin, with a higher percentage of extraction yield, compared to commercial pectin (Merck) is shown in **Table 4.**

Table 4. Characterization results of the pectin obtained.

Physic-Chemical Properties	Standard Pectin	Extracted Pectin
Ash (%)	3.77±3.39 (M)	1.29
Ash Alkalinity (%)	2.34±2.90 (M)	0.40
Free Acidity (meq/g)	0.78±0.46 (M)	0.96
Methoxyl content (%)	≥6.70 (USP)	10.33
Esterification Degree (%)	≥ 74 (USP) ≥65 (FAO/WHO	77.67
Galacturonic Acid (%)	≥81.50 (USP	82.87

(M) Merck, (USP) United States Pharmacopeia, (FAO) Food and Agriculture Organization of the United States, (WHO) World Health Organization

It can be observed that the samples had low ash content, even below the limit established by their commercial counterpart, indicating that they contained few solids and impurities [15]. High methoxyl pectins were obtained and they were classified as fast gelling pectin, because of their esterification degree [16,4]. Finally, the galacturonic acid content was above the established value, which indicates that the pectin had high-quality.

FTIR Analysis

The infrared spectra of the pectin obtained by hydrochloric acid hydrolysis, at pH 2.0, for 75 minutes and with a liquid-solid ratio of 1:25 g / ml is shown in figure 5. The characteristic functional groups of pectin were shown and were compared with a sample of commercial pectin. Due to that, we can say that they are chemically similar.

Figure 5. FTIR analysis for pectin obtained.

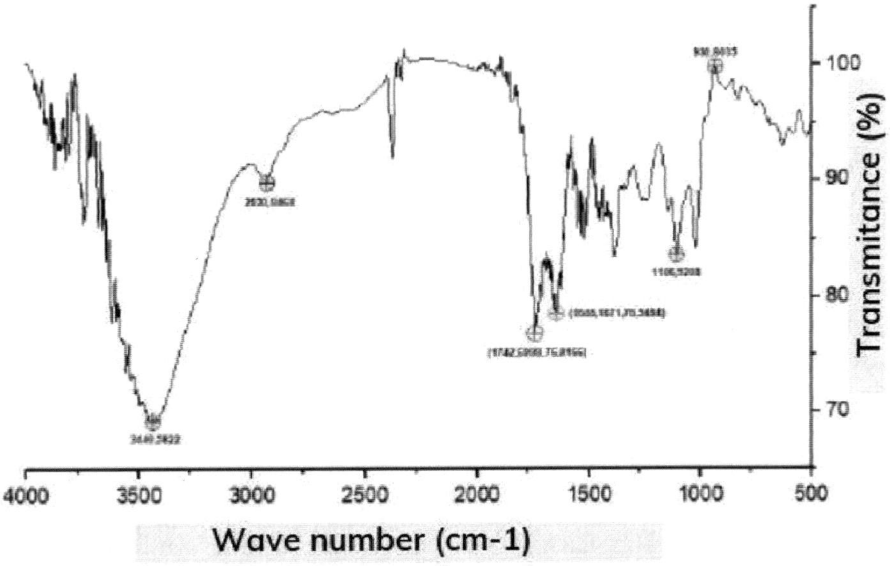

Pectin was again classified as rapid-set or fast gelling, since a greater elongation of the band corresponding to the carbonyl group of the ester is observed, compared with the corresponding band of the carbonyl group of the acid [15].

Conclusions

The best yield of pectin extracted with hydrochloric acid, was obtained under conditions of pH of 2.0, extraction time of 75 minutes, and liquid/solid ratio 1:25 g / ml. Which had a methoxyl content of 10.3%, a degree of esterification of 77.67%, and galacturonic acid content of 82.87%. Thus making it very competitive pectin at a commercial level. Also, the extraction time and pH were the variables that significantly influenced the extraction performance. Finally, the study carried out allows proposing the extraction of pectin as an attractive alternative for the recovery of this waste and the reduction of its environmental impact.

Acknowledgments

The authors acknowledge the Universidad del Atlántico in Barranquilla, Colombia for all the support and knowledge delivered.

References

[1] Cohen-Manrique, Carlos, et al. (2018), Modelado del Microclima de un Cultivo de Sandía (Citrullus lanatus) en la Sub-región Sabana del Departamento de Sucre, pp. 335–344.

[2] Durán, Ricardo, et al. (2017), Caracterización y extracción de citrulina de la corteza de la sandía (k consumida en Valledupar, *Temas agrarios,* pp. 62-69.

[3] Vargas, Yuri and Perez, Liliana (2018), Aprovechamiento de Residuos Agroindustriales en el mejoramiento de la calidad del ambiente, *Revista Facultad de Ciencias Básicas,* pp. 59-72.

[4] Zegada, Vanesa (2015), Extracción de Pectina de residuos de cáscara de naranja por hidrólisis ácida asistida por microondas (HMO), *Investigación & Desarrollo*, pp. 65-76.

[5] Garcia, Maribel, et al. (2010), Determinación de la composición química y actividad antioxidante en distintos estados de madurez de frutas de consumo habitual en Colombia, mora (Rubus glaucus B.), maracuyá (Passiflora edulis S.), guayaba (Psidium guajava L.) y papayuela (Carica cundinamarcensis J.), *Revista Alimentos*, pp.25-42

[6] Urango-Anaya, Katty (2018), Extracción Rápida de Pectina a Partir de Cáscara de Maracuyá (Passiflora edulis flavicarpa) empleando Microondas, *Información técnologica*, pp. 129-136.

[7] Sindoni, María, et al. (2014), Efecto de dos fases de maduración sobre la cantidad de pectina obtenida en dos varidades de parchita (Passiflora edulis F. Flavicarpa degener) de diferente procedencia. *Revista Iberoamericana de tecnología Postcosecha*. pp. 93-100.

[8] Paredes, Jorge, et al. (2015), Effect of degree of maturity on the physical and chemical properties of helmets pectins from guava (Psidium guajava L.). *Idesia,* pp. 35-41.

[9] Chasquibol, Nancy (2008), Extracción y caracterización de pectinas obtenidas a partir de frutos de la biodiversidad peruana. *Ingeniería Industrial*, pp. 176-199.

[10] Muñoz, Francisco (2011). Extracción y caracterización de la pectina obtenida a partir del fruto de dos ecotipos de cocona (solanum sessiliflorum), en diferentes grados de madurez; a nivel de planta piloto. *Facultad de Ingeniería Universidad Nacional de Colombia, Colombia*, p. 9.

[11] García, Jessica (2017), Evaluación del rendimiento de extracción de pectina en aguas mieles del beneficiado de café procedentes del desmucilaginado mecánico, *Facultad de Ingeniería y Arquitectura Universidad del Salvador, San Salvador,* p. 37.

[12] Jafari, Enrique, et al. (2017), Pectin from carrot pomace: Optimization of extraction and physico chemical properties. *Carbohydrate Polymers*. pp. 1315-1322.

[13] Bendahou, Abdelkader. et al (2007). Isolation and structural characterization of hemicelluloses from palm of Phoenix dactylifera L. *Carbohydrate Polymers*, pp. 601–608.

[14] Maran, Prakash, et al. (2014) Microwave assisted extraction of pectin from waste *Citrullus Lanatus* fruit rinds. *Carbohydrate Polymers*, p. 786.

[15] Cabarcas, Esteban, et al. (2012), Extracción y caracterización de pectina a partir de cáscaras de plátano para desarrollar un diseño general del proceso de producción, *Facultad de Ingeniería Universidad de Cartagena, Colombia*, pp. 44-48

[16] Acevedo, Vanesa and Ramírez, Diana (2011), Análisis técnico y económico de la pectina, a partir de la cáscara de la naranja (*Citrus sinesis*). *Facultad de Ingeniería Universidad de San Buenaventura, Colombia,* p. 33.

607885 Continuous Intensified Solvent Extraction of Biomolecules Using Hydrophobic Ionic Liquids

Abstract:

In pharmaceutical processes the high cost associated with downstream processing aimed at the purification and recovery of target products is one of the major issues limiting the widespread use of many bio-base products [1]. Separation processes and purification stages usually require numerous steps associated with high energy and chemicals consumption and typically represent a large percentage of the cost of the final product. Modern pharmaceutical manufacturing organizations would like separations that can integrate with continuous upstream flow synthesis and downstream isolation procedures, replacing current batch systems. Continuous manufacturing offers a range of advantages over batch, including speed, flexibility and safety. Solvent extraction remains one of the most attractive separation techniques due to its versatility, simplicity and effectiveness [2]. Continuous solvent extraction can be intensified with the application of small-scale two-phase contactors, where the reduction of the apparatus size leads to a range of benefits, such as high mass transfer rates and reduced solvent use. Small channel contactors have been successfully employed in the past for metal extractions [3].

Hydrophilic pharmaceutical molecules are difficult to extract from aqueous media with traditional organic solvents. The choice of solvent is often limiting as it needs to be inert for medical purposes. Ionic liquids have recently been proposed as alternatives, because of their numerous benefits including, low volatility and high chemical stability [4]. In this work we investigate the separation of hydrophilic amino acids, with hydrophobic imidazolium based ionic liquids using macrocyclic compounds as extractants. For amino acid L-tryptophan, the highest partition coefficients were obtained at pH values less than 2.38, where the amino acid was in its cationic form. Continuous extraction experiments were carried out in intensified small channels. The flow patterns of the aqueous and the ionic liquid phases were studied for total volumetric flowrates from 0.001 to 0.1 m^3/s and volume fraction of the two phases ranging in a 0.5 mm channel. For the conditions were plug flow was established, mass transfer experiments were carried out for different residence times.

Acknowledgements: Y.V. Phakoukaki would like to thank EPSRC and Johnson Matthey for her studentship

References:

[1] Behme, S. (2015). Manufacturing of pharmaceutical proteins. Weinheim: Wiley-VCH, pp.68-95.

[2] Poole, C. and Poole, S., 2010. Extraction of organic compounds with room temperature ionic liquids. *Journal of Chromatography A*, 1217(16), pp.2268-2286.

[3] Tsaoulidis, D. and Angeli, P. (2015). Effect of channel size on mass transfer during liquid–liquid plug flow in small scale extractors. Chemical Engineering Journal, 262, pp.785-793.

[4] Ghandi, K., 2014. A Review of Ionic Liquids, Their Limits and Applications. *Green and Sustainable Chemistry*, 04(01), pp.44-53.

599937 Continuous CC3 Porous Organic Cage Membranes

Abstract:

Porous organic cages (POCs) have emerged as a novel type of crystalline microporous materials which combine highly desirable properties, such as uniform micropores, high surface areas, and thermal and chemical stability, making them highly appealing candidates for diverse functional applications, including molecular gas separations. POCs consist of covalently bonded organic cages that can assemble into crystalline microporous materials displaying three-dimensional connectivity and uniform pore size. Herein, I discuss the successful synthesis of continuous CC3 membranes grown on alumina porous tubes, a geometry that is highly amenable for potential scale-up. The resultant membranes were evaluated for the separation of Xenon from light gases. Effectively separating Xenon from light gases can lead to a considerable reduction in its storage costs, and in potential revenue generated from its sale.

610114 Nanocomposite Molecular Sieve Membranes for the Separation of C4 Olefins

Abstract:

C4 mono and di-olefins (e.g., 1-butene, isobutene and 1,3-butadiene) are important chemicals for the manufacture of a wide range of polymers and materials. Their purification is currently performed by the energy-intensive extraction and distillation processes. Incorporating membranes into the existing separation schemes can potentially result in substantial energy savings. However, the membrane materials that have been explored for the separation of C4 olefins, which mainly include facilitated transport and polymer membranes, have not demonstrated adequate separation efficiency or stability. Here, we report the synthesis of nanocomposite molecular sieve membranes capable of separating C4 olefins based on the sub-ångström differences in their molecular sizes and shapes. The membrane structure consists of zeolitic imidazolate frameworks (ZIF) deposited inside an inorganic matrix through a vapor-phase synthesis approach. The ZIF phase possesses well-defined micropores and enhanced framework rigidity due to their confinement by the inorganic matrix. C4 single gas permeances through the membrane are strongly dependent on their kinetic diameters and/or molecular conformations, showing a diffusion-dominating permeation mechanism for the nanocomposite membranes. Due to the molecular sieving effect and enhanced structural stability, the membranes were found to be effective and stable in separating multi-component C4 hydrocarbon mixtures under varied conditions (e.g., feed pressure, feed composition, temperature, etc.). Adsorption results will be correlated with the membrane permeation data to further understand the separation mechanism.

611572 Preparation and Application of Zeolite and MOF Membranes from an Industrial viewpoint

Abstract:

Inorganic zeolites with well-defined pore network have received continuous worldwide interest, for fabricating membrane with molecular-sieving property. Despite the nearly 40 years of intense research on zeolite membranes, only the hydrophilic LTA membrane has been commercialized for industrial de-watering applications. The main features of state-of-the-art zeolite membranes are: 1. production capacity of a single plant is usually less than 10,000 tpy; 2. water content of the dewatered product is usually higher than 0.1 wt.%; 3. Only tubular type zeolite membrane is commercialized. To date, the annual market of zeolite membranes in China is around 15 million €. This is less than 1 percent of the market potential value. The main reasons are the high cost and the unsatisfied performance of the state-of-the-art zeolite membrane materials. A leap-forward development of zeolite membrane production is expected to be made in order to broaden the application of this technology.

Metal-organic frameworks (MOFs) possess not only zeolite-like topologies and properties, but also, more importantly, tunable pore sizes and well-designed functionalities, which makes them strong candidates for molecular sieve membranes. During the last 15 years, much progress in the development of MOF membranes has been witnessed. For example, numerous synthetic routes have been developed for preparing ZIF-8 membranes that demonstrate superior separation properties for propylene/propane mixture. However, the ZIF-8 membranes still face challenges in industrial applications. One of the biggest challenges is the unsatisfactory propylene flux when utilized in propylene/propane separation. Statistically, most of the reported ZIF-8 membranes have propylene flux of 0.1~0.2 $kgm^{-2}h^{-1}$, which is an order of magnitude lower than the water flux (1~5 $kgm^{-2}h^{-1}$) of commercialized LTA membrane for ethanol dehydration. Besides, the membrane stability under the real separation conditions needs further study.

In the presentation, we will report a continuous production line for zeolite membranes, with the aim to greatly increase the productivity and to reduce the membrane cost. Examples will be given in terms of ultra-deep dewatering of organic solvents. We will also talk about the possibility of continuous and scalable manufacture of ZIF-8 membranes. Compared with zeolite membranes, MOF membranes, endowed with simple preparation process, high separation efficiency and low manufacturing cost, offer enormous potential for scale-up preparation and commercial application. MOF membranes as a whole have an edge over zeolite membranes for commercialization in terms of the processing advantages. The application of MOF membranes in industrial gas separation can be reasonably expected for the next future.

612776 Development of Separation Technologies for the Reuse of Produced Water for Oil and Gas Operations

- Download/View Extended Abstract

Abstract:

Reuse of produced water for oil and gas extraction, or other beneficial reuse applications can reduce stress on freshwater resources. Produced water is the aqueous phase that is co-extracted with oil and gas from underground formations. It generally contains several constituents that require removal before reuse including particulate, oil and grease, dissolved organics, dissolved salts, and other constituents.

In one reuse application, water treatment processes are used to convert produced water into desalinated water to support steam flooding operations. Steam flooding is an enhance oil recovery (EOR) process in which steam is injected into oil and gas reservoirs to increase production. In some cases, reverse osmosis (RO) processes are used to remove the hardness and total dissolved solids in produced water to meet steam flooding boiler feed requirements. There are several potential challenges in desalinating produced water for steam flood due to the presence of foulants, high hardness, and high temperatures.

This presentation will examine challenges and R&D efforts in produced water treatment for beneficial reuse and disposal. Specifically, the talk will examine developments in removal of organics, high temperature reverse osmosis, and zero liquid discharge technologies.

605812 Membranes for Gas and Vapor Separations: The Innovations and the Challenges

Abstract:

Membrane technology has been a critical enabler for effective and efficient gas, vapor, and liquid separations for many years. Polymeric membrane-based industrial separation processes have evolved rapidly since the development of asymmetric integrally skinned polymeric membrane by Loeb and Sourirajan in 1960's and the development of thin film composite (TFC) membranes by Cadotte in 1980's.

Membranes for gas and vapor separations are of special interest to petroleum producers and refiners, chemical companies, and industrial gas suppliers. Several applications of gas separation membranes have achieved commercial success, such as nitrogen enrichment from air, natural gas upgrading, enhanced oil recovery, hydrogen purification and recovery, and recovery of organic vapors.

Innovations in membrane materials with long term stability, chemical resistance, and specific to applications, fabrication methods, module design, new process design, and process technology integration have been instrumental to the industrial adoption of membrane technology for gas separations. However, membrane technology is still facing challenges for large scale organic vapor separations, such as for olefin/paraffin, aromatic/non-aromatic, and aromatic/aromatic separations. Several recent successful cases and the strategies developed to achieve high and prolonged stable membrane performance for gas separations will be discussed.

AUTHOR INDEX